The Science of Ballistics

The Science of Ballistics

Edited by Osmond Lynch

CLANRYE
INTERNATIONAL
www.clanryeinternational.com

Clanrye International,
750 Third Avenue, 9th Floor,
New York, NY 10017, USA

ISBN: 978-1-63240-737-5

Cataloging-in-Publication Data

The science of ballistics / edited by Osmond Lynch.
p. cm.
Includes bibliographical references and index.
ISBN 978-1-63240-737-5
1. Ballistics. 2. Gunnery. 3. Projectiles. I. Lynch, Osmond.
UF820 .S35 2018
623.51--dc23

For information on all Clanrye International publications
visit our website at www.clanryeinternational.com

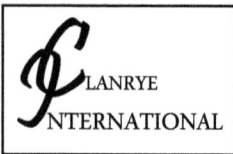

$\mathcal{CL}$ANRYE
INTERNATIONAL

Contents

Preface VII

Chapter 1   **An Overview of Ballistics** 1

Chapter 2   **Internal Ballistics: An Essential Aspect** 17
    a.   Internal Ballistics 17
    b.   Muzzle Energy 31
    c.   Ballistic Conduction 34
    d.   Gunpowder 39
    e.   Gunshot Residue 60
    f.   Rimfire (Firearms) 62

Chapter 3   **Transitional Ballistics: An Integrated Study** 68
    a.   Transitional Ballistics 68
    b.   Suppressor 70
    c.   Flash Suppressor 84
    d.   Muzzle Brake 88

Chapter 4   **Understanding External Ballistics** 95
    a.   External Ballistics 95
    b.   Propellant 122
    c.   Gun Barrel 126
    d.   Point-blank Range 127

Chapter 5   **An Overview of Projectile** 131
    a.   Projectile 131
    b.   Range of a Projectile 134
    c.   Projectile Motion 139
    d.   Trajectory 145
    e.   Trajectory of a Projectile 151
    f.   Muzzle Velocity 167
    g.   Catapult 169

Chapter 6   **Diverse Aspects of Ballistics** 176

    a.   Ballistic Coefficient 176

    b.   Deflection (Ballistics) 187

    c.   Elevation (Ballistics)..188

    d.   Ballistic Pendulum...189

    e.   Handgun Effectiveness 194

    f.   Rifleman's Rule 197

Permissions

Index

Preface

Ballistics deals with the mechanics of projectiles. A primary aspect of study under this field are bullets. It delves into the launch, effects and behavior of bullets. The designing of projectiles is also a significant aspect of this subject. Ballistics has four sub-fields namely terminal ballistics, internal ballistics, transitional ballistics, and external ballistics. This book attempts to understand the multiple branches that fall under the discipline of ballistics and how such concepts have practical applications. It elucidates the modern aspects and innovative models around prospective developments with respect to ballistics. This textbook is meant for students who are looking for an elaborate reference text on ballistics.

To facilitate a deeper understanding of the contents of this book a short introduction of every chapter is written below:

Chapter 1- The science of the behavior and effects of projectiles such as bullets, bombs and rockets is known as ballistics. Many weapons are designed based on this concept. This chapter is an overview of the subject matter incorporating all the major aspects of ballistics.

Chapter 2- Internal ballistics studies the propulsion of a projectile. Studying internal ballistics is important for the making of firearms. Muzzle energy, ballistic conduction, gunpowder, gunshot residue and rimfire are some of the significant and important topics related to internal ballistics. The following chapter unfolds its crucial aspects in a critical yet systematic manner.

Chapter 3- Traditional ballistics studies the behavior of the projectile from the moment it leaves the muzzle till the moment of equalization of force that provides motion. The two main devices that come under transitional ballistics realm are flash suppressors and sound suppressors. This chapter elucidates the crucial theories and principles of transitional ballistics.

Chapter 4- External ballistics is a part of ballistics that studies the performance of projectile when it is in motion. The projectile can either be guided, unguided, powered or un-powered. The topics discussed in the chapter are of great importance to broaden the existing knowledge on external ballistics.

Chapter 5- Any object that is thrown by the exertion of force is known as projectile. Range of a projectile, projectile motion, projectile point, trajectory, trajectory of a projectile, muzzle velocity and catapult are some of the topics related to the subject. Projectile is best understood in confluence with the major topics listed in the following chapter.

Chapter 6- The diverse aspects of ballistics include ballistic coefficient, deflection, elevation, ballistic pendulum, handgun effectiveness and Rifleman's rule. Ballistic coefficient is the capability of any object to overcome air resistance in flight whereas defection is the method that is used in pushing a projectile at a target which is moving at any speed.

Finally, I would like to thank the entire team involved in the inception of this book for their valuable time and contribution. This book would not have been possible without their efforts. I would also like to thank my friends and family for their constant support.

Editor

An Overview of Ballistics

The science of the behavior and effects of projectiles such as bullets, bombs and rockets is known as ballistics. Many weapons are designed based on this concept. This chapter is an overview of the subject matter incorporating all the major aspects of ballistics.

Ballistics

Ballistics is the science of mechanics that deals with the launching, flight, behavior, and effects of projectiles, especially bullets, unguided bombs, rockets, or the like; the science or art of designing and accelerating projectiles so as to achieve a desired performance.

A ballistic body is a body with momentum which is free to move, subject to forces, such as the pressure of gases in a gun or a propulsive nozzle, by rifling in a barrel, by gravity, or by air drag.

A ballistic missile is a missile only guided during the relatively brief initial powered phase of flight, whose trajectory is subsequently governed by the laws of classical mechanics, in contrast (for example) to a cruise missile which is aerodynamically guided in powered flight.

History and Prehistory

The earliest known ballistic projectiles were stones and spears, and the throwing stick.

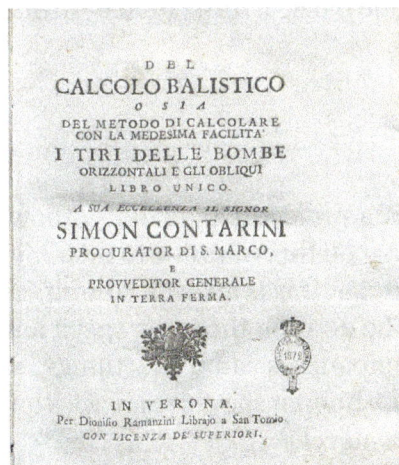

Gaetano Marzagaglia, *Del calcolo balistico*, 1748

The oldest evidence of stone-tipped projectiles, which may or may not have been propelled by a bow (c.f. atlatl), dating to c. 64,000 years ago, were found in Sibudu Cave, present day-South Africa. The oldest evidence of the use of bows to shoot arrows dates to about 10,000 years ago; it is based on pinewood arrows found in the Ahrensburg valley north of Hamburg. They had shallow grooves on the base, indicating that they were shot from a bow. The oldest bow so far recovered is about 8,000 years old, found in the Holmegård swamp in Denmark.

Archery seems to have arrived in the Americas with the Arctic small tool tradition, about 4,500 years ago.

The first devices identified as guns appeared in China around 1000 AD, and by the 12th century the technology was spreading through the rest of Asia, and into Europe by the 13th century.

After millennia of empirical development, the discipline of ballistics was initially studied and developed by Italian mathematician Niccolò Tartaglia in 1531 and went on to be put on a solid scientific and mathematical basis with the publication of Philosophiæ Naturalis Principia Mathematica in 1687 which gave mathematical laws of motion and gravity which for the first time made it possible to successfully predict trajectories.

The word *ballistics* comes from the Greek *ballein*, meaning "to throw".

Projectiles

A projectile is any object projected into space (empty or not) by the exertion of a force. Although any object in motion through space (for example a thrown baseball) is a projectile, the term most commonly refers to a ranged weapon. Mathematical equations of motion are used to analyze projectile trajectory.

Examples of projectiles include balls, arrows, bullets, artillery shells, rockets, etc.

Projectile Launchers

Throwing

Throwing is the launching of a projectile by hand. Although some other animals can throw, humans are unusually good throwers due to their high dexterity and good timing capabilities, and it is believed that this is an evolved trait. Evidence of human throwing dates back 2 million years. The 90 mph throwing speed found in many athletes far exceeds the speed at which chimpanzees can throw things, which is about 20 mph. This ability reflects the ability of the human shoulder muscles and tendons to store elasticity until it is needed to propel an object.

Baseball throws can exceed 100 mph

Sling

A sling is a projectile weapon typically used to throw a blunt projectile such as a stone, clay or lead "sling-bullet".

A sling has a small cradle or *pouch* in the middle of two lengths of cord. The *sling stone* is placed in the pouch. The middle finger or thumb is placed through a loop on the end of one cord, and a tab at the end of the other cord is placed between the thumb and forefinger. The sling is swung in an arc, and the tab released at a precise moment. This frees the projectile to fly to the target.

Bow

A bow is a flexible piece of material which shoots aerodynamic projectiles called arrows. A string joins the two ends and when the string is drawn back, the ends of the stick are flexed. When the string is released, the potential energy of the flexed stick is transformed into the velocity of the arrow. Archery is the art or sport of shooting arrows from bows.

Catapult

A catapult is a device used to launch a projectile a great distance without the aid of explosive devices — particularly various types of ancient and medieval siege engines. The catapult has been used since ancient times, because it was proven to be one of the most effective mechanisms during warfare. The word "catapult" comes from the Latin "catapulta", which in turn comes from the Greek (*katapeltēs*), itself from (*kata*), "downwards" and (*pallō*), "to toss, to hurl". Catapults were invented by the ancient Greeks.

Catapult 1 Mercato San Severino

Gun

USS *Iowa* (BB-61) fires a full broadside, 1984

A gun is a normally tubular weapon or other device designed to discharge projectiles or other material. The projectile may be solid, liquid, gas, or energy and may be free, as with bullets and artillery shells, or captive as with Taser probes and whaling harpoons. The means of projection varies according to design but is usually effected by the action of gas pressure, either produced through the rapid combustion of a propellant or compressed and stored by mechanical means, operating on the projectile inside an open-ended tube in the fashion of a piston. The confined gas accelerates the movable projectile down the length of the tube imparting sufficient velocity to sustain the projectile's travel once the action of the gas ceases at the end of the tube or muzzle. Alternatively, acceleration via electromagnetic field generation may be employed in which case the tube may be dispensed with and a guide rail substituted.

Rocket

SpaceX's Falcon 9 Full Thrust rocket, 2017

A rocket is a missile, spacecraft, aircraft or other vehicle that obtains thrust from a rocket engine. Rocket engine exhaust is formed entirely from propellants carried within the rocket before use. Rocket engines work by action and reaction. Rocket engines push rockets forward simply by throwing their exhaust backwards extremely fast.

While comparatively inefficient for low speed use, rockets are relatively lightweight and powerful, capable of generating large accelerations and of attaining extremely high speeds with reasonable efficiency. Rockets are not reliant on the atmosphere and work very well in space.

Rockets for military and recreational uses date back to at least 13th century China. Significant scientific, interplanetary and industrial use did not occur until the 20th century, when rocketry was the enabling technology for the Space Age, including setting foot on the Moon. Rockets are now used for fireworks, weaponry, ejection seats, launch vehicles for artificial satellites, human spaceflight, and space exploration.

Chemical rockets are the most common type of high performance rocket and they typically create their exhaust by the combustion of rocket propellant. Chemical rockets store a large amount of energy in an easily released form, and can be very dangerous. However, careful design, testing, construction and use minimizes risks.

Subfields

Ballistics is often broken down into the following four categories:

- *Internal ballistics* the study of the processes originally accelerating projectiles

- *Transition ballistics* the study of projectiles as they transition to unpowered flight

- *External ballistics* the study of the passage of the projectile (the trajectory) in flight

- *Terminal ballistics* the study of the projectile and its effects as it ends its flight

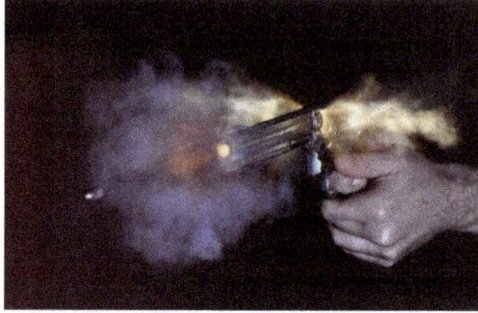

Ballistics can be studied using high-speed photography or high-speed cameras. A photo of a Smith & Wesson firing, taken with an ultra high speed air-gap flash. Using this sub-microsecond flash, the bullet can be imaged without motion blur.

Internal Ballistics

Internal ballistics (also interior ballistics), a sub-field of ballistics, is the study of the propulsion of a projectile.

In guns internal ballistics covers the time from the propellant's ignition until the projectile exits the gun barrel. The study of internal ballistics is important to designers and users of firearms of all types, from small-bore rifles and pistols, to high-tech artillery.

For rocket propelled projectiles, internal ballistics covers the period during which a rocket engine is providing thrust.

Transitional Ballistics

Transitional ballistics, also known as intermediate ballistics, is the study of a projectile's behavior from the time it leaves the muzzle until the pressure behind the projectile is equalized, so it lies between internal ballistics and external ballistics.

External Ballistics

External ballistics is the part of the science of ballistics that deals with the behaviour of a non-powered projectile in flight.

External ballistics is frequently associated with firearms, and deals with the unpowered free-flight phase of the bullet after it exits the gun barrel and before it hits the target, so it lies between transitional ballistics and terminal ballistics.

However, external ballistics is also concerned with the free-flight of rockets and other projectiles, such as balls, arrows etc.

Terminal Ballistics

Terminal ballistics is the study of the behavior and effects of a projectile when it hits its target.

Terminal ballistics is relevant both for small caliber projectiles as well as for large caliber projectiles (fired from artillery). The study of extremely high velocity impacts is still very new and is as yet mostly applied to spacecraft design.

Applications

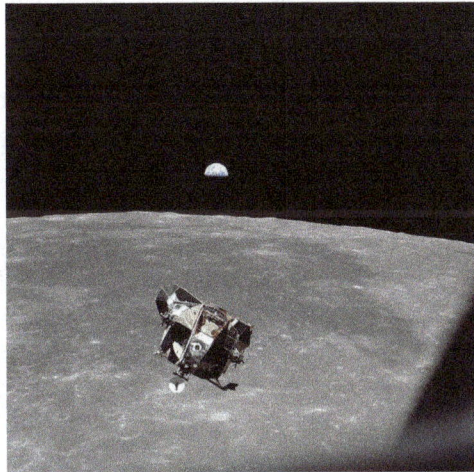

Apollo 11 — Astrodynamic calculations have permitted spacecraft to travel to and return from the Moon.

Forensic Ballistics

Forensic ballistics involves analysis of bullets and bullet impacts to determine information of use to a court or other part of a legal system. Separately from ballistics information, firearm and tool mark examinations ("ballistic fingerprinting") involve analyzing firearm, ammunition, and tool mark evidence in order to establish whether a certain firearm or tool was used in the commission of a crime.

Astrodynamics

Astrodynamics is the application of ballistics and celestial mechanics to the practical problems concerning the motion of rockets and other spacecraft. The motion of these objects is usually calculated from Newton's laws of motion and Newton's law of universal gravitation. It is a core discipline within space mission design and control.

Terminal Ballistics

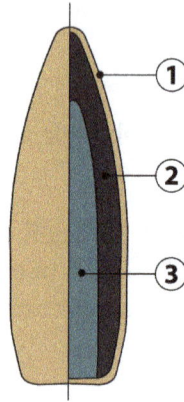

Bullet parts: 1 Metal Jacket, 2 Lead Core, 3 Steel penetrator.

Terminal ballistics, (also known as wound ballistics) a sub-field of ballistics, is the study of the behavior and effects of a projectile when it hits its target and transfers its energy to the target. Bullet design and the velocity of impact determine the effectiveness of its impact.

General

An early result is due to Newton; the impact depth of any projectile is the depth that a projectile will reach before stopping in a medium; in Newtonian mechanics, a projectile stops when it has transferred its momentum to an equal mass of the medium. If the impactor and medium have similar density this happens at an impact depth equal to the length of the impactor.

For this simple result to be valid, the arresting medium is considered to have no integral shear strength. Note that even though the projectile has stopped, the momentum is still transferred, and in the real world spalling and similar effects can occur.

Firearm Projectiles

Classes of Bullet

There are three basic classes of bullets:

- Those designed for maximum accuracy at varying ranges.
- Those designed to maximize damage to a target by penetrating as deeply as possible.
- Those designed to avoid over-penetration of a target, by deforming to control the depth to which the bullet penetrates, which as a by-product, deals more damage inside the wound.

The third class may limit penetration by expanding or fragmenting.

Target Shooting

.32 ACP full metal jacket, .32 S&W Long wadcutter,
380 ACP jacketed hollow point.

For short range target shooting on ranges up to 50 meters (55 yd), aerodynamics are relatively unimportant and velocities are low. As long as the bullet is balanced so it does not tumble, the aerodynamics are unimportant. For shooting at paper targets, the best bullet is one that will punch a perfect hole through the target. These bullets are called wadcutters. They have a very flat front, often with a relatively sharp edge along the perimeter. The flat front punches out a large hole in the paper, close to, if not equal to, the full diameter of the bullet.

This allows for easy, unambiguous scoring of the target. Since cutting the edge of a target ring will result in scoring the higher score, fractions of an inch are important. Magazine-fed pistols may not reliably feed wadcutters because of the angular shape. To address this, the semiwadcutter is used. The semiwadcutter consists of a conical section that comes to a smaller flat, and a thin sharp shoulder at the base of the cone. The flat point punches a clean hole, and the shoulder opens the hole up cleanly. For steel targets, the concern is to provide enough force to knock over the target while minimizing the damage to the target. A soft lead bullet, or a jacketed hollow-point bullet or soft-point bullet will flatten out on impact (if the velocity at impact is sufficient to make it deform), spreading the impact over a larger area of the target, allowing more total force to be applied without damaging the steel target.

There are also specialized bullets designed for use in long range precision target shooting with high-powered rifles; the designs vary somewhat from manufacturer to manufacturer. Research in the 1950s by the U.S. Air Force discovered that bullets are more stable in flight for longer distances and more resistant to crosswinds if the center of gravity is somewhat to the rear of the center of pressure The MatchKing bullet (which is still in wide use and holds many records) is a hollow point design with a tiny aperture in the jacket at the point of the bullet and a hollow air space under the point of the bullet, where previous conventional bullets had had a lead core that went all the way up to the point.

The U.S. military now issues ammunition to snipers that use bullets of this type. In 7.62×51mm NATO, M852 Match and M118LR ammunition are issued, both of which use Sierra MatchKing bullets; in 5.56×45mm NATO, those U.S. Navy and U.S. Marine snipers who use accurized M16-type rifles are issued the Mk 262 Mod 0 cartridge developed jointly by Black Hills Ammunition and Crane Naval Surface Warfare Center, using a bullet manufactured by Sierra Bullets that was cannelured according to military specifications for this project.

For ultra long range precision target shooting with high-powered rifles and military sniping, radically designed very-low-drag (VLD) bullets are available that are generally produced out of rods of mono-metal alloys on CNC lathes. The driving force behind these projectiles is the wish to enhance the practical maximum effective range beyond normal standards. To achieve this, the bullets have to be very long and normal cartridge overall lengths often have to be exceeded. Common rifling twist rates also often have to be tightened to stabilize very long projectiles. Such commercially nonexistent cartridges are termed "wildcats". The use of a wildcat based (ultra) long-range cartridge demands the use of a custom or customized rifle with an appropriately cut chamber and a fast-twist bore.

Maximum Penetration

For use against armored targets, or large, tough game animals, penetration is the most important consideration. Focusing the largest amount of momentum on the smallest possible area of the target provides the greatest penetration. Bullets for maximum penetration are designed to resist deformation on impact, and usually are made of lead that is covered in a copper, brass, or mild steel jacket (some are even solid copper or bronze alloy). The jacket completely covers the front of the bullet, although often the rear is left with exposed lead (this is a manufacturing consideration: the jacket is formed first, and the lead is swaged in from the rear).

For penetrating substances significantly harder than jacketed lead, the lead core is supplemented with or replaced with a harder material, such as hardened steel. Military armor-piercing small arms ammunition is made from a copper-jacketed steel core; the steel resists deformation better than the usual soft lead core leading to greater penetration. The current NATO 5.56mm SS109 (M855) bullet uses a steel-tipped lead core to improve penetration, the steel tip providing resistance to deformation for armor piercing, and the heavier lead core (25% heavier than the previous bullet, the M193) providing increased sectional density for better penetration in soft targets. For larger, higher-velocity calibers, such as tank guns, hardness is of secondary importance to density, and are normally sub-caliber projectiles made from tungsten carbide, tungsten hard alloy or depleted uranium fired in a light aluminum or magnesium alloy (or carbon fibre in some cases) sabot.

Many modern tank guns are smoothbore, not rifled, because practical rifling twists can only stabilize projectiles, such as an armour-piercing fin-stabilized discarding sabot

(APFSDS), with a length-to-diameter ratio of up to about 5:1, the spin imposed by rifling interferes with shaped-charge rounds, and also because the rifling adds friction and reduces the velocity it is possible to achieve. To get the maximum force on the smallest area, anti-tank rounds have aspect ratios of 10:1 or more. Since these cannot be stabilized by rifling, they are built instead like large darts, with fins providing the stabilizing force, negating the need for rifling. These subcaliber rounds are held in place in the bore by sabots. The sabot is a light material that transfers the pressure of the charge to the penetrator, then is discarded when the round leaves the barrel.

Controlled Penetration

The final category of bullets is that intended to control penetration so as not to harm anything behind the target. Such bullets are used primarily for hunting and civilian antipersonnel use; they are not generally used by the military, since the use of expanding bullets in international conflicts is prohibited by the Hague Convention and because these bullets have less chance of penetrating modern body armor. These bullets are designed to increase their surface area on impact, thus creating greater drag and limiting the travel through the target. A desirable side effect is that the expanded bullet makes a larger hole, increasing tissue disruption and speeding incapacitation.

In some applications, preventing exit from the rear of the target is also desirable. A bullet that penetrates through-and-through tends to cause more profuse bleeding, allowing a game animal to be bloodtrailed more easily. On the other hand, a perforating bullet can then continue on (likely not coaxial to the original trajectory due to target deflection) and might cause unintended damage or injury. Frangible bullets, made of tiny fragments held together by a weak binding, are often sold as an "ultimate" expanding bullet, as they will increase their effective diameter by an order of magnitude. When they work, they work extremely well, causing massive trauma to the target. On the other hand, when they fail, it is due to underpenetration, and the damage to the target is superficial and leads to very slow incapacitation.

Flat Point

The simplest maximum disruption bullet is one with a wide, flat tip. This increases the effective surface area, as rounded bullets can allow tissues to "flow" around the edges. It also increases drag during flight, which decreases the depth to which the bullet penetrates. Older centerfire rifles with tube magazines were designed to be used with flat-point bullets. Flat-point bullets, with fronts of up to 90% of the overall bullet diameter, are usually designed for use against large or dangerous game. They are often made of unusually hard alloys, are longer and heavier than normal for their caliber, and even include exotic materials such as tungsten to increase their sectional density.

These bullets are designed to penetrate deeply through muscle and bone, while causing a wound channel of nearly the full diameter of the bullet. These bullets are designed to penetrate deeply enough to reach vital organs from any shooting angle and at a far enough range. One of the hunting applications of the flat point bullet is large game such as bear hunted with a handgun in a .44 Magnum or larger caliber. More common than hunting is its use in a defensive "bear gun" carried by outdoorsmen. The disadvantage of flat point bullets is the reduction in aerodynamic performance; the flat point induces much drag, leading to significantly reduced velocities at long range.

Expanding

More effective on lighter targets are the expanding bullets, the hollow point bullet and the soft point bullet. These are designed to use the hydraulic pressure of muscle tissue to expand the bullet. The hollow point peels back into eight or nine connected pieces causing it to expand the damaged area. The hollow point fills with body water on impact, then expands as the bullet continues to have water pushed into it. This process is called mushrooming, as the ideal result is a shape that resembles a mushroom—a cylindrical base, topped with a wide surface where the tip of the bullet has peeled back to expose more area to create more drag while traveling through a body. A copper-plated hollowpoint loaded in a .44 Magnum, for example, with an original weight of 240 grains (15.55 g) and a diameter of 0.43 inch (11 mm) might mushroom on impact to form a rough circle with a diameter of 0.70 inch (18 mm) and a final weight of 239 grains (15.48 g).

This is excellent performance; almost the entire weight is retained, and the frontal surface area increased 63%. Penetration of the hollowpoint would be less than half that of a similar nonexpanding bullet, and the resulting wound or permanent cavity would be much wider.

Fragmenting

Example photo of the over-penetration of a fragmenting projectile.

This class of projectile is designed to break apart on impact, causing an effect similar to that of a frangible projectile, whilst being of a construction more akin to that of an expanding bullet. Fragmenting bullets are usually constructed like the hollowpoint projectiles described above, but with deeper and larger cavities. They may also have thinner copper jackets in order to reduce their overall integrity. For the purposes of aerodynamic efficiency the tip of the hollowpoint will often be tipped with a pointed polymer 'nose'. These bullets are typically fired at high velocities to maximize their fragmentation upon impact. In contrast to a hollowpoint which attempts to stay in one large piece retaining as much weight as possible whilst presenting the most surface area to the target, a fragmenting bullet is intended to break up into many small pieces almost instantly.

This means that all the kinetic energy from the bullet is transferred into the target in a very short period of time. The most common application of this bullet is the shooting of small vermin, such as prairie dogs. The effect of these bullets is quite dramatic, often resulting in the animal being blown apart upon impact. However, on larger game fragmenting ammunition provides inadequate penetration of vital organs to ensure a clean kill; instead, a "splash wound" may result. This also limits practical use of these rounds to supersonic (rifle) rounds, which have a high enough kinetic energy to ensure a lethal hit. The two main advantages of this ammunition are that it is very humane, as a hit almost anywhere on most small vermin will ensure an instant kill, and that instead of dangerously and uncontrollably ricocheting off surfaces, the bullet harmlessly breaks apart. Fragmenting bullets should not be confused with frangible bullets.

Frangible

The last category of expanding bullets is frangible bullets. These bullets are designed to break up on impact, which results in a huge increase in surface area. The most common of these bullets are made of small diameter lead pellets, placed in a thin copper shell and held in place by an epoxy or similar binding agent. On impact, the epoxy shatters and the copper shell opens up, much like a hollowpoint. The individual lead balls then spread out in a wide pattern, and due to their low mass to surface area ratio, stop very quickly. Similar bullets are made out of sintered metals, which turn to powder upon impact. These bullets are usually restricted to pistol cartridges, as the nonhomogenous cores tend to cause inaccuracies that, while acceptable at short pistol ranges, are not acceptable for the typical range at which rifles are used.

One interesting use of the sintered metal rounds is in shotguns in hostage rescue situations; the sintered metal round is used at near-contact range to shoot the lock mechanism out of doors. The resulting metal powder will immediately disperse after knocking out the door lock, and cause little or no damage to occupants of the room. Frangible rounds are also used by armed security agents on aircraft. The concern is not depressurization (a bullet hole will not depressurise an airliner), but over penetration and damage to vital electrical or hydraulic lines, or injury to an innocent bystander by a bullet that travels through a target's body completely instead of stopping in the body.

Also used are bullets similar to hollowpoint bullets or soft point bullets whose cores and/or jackets are deliberately weakened to cause deformation or fragmentation upon impact. The Warsaw Pact 5.45×39mm M74 assault rifle round exemplifies a trend that is becoming common in the era of high velocity, small caliber military rounds. The 5.45×39mm uses a steel-jacketed bullet with a two-part core, the rear being lead and the front being steel with an air pocket fore most. Upon impact, the unsupported tip deforms, bending the bullet nose into a slight "L" shape. This causes the bullet to tumble in the tissue, thus increasing its effective frontal surface area by traveling sideways more often than not.

This does not violate the Hague Convention, as it specifically mentions bullets that expand or flatten in the body. The NATO SS109 also tends to bend at the steel/lead junction, but with its weaker jacket, it fragments into many dozens of pieces. NATO 7.62 mm ball manufactured by some countries, such as Germany and Sweden, are also known to fragment due to jacket construction.

Other bullets in use by militaries are quite back heavy, due to a long, sharp point created in an attempt to get the maximum ballistic coefficient. These bullets will flip over after impact, then settle into a stable, back first orientation before stopping. The Swiss military actually redesigned their 5.56mm assault rifle bullet to prevent this, to more fully comply with the spirit of the Hague Convention, though according to some sources the present GP90 5.56×45mm Swiss assault rifle ammunition was actually designed as an armor-piercing bullet, because, in the 1980s, it was perceived that the Soviets and their Warsaw Pact allies were going to issue soft body armor to infantry units on a wide basis, but after the end of the Cold War, the Bofors corporation, having spent a great deal of money on developing the new bullet, changed the sales pitch in order to sell it to the Swiss government.

It might seem that if the whole purpose of a maximum disruption round is to expand to a larger diameter, it would make more sense to start out with the desired diameter rather than relying on the somewhat inconsistent results of expansion on impact. While there is merit to this (there is a strong following of the .45 ACP, as compared to the .40 S&W and 0.355 in diameter 9×19mm, for just this reason) there are also significant downsides. A larger diameter bullet is going to have significantly more drag than a smaller diameter bullet of the same mass, which means long range performance will be significantly degraded. A larger diameter bullet also means more space is required to store the ammunition, which means either bulkier guns or smaller magazine capacities. The common trade-off when comparing .45 ACP, .40 S&W, and 9×19mm pistols is a 7- to 14-round capacity in the .45 ACP vs. a 10- to 16-round capacity in the .40 S&W vs. a 13- to 19-round capacity in the 9×19mm.

Although several .45-caliber pistols are available with high-capacity magazines (Para Ordnance being one of the first in the late 1980s) many people find the wide grip required uncomfortable and difficult to use. Especially where the military requirement of a nonexpanding round is concerned, there is fierce debate over whether it is better to have fewer, larger bullets for enhanced terminal effects, or more, smaller bullets for increased number of potential target hits.

Large Caliber

The purpose of firing a large calibre projectile is not always the same. For example, one might need to create disorganisation within enemy troops, create casualties within enemy troops, eliminate the functioning of an enemy tank, or destroy an enemy bunker. Different purposes of course require different projectile designs.

Many large calibre projectiles are filled with a high explosive which, when detonated, shatters the shell casing, producing thousands of high velocity fragments and an accompanying sharply rising blast overpressure. More rarely, others are used to release chemical or biological agents, either on impact or when over the target area; designing an appropriate fuse is a difficult task which lies outside the realm of terminal ballistics.

Other large-calibre projectiles use bomblets (sub-munitions), which are released by the carrier projectile at a required height or time above their target. For US artillery ammunition, these projectiles are called Dual-Purpose Improved Conventional Munition (DPICM), a 155 mm M864 DPICM projectile for example contains a total of 72 shaped-charge fragmentation bomblets. The use of multiple bomblets over a single HE projectile allows for a denser and less wasteful fragmentation field to be produced. If a bomblet strikes an armoured vehicle, there is also a chance that the shaped charge will (if used) penetrate and disable the vehicle. A negative factor in their use is that any bomblets that fail to function go on to litter the battlefield in a highly sensitive and lethal state, causing casualties long after the cessation of conflict. International conventions tend to forbid or restrict the use of this type of projectile.

Some anti-armour projectiles use what is known as a shaped charge to defeat their target. Shaped charges have been used ever since it was discovered that a block of high explosives with letters engraved in it created perfect impressions of those letters when detonated against a piece of metal. A shaped charge is an explosive charge with a hollow lined cavity at one end and a detonator at the other. They operate by the detonating high explosive collapsing the (often copper) liner into itself. Some of the collapsing liner goes on to form a constantly stretching jet of material travelling at hypersonic speed. When detonated at the correct standoff to the armour, the jet violently forces its way through the target's armour.

Contrary to popular belief, the jet of a copper-lined shaped charge is not molten, although it is heated to about 500 °C. This misconception is due to the metal's fluid-like behaviour, which is caused by the massive pressures produced during the explosives detonation causing the metal to flow plastically. When used in the anti-tank role, a projectile that uses a shaped-charge warhead is known by the acronym HEAT (high-explosive anti-tank).

Shaped charges can be defended against by the use of explosive reactive armour (ERA), or complex composite armour arrays. ERA uses a high explosive sandwiched between two, relatively thin, (normally) metallic plates. The explosive is detonated when struck by the shaped charge's jet, the detonating explosive sandwich forces the two plates apart, lowering the jets' penetration by interfering with, and disrupting it. A disadvantage of using ERA is that each plate can protect against a single strike, and the resulting explosion can be extremely dangerous to nearby personnel and lightly armoured structures.

Tank fired HEAT projectiles are slowly being replaced for the attack of heavy armour by so-called "kinetic energy" penetrators. Ironically, it is the most primitive (in-shape) projectiles that are hardest to defend against. A KE penetrator requires an enormous thickness of steel, or a complex armour array to protect against. They also produce a much larger diameter hole in comparison to a shaped charge and hence produce a far more extensive behind armour effect. KE penetrators are most effective when constructed of a dense tough material that is formed into a long, narrow, arrow/dart like projectile.

Tungsten and depleted uranium alloys are often used as the penetrator material. The length of the penetrator is limited by the ability of the penetrator to withstand launch forces whilst in the bore and shear forces along its length at impact.

Internal Ballistics: An Essential Aspect

Internal ballistics studies the propulsion of a projectile. Studying internal ballistics is important for the making of firearms. Muzzle energy, ballistic conduction, gunpowder, gunshot residue and rimfire are some of the significant and important topics related to internal ballistics. The following chapter unfolds its crucial aspects in a critical yet systematic manner.

Internal Ballistics

Internal ballistics (also interior ballistics), a subfield of ballistics, is the study of the propulsion of a projectile.

In guns internal ballistics covers the time from the propellant's ignition until the projectile exits the gun barrel. The study of internal ballistics is important to designers and users of firearms of all types, from small-bore rifles and pistols, to high-tech artillery.

For rocket-propelled projectiles, internal ballistics covers the period during which a rocket motor is providing thrust.

Parts and Equations

Hatcher breaks the duration of interior ballistics into 3 parts:

- Lock time, the time from sear release until the primer is struck

- Ignition time, the time from when the primer is struck until the projectile starts to move

- Barrel time, the time from when the projectile starts to move until it exits the barrel.

There are many processes that are significant. The source of energy is the burning propellant. It generates hot gases that raise the chamber pressure. That pressure pushes on the base of the projectile, and causes the projectile to accelerate. The chamber pressure depends on many factors. The amount of propellant that has burned, the temperature of the gases, and the volume of the chamber. The burn rate of the propellant depends not only on the chemical make up, but also on the shape of the propellant grains. The temperature depends not only on the energy released, but also the heat lost to the sides of the barrel and chamber. The volume of the chamber is continuously changing: as the propellant burns, there is more volume for the gas to occupy. As the projectile travels down the barrel, the volume behind the projectile also increases. There are still other effects. Some energy is lost in deforming the projectile and causing it to spin. There are also frictional losses between the projectile and the barrel. The projectile, as it travels down the barrel, compresses the air in front of it.

Models have been developed for these processes. These processes affect the gun design. The breech and the barrel must resist the high-pressure gases without damage. Although the pressure initially rises to a high value, the pressure starts dropping when the projectile has traveled some distance down the barrel. Consequently, the muzzle end of the barrel does not need to be as strong as the chamber end.

There are five general equations used in interior ballistics:

1. The equation of state of the propellant

2. The equation of energy

3. The equation of motion

4. The burning rate equation

5. The equation of the form function

History

Prior to the mid-1800s, before the development of electronics and the necessary mathematics, and material science to fully understand pressure vessel design, internal ballistics did not have a lot of detailed objective information. Barrels and actions would simply be built strong enough to survive a known overload (Proof test), and muzzle velocity change could be surmised from the distance the projectile traveled.

In the 1800s test barrels began to be instrumented. Holes were drilled in the barrel, "crusher gauges" using copper pellets were attached, the gun was fired, and the pressure was measured indirectly by how much the copper pellet was deformed. But the measurement only indicated the maximum pressure that was reached at that point in the barrel. By the 1960s, piezoelectric strain gauges were used. They allow instantaneous pressures to be measured and did not need a pressure port drilled into the barrel. More recently, using advanced telemetry and acceleration-hardened sensors, instrumented projectiles were developed that could measure the pressure at the base of the projectile and its acceleration.

Priming Methods

Through the years, several methods of igniting the propellant have been developed.

Originally, a small hole (a touch hole) was drilled into the breech so that a fine propellant (Black powder, the same propellant used in the gun) could be poured in, and an external flame or spark was applied. Later, Percussion caps and self-contained cartridges, had primers that detonated after mechanical deformation, igniting the propellant. Electric current can also be used to ignite the propellant.

Propellants

Black Powder

Gunpowder (Black powder) is a finely ground, pressed and granulated mechanical pyrotechnic mixture of sulfur, charcoal, and potassium nitrate or sodium nitrate. It can be produced in a range of grain sizes. The size and shape of the grains can increase or decrease the relative surface area, and change the burning rate significantly. The burning rate of black powder is relatively insensitive to pressure, meaning it will burn quickly and predictably even without confinement, making it also suitable for use as a low explosive. It has a very slow decomposition rate, and therefore a very low brisance. It is not, in the strictest sense of the term, an explosive, but a "deflagrant", as it does not detonate but decomposes by deflagration due to its subsonic mechanism of flame-front propagation.

Nitrocellulose (Single-base Propellants)

Nitrocellulose or "guncotton" is formed by the action of nitric acid on cellulose fibers. It is a highly combustible fibrous material that deflagrates rapidly when heat is applied. It also burns very cleanly, burning almost entirely to gaseous components at high temperatures with little smoke or solid residue. *Gelatinised* nitrocellulose is a plastic, which can be formed into cylinders, tubes, balls, or flakes known as *single-base* propellants. The size and shape of the propellant grains can increase or decrease the relative surface area, and change the burn rate significantly. Additives and coatings can be added to the propellant to further modify the burn rate. Normally, very fast powders are used for light-bullet or low-velocity pistols and shotguns, medium-rate powders for magnum pistols and light rifle rounds, and slow powders for large-bore heavy rifle rounds.

Double-base Propellants

Nitroglycerin can be added to nitrocellulose to form "double-base propellants". Nitrocellulose desensitizes nitroglycerin to prevent detonation in propellant-sized grains, and the nitroglycerin gelatinises the nitrocellulose and increases the energy. Double-base powders burn faster than single-base powders of the same shape, though not as cleanly, and burn rate increases with nitroglycerin content.

In artillery, Ballistite or Cordite has been used in the form of rods, tubes, slotted-tube, perforated-cylinder or multi-tubular; the geometry being chosen to provide the required burning characteristics. (Round balls or rods, for example, are "degressive-burning" because their production of gas decreases with their surface area as the balls or rods burn smaller; thin flakes are "neutral-burning," since they burn on their flat surfaces until the flake is completely consumed. The longitudally perforated or multi-perforated cylinders used in large, long-barreled rifles or cannon are "progressive-burning;" the burning surface increases as the inside diameter of the holes enlarges, giving sustained burning and a long, continuous push on the projectile to produce higher velocity without increasing the peak pressure unduly. Progressive-burning powder compensates somewhat for the pressure drop as the projectile accelerates down the bore and increases the volume behind it.)

Solid Propellants (Caseless Ammunition)

A recent topic of research has been in the realm of "caseless ammunition". In a caseless cartridge, the propellant is cast as a single solid grain, with the priming compound placed in a hollow at the base, and the bullet attached to the front. Since the single propellant grain is so large (most smokeless powders have grain sizes around 1 mm, but a caseless grain will be perhaps 7 mm diameter and 15 mm long), the relative burn rate must be much higher. To reach this rate of burning, caseless propellants often use moderated explosives, such as RDX. (Caseless ammunition might be considered a return to the mid-19th century, since the first practical cartridge repeater, the "Volcanic" pistol, used a charge of black powder in a cavity in the bullet base. This weapon was the direct ancestor of the Henry and Winchester rifles, though they switched to metal-cased ammunition. Some early rifles and revolvers also used combustible-paper cartridges, but they required a separate ignition system.) The major advantages of a successful caseless round would be elimination of the need to extract and eject the spent cartridge case, permitting higher rates of fire and a simpler mechanism, and also reduced ammunition weight by eliminating the weight (and cost) of the brass or steel case.

While there is at least one experimental military rifle (the H&K G11), and one commercial rifle (the Voere VEC-91), that use caseless rounds, they have met with little success. One other commercial rifle was the Daisy VL rifle made by the Daisy Air Rifle Co. and chambered for .22 caliber caseless ammunition that was ignited by a hot blast of compressed air from the lever used to compress a strong spring like for an air rifle. The caseless ammunition is of course not reloadable, since there is no casing left after firing the bullet, and the exposed propellant makes the rounds less durable. Also, the case in a standard cartridge serves as a seal, keeping gas from escaping the breech. Caseless arms must use a more complex self-sealing breech, which increases the design and manufacturing complexity. Another unpleasant problem, common to all rapid-firing arms but particularly problematic for those firing caseless rounds, is the problem of rounds "cooking off". This problem is caused by residual heat from the chamber heating the round in the chamber to the point where it ignites, causing an unintentional discharge.

To minimize the risk of cartridge cook-off, machineguns can be designed to fire from an open bolt, with the round not chambered until the trigger is pulled, and so there is no chance for the round to cook off before the operator is ready. Such weapons could use caseless ammunition effectively. Open-bolt designs are generally undesirable for anything but machine guns; the mass of the bolt moving forward causes the gun to lurch in reaction, which significantly reduces the accuracy of the gun, which is generally not an issue for machinegun fire.

Propellant Charge

Load Density and Consistency

Load density is the percentage of the space in the cartridge case that is filled with powder. In general, loads close to 100% density (or even loads where seating the bullet in the case, compresses the powder) ignite and burn more consistently than lower-density loads. In cartridges surviving from the black-powder era (examples being .45 Colt, .45-70 Government), the case is much larger than is needed to hold the maximum charge of high-density smokeless powder. This extra room allows the powder to shift in the case, piling up near the front or back of the case and potentially causing significant variations in burning rate, as powder near the rear of the case will ignite rapidly but powder near the front of the case will ignite later. This change has less impact with fast powders. Such high-capacity, low-density cartridges generally deliver best accuracy with the fastest appropriate powder, although this keeps the total energy low due to the sharp high-pressure peak.

Magnum pistol cartridges reverse this power/accuracy tradeoff by using lower-density, slower-burning powders that give high load density and a broad pressure curve. The downside is the increased recoil and muzzle blast from the high powder mass, and high muzzle pressure.

Most rifle cartridges have a high load density with the appropriate powders. Rifle cartridges tend to be bottlenecked, with a wide base narrowing down to a smaller diameter, to hold a light, high-velocity bullet. These cases are designed to hold a large charge of low-density powder, for an even broader pressure curve than a magnum pistol cartridge. These cases require the use of a long rifle barrel to extract their full efficiency, although they are also chambered in rifle-like pistols (single-shot or bolt-action) with barrels of 10 to 15 inches (25 to 38 cm).

One unusual phenomenon occurs when dense, low-volume powders are used in large-capacity rifle cases. Small charges of powder, unless held tightly near the rear of the case by wadding, can apparently detonate when ignited, sometimes causing catastrophic failure of the firearm. The mechanism of this phenomenon is not well known, and generally it is not encountered except when loading low recoil or low-velocity subsonic rounds for rifles. These rounds generally have velocities of under 1100 ft/s (320 m/s), and are used for indoor shooting, in conjunction with a suppressor or for pest control, where the power and muzzle blast of a full-power round is not needed or desired.

Chamber

Straight vs. Bottleneck

Straight walled cases were the standard from the beginnings of cartridge arms. With the low burning speed of black powder, the best efficiency was achieved with large, heavy bullets, so the bullet was the largest practical diameter. The large diameter allowed a short, stable bullet with high weight, and the maximum practical bore volume to extract the most energy possible in a given length barrel. There were a few cartridges that had long, shallow tapers, but these were generally an attempt to use an existing cartridge to fire a smaller bullet with a higher velocity and lower recoil. With the advent of smokeless powders, it was possible to generate far higher velocities by using a slow smokeless powder in a large volume case, pushing a small, light bullet. The odd, highly tapered 8 mm Lebel, made by necking down an older 11 mm black-powder cartridge, was introduced in 1886, and it was soon followed by the 7.92×57mm Mauser and 7×57mm Mauser military rounds, and the commercial .30-30 Winchester, all of which were new designs built to use smokeless powder. All of these have a distinct shoulder that closely resembles modern cartridges, and with the exception of the Lebel they are still chambered in modern firearms even though the cartridges are over a century old.

Aspect Ratio and Consistency

When selecting a rifle cartridge for maximum accuracy, a short, fat cartridge with very little case taper may yield higher efficiency and more consistent velocity than a long, thin cartridge with a lot of case taper (part of the reason for a bottle-necked design). Given current trends towards shorter and fatter cases, such as the new Winchester Super Short Magnum cartridges, it appears the ideal might be a case approaching spherical inside. Target and varmint hunting rounds require the greatest accuracy, so their cases tend to be short, fat, and nearly untapered with sharp shoulders on the case. Short, fat cases also allow short-action weapons to be made lighter and stronger for the same level of performance. The trade-off for this performance is fat rounds which take up more space in a magazine, sharp shoulders that do not feed as easily out of a magazine, and less reliable extraction of the spent round. For these reasons, when reliable feeding is more important than accuracy, such as with military rifles, longer cases with shallower shoulder angles are favored. There has been a long-term trend however, even among military weapons, towards shorter, fatter cases. The current 7.62×51mm NATO case replacing the longer .30-06 Springfield is a good example, as is the new 6.5 Grendel cartridge designed to increase the performance of the AR-15 family of rifles and carbines. Nevertheless, there is significantly more to accuracy and cartridge lethality than the length and diameter of the case, and the 7.62x51mm NATO has a smaller case capacity than the .30-06 Springfield, reducing the amount of propellant that can be used, directly reducing the bullet weight and muzzle velocity combination that contributes to lethality, (as detailed in the published cartridge specifications linked herein for comparison). The 6.5 Grendel, on the other hand, is capable of firing a significantly heavier bullet than the 5.56 NATO out of the AR-15 family of weapons, with only a slight decrease in muzzle velocity, perhaps providing a more advantageous performance tradeoff.

Friction and Inertia

Static Friction and Ignition

Since the burning rate of smokeless powder varies directly with the pressure, the initial pressure buildup,(i.e. "the shot-start pressure"), has a significant effect on the final velocity, especially in large cartridges with very fast powders and relatively light weight projectiles. In small caliber firearms, the friction holding the bullet in the case, determines how soon after ignition the bullet moves, and since the motion of the bullet increases the volume and drops the pressure, a difference in friction can change the slope of the pressure curve. In general, a tight fit is desired, to the extent of crimping the bullet into the case. In straight-walled rimless cases, such as the .45 ACP, an aggressive crimp is not possible, since the case is held in the chamber by the mouth of the case, but sizing the case to allow a tight interference fit with the bullet, can give the desired result. In larger caliber firearms, the shot start pressure is often determined by the force required to initially engrave the projectile driving band into the start of the barrel rifling; smoothbore guns, which do not have rifling, achieve shot start pressure by initially driving the projectile into a "forcing cone" that provides resistance as it compresses the projectile obturation ring.

Kinetic Friction

The bullet must tightly fit the bore to seal the high pressure of the burning gunpowder. This tight fit results in a large frictional force. The friction of the bullet in the bore does have a slight impact on the final velocity, but that is generally not much of a concern. Of greater concern is the heat that is generated due to the friction. At velocities of about 300 m/s (980 ft/s), lead begins to melt, and deposit in the bore. This lead build-up constricts the bore, increasing the pressure and decreasing the accuracy of subsequent rounds, and is difficult to scrub out without damaging the bore. Rounds, used at velocities up to 460 m/s (1,500 ft/s), can use wax lubricants on the bullet to reduce lead build-up. At velocities over 460 m/s (1,500 ft/s), nearly all bullets are jacketed in copper, or a similar alloy that is soft enough not to wear on the barrel, but melts at a high enough temperature to reduce build-up in the bore. Copper build-up does begin to occur in rounds that exceed 760 m/s (2,500 ft/s), and a common solution is to impregnate the surface of the bullet with molybdenum disulfide lubricant. This reduces copper build-up in the bore, and results in better long-term accuracy. Large caliber projectiles also employ copper driving bands for rifled barrels for spin-stabilized projectiles; however, fin-stabilized projectiles fired from both rifle and smoothbore barrels, such as the APFSDS anti-armor projectiles, employ nylon obturation rings that are sufficient to seal high pressure propellant gasses and also minimize in-bore friction, providing a small boost to muzzle velocity.

The Role of Inertia

In the first few centimeters of travel down the bore, the bullet reaches a significant percentage of its final velocity, even for high-capacity rifles, with slow burning powder. The acceleration is on the order of tens of thousands of gravities, so even a projectile as light as 40 grains (2.6 g) can provide over 1,000 newtons (220 lbf) of resistance due to inertia. Changes in bullet mass, therefore, have a huge impact on the pressure curves of smokeless powder cartridges, unlike black-powder cartridges. The loading or reloading of smokeless cartridges thus requires high-precision equipment, and carefully measured tables of load data for given cartridges, powders, and bullet weights.

Pressure-velocity Relationships

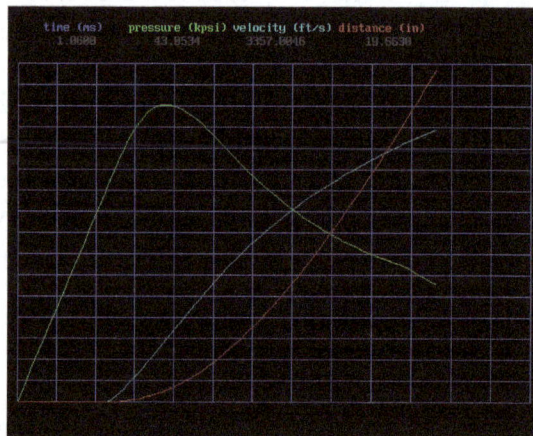

This is a graph of a simulation of the 5.56 mm NATO round, being fired from a 20-inch (510 mm) barrel. The horizontal axis represents time, the vertical axis represents pressure (green line), bullet travel (red line), and bullet velocity (light blue line). The values shown at top are peak values.

Energy is imparted to the bullet in a firearm by the pressure of gases produced by burning propellant. While higher pressures produce higher velocities, pressure duration is also important. Peak pressure may represent only a small fraction of the time the bullet is accelerating. The entire duration of the bullet's travel through the barrel must be considered.

Peak vs. Area

Energy is defined as the ability to do work on an object; for example, the work required to lift a one-pound weight, one foot against the pull of gravity defines a foot-pound of energy (One joule is equal to the energy needed to move a body over a distance of one meter using one newton of force). If we were to modify the graph to reflect force (the pressure exerted on the base of the bullet multiplied by the area of the base of the bullet) as a function of distance, the area under that curve would be the total energy imparted to the bullet. Increasing the energy of the bullet requires increasing the area under that curve, either by raising the average pressure, or increasing the distance the bullet travels under pressure. Pressure is limited by the strength of the firearm, and duration is limited by barrel length.

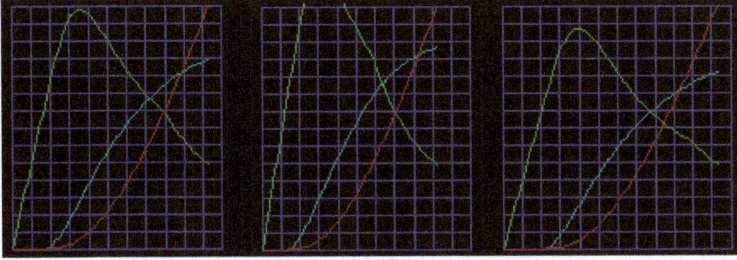

This graph shows different pressure curves for powders with different burn rates. The leftmost graph is the same as the large graph above. The middle graph shows a powder with a 25% faster burn rate, and the rightmost graph shows a powder with a 20% slower burn rate.

Propellant Design

Propellants are carefully matched to firearm strength, chamber volume and barrel length, and to bullet material, weight and dimensions. The rate of gas generation is proportional to the surface area of burning propellant grains in accordance with Piobert's Law. Progression of burning from the surface into the grains is attributed to heat transfer from the surface of energy necessary to initiate the reaction. Smokeless propellant reactions occur in a series of zones or phases as the reaction proceeds from the surface into the solid. The deepest portion of the solid experiencing heat transfer melts and begins phase transition from solid to gas in a *foam zone*. The gaseous propellant decomposes into simpler molecules in a surrounding *fizz zone*. Endothermic transformations in the *foam zone* and *fizz zone* require energy initially provided by the primer and subsequently released in a luminous outer *flame zone* where the simpler gas molecules react to form conventional combustion products like steam and carbon monoxide.

The heat transfer rate of smokeless propellants increases with pressure; so the rate of gas generation from a given grain surface area increases at higher pressures. Accelerating gas generation from fast burning propellants may rapidly create a destructively high pressure spike before bullet movement increases reaction volume. Conversely, propellants designed for a minimum heat transfer pressure may cease decomposition into gaseous reactants if bullet movement decreases pressure before a slow burning propellant has been consumed. Unburned propellant grains may remain in the barrel if the energy-releasing *flame zone* cannot be sustained in the resultant absence of gaseous reactants from the inner zones.

Propellant Burnout

Another issue to consider, when choosing a powder burn rate, is the time the powder takes to completely burn vs. the time the bullet spends in the barrel. Looking carefully at the left graph, there is a change in the curve, at about 0.8 ms. This is the point at which the powder is completely burned, and no new gas is created. With a faster powder, burnout occurs earlier, and with the slower powder, it occurs later. Propellant that is unburned when the bullet reaches the muzzle is wasted — it adds no energy to the bullet, but it does add to the recoil and muzzle blast. For maximum power, the powder should burn until the bullet is just short of the muzzle.

Since smokeless powders burn, not detonate, the reaction can only take place on the surface of the powder. Smokeless powders come in a variety of shapes, which serve to determine how fast they burn, and also how the burn rate changes as the powder burns. The simplest shape is a ball powder, which is in the form of round or slightly flattened spheres. Ball powder has a comparatively small surface-area-to-volume ratio, so it burns comparatively slowly, and as it burns, its surface area decreases. This means as the powder burns, the burn rate slows down.

To some degree, this can be offset by the use of a retardant coating on the surface of the powder, which slows the initial burn rate and flattens out the rate of change. Ball powders are generally formulated as slow pistol powders, or fast rifle powders.

Flake powders are in the form of flat, round flakes which have a relatively high surface-area-to-volume ratio. Flake powders have a nearly constant rate of burn, and are usually formulated as fast pistol or shotgun powders. The last common shape is an extruded powder, which is in the form of a cylinder, sometimes hollow. Extruded powders generally have a lower ratio of nitroglycerin to nitrocellulose, and are often progressive burning — that is, they burn at a faster rate as they burn. Extruded powders are generally medium to slow rifle powders.

Muzzle Pressure Concerns

From the pressure graphs, it can be seen that the residual pressure in the barrel as the bullet exits is quite high, in this case over 16 kpsi / 110 MPa / 1100 bar. While lengthening the barrel or reducing the amount of propellant gas will reduce this pressure, that often is not possible due to issues of firearm size and minimum required energy. Short-range target guns usually are chambered for .22 Long Rifle or .22 Short, which have very tiny powder capacities and little residual pressure. When higher energies are required for long-range shooting, hunting or anti-personnel use, high muzzle pressures are a necessary evil. With these high muzzle pressures come increased flash and noise from the muzzle blast, and, due to the large powder charges used, higher recoil. Recoil includes the reaction caused not just by the bullet, but also by the powder mass and speed (with the residual gases acting as a rocket exhaust). However, for a muzzle brake to be effective there must be significant muzzle pressure.

General Concerns

Bore Diameter and Energy Transfer

A firearm, in many ways, is like a piston engine on the power stroke. There is a certain amount of high-pressure gas available, and energy is extracted from it by making the gas move a piston — in this case, the projectile is the piston. The swept volume of the piston determines how much energy can be extracted from the given gas. The more volume that is swept by the piston, the lower is the exhaust pressure (in this case, the muzzle pressure). Any remaining pressure at the muzzle or at the end of the engine's power stroke represents lost energy.

To extract the maximum amount of energy, then, the swept volume is maximized. This can be done in one of two ways — increasing the length of the barrel or increasing the diameter of the projectile. Increasing the barrel length will increase the swept volume linearly, while increasing the diameter will increase the swept volume as the square of the diameter. Since barrel length is limited by practical concerns to about arm's length for a rifle and much shorter for a handgun, increasing bore diameter is the normal way to increase the efficiency of a cartridge. The limit to bore diameter is generally the sectional density of the projectile. Larger-diameter bullets of the same weight have much more drag, and so they lose energy more quickly after exiting the barrel. In general, most handguns use bullets between .355 (9 mm) and .45 (11.5 mm) caliber, while most rifles generally range from .223 (5.56 mm) to .32 (8 mm) caliber. There are many exceptions, of course, but bullets in the given ranges provide the best general-purpose performance. Handguns use the larger-diameter bullets for greater efficiency in short barrels, and tolerate the long-range velocity loss since handguns are seldom used for long-range shooting. Handguns designed for long-range shooting are generally closer to shortened rifles than to other handguns.

Ratio of Propellant to Projectile Mass

Another issue, when choosing or developing a cartridge, is the issue of recoil. The recoil is not just the reaction from the projectile being launched, but also from the powder gas, which will exit the barrel with a velocity even higher than that of the bullet. For handgun cartridges, with heavy bullets and light powder charges (a 9×19mm, for example, might use 5 grains (320 mg) of powder, and a 115 grains (7.5 g) bullet), the powder recoil is not a significant force; for a rifle cartridge (a .22-250 Remington, using 40 grains (2.6 g) of powder and a 40 grains (2.6 g) bullet), the powder can be the majority of the recoil force.

There is a solution to the recoil issue, though it is not without cost. A muzzle brake or recoil compensator is a device which redirects the powder gas at the muzzle, usually up and back. This acts like a rocket, pushing the muzzle down and forward. The forward push helps negate the feel of the projectile recoil by pulling the firearm forwards. The downward push, on the other hand, helps counteract the rotation imparted by the fact that most firearms have the barrel mounted above the center of gravity. Overt combat guns, large-bore high-powered rifles, long-range handguns chambered for rifle ammunition, and action-shooting handguns designed for accurate rapid fire, all benefit from muzzle brakes.

The high-powered firearms use the muzzle brake mainly for recoil reduction, which reduces the battering of the shooter by the severe recoil. The action-shooting handguns redirect all the energy up to counteract the rotation of the recoil, and make following shots faster by leaving the gun on target. The disadvantage of the muzzle brake is a longer, heavier barrel, and a large increase in sound levels and flash behind the muzzle of the rifle. Shooting firearms without muzzle brakes and without hearing protection can eventually damage the operator's hearing; however, shooting rifles with muzzle brakes - with or without hearing protection - causes permanent ear damage.

Powder-to-projectile-weight ratio also touches on the subject of efficiency. In the case of the .22-250 Remington, more energy goes into propelling the powder gas than goes into propelling the bullet. The .22-250 pays for this by requiring a large case, with lots of powder, all for a fairly small gain in velocity and energy over other .22 caliber cartridges.

Accuracy and Bore Characteristics

Nearly all small bore firearms, with the exception of shotguns, have rifled barrels. The rifling imparts a spin on the bullet, which keeps it from tumbling in flight. The rifling is usually in the form of sharp edged grooves cut as helices along the axis of the bore, anywhere from 2 to 16 in number. The areas between the grooves are known as lands.

Another system, polygonal rifling, gives the bore a polygonal cross section. Polygonal rifling is not very common, used by only a few European manufacturers as well as the American gun manufacturer Kahr Arms. The companies that use polygonal rifling claim greater accuracy, lower friction, and less lead and/or copper buildup in the barrel. Traditional land and groove rifling is used in most competition firearms, however, so the advantages of polygonal rifling are unproven.

There are three common ways of rifling a barrel, and one emerging technology:

- The most basic is to use a single point cutter, drawn down the bore by a machine that carefully controls the rotation of the cutting head relative to the barrel. This is the slowest process, but as it requires the simplest equipment, it is often used by custom gunsmiths, and can result in superbly accurate barrels.

- The next method is button rifling. This method uses a die with a negative image of the rifling cut on it. This die is drawn down the barrel while carefully rotated, and it swages the inside of the barrel. This "cuts" all the grooves at once (it does not really cut metal), and so is faster than cut rifling. Detractors claim that the process leaves considerable residual stress in the barrel, but world records have been set with button-rifled barrels, so again there is no clear disadvantage.

- The last common method used is hammer forging. In this process, a slightly oversized, bored barrel is placed around a mandrel that contains a negative image of the entire length of the rifled barrel. The barrel and mandrel are rotated and hammered by power hammers, which forms the inside of the barrel all at once. This is the fastest (and in the long run, cheapest) method of making a barrel, but the equipment is prohibitively expensive for all but the largest gun makers. Hammer-forged barrels are strictly mass-produced, so they are generally not capable of top accuracy as produced, but with some careful hand work, they can be made to shoot far better than most shooters are capable of.

- A new technique being applied to barrel manufacture is electrical machining, in the form of Electrical discharge machining (EDM) or Electro chemical machining (ECM). These processes use electricity to erode away material, a process produces a highly consistent diameter and very smooth finish, with less stress than other rifling methods. EDM is very costly and primarily used in large bore, long barrel cannon, where traditional methods are very difficult, while ECM is used by some smaller barrel makers.

The purpose of the barrel is to provide a consistent seal, allowing the bullet to accelerate to a consistent velocity. It must also impart the right spin, and release the bullet consistently, perfectly concentric to the bore. The residual pressure in the bore must be released symmetrically, so that no side of the bullet receives any more or less push than the rest. The muzzle of the barrel is the most critical part, since that is the part that controls the release of the bullet. Some rimfires and airguns actually have a slight constriction, called a choke, in the barrel at the muzzle. This guarantees that the bullet is held securely just before release.

To keep a good seal, the bore must be a very precise, constant diameter, or have a slight decrease in diameter from breech to muzzle. Any increase in bore diameter will allow the bullet to shift. This can cause gas to leak past the bullet, affecting the velocity, or cause the bullet to tip, so that it is no longer perfectly coaxial with the bore. High quality barrels are lapped to remove any constrictions in the bore which will cause a change in diameter.

A lapping process known as "fire lapping" uses a lead "slug" that is slightly larger than the bore and covered in fine abrasive compound to cut out the constrictions. The slug is passed from breech to muzzle, so that as it encounters constrictions, it cuts them away, and does no cutting on areas that are larger than the constriction. Many passes are made, and as the bore becomes more uniform, finer grades of abrasive compound are used. The final result is a barrel that is mirror-smooth, and with a consistent or slightly tapering bore. The hand-lapping technique uses a wooden or soft metal rod to pull or push the slug through the bore, while the newer fire-lapping technique uses specially loaded, low-power cartridges to push abrasive-covered soft-lead bullets down the barrel.

Another issue that has an effect on the barrel's hold on the bullet is the rifling. When the bullet is fired, it is forced into the rifling, which cuts or "engraves" the surface of the bullet. If the rifling is a constant twist, then the rifling rides in the grooves engraved in the bullet, and everything is secure and sealed. If the rifling has a decreasing twist, then the changing angle of the rifling in the engraved grooves of the bullet causes the rifling to become narrower than the grooves. This allows gas to blow by, and loosens the hold of the bullet on the barrel. An increasing twist, however, will make the rifling become wider than the grooves in the bullet, maintaining the seal. When a rifled-barrel blank is selected for a gun, careful measurement of the inevitable variations in manufacture can determine if the rifling twist varies, and put the higher-twist end at the muzzle.

The muzzle of the barrel is the last thing to touch the bullet before it goes into ballistic flight, and as such has the greatest potential to disrupt the bullet's flight. The muzzle must allow the gas to escape the barrel symmetrically; any asymmetry will cause an uneven pressure on the base of the bullet, which will disrupt its flight. The muzzle end of the barrel is called the "crown", and it is usually either beveled or recessed to protect it from bumps or scratches that might affect accuracy. A sign of a good crown will be a symmetric, star-shaped pattern on the muzzle end of the barrel, formed by soot deposited, as the powder gases escape the barrel. If the star is uneven, then it is a sign of an uneven crown, and an inaccurate barrel.

Before the barrel can release the bullet in a consistent manner, it must grip the bullet in a consistent manner. The part of the barrel between where the bullet exits the cartridge, and engages the rifling, is called the "throat", and the length of the throat is the freebore. In some firearms, the freebore is all but nonexistent — the act of chambering the cartridge forces the bullet into the rifling. This is common in low-powered rimfire target rifles. The placement of the bullet in the rifling ensures that the transition between cartridge and rifling is quick and stable. The downside is that the cartridge is firmly held in place, and attempting to extract the unfired round can be difficult, to the point of even pulling the bullet from the cartridge in extreme cases.

With high-powered cartridges, there is an additional disadvantage to a short freebore. A significant amount of force is required to engrave the bullet, and this additional resistance can raise the pressure in the chamber by quite a bit. To mitigate this effect, higher-powered rifles tend to have more freebore, so that the bullet is allowed to gain some momentum, and the chamber pressure is allowed to drop slightly, before the bullet engages the rifling. The downside is that the bullet hits the rifling when already moving, and any slight misalignment can cause the bullet to tip as it engages the rifling. This will, in turn, mean that the bullet does not exit the barrel coaxially. The amount of freebore is a function of both the barrel and the cartridge. The manufacturer or gunsmith who cuts the chamber will determine the amount of space between the cartridge case mouth and the rifling. Setting the bullet further forward or back in the cartridge can decrease or increase the amount of freebore, but only within a small range. Careful testing by the ammunition loader can optimize the amount of freebore to maximize accuracy, while keeping the peak pressure within limits.

Revolver-specific Issues

The defining characteristic of a revolver is the revolving cylinder, separate from the barrel, that contains the chambers. Revolvers typically have 5 to 10 chambers, and the first issue is ensuring consistency among the chambers, because if they aren't consistent then the point of impact will vary from chamber to chamber. The chambers must also align consistently with the barrel, so the bullet enters the barrel the same way from each chamber.

The throat in a revolver is part of the cylinder, and like any other chamber, the throat should be sized so that it is concentric to the chamber and very slightly over the bullet diameter. At the end of the throat, however, things change. First, the throat in a revolver is at least as long as the maximum overall length of the cartridge, otherwise the cylinder cannot revolve. The next step is the cylinder gap, the space between the cylinder and barrel. This must be wide enough to allow free rotation of the cylinder even when it becomes fouled with powder residue, but not so large that excessive gas is released. The next step is the forcing cone. The forcing cone is where the bullet is guided from the cylinder into the bore of the barrel. It should be concentric with the bore, and deep enough to force the bullet into the bore without significant deformation. Unlike rifles, where the threaded portion of the barrel is in the chamber, revolver barrels threads surround the breech end of the bore, and it is possible that the bore will be compressed when the barrel is screwed into the frame. Cutting a longer forcing cone can relieve this "choke" point, as can lapping of the barrel after it is fitted to the frame.

Muzzle Energy

$$E_k = \frac{1}{2}\,mv^2$$

Pellet exiting muzzle, with formula for energy overlaid.

Muzzle energy is the kinetic energy of a bullet as it is expelled from the muzzle of a firearm. It is often used as a rough indication of the destructive potential of a given firearm or load. The heavier the bullet and especially the faster it moves, the higher its muzzle energy and the more damage it will do.

The general formula for the kinetic energy is

$$E_k = \frac{1}{2}mv^2$$

where

v is the velocity of the bullet

m is the mass of the bullet.

A subsonic variant of ammunition that would otherwise be supersonic has its velocity limited to less than the speed of sound, i.e. v is fixed at around 300 metres per second (980 ft/s). For ammunition with this limitation the muzzle energy is variable only with respect to the bullet mass m.

Calculating Muzzle Energy

- In the SI system the above E_k will be in unit joule if the mass, m, is in kilogram, and the speed, v, is in metre per second.

- In United States engineering units, particular care must be taken to ensure that consistent units are used.

  - Mass, m, is usually given in grains and the speed, v, in feet per second but kinetic energy, E_k, is typically given in foot-pound force (abbreviated ft-lbf). Most sporting arms publications within the United States report muzzle energies in foot-pound force. If m is specified in grains and v in feet per second, the following formula can be used, which gives the energy in foot-pound force:

$$E_k = \frac{1}{2}mv^2 \times \left(\frac{1 \text{ ft} \cdot \text{lbf}}{7000 \text{ gr} \times 32.1739 \text{ ft/s}^2} \right)$$

  - When publishing kinetic energy tables for small arms ammunition, an acceleration due to gravity of 32.163 ft/s² rather than the standard of 32.1739 ft/s² is used. The formula therefore becomes

$$E_k = \frac{1}{2}mv^2 \times \left(\frac{1 \text{ ft} \cdot \text{lbf}}{7000 \text{ gr} \times 32.163 \text{ ft/s}^2} \right)$$

The bullet energy, remaining energy, down range energy, and impact energy of a projectile may also be calculated using the above formulas.

The first of the two US formulas can be simplified in conformance with US practice to:

$$E = (M \times V^2) \div K$$

where K = 450,435 and is derived from (2 x 32.1739 x 7000),

        M is the weight of the projectile, in grains,

        V is the velocity in feet per second and

        E is the energy in foot pounds.

Typical Muzzle Energies of Common Firearms and Cartridges

Example muzzle energy levels of different types of firearms			
Firearm (except listed air guns)	**Caliber**	**Muzzle energy**	
		ft-lbs	**joules**
air gun spring	.177	15	20
air gun mag spring	.22	22	30
air gun PCP	.22	30+	40+
pistol	.22LR	117	159
pistol	9 mm	383	519
pistol	.45 ACP	416	564
pistol	.357 Magnum	640	873
pistol	10mm	775	1,057
rifle	5.56×45 mm	1,325	1,796
rifle	7.62×39 mm	1,527	2,070
rifle	7.62 × 51 mm	2,802	3,799
rifle	.338 Lapua Magnum	4,893	6,634
heavy	.50 BMG	11,091	15,037
heavy	14.5 × 114 mm	23,744	32,000

Average muzzle energies for common pistol cartridges		
Cartridge	**Muzzle energy**	
	ft-lbf	**joules**
.380 ACP	199	270
.38 Special	310	420
9 mm Luger	350	470
.45 Colt	370	500
.45 GAP	400	540
.45 ACP	400	540

.40 S&W	425	576
.357 Sig	475	644
.357 Mag	550	750
10mm Auto	650	880
.41 Mag	910	1,230
.44 Mag	1,000	1,400
.50 AE	1,500	2,000
.454 Casull	1,900	2,600
.460 SW	2,400	3,300
.500 SW	2,600	3,500

It must be stressed that muzzle energy is dependent upon the factors previously listed and that even velocity is highly variable depending upon the length of the barrel a projectile is fired from. Also note that the muzzle energy is only an upper limit for how much energy is transmitted to the target and that the effects of a ballistic trauma depend on several other factors as well. While the above list mentions some averages, there is wide variation in commercial ammunition. A 180 grain bullet fired from .357 magnum handgun can achieve a muzzle energy of 580 foot-pounds. A 110 grain bullet fired from the same gun might only achieve 400 foot-pounds of muzzle energy, depending upon the manufacture of the cartridge. Some .45 Colt ammunition can produce 1,200 foot-pounds of muzzle energy, far in excess of the average listed above.

Legal Requirements on Muzzle Energy

Some jurisdictions stipulate minimum muzzle energies for hunting. For example, in Denmark rifle ammunition used for hunting the largest types of game there such as red deer must 100 m down range have a kinetic energy E_{100} of at least 2700 J and a bullet mass of at least 9 g or alternatively an E_{100} of at least 2000 J and a bullet mass of at least 10 g. Namibia specifies three levels of minimum muzzle energy for hunting, 1350 J for game such as springbok, 2700 J for game such as hartebeest and 5400 J for big game, together with a minimum caliber of 7 mm.

In Germany airsoft guns with a muzzle energy of no more than 0.5 J are exempt from the gun law, while air guns with a muzzle energy of no more than 7.5 J may be acquired without a firearms license.

Ballistic Conduction

Ballistic conduction or Ballistic transport is the transport of electrons in a medium having negligible electrical resistivity caused by scattering. Without scattering, electrons simply obey Newton's second law of motion at non-relativistic speeds.

In general, the resistivity exists because an electron, while moving inside a medium, is scattered by impurities, defects, the atoms/molecules composing the medium that simply oscillate around their equilibrium position (in a solid), or, generally, by any freely-moving atom/molecule composing the medium, in a gas or liquid.

For a given medium a moving electron can be ascribed a mean free path as being the average length that the electron can travel freely, i.e., before a collision, which could change its momentum. The mean free path can be increased by reducing the number of impurities in a crystal or by lowering its temperature.

Ballistic transport is observed when the mean free path of the electron is (much) longer than the dimension of the medium through which the electron travels. The electron alters its motion only upon collision with the *walls*. In the case of a wire suspended in air/vacuum the surface of the wire plays the role of the *box* reflecting the electrons and preventing them from exiting toward the empty space/open air. This is because there is an energy to be paid to extract the electron from the medium (work function).

For example, ballistic transport can be observed in a metal nanowire: this is simply because the wire is of the size of a nanometer (10^{-9} meters) and the mean free path can be longer than that in a metal

Ballistic conduction is the unimpeded flow of charge, or energy-carrying particles, over relatively long distances in a material. Normally, transport of electrons (or holes) is dominated by scattering events, which relax the carrier momentum in an effort to bring the conducting material to equilibrium. Thus, ballistic transport in a material is determined by how ballistically conductive that material is. Ballistic conduction differs from superconductivity due to the absence of the Meissner effect in the material. A ballistic conductor would stop conducting if the driving force is turned off, whereas in a superconductor current would continue to flow after the driving supply is disconnected.

Ballistic conduction is typically observed in quasi-1D structures, such as carbon nanotubes or silicon nanowires, because of extreme size quantization effects in these materials. Ballistic conduction is not limited to electrons (or holes) but can also apply to phonons. It is theoretically possible for ballistic conduction to be extended to other quasi-particles, but this has not been experimentally verified.

Theory

Scattering Mechanisms

In general, carriers will exhibit ballistic conduction when $L \leq \lambda_{MFP}$ where L is the length of the active part of the device (i.e., a channel in a MOSFET). λ_{MFP} is the mean scattering length for the carrier which can be given by Matthiessen's Rule, written here for electrons:

$$\frac{1}{\lambda_{MFP}} = \frac{1}{\lambda_{el-el}} + \frac{1}{\lambda_{ap}} + \frac{1}{\lambda_{op,ems}} + \frac{1}{\lambda_{op,abs}} + \frac{1}{\lambda_{impurity}} + \frac{1}{\lambda_{defect}} + \frac{1}{\lambda_{boundary}}$$

where λ_{el-el} is the electron-electron scattering length, λ_{ap} is the acoustic phonon (emission and absorption) scattering length, $\lambda_{op,ems}$ is the optical phonon emission scattering length, $\lambda_{op,abs}$ is the optical phonon absorption scattering length, $\lambda_{impurity}$ is the electron-impurity scattering length, λ_{defect} is the electron-defect scattering length, $\lambda_{boundary}$ is the electron scattering length with the boundary, and λ_{MFP} is the total electron mean free path (electron scattering length). In terms of scattering mechanisms, optical phonon emission normally dominates, depending on the material and transport conditions. There are also other scattering mechanisms which apply to different carriers that are not considered here (e.g. remote interface phonon scattering, umklapp scattering). To get these characteristic scattering rates, one would need to derive a Hamiltonian and solve Fermi's Golden Rule for the system in question.

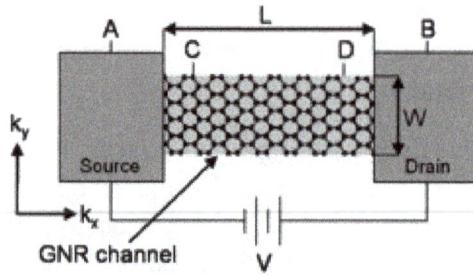

A graphene nanoribbon field-effect transistor (GNRFET). Here contacts A and B are at two different Fermi levels E_{F_A} and E_{F_B}.

Landauer-Büttiker Formalism

In 1957, Rolf Landauer proposed that conduction in a 1D system could be viewed as a transmission problem. For the 1D GNRFET on the right (where the graphene nanoribbon channel is assumed to be ballistic), the current from A to B (given by the Boltzmann transport equation) is

$$I_{AB} = \frac{g_s e}{h} \int_{E_{F_B}}^{E_{F_A}} M(E) f'(E) T(E) dE$$

where $g_s = 2$ due to spin degeneracy, e is the electron charge, h=Planck's constant, E_{F_A} and E_{F_B} are the Fermi levels of A and B, $M(E)$ is the number of propagating modes in the channel, $f'(E)$ is the deviation from the equilibrium electron distribution (perturbation), and $T(E)$ is the transmission probability (T=1 for ballistic). Based on the definition of conductance $G = \frac{I}{V}$ and the voltage separation between the Fermi levels is approximately $eV = E_{F_A} - E_{F_B}$, it follows that $G = \frac{2e^2}{h} MT$ where M is the number of modes in the transmission channel and spin is included. G is known as the quantized conductance. The contacts have a multiplicity of modes due to their larger size in comparison to the channel. Conversely, the quantum confinement in the 1D GNR channel constricts the number of modes to carrier degeneracy and restrictions from the material's energy dispersion relationship and Brillouin zone. For example, electrons in carbon nanotubes have two intervalley modes and two spin modes. Since the contacts and the GNR channel are connected by leads, the transmission probability is smaller at contacts A and B, $T \approx \dfrac{M}{M_{contact}}$. Thus the quantum conductance is approximately the same if measured at A and B or C and D.

Landauer-Buttiker formalism holds as long as the carriers are coherent (which means the length of the active channel is less than the phase-breaking mean free path) and the transmission functions can calculated from Schrödinger's equation or approximated by the WKB approximation. Therefore, even in the case of a perfect ballistic transport, there is a fundamental ballistic conductance which saturates the current of the device with a resistance of approximately $\dfrac{12.9k\Omega}{M}$ (spin degeneracy included). There is, however, a generalization of the Landauer-Büttiker formalism of transport applicable to time-dependent problems in the presence of dissipation.

Importance

Ballistic conduction enables use of quantum mechanical properties of electron wave functions. Ballistic transport is coherent in wave mechanics terms. Phenomena like double-split interference, spatial resonance (and other optical or microwave-like effects) could be exploited in electronic systems at nanoscale.

Optical Analogies

A comparison with light provides an analogy between ballistic and non-ballistic conduction. Ballistic electrons behave like light in a waveguide or a high-quality optical assembly. Non-ballistic electrons behave like light diffused in milk or reflected off a white wall or a piece of paper.

Electrons can be scattered several ways in a conductor. Electrons have several properties: wavelength (energy), direction, phase, and spin orientation. Different materials have different scattering probabilities which cause different incoherence rates (stochasticity). Some kinds of scattering can only cause a change in electron direction, others can cause energy loss.

Consider a coherent source of electrons connected to a conductor. Over a limited distance, the electron wave function will remain coherent. You still can deterministically predict its behavior (and use it for computation theoretically). After some greater distance, scattering causes each electron to have a slightly different phase and/or direction. But there is still almost no energy loss. Like monochromatic light passing through milk, electrons undergo elastic interactions. Information about the state of the electrons at the input is then lost. Transport becomes statistical and stochastic. From the resistance point of view, stochastic (not oriented) movement of electrons is useless even if they carry the same energy - they move thermally. If the electrons undergo inelastic interactions too, they lose energy and the result is a second mechanism of resistance. Electrons which undergo inelastic interaction are then similar to non-monochromatic light.

For correct usage of this analogy consideration of several facts is needed:

- Photons are bosons and electrons are fermions.

- There is coulombic repulsion between electrons.

Thus this analogy is good only for single-electron conduction because electron processes are strongly and nonlinear dependent on other electrons.

- It is more likely that an electron would lose more energy than a photon would, because of the electron's non-zero rest mass.

- Electron interactions with the environment, each other, and other particles are generally stronger than interactions with and between photons.

Examples

As mentioned, nanostructures such as carbon nanotubes or graphene nanoribbons are often considered ballistic, but these devices only very closely resemble ballistic conduction. Their ballisticity is nearly 0.9 at room temperature.

Carbon Nanotubes and Graphene Nanoribbon

The dominant scattering mechanism at room temperature is that of electrons emitting optical phonons. If electrons don't scatter with enough phonons (for example if the scattering rate is low), the mean free path tends to be very long ($\lambda_{MFP} \approx 1 \mu m$ m). So a nanotube or graphene nanoribbon could be a good ballistic conductor if the electrons in transit don't scatter with too many phonons and if the device is about 100 nm long. Such a transport regime has been found to depend on the nanoribbon edge structure and the electron energy.

Si Nanowires

It is often incorrectly thought that Si nanowires are quantum confined ballistic conductors. There are major differences between carbon nanotubes (which are hollow) and Si nanowires (which are solid). Nanowires are about 20-50 nm in diameter and are 3D solid while carbon nanotubes have diameters around the wavelength of the electrons (2-3 nm) and are essentially 1D conductors. However it is still possible to observe ballistic conduction in Si nanowires at very low temperatures (2-3 K).

Gunpowder

Black powder for muzzleloading firearms and pistols in FFFg granulation size. U.S. quarter (diameter 24 mm) for comparison.

Gunpowder, also known as black powder to distinguish it from modern gunpowder, is the earliest known chemical explosive. It consists of a mixture of sulfur, charcoal, and potassium nitrate (saltpeter). The sulfur and charcoal act as fuels while the saltpeter is an oxidizer. Because of its incendiary properties and the amount of heat and gas volume that it generates, gunpowder has been widely used as a propellant in firearms, rockets, fireworks, and as a blasting powder in quarrying, mining, and road building.

Gunpowder was invented in 9th century China and spread throughout most parts of Eurasia by the end of the 13th century. Most arguments on early gunpowder developments now revolve around how much Chinese advancements in gunpowder influenced gunpowder warfare in the Middle East and Europe, but controversy over gunpowder's precise origins continue to be debated today.

Gunpowder is classified as a low explosive because of its relatively slow decomposition rate and consequently low brisance. Low explosives deflagrate (i.e., burn) at *subsonic* speeds, whereas high explosives detonate, producing a supersonic wave. Ignition of powder packed behind a bullet must generate enough pressure to force it from the muzzle at high speed, but not enough to rupture the gun barrel. Gunpowder thus makes a good propellant, but is less suitable for shattering rock or fortifications. Gunpowder was widely used to fill artillery shells and in mining and civil engineering to blast rock until the second half of the 19th century, when the first high explosives were put into use. Gunpowder is no longer used in modern weapons nor is it used for industrial purposes due to its relatively inefficient cost compared to newer alternatives such as dynamite and ammonium nitrate/fuel oil (ANFO). Today gunpowder firearms are limited primarily to hunting, target shooting, and bulletless historical reenactments.

History of Gunpowder

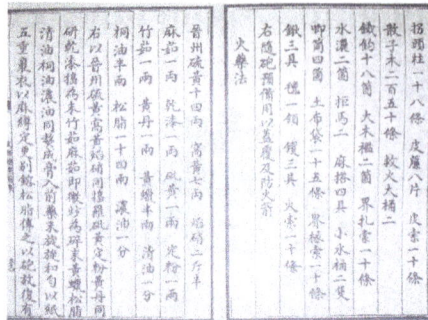

Earliest known written formula for gunpowder, from the *Wujing Zongyao* of 1044 AD.

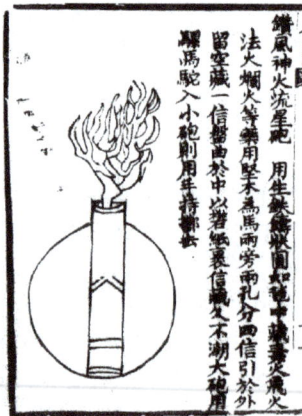

A 'magic fire meteor going against the wind' bomb as depicted in the *Huolongjing* ca. 1350.

Stoneware bombs, known in Japanese as *Tetsuhau* (iron bomb), or in Chinese as *Zhentianlei* (thunder crash bomb), excavated from the Takashima shipwreck, October 2011, dated to the Mongol invasions of Japan (1271–1284 AD).

The earliest chemical formula for gunpowder appeared in the 11th century Song dynasty text, *Wujing Zongyao*, however gunpowder had already been used for fire arrows since at least the 10th century. In the following centuries various gunpowder weapons such as bombs, fire lances, and the gun appeared in China. The technology possibly spread from China throughout Eurasia. The earliest Western accounts of gunpowder appear in texts written by English philosopher Roger Bacon in the 13th century.

A major problem confronting the study of early gunpowder history is ready access to sources close to the events described. Often the first records potentially describing use of gunpowder in warfare were written several centuries after the fact, and may well have been colored by the contemporary experiences of the chronicler. It is also difficult to accurately translate original Chinese alchemical texts, which tend to explain phenomena through metaphor, into modern scientific language with rigidly defined terminology in English. Translation difficulties have led to errors or loose interpretations bordering on artistic licence. Early texts potentially mentioning gunpowder are sometimes marked by a linguistic process where semantic change occurred. For instance, the Arabic word *naft* transitioned from denoting naphtha to denoting gunpowder, and the Chinese word *pào* changed in meaning from catapult to referring to a cannon. This has led to arguments on the exact origins of gunpowder based on etymological foundations. Science and technology historian Bert S. Hall makes the observation that:

It goes without saying, however, that historians bent on special pleading, or simply with axes of their own to grind, can find rich material in these terminological thickets.

— Bert S. Hall

China

Saltpeter was known to the Chinese by the mid-1st century AD and was primarily produced in the provinces of Sichuan, Shanxi, and Shandong. There is strong evidence of the use of saltpeter and sulfur in various medicinal combinations. A Chinese alchemical text dated 492 noted saltpeter burnt with a purple flame, providing a practical and reliable means of distinguishing it from other inorganic salts, thus enabling alchemists to evaluate and compare purification techniques; the earliest Latin accounts of saltpeter purification are dated after 1200.

The first reference to the incendiary properties of such mixtures is the passage of the *Zhenyuan miaodao yaoliie*, a Taoist text tentatively dated to the mid-9th century: "Some have heated together sulfur, realgar and saltpeter with honey; smoke and flames result, so that their hands and faces have been burnt, and even the whole house where they were working burned down." The word "gunpowder" in Chinese, means "Fire Medicine"; however this name only came into use some centuries after the mixture's discovery. In the following centuries a variety of gunpowder weapons such as rockets, bombs, and land mines appeared before the first metal barrel firearms were invented. Explosive weapons such as bombs have been discovered in a shipwreck off the shore of Japan dated from 1281, during the Mongol invasions of Japan.

The Chinese *Wujing Zongyao* (*Complete Essentials from the Military Classics*), written by Zeng Gongliang between 1040–1044, provides encyclopedia references to a variety of mixtures that included petrochemicals—as well as garlic and honey. A slow match for flame throwing mechanisms using the siphon principle and for fireworks and rockets is mentioned. The mixture formulas in this book do not contain enough saltpeter to create an explosive however; being limited to at most 50% saltpeter, they produce an incendiary. The *Essentials* was however written by a Song dynasty court bureaucrat, and there is little evidence that it had any immediate impact on warfare; there is no mention of gunpowder use in the chronicles of the wars against the Tanguts in the 11th century, and China was otherwise mostly at peace during this century.

However by 1083 the Song court was producing hundreds of thousands of fire arrows for their garrisons. Bombs and fire lances became prominent during the 12th century and were used by the Song during the Jin-Song Wars. The first proto-guns, known as "fire lances", were first recorded to have been used at the siege of De'an in 1132 by Song forces against the Jin. In the early 13th century the Jin utilized iron casing bombs. Projectiles were added to fire lances, re-usable fire lance barrels were developed, first out of hardened paper, and then metal. By 1257 some fire lances were firing wads of bullets. In the late 13th century metal fire lances became 'eruptors,' proto-cannons firing co-viative projectiles, and by 1287 at the latest, had become true guns, the hand cannon.

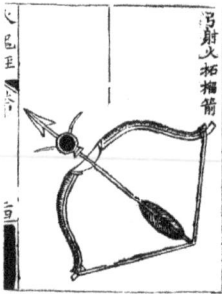

An arrow strapped with gunpowder ready to be shot from a bow. From the *Huolongjing* ca. 1350.

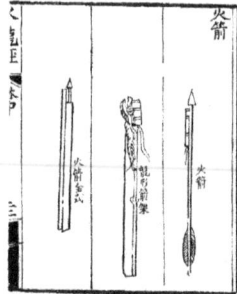

The oldest known depiction of rocket arrows, from the *Huolongjing*. The right arrow reads 'fire arrow,' the middle is an 'arrow frame in the shape of a dragon,' and the left is a 'complete fire arrow.'

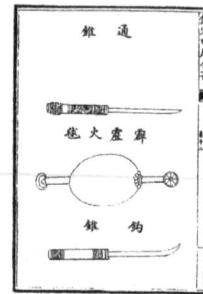

An illustration of a thunderclap bomb as depicted in the 1044 text *Wujing Zongyao*. Considered to be a pseudo-explosive. The top item is a through awl and the bottom one is a hook awl.

A fire lance as depicted in the *Huolongjing* ca. 1350.

The 'flying-cloud thunderclap-eruptor' cannon from the *Huolongjing* ca. 1350.

An organ gun known as the 'mother of a hundred bullets gun' from the *Huolongjing* ca. 1350.

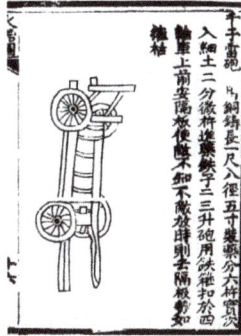

An illustration of a bronze "thousand ball thunder cannon" from the *Huolongjing* ca. 1350.

The 'self-tripped trespass land mine' from the *Huolongjing* ca. 1350.

Middle East

A picture of a 15th-century Granadian cannon from the book *Al-izz wal rifa'a*.

The Muslims acquired knowledge of gunpowder some time between 1240 and 1280, by which point the Syrian Hasan al-Rammah had written, in Arabic, recipes for gunpowder, instructions for the purification of saltpeter, and descriptions of gunpowder incendiaries. It is implied by al-Rammah's usage of "terms that suggested he derived his knowledge from Chinese sources" and his references to saltpeter as "Chinese snow" , fireworks as "Chinese flowers" and rockets as "Chinese arrows" that knowledge of gunpowder arrived from China. However, because al-Rammah attributes his material to "his father and forefathers", al-Hassan argues that gunpowder became prevalent in Syria and Egypt by "the end of the twelfth century or the beginning of the thirteenth". In Persia saltpeter was known as "Chinese salt" or "salt from Chinese salt marshes".

Hasan al-Rammah included 107 gunpowder recipes in his text *al-Furusiyyah wa al-Manasib al-Harbiyya* (*The Book of Military Horsemanship and Ingenious War Devices*), 22 of which are for rockets. If one takes the median of 17 of these 22 compositions for rockets (75% nitrates, 9.06% sulfur, and 15.94% charcoal), it is nearly identical to the modern reported ideal gunpowder recipe of 75% potassium nitrate, 10% sulfur, and 15% charcoal.

Al-Hassan claims that in the Battle of Ain Jalut of 1260, the Mamluks used against the Mongols in "the first cannon in history" gunpowder formula with near-identical ideal composition ratios for explosive gunpowder. Other historians urge caution regarding claims of Islamic firearms use in the 1204–1324 period as late medieval Arabic texts used the same word for gunpowder, *naft*, that they used for an earlier incendiary, naphtha.

Khan claims that it was invading Mongols who introduced gunpowder to the Islamic world and cites Mamluk antagonism towards early musketeers in their infantry as an example of how gunpowder weapons were not always met with open acceptance in the Middle East. Similarly, the refusal of their Qizilbash forces to use firearms contributed to the Safavid rout at Chaldiran in 1514.

The state-controlled manufacture of gunpowder by the Ottoman Empire through early supply chains to obtain nitre, sulfur and high-quality charcoal from oaks in Anatolia contributed significantly to its expansion between the 15th and 18th century. It was not until later in the 19th century when the syndicalist production of Turkish gunpowder was greatly reduced, which coincided with the decline of its military might.

Mainland Europe

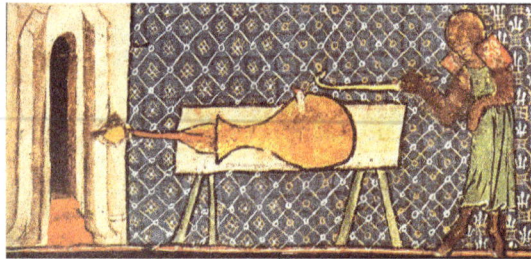

Earliest depiction of a European cannon, "De Nobilitatibus Sapientii Et Prudentiis Regum", Walter de Milemete, 1326.

Büchsenmeysterei : von Geschoß, Büchsen, Pulver, Salpeter und Feurwergken, 1531.

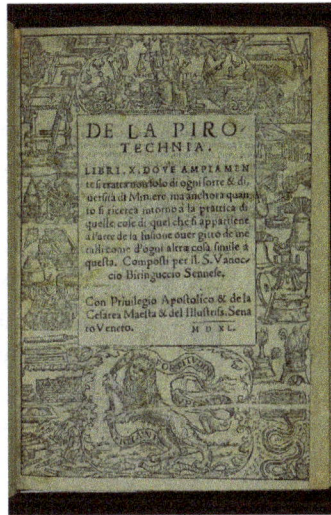

De la pirotechnia, 1540

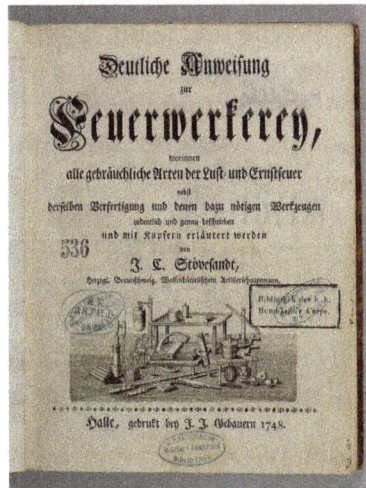

Deutliche Anweisung zur Feuerwerkerey, 1748

Several sources mention Chinese firearms and gunpowder weapons being deployed by the Mongols against European forces at the Battle of Mohi in 1241. Professor Kenneth Warren Chase credits the Mongols for introducing into Europe gunpowder and its associated weaponry. However there is no clear route of transmission, and while the Mongols are often pointed to as the likeliest vector, Timothy May points out that "there is no concrete evidence that the Mongols used gunpowder weapons on a regular basis outside of China."

In Europe, one of the first mentions of gunpowder use appears in a passage found in Roger Bacon's *Opus Maius* of 1267 and *Opus Tertium* in what has been interpreted as being firecrackers. The most telling passage reads: "We have an example of these things (that act on the senses) in [the sound and fire of] that children's toy which is made in many [diverse] parts of the world; i.e., a device no bigger than one's thumb. From the violence of that salt called saltpeter [together with sulfur and willow charcoal, combined into a powder] so horrible a sound is made by the bursting of a thing so small, no more than a bit of parchment [containing it], that we find [the ear assaulted by a noise] exceeding the roar of strong thunder, and a flash brighter than the most brilliant lightning." In the early 20th century, British artillery officer Henry William Lovett Hime proposed that another work tentatively attributed to Bacon, *Epistola de Secretis Operibus Artis et Naturae, et de Nullitate Magiae* contained an encrypted formula for gunpowder. This claim has been disputed by historians of science including Lynn Thorndike, John Maxson Stillman and George Sarton and by Bacon's editor Robert Steele, both in terms of authenticity of the work, and with respect to the decryption method. In any case, the formula claimed to have been decrypted (7:5:5 saltpeter:charcoal:sulfur) is not useful for firearms use or even firecrackers, burning slowly and producing mostly smoke.

The *Liber Ignium*, or *Book of Fires*, attributed to Marcus Graecus, is a collection of incendiary recipes, including some gunpowder recipes. Partington dates the gunpowder recipes to approximately 1300. One recipe for "flying fire" (*ingis volatilis*) involves saltpeter, sulfur, and colophonium, which, when inserted into a reed or hollow wood, "flies away suddenly and burns up everything." Another recipe, for artificial "thunder", specifies a mixture of one pound native sulfur, two pounds linden or willow charcoal, and six pounds of saltpeter. Another specifies a 1:3:9 ratio.

Some of the gunpowder recipes of *De Mirabilibus Mundi* of Albertus Magnus are identical to the recipes of the *Liber Ignium*, and according to Partington, "may have been taken from that work, rather than conversely." Partington suggests that some of the book may have been compiled by Albert's students, "but since it is found in thirteenth century manuscripts, it may well be by Albert." Albertus Magnus died in 1280.

A major advance in manufacturing began in Europe in the late 14th century when the safety and thoroughness of incorporation was improved by wet grinding; liquid, such as distilled spirits or perhaps the urine of wine-drinking bishops was added during the grinding-together of the ingredients and the moist paste dried afterwards. The principle of wet mixing to prevent the separation of dry ingredients, invented for gunpowder, is used today in the pharmaceutical industry.

It was also discovered that if the paste was rolled into balls before drying the resulting gunpowder absorbed less water from the air during storage and traveled better. The balls were then crushed in a mortar by the gunner immediately before use, with the old problem of uneven particle size and packing causing unpredictable results. If the right size particles were chosen, however, the result was a great improvement in power. Forming the damp paste into *corn*-sized clumps by hand or with the use of a sieve instead of larger balls produced a product after drying that loaded much better, as each tiny piece provided its own surrounding air space that allowed much more rapid combustion than a fine powder. This "corned" gunpowder was from 30% to 300% more powerful. An example is cited where 34 pounds of serpentine was needed to shoot a 47-pound ball, but only 18 pounds of corned powder. The optimum size of the grain depended on its use; larger for large cannon, finer for small arms. Larger cast cannons were easily muzzle-loaded with corned powder using a long-handled ladle. Corned powder also retained the advantage of low moisture absorption, as even tiny grains still had much less surface area to attract water than a floury powder.

During this time, European manufacturers also began regularly purifying saltpeter, using wood ashes containing potassium carbonate to precipitate calcium from their dung liquor, and using ox blood, alum, and slices of turnip to clarify the solution.

During the Renaissance, two European schools of pyrotechnic thought emerged, one in Italy and the other at Nuremberg, Germany. The German printer and publisher Christiaan Egenolff adapted an earlier work on pyrotechnics from manuscript to print form, publishing his *Büchsenmeysterei* in 1529 and reprinting it in 1531. Now extremely rare, the book discusses the manufacturing of gunpowder, the operation of artillery and the rules of conduct for the gunsmith.

In Italy, Vannoccio Biringuccio, born in 1480, was a member of the guild *Fraternita di Santa Barbara* but broke with the tradition of secrecy by setting down everything he knew in a book titled *De la pirotechnia*, written in vernacular. It was published posthumously in 1540, with 9 editions over 138 years, and also reprinted by MIT Press in 1966.

By the mid-17th century fireworks were used for entertainment on an unprecedented scale in Europe, being popular even at resorts and public gardens. With the publication of *Deutliche Anweisung zur Feuerwerkerey* (1748), methods for creating fireworks were sufficiently well-known and well-described that "Firework making has become an exact science." In 1774 Louis XVI ascended to the throne of France at age 20. After he discovered that France was not self-sufficient in gunpowder, a Gunpowder Administration was established; to head it, the lawyer Antoine Lavoisier was appointed. Although from a bourgeois family, after his degree in law Lavoisier became wealthy from a company set up to collect taxes for the Crown; this allowed him to pursue experimental natural science as a hobby.

Without access to cheap saltpeter (controlled by the British), for hundreds of years France had relied on saltpetermen with royal warrants, the *droit de fouille* or "right to dig", to seize nitrous-containing soil and demolish walls of barnyards, without compensation to the owners. This caused farmers, the wealthy, or entire villages to bribe the petermen and the associated bureaucracy to leave their buildings alone and the saltpeter uncollected. Lavoisier instituted a crash program to increase saltpeter production, revised (and later eliminated) the *droit de fouille*, researched best refining and powder manufacturing methods, instituted management and record-keeping, and established pricing that encouraged private investment in works. Although saltpeter from new

Prussian-style putrefaction works had not been produced yet (the process taking about 18 months), in only a year France had gunpowder to export. A chief beneficiary of this surplus was the American Revolution. By careful testing and adjusting the proportions and grinding time, powder from mills such as at Essonne outside Paris became the best in the world by 1788, and inexpensive.

Britain and Ireland

Gunpowder production in Britain appears to have started in the mid 14th century with the aim of supplying the English Crown. Records show that, in England, gunpowder was being made in 1346 at the Tower of London; a powder house existed at the Tower in 1461; and in 1515 three King's gunpowder makers worked there. Gunpowder was also being made or stored at other Royal castles, such as Portchester. By the early 14th century, according to N.J.G. Pounds's study *The Medieval Castle in England and Wales*, many English castles had been deserted and others were crumbling. Their military significance faded except on the borders. Gunpowder had made smaller castles useless.

Henry VIII of England was short of gunpowder when he invaded France in 1544 and England needed to import gunpowder via the port of Antwerp in what is now Belgium.

The English Civil War (1642–1645) led to an expansion of the gunpowder industry, with the repeal of the Royal Patent in August 1641.

Two British physicists, Andrew Noble and Frederick Abel, worked to improve the properties of black powder during the late 19th century. This formed the basis for the Noble-Abel gas equation for internal ballistics.

The introduction of smokeless powder in the late 19th century led to a contraction of the gunpowder industry. After the end of World War I, the majority of the United Kingdom gunpowder manufacturers merged into a single company, "Explosives Trades limited"; and a number of sites were closed down, including those in Ireland. This company became Nobel Industries Limited; and in 1926 became a founding member of Imperial Chemical Industries. The Home Office removed gunpowder from its list of *Permitted Explosives*; and shortly afterwards, on 31 December 1931, the former Curtis & Harvey's Glynneath gunpowder factory at Pontneddfechan, in Wales, closed down, and it was demolished by fire in 1932.

The last remaining gunpowder mill at the Royal Gunpowder Factory, Waltham Abbey was damaged by a German parachute mine in 1941 and it never reopened. This was followed by the closure of the gunpowder section at the Royal Ordnance Factory, ROF Chorley, the section was closed and demolished at the end of World War II; and ICI Nobel's Roslin gunpowder factory, which closed in 1954.

This left the sole United Kingdom gunpowder factory at ICI Nobel's Ardeer site in Scotland; it too closed in October 1976. Since then gunpowder has been imported into the United

Kingdom. In the late 1970s/early 1980s gunpowder was bought from eastern Europe, particularly from what was then the German Democratic Republic and former Yugoslavia.

India

In the year 1780 the British began to annex the territories of the Sultanate of Mysore, during the Second Anglo-Mysore War. The British battalion was defeated during the Battle of Guntur, by the forces of Hyder Ali, who effectively utilized Mysorean rockets and rocket artillery against the closely massed British forces.

Mughal Emperor Shah Jahan, hunting deer using a matchlock as the sun sets in the horizon.

Gunpowder and gunpowder weapons were transmitted to India through the Mongol invasions of India. The Mongols were defeated by Alauddin Khilji of the Delhi Sultanate, and some of the Mongol soldiers remained in northern India after their conversion to Islam. It was written in the *Tarikh-i Firishta* (1606–1607) that Nasir ud din Mahmud the ruler of the Delhi Sultanate presented the envoy of the Mongol ruler Hulegu Khan with a dazzling pyrotechnics display upon his arrival in Delhi in 1258. Nasir ud din Mahmud tried to express his strength as a ruler and tried to ward off any Mongol attempt similar to the Siege of Baghdad (1258). Firearms known as *top-o-tufak* also existed in many Muslim kingdoms in India by as early as 1366. From then on the employment of gunpowder warfare in India was prevalent, with events such as the "Siege of Belgaum" in 1473 by Sultan Muhammad Shah Bahmani.

The shipwrecked Ottoman Admiral Seydi Ali Reis is known to have introduced the earliest type of matchlock weapons, which the Ottomans used against the Portuguese during the Siege of Diu (1531). After that, a diverse variety of firearms, large guns in particular, became visible in Tanjore, Dacca, Bijapur, and Murshidabad. Guns made of bronze were recovered from Calicut (1504)- the former capital of the Zamorins

The Mughal emperor Akbar mass-produced matchlocks for the Mughal Army. Akbar is personally known to have shot a leading Rajput commander during the Siege of Chittorgarh. The Mughals began to use bamboo rockets (mainly for signalling) and employ sappers: special units that undermined heavy stone fortifications to plant gunpowder charges.

The Mughal Emperor Shah Jahan is known to have introduced much more advanced matchlocks, their designs were a combination of Ottoman and Mughal designs. Shah Jahan also countered the British and other Europeans in his province of Gujarāt, which supplied Europe saltpeter for use in gunpowder warfare during the 17th century. Bengal and Mālwa participated in saltpeter production. The Dutch, French, Portuguese, and English used Chhapra as a center of saltpeter refining.

Ever since the founding of the Sultanate of Mysore by Hyder Ali, French military officers were employed to train the Mysore Army. Hyder Ali and his son Tipu Sultan were the first to introduce modern cannons and muskets, their army was also the first in India to have official uniforms. During the Second Anglo-Mysore War Hyder Ali and his son Tipu Sultan unleashed the Mysorean rockets at their British opponents effectively defeating them on various occasions. The Mysorean rockets inspired the development of the Congreve rocket, which the British widely utilized during the Napoleonic Wars and the War of 1812.

Indonesia

The Javanese Majapahit Empire was arguably able to encompass much of modern-day Indonesia due to its unique mastery of bronze-smithing and use of a central arsenal fed by a large number of cottage industries within the immediate region. Documentary and archeological evidence indicate that Arab traders introduced gunpowder, gonnes, muskets, blunderbusses, and cannons to the Javanese, Acehnese, and Batak via long established commercial trade routes around the early to mid 14th century. Portuguese and Spanish invaders were unpleasantly surprised and even outgunned on occasion. The resurgent Singhasari Empire overtook Sriwijaya and later emerged as the Majapahit whose warfare featured the use of fire-arms and cannonade. Circa 1540, the Javanese, always alert for new weapons found the newly arrived Portuguese weaponry superior to that of the locally made variants. Javanese bronze breech-loaded swivel-guns, known as meriam, or erroneously as lantaka, was used widely by the Majapahit navy as well as by pirates and rival lords. The demise of the Majapahit empire and the dispersal of disaffected skilled bronze cannon-smiths to Brunei, modern Sumatra, Malaysia and the Philippines lead to widespread use, especially in the Makassar Strait.

Saltpeter harvesting was recorded by Dutch and German travelers as being common in even the smallest villages and was collected from the decomposition process of large dung hills specifically piled for the purpose. The Dutch punishment for possession of non-permitted gunpowder appears to have been amputation. Ownership and manufacture of gunpowder was later prohibited by the colonial Dutch occupiers. According to a colonel McKenzie quoted in Sir Thomas Stamford Raffles, *The History of Java* (1817), the purest sulfur was supplied from a crater from a mountain near the straits of Bali.

Manufacturing Technology

Edge-runner mill in a restored mill, at The Hagley Museum

Gunpowder storing barrels at Martello tower in Point Pleasant Park

For the most powerful black powder, meal powder, a wood charcoal, is used. The best wood for the purpose is Pacific willow, but others such as alder or buckthorn can be used. In Great Britain between the 15th and 19th centuries charcoal from alder buckthorn was greatly prized for gunpowder manufacture; cottonwood was used by the American Confederate States. The ingredients are reduced in particle size and mixed as intimately as possible. Originally, this was with a mortar-and-pestle or a similarly operating stamping-mill, using copper, bronze or other non-sparking materials, until supplanted by the rotating ball mill principle with non-sparking bronze or lead. Historically, a marble or limestone edge runner mill, running on a limestone bed, was used in Great Britain; however, by the mid 19th century this had changed to either an iron-shod stone wheel or a cast iron wheel running on an iron bed. The mix was dampened with alcohol or water during grinding to prevent accidental ignition. This also helps the extremely soluble saltpeter to mix into the microscopic nooks and crannies of the very high surface-area charcoal.

Around the late 14th century, European powdermakers first began adding liquid during grinding to improve mixing, reduce dust, and with it the risk of explosion. The powder-makers would then shape the resulting paste of dampened gunpowder, known as mill cake, into corns, or grains, to dry. Not only did corned powder keep better because of its reduced surface area, gunners also found that it was more powerful and easier to load into guns. Before long, powder-makers standardized the process by forcing mill cake through sieves instead of corning powder by hand.

The improvement was based on reducing the surface area of a higher density composition. At the beginning of the 19th century, makers increased density further by static pressing. They shoveled damp mill cake into a two-foot square box, placed this beneath a screw press and reduced it to $\frac{1}{2}$ its volume. "Press cake" had the hardness of slate. They broke the dried slabs with hammers or rollers, and sorted the granules with sieves into different grades. In the United States, Eleuthere Irenee du Pont, who had learned the trade from Lavoisier, tumbled the dried grains in rotating barrels to round the edges and increase durability during shipping and handling. (Sharp grains rounded off in transport, producing fine "meal dust" that changed the burning properties.)

Another advance was the manufacture of kiln charcoal by distilling wood in heated iron retorts instead of burning it in earthen pits. Controlling the temperature influenced the power and consistency of the finished gunpowder. In 1863, in response to high prices for Indian saltpeter, DuPont chemists developed a process using potash or mined potassium chloride to convert plentiful Chilean sodium nitrate to potassium nitrate.

The following year (1864) the Gatebeck Low Gunpowder Works in Cumbria (Great Britain) started a plant to manufacture potassium nitrate by essentially the same chemical process. This is nowadays called the 'Wakefield Process', after the owners of the company. It would have used potassium chloride from the Staßfurt mines, near Magdeburg, Germany, which had recently become available in industrial quantities.

During the 18th century, gunpowder factories became increasingly dependent on mechanical energy. Despite mechanization, production difficulties related to humidity control, especially during the pressing, were still present in the late 19th century. A paper from 1885 laments that "Gunpowder is such a nervous and sensitive spirit, that in almost every process of manufacture it changes under our hands as the weather changes." Pressing times to the desired density could vary by a factor of three depending on the atmospheric humidity.

Composition and Characteristics

The term *black powder* was coined in the late 19th century, primarily in the United States, to distinguish prior gunpowder formulations from the new smokeless powders and semi-smokeless powders, in cases where these are not referred to as cordite. Semi-smokeless powders featured bulk volume properties that approximated black powder, but had significantly reduced amounts of smoke and combustion products. Smokeless powder has different burning properties (pressure vs. time) and can generate higher pressures and work per gram. This can rupture older weapons designed for black powder. Smokeless powders ranged in color from brownish tan to yellow to white. Most of the bulk semi-smokeless powders ceased to be manufactured in the 1920s.

Black powder is a granular mixture of

- a nitrate, typically potassium nitrate (KNO_3), which supplies oxygen for the reaction;

- charcoal, which provides carbon and other fuel for the reaction, simplified as carbon (C);

- sulfur (S), which, while also serving as a fuel, lowers the temperature required to ignite the mixture, thereby increasing the rate of combustion.

Potassium nitrate is the most important ingredient in terms of both bulk and function because the combustion process releases oxygen from the potassium nitrate, promoting the rapid burning of the other ingredients. To reduce the likelihood of accidental ignition by static electricity, the granules of modern black powder are typically coated with graphite, which prevents the build-up of electrostatic charge.

Charcoal does not consist of pure carbon; rather, it consists of partially pyrolyzed cellulose, in which the wood is not completely decomposed. Carbon differs from ordinary charcoal. Whereas charcoal's autoignition temperature is relatively low, carbon's is much greater. Thus, a black powder composition containing pure carbon would burn similarly to a match head, at best.

The current standard composition for the black powders that are manufactured by pyrotechnicians was adopted as long ago as 1780. Proportions by weight are 75% potassium nitrate (known as saltpeter or saltpetre), 15% softwood charcoal, and 10% sulfur. These ratios have varied over the centuries and by country, and can be altered somewhat depending on the purpose of the powder. For instance, power grades of black powder, unsuitable for use in firearms but adequate for blasting rock in quarrying operations, are called blasting powder rather than gunpowder with standard proportions of 70% nitrate, 14% charcoal, and 16% sulfur; blasting powder may be made with the cheaper sodium nitrate substituted for potassium nitrate and proportions may be as low as 40% nitrate, 30% charcoal, and 30% sulfur. In 1857, Lammot du Pont solved the main problem of using cheaper sodium nitrate formulations when he patented DuPont "B" blasting powder. After manufacturing grains from press-cake in the usual way, his process tumbled the powder with graphite dust for 12 hours. This formed a graphite coating on each grain that reduced its ability to absorb moisture.

Neither the use of graphite nor sodium nitrate was new. Glossing gunpowder corns with graphite was already an accepted technique in 1839, and sodium nitrate-based blasting powder had been made in Peru for many years using the sodium nitrate mined at Tarapacá (now in Chile). Also, in 1846, two plants were built in south-west England to make blasting powder using this sodium nitrate. The idea may well have been brought from Peru by Cornish miners returning home after completing their contracts. Another suggestion is that it was William Lobb, the planthunter, who recognised the possibilities of sodium nitrate during his travels in South America. Lammot du Pont would have known about the use of graphite and probably also knew about the plants in south-west England. In his patent he was careful to state that his claim was for the combination of graphite with sodium nitrate-based powder, rather than for either of the two individual technologies.

French war powder in 1879 used the ratio 75% saltpeter, 12.5% charcoal, 12.5% sulfur. English war powder in 1879 used the ratio 75% saltpeter, 15% charcoal, 10% sulfur. The British Congreve rockets used 62.4% saltpeter, 23.2% charcoal and 14.4% sulfur, but the British Mark VII gunpowder was changed to 65% saltpeter, 20% charcoal and 15% sulfur. The explanation for the wide variety in formulation relates to usage. Powder used for rocketry can use a slower burn rate since it accelerates the projectile for a much longer time—whereas powders for weapons such as flintlocks, cap-locks, or matchlocks need a higher burn rate to accelerate the projectile in a much shorter distance. Cannons usually used lower burn rate powders, because most would burst with higher burn rate powders.

Serpentine

The original dry-compounded powder used in 15th-century Europe was known as "Serpentine", either a reference to Satan or to a common artillery piece that used it. The ingredients were ground together with a mortar and pestle, perhaps for 24 hours, resulting in a fine flour. Vibration during transportation could cause the components to separate again, requiring remixing in the field. Also if the quality of the saltpeter was low (for instance if it was contaminated with highly hygroscopic calcium nitrate), or if the powder was simply old (due to the mildly hygroscopic nature of potassium nitrate), in humid weather it would need to be re-dried. The dust from "repairing" powder in the field was a major hazard.

Loading cannons or bombards before the powder-making advances of the Renaissance was a skilled art. Fine powder loaded haphazardly or too tightly would burn incompletely or too slowly. Typically, the breech-loading powder chamber in the rear of the piece was filled only about half full, the serpentine powder neither too compressed nor too loose, a wooden bung pounded in to seal the chamber from the barrel when assembled, and the projectile placed on. A carefully determined empty space was necessary for the charge to burn effectively. When the cannon was fired through the touchhole, turbulence from the initial surface combustion caused the rest of the powder to be rapidly exposed to the flame.

The advent of much more powerful and easy to use *corned* powder changed this procedure, but serpentine was used with older guns into the 17th century.

Corning

For gunpowder to explode effectively, the combustible ingredients must be reduced to the smallest possible particle sizes, and be as thoroughly mixed as possible. Once mixed, however, for better results in a gun, makers discovered that the final product should be in the form of individual dense grains that spread the fire quickly from grain to grain, much as straw or twigs catch fire more quickly than a pile of sawdust.

Primarily for safety reasons, size reduction and mixing is done while the ingredients are damp, usually with water. After 1800, instead of forming grains by hand or with sieves, the damp *mill-cake* was pressed in molds to increase its density and extract the liquid, forming *press-cake*. The pressing took varying amounts of time, depending on conditions such as atmospheric humidity. The hard, dense product was broken again into tiny pieces, which were separated with sieves to produce a uniform product for each purpose: coarse powders for cannons, finer grained powders for muskets, and the finest for small hand guns and priming. Inappropriately fine-grained powder often caused cannons to burst before the projectile could move down the barrel, due to the high initial spike in pressure. *Mammoth* powder with large grains, made for Rodman's 15-inch cannon, reduced the pressure to only 20 percent as high as ordinary cannon powder would have produced.

In the mid-19th century, measurements were made determining that the burning rate within a grain of black powder (or a tightly packed mass) is about 6 cm/s (0.20 feet/s), while the rate of ignition propagation from grain to grain is around 9 m/s (30 feet/s), over two orders of magnitude faster.

Modern Types

Modern corning first compresses the fine black powder meal into blocks with a fixed density (1.7 g/cm³). In the United States, gunpowder grains were designated F (for fine) or C (for coarse). Grain diameter decreased with a larger number of Fs and increased with a larger number of Cs, ranging from about 2 mm (0.08 in) for 7F to 15 mm (0.6 in) for 7C. Even larger grains were produced for artillery bore diameters greater than about 17 cm (6.7 in). The standard DuPont *Mammoth* powder developed by Thomas Rodman and Lammot du Pont for use during the American Civil War had grains averaging 0.6 inches (15 mm) in diameter with edges rounded in a glazing barrel. Other versions had grains the size of golf and tennis balls for use in 20-inch (51 cm) Rodman guns. In 1875 DuPont introduced *Hexagonal* powder for large artillery, which was pressed using shaped plates with a small center core—about 1.5 inches (3.8 cm) diameter, like a wagon wheel nut, the center hole widened as the grain burned. By 1882 German makers also produced hexagonal grained powders of a similar size for artillery.

By the late 19th century manufacturing focused on standard grades of black powder from Fg used in large bore rifles and shotguns, through FFg (medium and small-bore arms such as muskets and fusils), FFFg (small-bore rifles and pistols), and FFFFg (extreme small bore, short pistols and most commonly for priming flintlocks). A coarser grade for use in military artillery blanks was designated A-1. These grades were sorted on a system of screens with oversize retained on a mesh of 6 wires per inch, A-1 retained on 10 wires per inch, Fg retained on 14, FFg on 24, FFFg on 46, and FFFFg on 60. Fines designated FFFFFg were usually reprocessed to minimize explosive dust hazards. In the United Kingdom, the main service gunpowders were classified RFG (rifle grained fine) with diameter of one or two millimeters and RLG (rifle grained large) for grain diameters between two and six millimeters. Gunpowder grains can alternatively be categorized by mesh size: the BSS sieve mesh size, being the smallest mesh size, which retains no grains. Recognized grain sizes are Gunpowder G 7, G 20, G 40, and G 90.

Owing to the large market of antique and replica black-powder firearms in the US, modern gunpowder substitutes like Pyrodex, Triple Seven and Black Mag3 pellets have been developed since the 1970s. These products, which should not be confused with smokeless powders, aim to produce less fouling (solid residue), while maintaining the traditional volumetric measurement system for charges. Claims of less corrosiveness of these products have been controversial however. New cleaning products for black-powder guns have also been developed for this market.

Other Types of Gunpowder

Besides black powder, there are other historically important types of gunpowder. "Brown gunpowder" is cited as composed of 79% nitre, 3% sulfur, and 18% charcoal per 100 of dry powder, with about 2% moisture. Prismatic Brown Powder is a large-grained product the Rottweil Company introduced in 1884 in Germany, which was adopted by the British Royal Navy shortly thereafter. The French navy adopted a fine, 3.1 millimeter, not prismatic grained product called *Slow Burning Cocoa* (SBC) or "cocoa powder". These brown powders reduced burning rate even further by using as little as 2 percent sulfur and using charcoal made from rye straw that had not been completely charred, hence the brown color.

Lesmok powder was a product developed by DuPont in 1911, one of several semi-smokeless products in the industry containing a mixture of black and nitrocellulose powder. It was sold to Winchester and others primarily for .22 and .32 small calibers. Its advantage was that it was believed at the time to be less corrosive than smokeless powders then in use. It was not understood in the U.S. until the 1920s that the actual source of corrosion was the potassium chloride residue from potassium chlorate sensitized primers. The bulkier black powder fouling better disperses primer residue. Failure to mitigate primer corrosion by dispersion caused the false impression that nitrocellulose-based powder caused corrosion. Lesmok had some of the bulk of black powder for dispersing primer residue, but somewhat less total bulk than straight black powder, thus requiring less frequent bore cleaning. It was last sold by Winchester in 1947.

Sulfur-free Gunpowder

Burst barrel of a muzzle loader pistol replica, which was loaded with nitrocellulose powder instead of black powder and could not withstand the higher pressures of the modern propellant.

The development of smokeless powders, such as cordite, in the late 19th century created the need for a spark-sensitive priming charge, such as gunpowder. However, the sulfur content of traditional gunpowders caused corrosion problems with Cordite Mk I and this led to the introduction of a range of sulfur-free gunpowders, of varying grain sizes. They typically contain 70.5 parts of saltpeter and 29.5 parts of charcoal. Like black powder, they were produced in different grain sizes. In the United Kingdom, the finest grain was known as *sulfur-free mealed powder* (*SMP*). Coarser grains were numbered as sulfur-free gunpowder (SFG n): 'SFG 12', 'SFG 20', 'SFG 40' and 'SFG 90', for example; where the number represents the smallest BSS sieve mesh size, which retained no grains.

Sulfur's main role in gunpowder is to decrease the ignition temperature. A sample reaction for sulfur-free gunpowder would be

$$6\ KNO_3 + C_7H_4O \rightarrow 3\ K_2CO_3 + 4\ CO_2 + 2\ H_2O + 3\ N_2$$

Combustion Characteristics

Chemical Reaction

Gunpowder does not burn as a single reaction, so the byproducts are not easily predicted. One study showed that it produced (in order of descending quantities) 55.91% solid products: potassium carbonate, potassium sulfate, potassium sulfide, sulfur, potassium nitrate, potassium thiocyanate, carbon, ammonium carbonate and 42.98% gaseous products: carbon dioxide, nitrogen, carbon monoxide, hydrogen sulfide, hydrogen, methane, 1.11% water.

However, simplified equations have been cited.

A simple, commonly cited, chemical equation for the combustion of black powder is

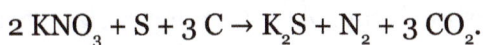

$$2\ KNO_3 + S + 3\ C \rightarrow K_2S + N_2 + 3\ CO_2.$$

A balanced, but still simplified, equation is:

$$10\ KNO_3 + 3\ S + 8\ C \rightarrow 2\ K_2CO_3 + 3\ K_2SO_4 + 6\ CO_2 + 5\ N_2.$$

Both previous equation are based on the assumption that charcoal is pure carbon, while in real life charcoal's chemical formula varies, but it can be summed up by its empirical formula: C_7H_4O. Therefore, a more accurate equation of the decomposition of regular black powder with sulfur is:

$$6 \text{ KNO}_3 + C_7H_4O + 2 \text{ S} \rightarrow K_2CO_3 + K_2SO_4 + K_2S + 4 \text{ CO}_2 + 2 \text{ CO} + 2 \text{ H}_2O + 3 \text{ N}_2$$

Likewise, black powder without sulfur gives:

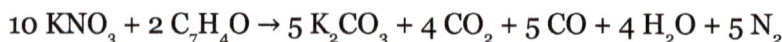

$$10 \text{ KNO}_3 + 2 \text{ C}_7H_4O \rightarrow 5 \text{ K}_2CO_3 + 4 \text{ CO}_2 + 5 \text{ CO} + 4 \text{ H}_2O + 5 \text{ N}_2$$

Black powder made with less-expensive and more plentiful sodium nitrate (in appropriate proportions) works just as well, and previous equations apply, with sodium instead of potassium. However, it is more hygroscopic than powders made from potassium nitrate—popularly known as saltpeter. Because *corned* black powder grains made with saltpeter are less affected by moisture in the air, they can be stored unsealed without degradation by humidity. Muzzleloaders have been known to fire after hanging on a wall for decades in a loaded state, provided they remained dry. By contrast, black powder made with sodium nitrate must be kept sealed to remain stable.

The matchlock musket or pistol (an early gun ignition system), as well as the flintlock would often be unusable in wet weather, due to powder in the pan being exposed and dampened.

Energy

Gunpowder releases 3 megajoules per kilogram and contains its own oxidant. This is lower than TNT (4.7 megajoules per kilogram), or gasoline (47.2 megajoules per kilogram, but gasoline requires an oxidant, so an optimized gasoline and O_2 mixture contains 10.4 megajoules per kilogram). Black powder also has a low energy density compared to modern "smokeless" powders, and thus to achieve high energy loadings, large amounts of black powder are needed with heavy projectiles.

Effects

Gunpowder is a low explosive, that is, it does not detonate but rather deflagrates (burns quickly). This is an advantage in a propeller device, where you don't want a shock that would shatter the gun and potentially harm the operator, however it is a drawback when some explosion is wanted. In that case, gunpowder (and most importantly, gases produced by its burning) must be confined. Since it contains its own oxidizer and additionally burns faster under pressure, its combustion is capable of bursting containers such as shell, grenade, or improvised "pipe bomb" or "pressure cooker" casings to form shrapnel.

In quarrying, high explosives are generally preferred for shattering rock. However, because of its low brisance, black powder causes fewer fractures and results in more usable stone compared to other explosives, making black powder useful for blasting monumental stone such as granite and marble. Black powder is well suited for blank rounds, signal flares, burst charges, and rescue-line launches. Black powder is also used in fireworks for lifting shells, in rockets as fuel, and in certain special effects.

As seen above, combustion converts less than half the mass of black powder to gas, most of it turns into particulate matter. Some of it is ejected, wasting propelling power, fouling the air, and generally being a nuisance (giving off a soldier position, generating fog that hinders vision, etc.). Some of it ends up as a thick layer of soot inside the barrel, where it also is a nuisance for subsequent shots, and a cause of jamming an automatic weapon. Moreover this residue is hygroscopic, and with the addition of moisture absorbed from the air forms a corrosive substance. The soot contains potassium oxide or sodium oxide that turns into potassium hydroxide, or sodium hydroxide, which corrodes wrought iron or steel gun barrels. Black powder arms must therefore be well cleaned after use, both inside and out, to remove the residue.

Transportation Regulations

The United Nations Model Regulations on the Transportation of Dangerous Goods and national transportation authorities, such as United States Department of Transportation, have classified gunpowder (black powder) as a *Group A: Primary explosive substance* for shipment because it ignites so easily. Complete manufactured devices containing black powder are usually classified as *Group D: Secondary detonating substance, or black powder, or article containing secondary detonating substance*, such as firework, class D model rocket engine, etc., for shipment because they are harder to ignite than loose powder. As explosives, they all fall into the category of Class 1.

Other uses

Besides its use as an explosive, gunpowder has been occasionally employed for other purposes; after the Battle of Aspern-Essling (1809), the surgeon of the Napoleonic Army Larrey, lacking salt, seasoned a horse meat bouillon for the wounded under his care with gunpowder. It was also used for sterilization in ships when there was no alcohol.

Jack Tars (British sailors) used gunpowder to create tattoos when ink wasn't available, by pricking the skin and rubbing the powder into the wound in a method known as traumatic tattooing.

Christiaan Huygens experimented with gunpowder in 1673 in an early attempt to build an internal combustion engine, but he did not succeed. Modern attempts to recreate his invention were similarly unsuccessful.

Fireworks use gunpowder as lifting and burst charges, although sometimes other more powerful compositions are added to the burst charge to improve performance in small shells or provide a louder report. Most modern firecrackers no longer contain black powder.

Beginning in the 1930s, gunpowder or smokeless powder was used in rivet guns, stun guns for animals, cable splicers and other industrial construction tools. The "stud gun" drove nails or screws into solid concrete, a function not possible with hydraulic tools. Shotguns have been used to eliminate persistent material rings in operating rotary kilns (such as those for cement, lime, phosphate, etc.) and clinker in operating furnaces, and commercial tools make the method more reliable.

Near London in 1853, Captain Shrapnel demonstrated a method for crushing gold-bearing ores by firing them from a cannon into an iron chamber, and "much satisfaction was expressed by all present". He hoped it would be useful on the goldfields of California and Australia. Nothing came of the invention, as continuously-operating crushing machines that achieved more reliable comminution were already coming into use.

Gunshot Residue

Gunshot residue (GSR), also known as cartridge discharge residue (CDR), "gunfire residue" (GFR), or firearm discharge residue (FDR), is residue deposited on the hands and clothes of someone who discharges a firearm. It is principally composed of burnt and unburnt particles from the explosive primer, the propellant—and possibly fragments of the bullet, cartridge case, and the firearm.

Law enforcement investigators test the clothing and skin of people for GSR to determine if they were near a gun when it discharged. Gunshot residue can travel over 3–5 feet (0.9–1.5 meters) from the gun. At the farthest distance, only a few trace particles may be present.

History

In 1971 John Boehm presented some micrographs of GSR particles found during the examination of bullet entrance holes using a scanning electron microscope. If the scanning electron microscope is equipped with an energy-dispersive X-ray spectroscopy detector, the chemical elements present in such particles, mainly lead, antimony and barium, can be identified.

In 1979 Wolten et al. proposed a classification of GSR based on composition, morphology, and size. Four compositions were considered *characteristic*:

- Lead, antimony, and barium
- Barium, calcium, and silicon
- Antimony
- Barium

The authors proposed some rules about chemical elements that could also be present in these particles.

Wallace and McQuillan published a new classification of the GSR particles in 1984. They labeled as *unique* particles those that contain lead, antimony, and barium, or that contain antimony and barium. Wallace and McQuillan also maintained that these particles could contain only some chemical elements.

Current Practice

In the latest ASTM Standard Guide for GSR analysis by Scanning Electron Microscopy/ Energy Dispersive X-ray Spectrometry (SEM-EDX) particles containing lead, antimony and barium, and respecting some rules related to the morphology and to the presence of other elements are considered characteristic of GSR. The most definitive method to determine if a particle is characteristic of or consistent with GSR is by its elemental profile. An approach to the identification of particles characteristic of or consistent with GSR is to compare the elemental profile of the recovered particulate with that collected from case-specific known source items, such as the recovered weapon, Cartridge cases or victim-related items whenever necessary. This approach was called "case by case" by Romolo and Margot in an article published in 2001. In 2010 Dalby et al. published the latest review on the subject and concluded that the adoption of a "case by case" approach to GSR analysis must be seen as preferable, in agreement with Romolo and Margot.

In light of similar particles produced from extraneous sources, both Mosher et al. (1998) and Grima et al. (2012) presented evidence of pyrotechnic particles that can be mistakenly identified as GSR. Both publications highlight that certain markers of exclusion and reference to the general population of collected particulate can help the expert in designating GSR-similar particles as firework-sourced.

Particle analysis by scanning electron microscope equipped with an energy-dispersive X-ray spectroscopy detector is the most powerful forensic tool that investigators can use to determine a subject's proximity to a discharging firearm or contact with a surface exposed to GSR (firearm, spent cartridge case, target hole). Test accuracy requires procedures that avoid secondary gunshot residue transfer from police officers onto subjects or items to be tested, and that avoid contamination in the laboratory.

The two main groups of specialists currently active on gunshot residue analysis are the Scientific Working Group for Gunshot Residue (SWGGSR) based in USA and the ENFSI EWG Firearms/GSR Working Group based in Europe.

Results

A positive result for GSR from SEM-EDX analysis can mean many things. Mainly it indicates that the person sampled was either in the vicinity of a gun when it was fired, handled a gun after it was fired, or touched something that was around the gun when it was fired. (For example: When a person goes to the aid of a victim of a gunshot wound, some GSR particles can transfer from the victim.)

A negative result can mean that the person was nowhere near the gun when it was fired, or that they were near it but not close enough for GSR to land on them, or it can mean that the GSR deposited on them wore off. GSR is the consistency of flour and typically only stays on the hands of a living person for 4–6 hours. Wiping the hands on anything, even putting them in and out of pockets can transfer GSR off the hands. Victims don't always get GSR on them; even suicide victims can test negative for GSR.

Matching GSR to a Specific Source

If the ammunition used was specifically tagged in some way by special elements, it is possible to know the cartridge used to produce the GSR. Inference about the source of GSR can be based on the examination of the particles found on a suspect and the population of particles found on the victim, in the firearm or in the cartridge case, as suggested by the ASTM Standard Guide for GSR analysis by Scanning Electron Microscopy/Energy Dispersive X-ray Spectrometry. Advanced analytical techniques such as Ion Beam Analysis (IBA), carried out after Scanning Electron Microscopy, can support further information allowing to infer about the source of GSR particles. Christopher et al. showed as the grouping behaviour of different makes of ammunition can be determined using multivariate analysis. Bullets can be matched back to a gun using comparative ballistics.

Organic Gunshot Residue

Organic gunshot residue can be analysed by analytical techniques such as chromatography, capillary electrophoresis, and mass spectrometry.

Rimfire (Firearms)

Rimfire ammunition, from left to right, .22 Short, .22 Long Rifle, .22 WMR, .17 HM2, .17 HMR

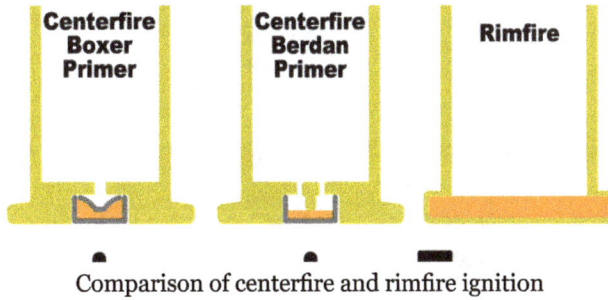

Comparison of centerfire and rimfire ignition

Rimfire is a method of ignition for metallic firearm cartridges as well as the cartridges themselves. It is called rimfire because the firing pin of a gun strikes and crushes the base's rim to ignite the primer. The rim of the rimfire cartridge is essentially an extended and widened percussion cap which contains the priming compound, while the cartridge case itself contains the propellant powder and the projectile (bullet). Once the rim of the cartridge has been struck and the bullet discharged, the cartridge cannot be reloaded, because the head has been deformed by the firing pin impact. While many other different cartridge priming methods have been tried since the 19th century, only rimfire technology and centerfire technology survive today in significant use.

Frenchman Louis-Nicolas Flobert invented the first rimfire metallic cartridge in 1845. His cartridge consisted of a percussion cap with a bullet attached to the top and the idea was to improve the safety of indoor shooting. Usually derived in the 6 mm and 9 mm calibres, it is since then called the Flobert cartridge or the Bosquette cartridge but it doesn't contain any powder, the only propellant substance contained in the cartridge is the percussion cap itself. In English-speaking countries, the Flobert cartridge corresponds to .22 BB and .22 CB ammunition.

Characteristics

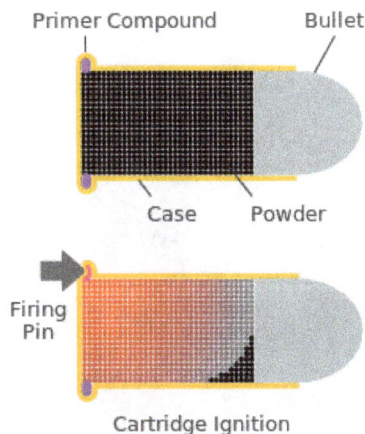

A schematic of a rimfire cartridge.

Fired rimfire (left) and centerfire cartridges. A rimfire firing pin produces a notch at the edge of the case; a centerfire pin produces a divot in the center of the primer.

Rimfire cartridges are limited to low pressures because they require a thin case so that the firing pin can crush the rim and ignite the primer. Rimfire cartridges of .44 caliber (actually .45 caliber) up to .56 caliber were once common when black powder was used as a propellant. However, modern rimfire cartridges use smokeless powder which generates much higher pressures and tend to be of .22 caliber (5.5 mm) or smaller. The low pressures necessitated by the rimfire design mean that rimfire firearms can be very light and inexpensive, which has helped lead to the continuing popularity of these small-caliber cartridges.

Economics

Rimfire cartridges are typically inexpensive, primarily due to the inherent cost-efficiency of the ability to manufacture the cartridges in large lots. The price of metals used in the cartridges (lead, copper and zinc) increased in 2002; the prices of the ammunition then further increased in 2012 possibly due to hoarding.

History

Cartridge .44-Henry-Flat

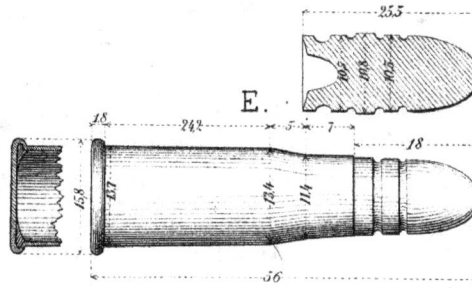

Cartridge Vetterli Ord. Suisse 1869

The idea of placing a priming compound in the rim of the cartridge evolved from an 1831 patent, which called for a thin case, coated all along the inside with priming compound.

By 1845, this had evolved into the Flobert .22 BB Cap, in which the priming compound is distributed just inside the rim. The .22 BB Cap is essentially just a percussion cap with a round ball pressed in the front, and a rim to hold it securely in the chamber. Intended for use in an indoor "gallery" target rifle, it used no gunpowder, but relied entirely on the priming compound for propulsion. Its velocities were very low, comparable to an airgun. The next rimfire cartridge was the .22 Short, developed for Smith & Wesson's first revolver, in 1857; it used a longer rimfire case and 4 grains (260 mg) of black powder to fire a conical bullet.

This led to the .22 Long, with the same bullet weight as the short, but with a longer case and 5 grains (320 mg) of black powder. This was followed by the .22 Extra Long with a case longer than the .22 Long and a heavier bullet. The .22 Long Rifle is a .22 Long case loaded with the heavier Extra Long bullet intended for better performance in the long barrel of a rifle. Larger rimfire calibers were used during the American Civil War in the Henry Repeater, the Spencer Repeater, the Ballard rifle and the Frank Wesson carbine. While larger rimfire calibers were made, such as the, .30 rimfire, .32 rimfire, .38 rimfire .41 Short, the .44 Henry Flat devised for the famous Winchester 1866 carbine, up to the .58 Miller, the larger calibers were quickly replaced by centerfire versions, and today the .22 caliber rimfires are all that survive of these early rimfire cartridges.

The early 21st century has seen a revival in interest in rimfire cartridges, with two new rimfires introduced, both in .17 caliber (4.5 mm) and the re-introduction of the 5mm Remington Rimfire Magnum in 2008.

A new and increasingly popular rimfire, the 17 HMR is based on a .22 WMR casing with a smaller formed neck which accepts a .17 bullet. The advantages of the 17 HMR over .22 WMR and other rimfire cartridges are its much flatter trajectory, and its highly frangible hollow point bullets (often manufactured with plastic "ballistic tips" that improve the bullet's external ballistics). The .17 HM2 [Hornardy Mach 2] is based on the .22 Long Rifle and offers similar performance advantages over its parent cartridge, at a significantly higher cost. While .17 HM2 sells for about four times the cost of .22

Long Rifle ammunition (per box of 50 rounds), it is still significantly cheaper than most centerfire ammunition, and somewhat cheaper than the .17 HMR.

A notable rimfire cartridge that is still in production in Europe, and is chambered by the Winchester Model 39 in the 1920s, is the 9 mm Flobert. This cartridge can fire a small ball, but is primarily loaded with a small amount of shot, and used in smoothbore guns as a miniature shotgun, or "garden gun". Its power and range are very limited, making it suitable only for pest control. An example of rare but modern 9 mm Flobert Rimfire among hunters in Europe is the 1.75" Brass Shotshell manufactured by Fiocchi in Lecco, Italy using a .25 oz shot of #8 shot with a velocity of 600 fps.

Below is a list of the most common current production rimfire ammunition:

- The powderless .22 Cap rounds (.22 CB), including BB Cap (.22 BB).

- .22 Short, used for target shooting and Olympic and ISSF 25 m Rapid Fire Pistol competition until 2005

- .22 Long (obsolete but available)

- .22 Long Rifle (.22 LR), the most common cartridge made

- .22 Stinger (a form of .22 Long Rifle with a slightly longer case and the same overall loaded length), its case is the basis for the .17 HM2

- .22 Winchester Rimfire (.22 WRF) AKA .22 Remington Special (obsolete but available)

- .22 Winchester Magnum Rimfire (.22 WMR)

- 5 mm Remington Rimfire Magnum (formerly discontinued but currently in commercial production by Aguila Ammunition/Centurion)

- .17 Hornady Magnum Rimfire (.17 HMR), a .17 caliber cartridge based on a modified .22 WMR case

- .17 Hornady Mach 2 (.17 HM2), a .17 caliber cartridge based on a modified .22 Stinger case

- .17 Winchester Super Magnum (.17 WSM) a .17 caliber cartridge based on a modified .27 caliber nail gun blank case

Shot

Some rimfire cartridges are loaded with a small amount of #11 or #12 shot (about 1/15th ounce). This "rat-shot" is only marginally effective in close ranges, and is usually used for shooting rats or other small animals. It is also useful for shooting birds inside storage buildings as it will not penetrate walls or ceilings. At a distance of about 10 feet

(3 m) the pattern is about 8 inches (20 cm) in diameter from a standard rifle, which is about the maximum effective range. Special smoothbore shotguns, such as Marlin's *Garden Gun* can produce effective patterns out to 15 or 20 yards using .22 WMR shotshells, which hold 1/8 oz. of #11 or #12 shot contained in a plastic capsule.

Shotshells will not feed reliably in some magazine fed firearms, due to the unusual shape of some cartridges that are crimped closed at the case mouth, and the relatively fragile plastic tips of other designs. Shotshells will also not produce sufficient power to cycle semiautomatic actions, because, unlike projectile ammunition, nothing forms to the lands and grooves of the barrel to create the pressure necessary to cycle the firearm's action.

Collectibility

Rimfire ammunition is popular with ammunition cartridge collectors, who base much of the collectibility value of rimfire cartridges on the rarity of the stamped mark on the head of the cartridge (the headstamp). There is a subcategory of collectible rimfire ammunition with headstamps commemorating certain persons who have worked in the industry, often issued in extremely small quantities on the occasion of that person's retirement. Often the majority of these cartridges are given to the retiring individual, leaving him or her to decide to whom they are traded or distributed. This increases their perceived value as collectibles.

References

- De Haas, Frank; Wayne Van Zwoll (2003). "Short Stature, Long Range". Bolt Action Rifles - 4th Edition. Krause Publications. pp. 636–643. ISBN 978-0-87349-660-5

- Kosanke, Bonnie J. (2002), "Selected Pyrotechnic Publications of K. L. and B. J. Kosanke: 1998 Through 2000", Journal of Pyrotechnics: 34–45, ISBN 978-1-889526-13-3

- Andrade, Tonio (2016), The Gunpowder Age: China, Military Innovation, and the Rise of the West in World History, Princeton University Press, ISBN 978-0-691-13597-7

- Al-Hassan, Ahmad Y. (2001), "Potassium Nitrate in Arabic and Latin Sources", History of Science and Technology in Islam, retrieved 23 July 2007

- Khan, Iqtidar Alam (1996), "Coming of Gunpowder to the Islamic World and North India: Spotlight on the Role of the Mongols", Journal of Asian History, 30: 41–5

- Kelly, Jack (2004), Gunpowder: Alchemy, Bombards, & Pyrotechnics: The History of the Explosive that Changed the World, Basic Books, ISBN 0-465-03718-6

- P.V. Mosher, M.J. McVicar, E.D. Randall, E.H. Sild, Gunshot residue-similar particles produced by fireworks, Journal of the Canadian Society of Forens. Sci. 31 (3)(1998) 157–168

- Perrin, Noel (1979), Giving up the Gun, Japan's reversion to the Sword, 1543–1879, Boston: David R. Godine, ISBN 0-87923-773-2

Transitional Ballistics: An Integrated Study

Traditional ballistics studies the behavior of the projectile from the moment it leaves the muzzle till the moment of equalization of force that provides motion. The two main devices that come under transitional ballistics realm are flash suppressors and sound suppressors. This chapter elucidates the crucial theories and principles of transitional ballistics.

Transitional Ballistics

Transitional ballistics, also known as intermediate ballistics, is the study of a projectile's behavior from the time it leaves the muzzle until the pressure behind the projectile is equalized, so it lies between internal ballistics and external ballistics.

The Transitional Period

Transitional ballistics is a complex field that involves a number of variables that are not fully understood; therefore, it is not an exact science. When the bullet reaches the muzzle of the barrel, the escaping gases are still, in many cases, at hundreds of atmospheres of pressure. Once the bullet exits the barrel, breaking the seal, the gases are free to move past the bullet and expand in all directions. This expansion is what gives gunfire its explosive sound (in conjunction with the sonic boom of the projectile), and is often accompanied by a bright flash as the gases combine with the oxygen in the air and finish combusting.

The propellant gases continue to exert force on the bullet and firearm for a short while after the bullet leaves the barrel. One of the essential elements of accurizing a firearm is to make sure that this force does not disrupt the bullet from its path. The worst case is a muzzle or muzzle device such as a flash-hider that is cut at a non-square angle, so that one side of the bullet leaves the barrel early; this will cause the gas to escape in an asymmetric pattern, and will push the bullet away from that side, causing shots to form a "string", where the shots cluster along a line rather than forming a normal Gaussian pattern.

Most firearms have muzzle velocities in excess of the ambient speed of sound, and even in subsonic cartridges the escaping gases will exceed the speed of sound, forming a shock wave. This wave will quickly slow as the expanding gas cools, dropping the speed

of sound within the expanding gas, but at close range this shockwave can be very damaging. The muzzle blast from a high powered cartridge can literally shred soft objects in
its vicinity, as careless benchrest pistol shooters occasionally find out when the muzzle
slips back onto their sandbag and the muzzle blast sends sand flying.

Initial Velocity Calculation

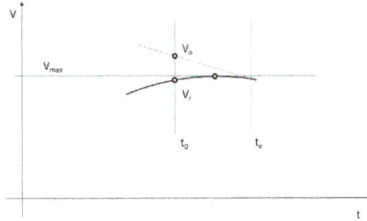

The initial velocity (Vo) and real muzzle velocity (Vr) difference

During the first part of the intermediate ballistics period the real velocity of the projectile increases. It is caused by the propellant gases exiting the muzzle. For that reason
the real maximum projectile velocity (Vmax) is higher than the real muzzle velocity
(Vr). The external ballistics uses so-called initial velocity Vo, which is not the same as
the real muzzle velocity. The initial velocity Vo is calculated via an extrapolation of the
decaying part of velocity curve to the position of the muzzle (to). The difference between these two velocities is visible in the chart.

Altering Transitional Ballistics

In addition to the process of "crowning" a barrel to ensure a clean and accurate exit of
the bullet, there are a number of devices that attempt to harness the muzzle blast for
various reasons.

Suppressing the Blast

Flash suppressors and sound suppressors are the most obvious devices that operate in
the transitional ballistics realm. These both alter the flow of the escaping gas to reduce
the effects of the muzzle blast. Flash suppressors introduce turbulence into the mixing
of fuel-laden hot gases escaping from the muzzle and the surrounding oxygen-rich air,
reducing combustion efficiency and thus reducing the size and brightness of the flash.
Sound suppressors slow the expansion of gases, allowing it to cool and reducing the
rate at which it escapes to prevent a shockwave from forming.

A *recoil compensator* is designed to direct the gases upwards at roughly a right angle
to the bore, in essence making it a small rocket that pushes the muzzle downwards, and
counters the "flip", or rise of the muzzle caused by the high bore line of most firearms.
These are often found on "raceguns" used for action shooting and in heavy, rifle caliber
handguns used in metallic silhouette shooting. In the former case, the compensator
serves to keep the sights down on target for a quick follow-up shot, while in the latter

case they keep the heavy recoil directed backwards, preventing the pistol from trying to twist out of the shooter's grip.

A *muzzle brake* is designed to redirect the muzzle blast backwards, and therefore counter the recoil of the bullet. Muzzle brakes tend to be found on larger firearms, such as magnum rifles and artillery. A well designed muzzle brake can significantly reduce recoil, turning a rifle that would otherwise be punishing to shoot into a far more tolerable experience. A good example may be seen on the M82 Barrett sniper rifle.

There are downsides to both recoil compensators and muzzle brakes. They direct more of the muzzle flash to the sides or back towards the shooter—this is especially true of muzzle brakes. While eye and ear protection should always be used when shooting, this is even more essential with the muzzle blast directed back towards the shooter. Brakes and compensators are often quite bulky, adding length, diameter, and mass to the muzzle end of the firearm where it will affect the firearm's handling worst. While a simple slot milled in the barrel, such as those used in Magna-Porting, will provide some benefit, efficient redirection of the gas flow requires large ports and baffles to deflect as much gas as possible. It is also highly inadvisable to fire sabot rounds like shotgun slugs or APDS rounds through a muzzle brake not designed for them.

Suppressor

Several firearms with detachable suppressors, from top to bottom: an Uzi, an AR-15, a Heckler & Koch USP Tactical, a Beretta 92FS, and a SIG Mosquito.

CZ 452 bolt-action rimfire rifle with suppressor

AWM-F sniper rifle with attached suppressor

Integral suppressor on VSS Vintorez sniper rifle and AS Val assault rifle.

A suppressor, sound suppressor, sound moderator, or silencer is a device attached to or part of the barrel of a firearm or air gun which reduces the amount of noise and visible muzzle flash generated by firing. Suppressors are typically constructed of a metal cylinder with internal mechanisms to reduce the sound of firing by slowing the escaping propellant gas. (A flash suppressor specifically cools or disperses burning gases typically exiting from the muzzle of a carbine-length weapon, without reference to sound reduction.)

In most countries, silencers are regulated by firearm legislation to varying degrees. While some have allowed for sporting use of silencers (especially to mitigate the costs of hearing loss and noise pollution), other governments have opted to ban them from civilian use.

History

American inventor Hiram Percy Maxim, the son of Maxim gun inventor Hiram Stevens Maxim, is usually credited with inventing and selling the first commercially successful models circa 1902 (patented March 30, 1909). Maxim gave his device the trademarked name *Maxim Silencer*, and they were regularly advertised in sporting goods magazines. The muffler for internal combustion engines was developed in parallel with the firearm suppressor by Maxim in the early 20th century, using many of the same techniques to provide quieter-running engines (in many English-speaking countries automobile mufflers are called silencers). Former president of the United States Theodore Roosevelt was known to purchase and use Maxim Silencers.

Suppressors were regularly used by agents of the United States Office of Strategic Services, who favored the newly designed High Standard HDM .22 Long Rifle pistol during World War II. OSS Director William Joseph "Wild Bill" Donovan demonstrated the pistol for President Franklin D. Roosevelt at the White House. According to OSS research chief Stanley Lovell, Donovan (an old and trusted friend of the President) was waved into the Oval Office, where Roosevelt was dictating a letter. While Roosevelt finished his message, Donovan turned his back and fired ten shots into a sandbag he had brought with him, announced what he had done and handed the smoking gun to the astonished president. The British SOE (Special Operations Executive) Welrod pistol with an integral suppressor was also used by the American OSS on clandestine operations in Nazi occupied Europe during the Second World War.

Terminology

Both the United States Department of Justice and the ATF (Bureau of Alcohol, Tobacco, Firearms and Explosive) refer to suppressors as "silencers". Additionally, Hiram Percy Maxim (the original inventor of the device) marketed them as "Maxim Silencers". The earliest use of the term "suppressor" to refer to firearm noise reduction is in US Patent 4530417, July 23, 1985 "A suppressor for reducing the muzzle blast of firearms or the like.".

Firearm Noise Anatomy

When a firearm is discharged, there are three ways sound is produced. Part of it can be managed; however, some of it is beyond the ability of the operator or manufacturers to eliminate.

In order of importance, the three ways a firearm generates sound are:

- Muzzle blast (high-temperature, high-pressure gases escaping after bullet)

- Sonic boom (sound associated with shock waves created by an object exceeding the speed of sound)

- Mechanical noise (moving parts of the firearm)

A suppressor can only affect the noise generated by the two primary sources—muzzle blast and sonic boom—and in most cases only the former. While subsonic ammunition can negate the sonic boom, mechanical noise can be mitigated but is nearly impossible to eliminate. For these reasons, it is difficult to completely silence any firearm, or achieve an acceptable level of noise suppression in revolvers that function under standard operating principles. Some revolvers have technical features that enable suppression and include the Russian Nagant M1895 and OTs-38 revolvers, and the S&W QSPR.

Muzzle blast generated by discharge is directly proportional to the amount of propellent contained within the cartridge. Therefore, the greater the case capacity the larger muzzle blast and consequently a more efficient or larger system is required. A gunshot (the combination of the sonic boom, the vacuum release, and hot gases) will almost always be louder than the sound of the action cycling of an auto-loading firearm. Alan C. Paulson, a renowned firearms specialist, claimed to have encountered an integrally suppressed .22 LR that had such a quiet report, although this is somewhat uncommon. Properly evaluating the sound generated by a firearm can only be done using a decibel meter in conjunction with a frequency spectrum analyser during live tests.

Design and Construction

BR Tuote Ky Reflex Rifle Suppressor Cross-section	
Cross-section drawing of a BR Tuote rifle suppressor, showing expansion chamber "reflexed" (going back around) the rifle barrel, and four baffles. The diffractor and baffles are carefully shaped to deflect gas.	Cross section diagram of Heckler & Koch MP5SD early model suppressor, from H&K 1971 patent. Vented barrel surrounded by metal mesh packing in expansion chambers, followed by conical baffles in forward chambers.
US Navy Hush Puppy Pistol Suppressor Cross-section	Vaime .22 suppressor Cross-section
Cross-section drawing of a US Navy Hush Puppy Mk 2 pistol suppressor, showing expansion chamber wrapped around inner suppressor assembly, and four wipes. The bullet pushes a bullet-diameter hole through the wipes, trapping propellant gas behind it entirely until the bullet has passed through the wipe completely.	Cross-section drawing of a Vaime .22 caliber rifle suppressor, showing short expansion chamber and thirteen plastic baffles. These baffles use alternating angled flat surfaces to repeatedly deflect gas expanding through the suppressor. In the actual suppressor, the baffles are oriented at 90 degrees to each other about the axis of bullet travel (the illustration cannot demonstrate this well).

The suppressor is typically a hollow metal tube manufactured from steel, aluminium, or titanium and contains expansion chambers. This device, typically cylindrical in shape, attaches to the muzzle of a pistol, submachine gun, or rifle. Some "can"-type suppressors (so-called as they often resemble a beverage can), may be detached by the user and attached to a different firearm. Another type is the "integral" suppressor, which typically consists of an expansion chamber or chambers surrounding the barrel. The barrel has openings or "ports" which bleed off gases into the chambers. This type of suppressor is part of the firearm (thus the term "integral"), and maintenance of the suppressor requires that the firearm be at least partially disassembled.

Both types of suppressors reduce noise by allowing the rapidly expanding gases from the firing of the cartridge to be decelerated and cooled through a series of hollow chambers. The trapped gas exits the suppressor over a longer period of time and at a greatly reduced velocity, producing less noise signature. The chambers are divided by either *baffles* or *wipes*. There are typically at least four and up to perhaps fifteen chambers in a suppressor, depending on the intended use and design details. Often, a single, larger expansion chamber is located at the muzzle end of a can-type suppressor, which allows the propellant gas to expand considerably and slow down before it encounters the baffles or wipes. This larger chamber may be "reflexed" toward the rear of the barrel to minimize the overall length of the combined firearm and suppressor, especially with longer weapons such as rifles.

Huntertown Kestrel 5.56 firearm suppressor disassembled to show blast chamber, baffles, and sections of the outer tube.

Suppressors vary greatly in size and efficiency. One disposable type developed in the 1980s by the U.S. Navy for 9×19mm pistols was 150 mm (5.9 in) long and 45 mm (1.8 in) in outside diameter, and was designed for six shots with standard ammunition or up to thirty shots with subsonic (slower than the speed of sound) ammunition. In contrast, one suppressor designed for rifles firing the powerful .50 caliber (BMG) cartridge is 509 mm (20.0 in) long and 76 mm (3.0 in) in diameter.

Two ancillary advantages to the suppressor are recoil reduction and flash suppression. Muzzle flash is reduced by both being contained in the suppressor and through the arresting of unburned powder that would normally burn in the air, adding to the flash. Recoil reduction results from the slowing of propellant gasses, which can contribute 30–50% of recoil velocity. The weight of suppressor and the location of that additional weight at the muzzle reduce recoil through basic mass as well as muzzle flip due to the location of this mass.

Components

Baffles and Spacers

Baffles are usually circular metal dividers which separate the expansion chambers. Each baffle has a hole in its center to permit the passage of the bullet through the suppressor and toward the target. The hole is typically at least 1 mm larger than the bullet caliber to minimize the risk of the bullet hitting the baffle in what is known as a "baffle strike". Baffles are typically made of stainless steel, aluminium, titanium, or alloys such as Inconel, and are either machined out of solid metal or stamped out of sheet metal. A few suppressors for low-powered cartridges such as the .22 Long Rifle have successfully used plastic baffles (certain models by Vaime and others.)

There are several unique baffle designs. M, K, Z, monolithic core and Ω (Omega) are the most prevalent. M-type is the crudest and comprises an inverted cone. K forms slanted obstructions diverging from the sidewalls, creating turbulence across the boreline. Z is expensive to machine and includes "pockets" of dead airspace along the sidewalls which trap expanded gases and hold them thereby lengthening the time that the gases cool before exiting. Omega forms a series of spaced cones drawing gas away from the boreline, incorporates a scalloped mouth creating cross-bore turbulence, which is in turn directed to a "mouse-hole" opening between the baffle stack and sidewall.

Propellant gas heats and erodes the baffles, causing wear, which is worsened by high rates of fire. Aluminium baffles are seldom used with fully automatic weapons, because service life is unacceptably short. Some modern suppressors using steel or high-temperature alloy baffles can endure extended periods of fully automatic fire without damage. The highest-quality rifle suppressors available today have a claimed service life of greater than 30,000 rounds. Baffles have not been given any specific angles, a specific size, or weight to meet any standards; they are created on a trial and error basis.

Spacers separate baffles and keep them aligned at a specified distance from each other inside the suppressor. Many baffles and spacers are manufactured as a single assembly and several suppressor designs have all the baffles attached together with spacers as a one-piece helical baffle stack. Modern baffles are usually carefully shaped to divert the propellant gases effectively into the chambers. This shaping can be a slanted flat surface, canted at an angle to the bore, or a conical or otherwise curved surface. One popular technique is to have alternating angled surfaces through the stack of baffles.

Wipes and Packing Material

Wipes are inner dividers intended to touch the bullet as it passes through the suppressor, and are typically made of rubber, plastic, or foam. Each wipe may either have a hole drilled in it before use, a pattern stamped into its surface at the point where the bullet will strike it, or it may simply be punched through by the bullet. Wipes typically last for a small number of firings (perhaps no more than five) before their performance is significantly degraded. While many suppressors used wipes in the Vietnam War era, most modern suppressors do not use them as anything that touches the projectile has significant accuracy implications. All "wipes" deteriorate quickly and require disassembly and spare parts replacement.

"Wet" suppressors or "wet cans" use a small quantity of water, oil, grease, or gel in the expansion chambers to cool the propellant gases and reduce their volume. The coolant lasts only a few shots before it must be replenished, but can greatly increase the effectiveness of the suppressor. Water is most effective, due to its high heat of vaporization, but it can run or evaporate out of the suppressor. Grease, while messier and less effective than water, can be left in the suppressor indefinitely without losing effectiveness. Oil is the least effective and least preferable, as it runs while being as messy as grease, and leaves behind a fine mist of aerosolized oil after each shot. Water-based gels, such as wire-pulling lubricant gel, are a good compromise; they offer the efficacy of water with less mess, as they do not run or drip. However, they take longer to apply, as they must be cleared from the bore of the suppressor to ensure a clear path for the bullet (grease requires this step as well). Generally, only pistol suppressors are shot wet, as rifle suppressors handle such high pressure and heat that the liquid is gone within 1–3 shots. Many manufacturers will not warranty their rifle suppressors for "wet" fire, as some feel this may even result in a dangerous over-pressurization of the silencer.

Packing materials such as metal mesh, steel wool, or metal washers may be used to fill the chambers and further dissipate and cool the gases. These are somewhat more effective than empty chambers, but less effective than wet designs. Metal mesh, if properly used, may last for hundreds or thousands of shots of spaced semi-automatic fire, however steel wool usually degrades within ten shots with stainless wool lasting longer than regular steel wool. Like wipes, packing materials are rarely found in modern suppressors.

Wipes, packing materials and purpose-designed wet cans have been generally abandoned in 21st-century suppressor design because they decrease overall accuracy and require excessive cleaning and maintenance. The instructions from several manufacturers state that their suppressors need not be cleaned at all. Furthermore, legal changes in the United States during the 1980s and 1990s made it much more difficult for end-users to legally replace internal silencer parts, and the newer designs reflect this reality.

Attachment

Suppressed 12 gauge handgun and 7.62x39mm rifle

Apart from integral suppressors which are integrated as a part of the firearm's barrel, most suppressors have a "female" threaded end, which attaches to "male" threads cut into the exterior of the barrel. These types of suppressors are mostly used on handguns and rifles chambered in .22LR. More powerful rifles may use this type of attachment, but harsh recoil may cause the suppressor to over-tighten to the barrel and the suppressor can become difficult to remove. SilencerCo's Salvo suppressor for shotguns attaches via internal barrel threading normally used to mount removable chokes.

Military rifles such as the M16 or M14 often utilize "quick-detach" suppressors which use coarser than normal threads and are installed over an existing muzzle device such as a flash suppressor and may include a secondary locking mechanism to allow the shooter to quickly and safely add or remove a sound suppressor based on individual needs.

Advanced Types

SilencerCo Osprey .45 suppressor on a Springfield pistol

In addition to containing and slowly releasing the gas pressure associated with muzzle blast or reducing pressure through the use of coolant mediums, advanced suppressor designs attempt to modify the properties of the sound waves generated by the muzzle blast. In these designs, effects known as frequency shifting and phase cancellation (or destructive interference) are used in an attempt to make the suppressor quieter. These effects are achieved by separating the flow of gases and causing them to collide with each other or by venting them through precision-made holes. The intended effect of frequency shifting is to shift audible sound waves frequencies into ultrasound (above 20 kHz), beyond the range of human hearing. The Russian AN-94 assault rifle features a muzzle attachment that claims apparent noise reduction by venting some gases through a "dog-whistle" type channel. Phase cancellation occurs when similar sound waves encounter each other 180° out of phase, cancelling the amplitude of the wave and eliminating the pressure variations perceived as sound.

Using either property to advantage requires that the suppressor be designed within the specification of the muzzle blast in mind. For example, the velocity of the sound waves is a major factor. This figure can change significantly between different cartridges and barrel lengths.

However, these concepts are controversial because muzzle blast creates broadband noise rather than pure tones, and phase cancellation in particular is therefore extremely difficult (if not impossible) to achieve. Some suppressor manufacturers claim to use phase cancellation in their designs.

From the practical perspective, supersonic cartridge loads are impractical to suppress past the levels that are merely hearing-safe for the shooter due to the sonic boom emitted by the bullet, and cartridges such as .22 LR and .45 ACP have long been recognized as the easiest to suppress even if using technology dating back to the 1940s.

Improvised Silencers

Improvised silencers have been made from a variety of materials. In 2015, Los Angeles County Sheriff deputies recovered a ZB vz. 26 light machine gun with an automobile oil filter attached. PVC pipes, plastic water bottles, and foam-filled pillows are also used. In the United States, improvised silencers are governed by the same laws as manufactured ones.

Characteristics

Rear of a suppressor with the Nielsen device protruding (completely assembled).

Retaining ring unscrewed and Nielsen device partially removed.

Nielsen device completely removed and disassembled.

Rear of suppressor showing the rotational indexing system incorporated into some Nielsen devices.

Functionally, a suppressor is meant to diminish the report of a discharged round, or make its sound unrecognizable. Other sounds emanating from the weapon remain unchanged. Even subsonic bullets make distinct sounds by their passage through the air and striking targets, and supersonic bullets produce a small sonic boom, resulting in a "ballistic crack". Semi-automatic and fully automatic firearms also make distinct noises as their actions cycle, ejecting the fired cartridge case and loading a new round.

Aside from reductions in volume, suppressors tend to alter the sound to something that is not identifiable as a gunshot. This reduces or eliminates attention drawn to the shooter. A Finnish expression dating from the Winter War says that "A silencer does not make a soldier silent, but it does make him invisible." Suppressors are particularly useful in enclosed spaces where the sound, flash and pressure effects of a weapon being fired are amplified. Such effects may disorient the shooter, affecting situational awareness, concentration and accuracy, and can permanently damage hearing very quickly.

As the suppressed sound of firing is overshadowed by ballistic crack, observers can be deceived as to the location of the shooter, often from 90 to 180 degrees from his actual location. However, counter-sniper tactics can include gunfire locators, where sensitive microphones are coupled to computers running algorithms, and use the ballistic crack to detect and localize the origin of the shot. The U.S. Boomerang system is one such example.

There are many advantages in using a suppressor that are not related to the sound.

Hunters using centerfire rifles find suppressors bring various important benefits that outweigh the extra weight and resulting change in the firearm's center of gravity. The most important advantage of a suppressor is the hearing protection for the shooter as well as their companions. Many hunters have suffered permanent hearing damage due to someone else firing a high-caliber gun too closely without warning. By reducing noise, recoil and muzzle-blast, it also enables the firer to follow through calmly on their first shot and fire a further carefully aimed shot without delay if necessary. Wildlife of all kinds are often confused as to the direction of the source of a well-suppressed shot. In the field, however, the comparatively large size of a centerfire rifle suppressor can cause unwanted noise if it bumps or rubs against vegetation or rocks, so many users cover them with neoprene sleeves.

Suppressors reduce firing recoil significantly, primarily by diverting and trapping the propellant gas. The gas generally has much less mass than the projectile, but it exits the muzzle at multiples of the projectile velocity, so reducing the speed and quantity of the gas expelled can significantly reduce the total momentum of the matter (gas and projectile) leaving the barrel, the negation of which, because momentum is conserved, is transferred to the gun as recoil. Paulson *et al.*, discussing low-velocity pistol calibers, suggest the recoil reduction is around 15%. With high-velocity calibers, recoil reduction runs in the range of 20–30%. The added mass of the suppressor—normally 300 to 500 grams—also helps to manage the recoil.

A suppressor also cools the hot gases coming out of the barrel enough that most of the lead-laced vapor that leaves the barrel condenses inside the suppressor, reducing the amount of lead that might be inhaled by the shooter and others around them. However, in auto-loading actions this might be offset by increased back pressure which results in propellant gas blowing back into a shooter's face through the chamber during case ejection.

Subsonic Ammunition

In weapons firing supersonic ammunition, the bullet itself produces a loud and very sharp sound as it leaves the muzzle in excess of the speed of sound and gradually reducing speed as it travels downrange. This is a small sonic boom, and is referred to in the firearm field as "ballistic crack" or "sonic signature". Subsonic ammunition eliminates this sound, but at the cost of lower velocity, resulting in decreased range and much decreased muzzle energy, thus lessening effectiveness on the target. For example, if the muzzle velocity is reduced from 820 m/s (common for f.ex. .308 Winchester) to a subsonic 290 m/s, the muzzle energy is reduced by a factor of 8. Military marksmen and police units may use this ammunition to maximize the effectiveness of their suppressed rifles. While the range may be decreased when using subsonic rounds, this may be acceptable for specialized situations, where the absolute minimum amount of noise is required.

However, the numeric effectiveness of subsonic rounds is, again, misrepresented by media. Independent testing of commercially available firearm suppressors with commercially available subsonic rounds has found that .308 subsonic rounds decreased the volume at the muzzle 10 to 12 dB when compared to the same caliber of suppressed supersonic ammunition. When combined with suppressors, the subsonic .308 rounds metered between 121 and 137 dB.

This ballistic crack depends on the speed of sound, which in turn depends mainly on air temperature. At sea level, an ambient temperature of 70 °F (21 °C), and under normal atmospheric conditions, the speed of sound is approximately 1,140 feet per second (350 m/s). Bullets that travel near the speed of sound are considered transonic, which means that the airflow over the surface of the bullet, which at points travels faster than the bullet itself, can break the speed of sound. Pointed bullets which gradually displace air can get closer to the speed of sound than round nosed bullets before becoming transonic.

Special cartridges have been developed for use with a suppressor. These cartridges use very heavy bullets to make up for the energy lost by keeping the bullet subsonic. A good example of this is the .300 Whisper cartridge, which is formed from a necked-up .221 Remington Fireball cartridge case. The subsonic .300 Whisper fires up to a 250 grains (16 g), .30 caliber bullet at about 980 feet per second (300 m/s), generating about 533 foot-pounds force (723 J) of energy at the muzzle. While this is similar to the energy available from the .45 ACP pistol cartridge, the reduced diameter and streamlined shape of the heavy .30 caliber bullet provides far better external ballistic performance, improving range substantially.

9×19mm Parabellum, a very popular caliber for suppressed shooting, can use almost any factory-loaded 147 grains (9.5 g) weight round to achieve subsonic performance. These 147 gr weight bullets typically have a velocity of 900–980 feet per second (270–300 m/s), which is less than the common 1,140 feet per second (350 m/s) speed of sound.

The Soviet / Russian armor-piercing 9×39mm ammunition used in for example the AS Val rifle has a high subsonic ballistic coefficient, high retained downrange energy, high sectional density, and moderate recoil. All elements combined make this a very attractive choice for close-quarters combat (CQC) firearms.

Instead of using subsonic ammunition, one can lower the muzzle velocity of a supersonic bullet before it leaves the barrel. Some suppressor designs, referred to as "integrals", do this by allowing gas to bleed off along the length of the barrel before the projectile exits. The MP5SD is the best example of this with holes right after the chamber of the barrel used to reduce a regular 115 or 124 gr ammunition to subsonic velocities.

Effectiveness

Firearm suppressors including the SilencerCo Osprey 9, SWR Octane 45, and SilencerCo Saker 5.56.

Live tests by independent reviewers of numerous commercially available suppressors find that even low-power, unsuppressed .22 LR handguns produce gunshots over 160 decibels. In testing, most of the suppressors reduced the volume to between 130 and 145 dB, with the quietest suppressors metering at 117 dB. The actual suppression of sound ranged from 14.3 to 43 dB, with most data points around the 30 dB mark. The De Lisle carbine, a British World War II suppressed rifle used in small numbers by Special Forces, was recorded at 85.5 dB in official firing tests.

Comparatively, ear protection commonly used while shooting provides 18 to 32 dB of sound reduction at the ear. Further, chainsaws, rock concerts, rocket engines, pneumatic drills, small firecrackers, and ambulance sirens are rated at 100 to 140 dB.

While some consider the noise reduction of a suppressor significant enough to permit safe shooting without hearing protection ("hearing safe"), noise-induced hearing loss may occur at 85 time-weighted-average decibels or above if exposed for a prolonged period, and suppressed gunshots regularly meter above 130 dB. However, the U.S. Occupational Safety and Health Administration uses 140 dB as the "safety cutoff" for impulsive noise, which has led most U.S. manufacturers to advertise sub-140 dB suppressors as "hearing safe". Current OSHA standards would allow no more than sub-single-second exposure to impact noise over 130 dB per 24 hours. That would equate to a single .308 round fired through a very efficient suppressor. This result effectively requires all users of suppressors to wear additional ear protection.

Decibel testing measures only the peak sound "pressure" noise, not duration or frequency. Limitations of dB testing become apparent in a comparison of sound between a .308 caliber rifle and a .300 Winchester Magnum rifle. The dB meter will show that both rifles produce the same decibel level of noise. Upon firing these rifles, however, it is clear that the .300 Winchester Magnum sounds much louder. What a dB meter does not show is that, although both rifles produce the same peak sound pressure level (SPL), the .300 Winchester Magnum holds its peak duration longer—meaning that the .300 Winchester Magnum sound remains at full value longer, while the .308 peaks and falls off more quickly. Decibel meters fail in this and other regards when being used as the principal means to determine suppressor capability. In a physical sense, dB meters essentially take a short-time average (RMS intensity of a sonic signal or impulse) over a specified period of time (sampling rate), and do not take into account the rate of increase of the sound wave packet (first derivative of packet envelope), which would in practice provide a better sense of the human perception of sound.

Alan C. Paulson, of Paladin Press, writes that suppressors are considered a great help, along with hearing protection, to preserve the hearing of the user and any onlookers.

Regulation

Legal regulation of suppressors varies widely around the world. In some nations, such as Finland and France, some or all types of suppressor are essentially unregulated and are sold through retail stores or by mail-order. In other countries their possession or use is more restricted.

Europe

- In Denmark, the Danish Weapons And Explosives Law makes the unlicensed possession of a suppressor illegal. It is legal to buy suppressors in Denmark if you have a valid gun license. As of 7 May 2014 it is legal to own and use suppressors for hunting.

- In Finland, a firearm suppressor is classified as a firearm part by law. Purchasing a suppressor requires a firearm ownership permit, which is to be shown to the vendor at the moment of purchase.

- Suppressors for rimfire pistols are sold without government oversight in France.

- In the Russian Federation, usage of firearm suppressors (legally defined as "devices for noiseless shooting") by civilians is prohibited, and the dealers are prohibited from selling them, but there is no penalty for purchasing or possession of such devices.

- In Sweden, suppressors for specified calibers are legal for hunting purposes and a license is required.

- In the United Kingdom, firearm certificate (FAC) will need to show permission for the purchase of a "sound moderator" and also the firearm for which it is intended. All firearms certificates have the firearm and caliber approved by the police and annotated to the document before a suppressor may be purchased. Applicants must show a "good reason" for needing the accessory.

- In Germany, a suppressor is treated the same in the eyes of the law as the weapon it is designed for. Accordingly, suppressors for air guns, which can be purchased by anyone over 18 years of age, can as well be purchased by anyone over said age. Since, amongst other things "good cause" must be shown to be issued a license to own a firearm in Germany, the same "good cause" requirement exists for suppressors for these firearms. This requirement is handled very varyingly across the States of Germany. The State of Bavaria accepts the possession of a valid hunter's license as "good cause" to own an unlimited number of suppressors, while North Rhine-Westphalia does not accept hunting as a "good cause" at all. Baden-Württemberg accepts "active exercise of hunting" as "good cause", but only allows the purchase of one suppressor.

North America

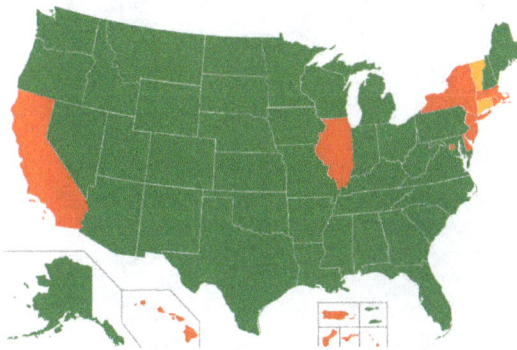

Legality of firearm suppressors by US jurisdiction

■ Suppressors legal to own and hunt with

■ Suppressors legal but illegal to hunt with

■ Suppressors illegal

- In Canada, a device to muffle or stop the sound of a firearm is a "prohibited device" under the Criminal Code. A prohibited device is not inherently illegal in Canada but it does require an uncommon and very specific prohibited device license for its possession, use, and transport. Suppressors cannot be imported into the country by civilians.

- In the United States, taxes and strict regulations affect the manufacture and sale of suppressors under the National Firearms Act. They are legal for individuals to possess and use for lawful purposes in 42 of the 50 states. However, a prospective owner must go through an application process administered by the Bureau of Alcohol, Tobacco, Firearms and Explosives (ATF), which requires a federal tax payment of $200 and a thorough criminal background check. The tax payment buys a revenue stamp, which is the legal document allowing possession of a suppressor. The 8 states that have explicitly banned any civilian from possessing a suppressor are: California, Delaware, Hawaii, Illinois, Massachusetts, New Jersey, New York, Rhode Island, and the District of Columbia. The states of Connecticut and Vermont allow for suppressor ownership but prohibit using suppressors while hunting. The federal legal requirements to manufacture a suppressor in the United States are enumerated in Title 26, Chapter 53 of the United States Code. The individual states and several municipalities also have their specific requirements. Federal law provides severe penalties for crimes of violence committed using firearms equipped with silencers: a minimum prison sentence of 30 years.

Flash Suppressor

Bullet exiting an A2-style flash suppressor, photographed with a high-speed air-gap flash.

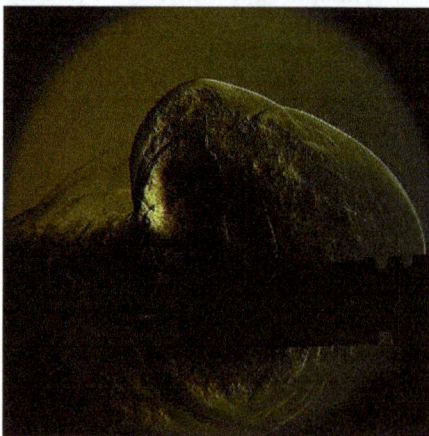

This image was taken from a high-speed Schlieren video of a flash suppressor.
Schlieren imaging reveals the heat and pressure gradients invisible to standard imaging.

Detail of the birdcage-type flash suppressor on a SIG SG 550.

A flash suppressor, also known as a flash guard, flash eliminator, flash hider, or flash cone, is a device attached to the muzzle of a rifle that reduces its visible signature while firing by cooling or dispersing the burning gases that exit the muzzle, a phenomenon typical of carbine-length weapons. Its primary intent is to reduce the chances that the shooter will be blinded in low-light shooting conditions. Contrary to popular belief, it is only a minor secondary benefit if a flash suppressor reduces the intensity of the flash visible to the enemy.

A flash suppressor is different from a muzzle brake, although they are typically mounted in the same position and sometimes confused with each other. While the former is intended to reduce visible flash, a muzzle brake is designed to reduce recoil inherent to large cartridges and typically does not reduce visible flash.

Rationale

Pre-20th century rifle designs tended to have longer barrels than modern rifles. A beneficial side effect of the long barrel is that the propellant is completely burned before the bullet leaves the barrel, usually resulting only in a puff of smoke from the muzzle. However, if the same weapon's barrel is "cut down" (shortened), as is common in cavalry and jungle-combat adapted versions, the bullet would often leave the barrel before the powder was completely consumed, resulting in a bright flash from the muzzle.

When barrel lengths were dramatically decreased with the introduction of various shorter-barreled rifles and carbines, the flash became a serious problem during nighttime combat, as the flash would imperil the shooter's night vision and would also make the shooter's position more apparent. Originally limited to "special purpose" roles, it was now expected that all infantry weapons with shorter barrels would experience this problem, and thereby be of limited use in low-light situations. Flash suppressors became common on late-World War II and later assault rifle designs, and are almost universal on these weapons today. Some designs such as those found on the AKS-74U serve a dual-role as a gas expansion chamber helping the shortened weapon's gas system function properly.

Military flash suppressors are designed to reduce the muzzle flash from the weapon to preserve the shooter's night vision, usually by diverting the incandescent gases to the sides, away from the line of sight of the shooter, and to secondarily reduce the flash visible to the enemy. Military forces engaging in night combat are still visible when firing, especially with night vision gear, and must move quickly after firing to avoid receiving return fire.

Limiting the amount of powder to what the length of a barrel can burn is one possible solution, but differences between individual cartridges mean that some cartridges will always have too much powder to be completely consumed, and the reduced powder load produces a lower projectile velocity. Muzzle flash can be controlled by using cartridges with a faster-burning propellant, so that the propellant gases will already have begun to cool by the time they exit the barrel, reducing flash intensities. Faster-burning powders, however, produce less projectile velocity, which reduces the accuracy due to introducing a more parabolic bullet flight path in place of a "flat" trajectory while also reducing lethality of the weapon by reducing the energy delivered on target.

Flash suppressors reduce, or in some cases eliminate, the flash by rapidly cooling the gases as they leave the end of the barrel. Although the overall amount of burning propellant is unchanged, the density and temperature are greatly reduced, as is the brightness of the flash.

Types

A number of different flash suppressing designs have been used over the years. The simplest is a cone placed on the end of the barrel, which was used on the late-World War II jungle-combat versions of the Lee–Enfield, the No. 5 variant, intended for use in the Pacific (the jungles of Malaya). More modern solutions tend to use a "basket" with several slits or holes cut in it, as seen on the M16 and other small-bore weapons. Cone-shaped flash eliminators are also evident on the ZB vzor 26 machine gun, and on the turret-mounted aircraft machine guns of British WWII heavy bombers, which were used mostly at night.

Duckbill flash suppressors have upper and lower "prongs" and direct gases to the sides. Early M60 machine guns and some early M16 models featured this type of flash suppressor. One disadvantage is that the prongs can become entangled with vines or other natural vegetation and objects in the field.

"Birdcage type" flash suppressors still have prongs, but feature a ring on the front to prevent vegetation entanglement between the prongs. The closed bottom port of the M16A2 design makes the device function as a compensator. Both designs require indexing with a crush washer which unfortunately contributes to flash signature.

The Vortex Flash Hider is a design developed in 1984, with a patent secured in 1995. The Vortex is somewhat reminiscent of the original "three-prong flash hider" found on the original Vietnam-era M-16. However, the Vortex is more robust and makes use of four solid tines, which are four equally spaced and angled 6° from a centerline, while the slots of the body incorporates a 5-, 10- and 15-degree twisted helix design, which eliminates up to 99% of visible muzzle flash by having the flash break up at multiple locations and angles.

The Noveske KX-3 is a flash suppressor intended for use on shorter barreled rifles and aids in reliability. The back pressure generated through this type of flash suppressor helps to cycle the rifle. Noveske patterned this design on the Krinkov brake found on the Soviet AKS-74U carbine, where it was explicitly used for this purpose. Essentially it is the cone-shaped suppressor of the AKS-74U within a chamber. Some other examples of cone-shaped hiders are found on the Bren machine gun, the .303 Rifle No 5 Mk 1 "Jungle Carbine" and some models of the RPK and German MG3.

The XM177 Commando variant of the M16 rifle used a unique flash suppressor sometimes called a flash or sound moderator for its 10-inch barrel. This device is 4.2 inches long and was designed primarily as a counterbalance measure as the shorter barrel made the weapon unwieldy. This device reduced flash signature greatly and sound signature slightly, making the normally louder short barreled rifle sound like a longer barreled M16A1. Although it has no internal baffles and does not completely reduce the sound signature to subsonic levels, because it alters the sound level of the weapon, the US Bureau of Alcohol Tobacco Firearms and Explosives has declared this device to be a sound suppressor and regulates its civilian purchase in the United States.

There are also devices referred to as hybrids that function as both flash suppressors and muzzle rise / recoil compensators such as the White Sound Defense FOSSA-556. The U.S. military A2 muzzle device is technically a hybrid device as it has vents that are biased upwardly to reduce muzzle rise.

Legality

New Zealand

Flash suppressors are seen as a "military" feature, and semi-automatic long guns with flash suppressors were defined as Military-Style Semi-Automatics in 1992, requiring a permit.

United States

Flash suppressors and barrel shrouds were seen as "military" features and were on the list of federally defined features that could cause a rifle to be defined as illegal, if the lower receiver was manufactured after the effective date of the Federal Assault Weapons Ban that went into effect in 1994 in the United States. In the context of that law, the National Rifle Association deemed telescoping stocks a "cosmetic feature". This ban expired in 2004, although some states such as California and New York have similar bans in place restricting the use of flash suppressors.

Germany & Australia

Flash suppressors and muzzle brakes can be legally acquired and used on all types of weapons, unless they are designed to significantly or predominantly reduce the sound.

Muzzle Brake

The muzzle brake of a 30mm Mauser autocannon.

The S&W Model 500 pistol features a muzzle brake.

The AKM rifle's slant-cut muzzle brake.

The prominent muzzle brake of the PGM Hecate II.

A muzzle brake or recoil compensator is a device connected to the muzzle of a firearm or cannon that redirects propellant gases to counter recoil and unwanted rising of the barrel during rapid fire. The concept was first introduced for artillery and was a common feature on many anti-tank guns, especially those mounted on tanks, in order to reduce the area needed to take up the strokes of recoil and kickback. They have been used in various forms for rifles and pistols to help control recoil and the rising of the barrel that normally occurs after firing. They are used on pistols for practical pistol competitions, and are usually called compensators in this context.

Rationale

The interchangeable terms muzzle rise, muzzle flip, or muzzle climb refer to the tendency of a handheld firearm's front end (the muzzle end of the barrel) to rise after firing. Firearms with less height from the grip line to the barrel centerline tend to experience less muzzle rise.

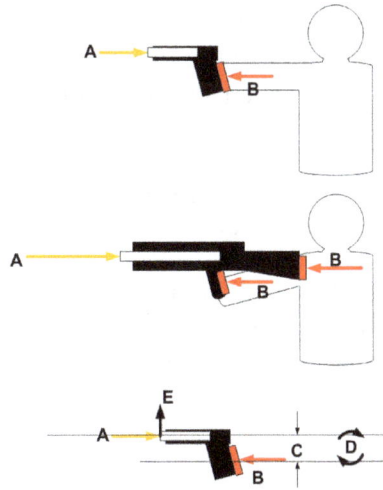

Illustration of forces in muzzle rise. Projectile and propellant gases act on barrel along barrel center line A. The shooter resists the forces by contact with the gun at grips and stock B. The height difference between barrel centerline and average point of contact is height C. The forces A and B operating over moment arm / height C create torque or moment D, which rotates the firearm's muzzle up as illustrated at E.

The muzzle rises primarily because, for most firearms, the centerline of the barrel is above the center of contact between the shooter and the firearm's grip and stock. The reactive forces from the fired bullet and propellant gases exiting the muzzle act directly down the centerline of the barrel. If that line of force is above the center of the contact points, this creates a moment or torque rotational force that makes the firearm rotate and the muzzle end rise upward. The M1946 Sieg automatic rifle had an unusual muzzle brake that made the rifle climb downward, but enabled the user to fire it with one hand in full automatic.

Design and Construction

Muzzle brakes are simple in concept, such as the one employed on the 90 mm M3 gun used on the M47 Patton tank. This consists of a small length of tubing mounted at right angles to the end of the barrel. Brakes most often utilize slots, vents, holes, baffles, and similar devices. The strategy of a muzzle brake is to redirect and control the burst of combustion gases that follows the departure of a projectile.

All muzzle brake designs share a basic principle: they partially divert combustion gases at a generally sideways angle, away from the muzzle end of the bore. The momentum of the diverted gases thus does not add to the recoil. The angle toward which the gases are directed will fundamentally affect how the brake behaves. If gases are directed upward, they will exert a downward force and counteract muzzle rise. Any device that is attached to the end of the muzzle will also add mass, increasing its inertia and moving its center of mass forward; the former will reduce recoil and the latter will reduce muzzle rise.

Construction of a muzzle brake or compensator can be as simple as a diagonal cut at the muzzle end of the barrel to direct some of the escaping gas upward. On the AKM assault rifle, the brake also angles slightly to the right to counteract the sideways movement of the rifle under recoil.

Another simple method is porting, where holes or slots are machined into the barrel near the muzzle to allow the gas to escape.

More advanced designs use baffles and expansion chambers to slow escaping gases. This is the basic principle behind a linear compensator. Ports are often added to the expansion chambers, producing the long, multi-chambered recoil compensators often seen on IPSC raceguns.

Venting Direction

Springfield-Armory Custom XD-40 V-10, showing the ported barrel and slide.

Most linear compensators redirect the gases forward. Since that is where the bullet is going, they typically work by allowing the gases to expand into the compensator, which surrounds the muzzle but only has holes facing forward; like any device which allows the gases to expand before leaving the firearm, they are effectively a type of muzzle shroud. They reduce muzzle rise similarly to the mechanism by which a sideways brake does: since all the gas is escaping in the same direction, any muzzle rise would need to alter the velocity of the gas, which costs kinetic energy. When the brake redirects the gases directly backward, instead, the effect is similar to the reverse thrust system on an aircraft jet engine; any blast energy coming back at the shooter is pushing "against" the recoil, effectively reducing the actual amount of recoil on the shooter. Of course, this also means the gases are directed toward the shooter.

When the gases are primarily directed upward, the braking is referred to as *porting*. Porting typically involves precision-drilled ports or holes in the forward top part of the barrel and slide on pistols. These holes divert a portion of the gases expelled prior to the departure of the projectile in a direction that reduces the tendency of the firearm to rise. The concept is an application of Newton's third law; the exhaust directed upward causes a reciprocal force downward. This is why firearms are never ported on the bottom of the barrel, as that would exacerbate muzzle rise, rather than mitigate it. Porting has the undesired consequences of shortening the effective barrel length and reducing muzzle velocity; while a muzzle brake is an extension added to the barrel and does not reduce muzzle velocity. Porting has the advantage for faster follow-up shots, especially for 3-round burst operation.

Effectiveness

Though there are numerous ways to measure the energy of a recoil impulse, in general a 10% to 50% reduction can be measured. Some muzzle brake manufacturers claim greater recoil reduction percentages. Muzzle brakes need sufficient propellant gas volume and high gas pressure at the muzzle of the firearm to achieve good measured recoil reduction percentages. This means cartridges with a small bore area to case volume ratio (overbore cartridges) combined with a high operating pressure benefit more from recoil reduction with muzzle brakes than smaller standard cartridges.

Besides reducing felt recoil, one of the primary advantages of a muzzle brake is the reduction of muzzle rise. This lets a shooter realign a weapon's sights more quickly. This is relevant for fully automatic weapons. Muzzle rise can theoretically be eliminated by an efficient design. Because the rifle moves rearward less, the shooter has little to compensate for. Muzzle brakes benefit rapid-fire, fully automatic fire, and large-bore hunting rifles. They are also common on small-bore vermin rifles, where reducing the muzzle rise lets the shooter see the bullet impact through a telescopic sight. A reduction in recoil also reduces the chance of undesired (painful) contacts between the shooter's head and the ocular of a telescopic sight or other aiming components that must be positioned near the shooter's eye (often referred to as "scope eye"). Another advantage of a muzzle brake is a reduction of recoil fatigue during extended practice sessions, enabling the shooter to consecutively fire more rounds accurately. Further, flinch (involuntary pre-trigger-release anxiety behaviour resulting in inaccurate aiming and shooting) caused by excessive recoil may be reduced or eliminated.

Disadvantages

There are advantages and disadvantages to muzzle brakes. Recoil may be perceived by different shooters as pain, movement of the sight line, rearward thrust, or some combination of the three. Recoil energy can be *sharp* if the impulse is fast or may be considered *soft* if the impulse is slower, even if the same total energy is transferred.

The advantages of brakes and compensators are not without downsides, however. The shooter, gun crew, or close bystanders may perceive an increase in sound pressure level as well as an increase in muzzle blast and lead exposure. This occurs because the sound, flash, pressure waves, and lead loaded smoke plume normally projected away from the shooter are now partially redirected outward to the side or sometimes at partially backward angles toward the shooter or gun crew. Standard eye and ear protection, important for all shooters, may not be adequate to avoid hearing damage with the muzzle blast partially vectored back toward the gun crew or spotters by arrowhead shaped reactive muzzle brakes found on sniper team fired anti-materiel rifles like the Barrett M82.

Measurements indicate that on a rifle, a muzzle brake adds 5 to 10 dB to the normal noise level perceived by the shooter, increasing total noise levels up to 160 dB(A) ± 3 dB. Painful discomfort occurs at approximately 120 to 125 dB(A), with some references claiming 133 dB(A) for the threshold of pain. Active ear muffs are available with electronic noise cancellation that can reduce direct path ear canal noise by approximately 17–33 dB, depending on the low, medium, or high frequency at which attenuation is measured. Passive ear plugs vary in their measured attenuation, ranging from 20 dB to 30 dB, depending on whether they are properly used, and if low pass mechanical filters are also being used. Using both ear muffs (whether passive or active) and ear plugs simultaneously results in maximum protection, but the efficacy of such combined protection relative to preventing permanent ear damage is inconclusive, with evidence indicating that a combined noise reduction ratio (NRR) of only 36 dB (C-weighted) is the maximum possible using ear muffs and ear plugs simultaneously, equating to only a 36 - 7 = 29 dB(A) protection. Some high-end, passive, custom-molded earplugs also have a mechanical filter inserted into the center of the earmolded plug, with a small opening facing to the outside; this design permits being able to hear range commands at a gun range, while still having full rating impulse noise protection. Such custom molded earplugs with low pass filter and mechanical valve typically have a +85 dB(A) mechanical clamp, in addition to having a lowpass filter response, thereby providing typically 30-31 dB attenuation to loud impulse noises, with only a 21 dB reduction under low noise conditions across the human voice audible frequency range (300–4000 Hz) (thereby providing low attenuation between shots being fired), to permit hearing range commands. Similar functions are also available in standardized ear plugs that are not custom molded. But, relative to a noise level of 160 dB(A), this means that even using ear muffs and ear plugs simultaneously cannot protect a shooter against permanent ear damage when using a muzzle brake, through leaving a shooter exposed to noise levels of approximately 131 dB(A) that is 11 dB above the point where permanent ear damage occurs.

Brakes and compensators also add length, diameter, and mass to the muzzle end of a firearm, where it most influences its handling and may interfere with accuracy as muzzle rise will occur when the brake is removed and shooting without the brake can throw off the strike of the round.

Another problem can occur when saboted ammunition is used as the sabot tends to break up inside the brake. The problem is particularly pronounced when armour-piercing fin-stabilized discarding-sabot (APFSDS), a type of long-rod penetrator (LRP) are used. Since these APFSDS rounds are the most common armour-piercing ammunition currently, virtually no modern main battle tank guns have muzzle brakes.

A serious tactical disadvantage of muzzle brakes on both small arms and artillery is that, depending on their designs, they may cause escaping gases to throw up dust and debris clouds that impair visibility and reveal one's position, not to mention posing a hazard to individuals without eye protection. Troops often wet the ground in front of antitank guns in defensive emplacements to prevent this, and snipers are specially trained in techniques for suppressing or concealing the magnified effects of lateral muzzle blast when firing rifles with such brakes. Linear compensators and suppressors do not have the disadvantages of a redirected muzzle blast; they actually reduce the blast by venting high pressure gas forward at reduced velocity.

The redirection of larger amounts escaping high pressure gas can cause discomfort caused by blast-induced sinus cavity concussion. Such discomfort can especially become a problem for anti-materiel rifle shooters due to the bigger than normal cartridges with accompanying large case capacities and propellant volumes these rifles use and can be a reason for promoting accelerated shooter fatigue and flinching. Furthermore, the redirected blast may direct pressure waves toward the eye, potentially leading to retinal detachment when repeated shooting is performed with anti-materiel and large caliber weapons.

US Legislation and Regulation

The State of California outlaws flash suppressors on semiautomatic rifles with detachable magazines, but allows muzzle brakes to be used instead.

The Bureau of Alcohol, Tobacco, Firearms, and Explosives (BATFE) made a regulatory determination in 2013 that the muzzle device of the SIG Sauer MPX Carbine, adapted from the baffle core of the integrally suppressed version's suppressor and claimed by SIG to be a muzzle brake, constituted a silencer and rendered the MPX-C a Title II NFA weapon. SIG Sauer, the rifle's maker, sued the ATF in 2014 to have the designation overturned. In September 2015, Federal Judge Paul Barbadora upheld the BATFE's ruling.

References

- Carlucci, Donald E; Sidney S. Jacobson (2007). Ballistics: Theory and Design of Guns and Ammunition. CRC Press. p. 3. ISBN 1-4200-6618-8

- Weingarten, Dean (November 27, 2015). "Homeless Man's Black Market Submachine Guns and Improvised Silencer".Retrieved 12 October 2016

- Webb, Brandon; Doherty, Glen (15 September 2010). 21st Century Sniper: The Complete Guide. Skyhorse Publishing Inc. pp. 71–72. ISBN 978-1-61608-001-3

- "Firearms and Hearing Protection | March 2007 | The Hearing Industry Resource". 2007-02-12. Retrieved 2009-03-09

- Carter, Gregg Lee (1 January 2002). Guns in American Society: An Encyclopedia of History, Politics, Culture, and the Law. ABC-CLIO. p. 512. ISBN 978-1-57607-268-4

- Dinan, Elizabeth (10 April 2014). "SIG Sauer sues ATF for calling its 'muzzle brake' a gun silencer". Retrieved 29 December 2014

- Carlucci, Donald E.; Jacobson, Sidney S. (11 December 2007). Ballistics: Theory and Design of Guns and Ammunition. CRC Press. p. 158. ISBN 978-1-4200-6619-7

- "Title 26, United States Code, Chapter 53 - Legal Information Institute, Cornell University Law School". Law.cornell.edu. 2008-09-24. Retrieved 2009-03-09

- Boatman, Robert H. (2004). Living with the Big .50: The Shooter's Guide to the World's Most Powerful Rifle. Paladin Press. p. 86. ISBN 978-1-58160-440-5

Understanding External Ballistics

External ballistics is a part of ballistics that studies the performance of projectile when it is in motion. The projectile can either be guided, unguided, powered or un-powered. The topics discussed in the chapter are of great importance to broaden the existing knowledge on external ballistics.

External Ballistics

External ballistics or exterior ballistics is the part of ballistics that deals with the behavior of a projectile in flight. The projectile may be powered or un-powered, guided or unguided, spin or fin stabilized, flying through an atmosphere or in the vacuum of space, but most certainly flying under the influence of a gravitational field.

Gun-launched projectiles may be unpowered, deriving all their velocity from the propellant's ignition until the projectile exits the gun barrel. However, exterior ballistics analysis also deals with the trajectories of rocket assisted gun-launched projectiles and gun launched rockets; and rockets that acquire all their trajectory velocity from the interior ballistics of their on-board propulsion system, either a rocket motor or air-breathing engine, both during their boost phase and after motor burnout. External ballistics is also concerned with the free-flight of other projectiles, such as balls, arrows etc.

Forces Acting on the Projectile

When in flight, the main or major forces acting on the projectile are gravity, drag, and if present, wind; if in powered flight, thrust; and if guided, the forces imparted by the control surfaces.

In small arms external ballistics applications, gravity imparts a downward acceleration on the projectile, causing it to drop from the line of sight. Drag, or the air resistance, decelerates the projectile with a force proportional to the square of the velocity. Wind makes the projectile deviate from its trajectory. During flight, gravity, drag, and wind have a major impact on the path of the projectile, and must be accounted for when predicting how the projectile will travel.

For medium to longer ranges and flight times, besides gravity, air resistance and wind, several intermediate or meso variables described in the external factors paragraph have to be taken into account for small arms. Meso variables can become significant for fire-arms users that have to deal with angled shot scenarios or extended ranges, but are seldom relevant at common hunting and target shooting distances.

For long to very long small arms target ranges and flight times, minor effects and forces such as the ones described in the long range factors paragraph become important and have to be taken into account. The practical effects of these minor variables are generally irrelevant for most firearms users, since normal group scatter at short and medium ranges prevails over the influence these effects exert on projectile trajectories.

At extremely long ranges, artillery must fire projectiles along trajectories that are not even approximately straight; they are closer to parabolic, although air resistance affects this. Extreme long range projectiles are subject to significantly deflections, depending on circumstances, from the line toward the target; and all external factors and long range factors must be taken into account when aiming.

In the case of ballistic missiles, the altitudes involved have a significant effect as well, with part of the flight taking place in a near-vacuum well above a rotating earth, steadily moving the target from where it was at launch time.

Stabilizing Non-spherical Projectiles During Flight

Two methods can be employed to stabilize non-spherical projectiles during flight:

- Projectiles like arrows or arrow like sabots such as the M829 Armor-Piercing, Fin-Stabilized, Discarding Sabot (APFSDS) achieve stability by forcing their center of pressure (CP) behind their center of gravity (CG) with tail surfaces. The CP behind the CG condition yields stable projectile flight, meaning the projectile will not overturn during flight through the atmosphere due to aerodynamic forces.

- Projectiles like small arms bullets and artillery shells must deal with their CP being in front of their CG, which destabilizes these projectiles during flight. To stabilize such projectiles the projectile is spun around its longitudinal (leading to trailing) axis. The spinning mass creates gyroscopic forces that keep the bullet's length axis resistant to the destabilizing overturning torque of the CP being in front of the CG.

Main Effects in External Ballistics

Projectile/Bullet Drop and Projectile Path

The effect of gravity on a projectile in flight is often referred to as projectile drop or bullet drop. It is important to understand the effect of gravity when zeroing the sighting components of a gun. To plan for projectile drop and compensate properly, one must understand parabolic shaped trajectories.

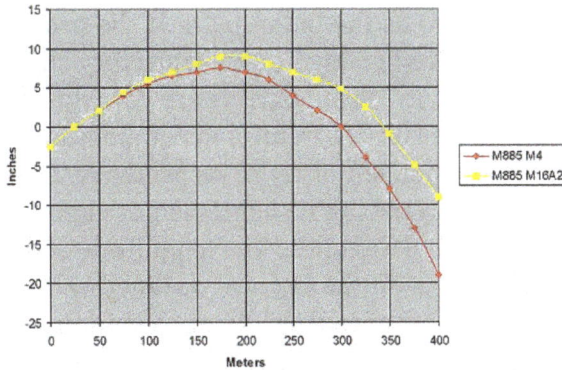

Typical trajectory graph for a M4 carbine and M16A2 rifle using identical M855 cartridges with identical projectiles. Though both trajectories have an identical 25 m near zero, the difference in muzzle velocity of the projectiles gradually causes a significant difference in trajectory and far zero. The 0 inch axis represents the line of sight or horizontal sighting plane.

Projectile/Bullet Drop

In order for a projectile to impact any distant target, the barrel must be inclined to a positive elevation angle relative to the target. This is due to the fact that the projectile will begin to respond to the effects of gravity the instant it is free from the mechanical constraints of the bore. The imaginary line down the center axis of the bore and out to infinity is called the line of departure and is the line on which the projectile leaves the barrel. Due to the effects of gravity a projectile can never impact a target higher than the line of departure. When a positively inclined projectile travels downrange, it arcs below the line of departure as it is being deflected off its initial path by gravity. Projectile/Bullet drop is defined as the vertical distance of the projectile below the line of departure from the bore. Even when the line of departure is tilted upward or downward, projectile drop is still defined as the distance between the bullet and the line of departure at any point along the trajectory. Projectile drop does not describe the actual trajectory of the projectile. Knowledge of projectile drop however is useful when conducting a direct comparison of two different projectiles regarding the shape of their trajectories, comparing the effects of variables such as velocity and drag behavior.

Projectile/Bullet Path

For hitting a distant target an appropriate positive elevation angle is required that is achieved by angling the line of sight from the shooter's eye through the centerline of the sighting system downward toward the line of departure. This can be accomplished by simply adjusting the sights down mechanically, or by securing the entire sighting system to a sloped mounting having a known downward slope, or by a combination of both. This procedure has the effect of elevating the muzzle when the barrel must be subsequently raised to align the sights with the target. A projectile leaving a muzzle at a given elevation angle follows a ballistic trajectory whose characteristics are dependent upon various factors such as muzzle velocity, gravity, and aerodynamic drag. This ballistic trajectory is referred to as the bullet path. If the projectile is spin stabilized, aerodynamic forces will also predictably arc the trajectory slightly to the right, if the rifling employs "right-hand twist." Some barrels are cut with left-hand twist, and the bullet will arc to the left, as a result. Therefore, to compensate for this path deviation, the sights also have to be adjusted left or right, respectively. A constant wind also predictably affects the bullet path, pushing it slightly left or right, and a little bit more up and down, depending on the wind direction. The magnitude of these deviations are also affected by whether the bullet is on the upward or downward slope of the trajectory, due to a phenomenon called "yaw of repose," where a spinning bullet tends to steadily and predictably align slightly off center from its point mass trajectory. Nevertheless, each of these trajectory perturbations are predictable once the projectile aerodynamic coefficients are established, through a combination of detailed analytical modeling and test range measurements.

Projectile/bullet path analysis is of great use to shooters because it allows them to establish ballistic tables that will predict how much vertical elevation and horizontal deflection corrections must be applied to the sight line for shots at various known distances. The most detailed ballistic tables are developed for long range artillery and are based on six-degree-of-freedom trajectory analysis, which accounts for aerodynamic behavior along the three axial directions—elevation, range, and deflection—and the three rotational directions—pitch, yaw, and spin. For small arms applications, trajectory modeling can often be simplified to calculations involving only four of these degrees-of-freedom, lumping the effects of pitch, yaw and spin into the effect of a yaw-of-repose to account for trajectory deflection. Once detailed range tables are established, shooters can relatively quickly adjust sights based on the range to target, wind, air temperature and humidity, and other geometric considerations, such as terrain elevation differences.

Projectile path values are determined by both the sight height, or the distance of the line of sight above the bore centerline, and the range at which the sights are zeroed, which in turn determines the elevation angle. A projectile following a ballistic trajectory has both forward and vertical motion. Forward motion is slowed due to air resistance, and in point mass modeling the vertical motion is dependent on a combination of the elevation angle and gravity. Initially, the projectile is rising with respect to the line of sight or the horizontal sighting plane. The projectile eventually reaches its apex (highest point in the trajectory parabola) where the vertical speed component decays to zero under the effect of gravity, and then begins to descend, eventually impacting the earth. The farther the distance to the intended target, the greater the elevation angle and the higher the apex.

The projectile path crosses the horizontal sighting plane two times. The point closest to the gun occurs while the bullet is climbing through the line of sight and is called the near zero. The second point occurs as the projectile is descending through the line of sight. It is called the far zero and defines the current sight in distance for the gun. Projectile path is described numerically as distances above or below the horizontal sighting plane at various points along the trajectory. This is in contrast to projectile drop which is referenced to the plane containing the line of departure regardless of the elevation angle. Since each of these two parameters uses a different reference datum, significant confusion can result because even though a projectile is tracking well below the line of departure it can still be gaining actual and significant height with respect to the line of sight as well as the surface of the earth in the case of a horizontal or near horizontal shot taken over flat terrain.

Maximum Point-blank Range and Battle Zero

Knowledge of the projectile drop and path has some practical uses to shooters even if it does not describe the actual trajectory of the projectile. For example, if the vertical projectile position over a certain range reach is within the vertical height of the target area the shooter wants to hit, the point of aim does not necessarily need to be adjusted over that range; the projectile is considered to have a sufficiently flat point-blank range trajectory for that particular target. Also known also as "battle zero", maximum point-blank range is also of importance to the military. Soldiers are instructed to fire at any target within this range by simply placing their weapon's sights on the center of mass of the enemy target. Any errors in range estimation are tactically irrelevant, as a well-aimed shot will hit the torso of the enemy soldier. The current trend for elevated sights and higher-velocity cartridges in assault rifles is in part due to a desire to extend the maximum point-blank range, which makes the rifle easier to use.

Drag Resistance

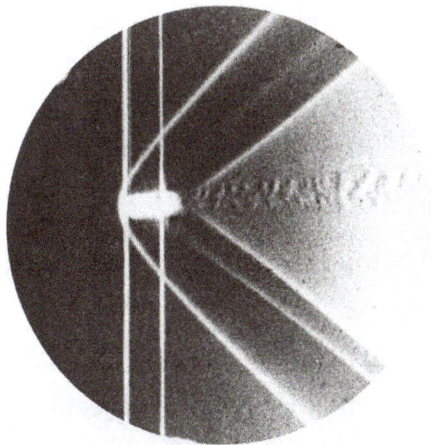

Schlieren photo/Shadowgraph of the detached shock or bow shockwave around
a bullet in supersonic flight, published by Ernst Mach in 1888.

Mathematical models, such as computational fluid dynamics, are used for calculating the effects of drag or air resistance; they are quite complex and not yet completely reliable, but research is ongoing. The most reliable method, therefore, of establishing the necessary projectile aerodynamic properties to properly describe flight tajectories is by empirical measurement.

Fixed Drag Curve Models Generated for Standard-shaped Projectiles

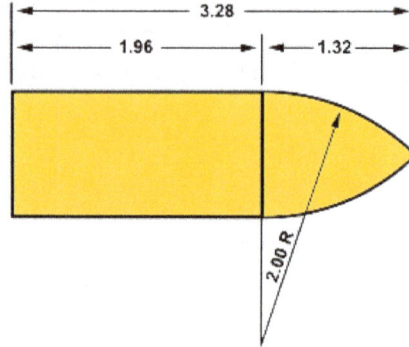

G1 shape standard projectile. All measurements in calibers/diameters.

Use of ballistics tables or ballistics software based on the Mayevski/Siacci method and G1 drag model, introduced in 1881, are the most common method used to work with external ballistics. Projectiles are described by a ballistic coefficient, or BC, which combines the air resistance of the bullet shape (the drag coefficient) and its sectional density (a function of mass and bullet diameter).

The deceleration due to drag that a projectile with mass m, velocity v, and diameter d will experience is proportional to $1/BC$, $1/m$, v^2 and d^2. The BC gives the ratio of ballistic efficiency compared to the standard G1 projectile, which is a fictitious projectile with a flat base, a length of 3.28 calibers/diameters, and a 2 calibers/diameters radius tangential curve for the point. The G1 standard projectile originates from the "C" standard reference projectile defined by the German steel, ammunition and armaments manufacturer Krupp in 1881. The G1 model standard projectile has a BC of 1. The French Gâvre Commission decided to use this projectile as their first reference projectile, giving the G1 name.

Sporting bullets, with a calibre d ranging from 0.177 to 0.50 inches (4.50 to 12.7 mm), have G1 BC's in the range 0.12 to slightly over 1.00, with 1.00 being the most aerodynamic, and 0.12 being the least. Very-low-drag bullets with BC's $\geq$ 1.10 can be designed and produced on CNC precision lathes out of mono-metal rods, but they often have to be fired from custom made full bore rifles with special barrels.

Sectional density is a very important aspect of a projectile or bullet, and is for a round projectile like a bullet the ratio of frontal surface area (half the bullet diameter squared, times pi) to bullet mass. Since, for a given bullet shape, frontal surface increases as the square of the calibre, and mass increases as the cube of the diameter, then sectional density grows linearly with bore diameter. Since BC combines shape and sectional density, a half scale model of the G1 projectile will have a BC of 0.5, and a quarter scale model will have a BC of 0.25.

Since different projectile shapes will respond differently to changes in velocity (particularly between supersonic and subsonic velocities), a BC provided by a bullet manufacturer will be an average BC that represents the common range of velocities for that bullet. For rifle bullets, this will probably be a supersonic velocity, for pistol bullets it will probably be subsonic. For projectiles that travel through the supersonic, transonic and subsonic flight regimes BC is not well approximated by a single constant, but is considered to be a function *BC(M)* of the Mach number M; here M equals the projectile velocity divided by the speed of sound. During the flight of the projectile the M will decrease, and therefore (in most cases) the BC will also decrease.

Most ballistic tables or software takes for granted that one specific drag function correctly describes the drag and hence the flight characteristics of a bullet related to its ballistics coefficient. Those models do not differentiate between wadcutter, flat-based, spitzer, boat-tail, very-low-drag, etc. bullet types or shapes. They assume one invariable drag function as indicated by the published BC.

Several drag curve models optimized for several standard projectile shapes are however available. The resulting fixed drag curve models for several standard projectile shapes or types are referred to as the:

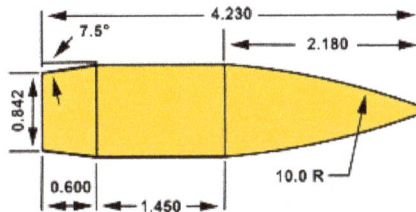

G7 shape standard projectile. All measurements in calibers/diameters.

- G1 or Ingalls (flatbase with 2 caliber (blunt) nose ogive - by far the most popular)

- G2 (Aberdeen J projectile)

- G5 (short 7.5° boat-tail, 6.19 calibers long tangent ogive)

- G6 (flatbase, 6 calibers long secant ogive)

- G7 (long 7.5° boat-tail, 10 calibers tangent ogive, preferred by some manufacturers for very-low-drag bullets)

- G8 (flatbase, 10 calibers long secant ogive)

- GL (blunt lead nose)

How different speed regimes affect .338 calibre rifle bullets can be seen in the .338 Lapua Magnum product brochure which states Doppler radar established G1 BC data. The reason for publishing data like in this brochure is that the Siacci/Mayevski G1 model can not be tuned for the drag behavior of a specific projectile whose shape significantly deviates from the used reference projectile shape. Some ballistic software designers, who based their programs on the Siacci/Mayevski G1 model, give the user the possibility to enter several different G1 BC constants for different speed regimes to calculate ballistic predictions that closer match a bullets flight behavior at longer ranges compared to calculations that use only one BC constant.

The above example illustrates the central problem fixed drag curve models have. These models will only yield satisfactory accurate predictions as long as the projectile of interest has the same shape as the reference projectile or a shape that closely resembles the reference projectile. Any deviation from the reference projectile shape will result in less accurate predictions. How much a projectile deviates from the applied reference projectile is mathematically expressed by the form factor (i). The form factor can be used to compare the drag experienced by a projectile of interest to the drag experienced by the employed reference projectile at a given velocity (range). The problem that the actual drag curve of a projectile can significantly deviate from the fixed drag curve of any employed reference projectile systematically limits the traditional drag resistance modeling approach. The relative simplicity however makes that it can be explained to and understood by the general shooting public and hence is also popular amongst ballistic software prediction developers and bullet manufacturers that want to market their products.

More Advanced Drag Models

Pejsa Model

Besides the traditional drag curve models for several standard projectile shapes or types other more advanced drag models exist. The most prominent alternative ballistic model is probably the model presented in 1980 by Dr. Arthur J. Pejsa. Mr. Pejsa claims on his website that his method was consistently capable of predicting (supersonic) rifle bullet trajectories within 2.5 mm (0.1 in) and bullet velocities within 0.3 m/s (1 ft/s) out to 914 m (1,000 yd) when compared to dozens of actual measurements. The Pejsa model is a closed-form solution that does not need to use any tables or fixed drag curves generated for standard-shaped projectiles.

The greatest strength of the Pejsa model is that any projectile within a given flight regime (for example the supersonic flight regime) can be mathematically modeled well with only two velocity measurements, a distance between said velocity measurements, and a slope or deceleration constant factor. The model allows the drag curve to change slopes or curvature at three different points. Down range velocity measurement data can be provided around key inflection points allowing for more accurate calculations of the projectile retardation rate. The Pejsa model allows the slope factor to be tuned to account for subtle differences in the retardation rate of different bullet shapes and sizes. It ranges from 0.1 (flat-nose bullets) to 0.9 (very-low-drag bullets). If this slope or deceleration constant factor is unknown a default value of 0.5 will predict the flight behavior of most modern spitzer-type rifle bullets quite well. With the help of test firing measurements the slope constant for a particular bullet/rifle system/shooter combination can be determined. These test firings should preferably be executed at 60% and for extreme long range ballistic predictions also at 80% to 90% of the supersonic range of the projectiles of interest, staying away from erratic transonic effects. With this the Pejsa model can easily be tuned for the specific drag behavior of a specific projectile, making significant better (supersonic) ballistic predictions for ranges beyond 500 m (547 yd) possible. A practical downside of the Pejsa model is that accurate projectile specific down range velocity measurements to provide these better predictions can not be easily performed by the vast majority of shooting enthusiasts.

An average retardation coefficient can be calculated for any given slope constant factor if velocity data points are known and distance between said velocity measurements is known. Obviously this is true only within the same flight regime. With velocity actual speed is meant, as velocity is a vector quantity and speed is the magnitude of the velocity vector. Because the power function does not have constant curvature a simple chord average cannot be used. The Pejsa model uses a weighted average retardation coefficient weighted at 0.25 range.The closer velocity is more heavily weighted. The retardation coefficient is measured in feet whereas range is measured in yards hence 0.25 * 3.0 = 0.75, in some places 0.8 rather than 0.75 is used. The 0.8 comes from rounding in order to allow easy entry on hand calculators. Since the Pejsa model does not use a simple chord weighted average, two velocity measurements are used to find the chord average retardation coefficient at midrange between the two velocity measurements points. In order to find the starting retardation coefficient Dr. Pejsa provides two separate equations in his two books. The first involves the power function. The second equation is identical to the one used to find the weighted average at R / 4; add N * (R/2) where R is the range in feet to the chord average retardation coefficient at midrange and where N is the slope constant factor. After the starting retardation coefficient is found the opposite procedure is used in order find the weighted average at R / 4; the starting retardation coefficient minus N * (R/4). In other words, N is used as the slope of the chord line. Dr. Pejsa states that he expanded his drop formula in a power series in order to prove that the weighted average retardation coefficient at R / 4 was a good approximation. For this Dr. Pejsa compared the power series expansion of his drop formula to some other unnamed drop formula's power expansion to reach his conclusions. The fourth term in both power series matched when the retardation coefficient at 0.25 range was used in Pejsa's drop formula. The fourth term was also the first term to use N. The higher terms involving N where insignificant and disappeared at N = 0.36, which according to Dr. Pejsa was a lucky coincidence making for an exceedingly accurate linear approximation, especially for N's around 0.36. If a retardation coefficient function is used exact average values for any N can be obtained because from calculus it is trivial to find the average of any integrable function. Dr. Pejsa states that the retardation coefficient can be modeled by $C * V^N$ where C is a fitting coefficient which disappears during the derivation of the drop formula and N the slope constant factor. The retardation coefficient equals the velocity squared divided by the retardation rate A. Using an average retardation coefficient allows the Pejsa model to be a closed-form expression within a given flight regime.

In order to allow the use of a G1 ballistic coefficient rather than velocity data Dr. Pejsa provided two reference drag curves. The first reference drag curve is based purely on the Siacci/Mayevski retardation rate function. The second reference drag curve is adjusted to equal the Siacci/Mayevski retardation rate function at a projectile velocity of 2600 fps (792.5 m/s) using a .30-06 Springfield Cartridge, Ball, Caliber .30 M2 152 grains (9.8 g) rifle spitzer bullet with a slope or deceleration constant factor of 0.5 in the supersonic flight regime. In other flight regimes the second Pejsa reference drag curve model uses slope constant factors of 0.0 or -4.0. These deceleration constant factors can be verified by backing out Pejsa's formulas (the drag curve segments fits the form $V^{(2-N)} / C$ and the retardation coefficient curve segments fits the form $V^2 / (V^{(2-N)} / C) = C * V^N$ where C is a fitting coefficient). The empirical test data Pejsa used to determine the exact shape of his chosen reference drag curve and pre-defined mathematical function that returns the retardation coefficient at a given Mach number was provided by the US military for the Cartridge, Ball, Caliber .30 M2 bullet. The calculation of the retardation coefficient function also involves air density, which Pejsa did not mention explicitly. The Siacci/Mayevski G1 model uses the following deceleration parametrization (60 °F, 30 inHg and 67% humidity, air density $\rho = 1.2209$ kg/m^3). Dr. Pejsa suggests using the second drag curve because the Siacci/Mayevski G1 drag curve does not provide a good fit for modern spitzer bullets. To obtain relevant retardation coefficients for optimal long range modeling Dr. Pejsa suggested using accurate projectile specific down range velocity measurement data for a particular projectile to empirically derive the average retardation coefficient rather than using a reference drag curve derived average retardation coefficient. Further he suggested using ammunition with reduced propellant loads to empirically test actual projectile flight behavior at lower velocities. When working with reduced propellant loads utmost care must be taken to avoid dangerous or catastrophic conditions (detonations) with can occur when firing experimental loads in firearms.

Some software developers offer commercial software which is based on the Pejsa drag model enhanced and improved with refinements to account for normally minor effects (Coriolis, gyroscopic drift, etc.) that come into play at long range. The developers of these Pejsa model derived programs applied enhancements to the original Pejsa algorithms, using iterative or recursive mathematic approaches for each segment of the retardation coefficient curve with a different slope factor N, for better valid ballistic predictions beyond 1,000 m (1,094 yd) where high powered very-low-drag rifle bullets tend to go transonic and eventually subsonic. These Pejsa model based programs may also allow the use of an average retardation coefficient derived from velocity data as Dr.

Pejsa's original ballistic software did. Further these other Pejsa model based ballistic software developers may have arbitrary chosen different reference projectile, velocity and/or air density variables than Dr. Pejsa used to alter the mathematical drag function or make their software suitable to allow the use of a non G1 ballistic coefficient.

Manges Model

Although not as well known as the Pejsa model, an additional alternative ballistic model was presented in 1989 by Colonel Duff Manges (U S Army Retired) at the American Defense Preparedness (ADPA) 11th International Ballistic Symposium held at the Brussels Congress Center, Brussels, Belgium, May 9–11, 1989. A paper titled "Closed Form Trajectory Solutions for Direct Fire Weapons Systems" appears in the proceedings, Volume 1, Propulsion Dynamics, Launch Dynamics, Flight Dynamics, pages 665-674. Originally conceived to model projectile drag for 120 mm tank gun ammunition, the novel drag coefficient formula has been applied subsequently to ballistic trajectories of center-fired rifle ammunition with results comparable to those claimed for the Pejsa model.

The Manges model uses a first principles theoretical approach that eschews "G" curves and "ballistic coefficients" based on the standard G1 and other similarity curves. The theoretical description has three main parts. The first is to develop and solve a formulation of the two dimensional differential equations of motion governing flat trajectories of point mass projectiles by defining mathematically a set of quadratures that permit closed form solutions for the trajectory differential equations of motion. A sequence of successive approximation drag coefficient functions is generated that converge rapidly to actual observed drag data. The vacuum trajectory, simplified aerodynamic, d'Antonio, and Euler drag law models are special cases. The Manges drag law thereby provides a unifying influence with respect to earlier models used to obtain two dimensional closed form solutions to the point-mass equations of motion. The third purpose of this paper is to describe a least squares fitting procedure for obtaining the new drag functions from observed experimental data. The author claims that results show excellent agreement with six degree of freedom numerical calculations for modern tank ammunition and available published firing tables for center-fired rifle ammunition having a wide variety of shapes and sizes.

A Microsoft Excel application has been authored that uses least squares fits of wind tunnel acquired tabular drag coefficients. Alternatively, manufacturer supplied ballistic trajectory data, or Doppler acquired velocity data can be fitted as well to calibrate the model. The Excel application then employs custom macroinstructions to calculate the trajectory variables of interest. A modified 4th order Runge-Kutta integration algorithm is used. Like Pejsa, Colonel Manges claims center-fired rifle accuracies to the nearest one tenth of an inch for bullet position, and nearest foot per second for the projectile velocity.

The Proceedings of the 11th International Ballistic Symposium are available through the National Defense Industrial Association (NDIA).

Six Degrees of Freedom Model

There are also advanced professional ballistic models like PRODAS available. These are based on six degrees of freedom (6 DoF) calculations. 6 DoF modeling accounts for x, y, and z position in space along with the projectiles pitch, yaw, and roll rates. 6 DoF modeling needs such elaborate data input, knowledge of the employed projectiles and expensive data collection and verification methods that it is impractical for non-professional ballisticians, but not impossible for the curious, computer literate, and mathematically inclined. Semi-empirical aeroprediction models have been developed that reduced extensive test range data on a wide variety of projectile shapes, normalizing dimensional input geometries to calibers; accounting for nose length and radius, body length, and boattail size, and allowing the full set of 6-dof aerodynamic coefficients to be estimated. Early research on spin-stabilized aeroprediction software resulted in the SPINNER computer program. The FINNER aeroprediction code calculates 6-dof inputs for fin stabilized projectiles. Solids modeling software that determines the projectile parameters of mass, center of gravity, axial and transverse moments of inertia necessary for stability analysis are also readily available, and simple to computer program. Finally, algorithms for 6-dof numerical integration suitable to a 4th order Runge-Kutta are readily available. All that is required for the amateur ballistician to investigate the finer analytical details of projectile trajectories, along with bullet nutation and precession behavior, is computer programming determination. Nevertheless, for the small arms enthusiast, aside from academic curiosity, one will discover that being able to predict trajectories to 6-dof accuracy is probably not of practical significance compared to more simplified point mass trajectories based on published bullet ballistic coefficients. 6 DoF is generally used by the aerospace and defense industry and military organizations that study the ballistic behavior of a limited number of (intended) military issue projectiles. Calculated 6 DoF trends can be incorporated as correction tables in more conventional ballistic software applications.

Though 6 DoF modeling and software applications are used by professional well equipped organizations for decades, the computing power restrictions of mobile computing devices like (ruggedized) personal digital assistants, tablet computers or smartphones impaired field use as calculations generally have to be done on the fly. In 2016 the Scandinavian ammunition manufacturer Nammo Lapua Oy released a 6 DoF calculation model based ballistic free software named Lapua Ballistics. The software is distributed as a mobile app only and available for Android and iOS devices. The employed 6 DoF model is however limited to Lapua bullets as a 6 DoF solver needs bullet specific drag coefficient (Cd)/Doppler radar data and geometric dimensions of the projectile(s) of interest. For other bullets the Lapua Ballistics solver is limited to and based on G1 or G7 ballistic coefficients and the Mayevski/Siacci method.

Artillery Software Suites

Military organizations have developed ballistic models like the NATO Armament Ballistic Kernel (NABK) for fire-control systems for artillery like the SG2 Shareable (Fire Control) Software Suite (S4) from the NATO Army Armaments Group (NAAG).

Doppler Radar-measurements

For the precise establishment of drag or air resistance effects on projectiles, Doppler radar measurements are required. Weibel 1000e or Infinition BR-1001 Doppler radars are used by governments, professional ballisticians, defence forces and a few ammunition manufacturers to obtain real-world data of the flight behavior of projectiles of their interest. Correctly established state of the art Doppler radar measurements can determine the flight behavior of projectiles as small as airgun pellets in three-dimensional space to within a few millimetres accuracy. The gathered data regarding the projectile deceleration can be derived and expressed in several ways, such as ballistic coefficients (BC) or drag coefficients (C_d). Because a spinning projectile experiences both precession and nutation about its center of gravity as it flies, further data reduction of doppler radar measurements is required to separate yaw induced drag and lift coefficients from the zero yaw drag coefficient, in order to make measurements fully applicable to 6-dof trajectory analysis.

Doppler radar measurement results for a lathe-turned monolithic solid .50 BMG very-low-drag bullet (Lost River J40 .510-773 grain monolithic solid bullet / twist rate 1:15 in) look like this:

Range (m)	500	600	700	800	900	1000	1100	1200
Ballistic coefficient	1.040	1.051	1.057	1.063	1.064	1.067	1.068	1.068

Range (m)	1300	1400	1500	1600	1700	1800	1900	2000
Ballistic coefficient	1.068	1.066	1.064	1.060	1.056	1.050	1.042	1.032

The initial rise in the BC value is attributed to a projectile's always present yaw and precession out of the bore. The test results were obtained from many shots not just a single shot. The bullet was assigned 1.062 for its BC number by the bullet's manufacturer Lost River Ballistic Technologies.

This tested bullet experiences its maximum drag coefficient when entering the transonic flight regime around Mach 1.200.

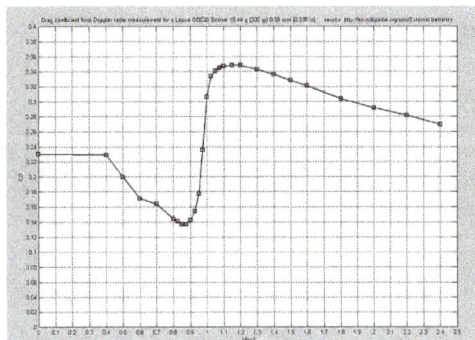

With the help of Doppler radar measurements projectile specific drag models can be established that are most useful when shooting at extended ranges where the bullet speed slows to the transonic speed region near the speed of sound. This is where the projectile drag predicted by mathematic modeling can significantly depart from the actual drag experienced by the projectile. Further Doppler radar measurements are used to study subtle in-flight effects of various bullet constructions.

Governments, professional ballisticians, defence forces and ammunition manufacturers can supplement Doppler radar measurements with measurements gathered by telemetry probes fitted to larger projectiles.

General Trends in Drag or Ballistic Coefficient

In general, a pointed projectile will have a better drag coefficient (C_d) or ballistic coefficient (BC) than a round nosed bullet, and a round nosed bullet will have a better C_d or BC than a flat point bullet. Large radius curves, resulting in a shallower point angle, will produce lower drags, particularly at supersonic velocities. Hollow point bullets behave much like a flat point of the same point diameter. Projectiles designed for supersonic use often have a slightly tapered base at the rear, called a boat tail, which reduces air resistance in flight. Cannelures, which are recessed rings around the projectile used to crimp the projectile securely into the case, will cause an increase in drag.

Analytical software has been developed that reduces actual test range data to parametric relationships for projectile drag coefficient prediction. Large caliber artillery also employ drag reduction mechanisms in addition to streamlining geometry. Rocket-assisted projectiles employ a small rocket motor that ignites upon muzzle exit providing additional thrust to overcome aerodynamic drag. Rocket assist is most effective with subsonic artillery projectiles. For supersonic long range artillery, where base drag dominates, base bleed is employed. Base bleed is a form of a gas generator that does not provide significant thrust, but rather fills the low-pressure area behind the projectile with gas pressure, effectively reducing the base drag and the overall projectile drag coefficient.

Transonic Problem

When the velocity of a projectile fired at supersonic muzzle velocity approaches the speed of sound it enters the transonic region (about Mach 1.2–0.8). In the transonic region, the centre of pressure (CP) of most non spherical projectiles shifts forward as the projectile decelerates. That CP shift affects the (dynamic) stability of the projectile. If the projectile is not well stabilized, it can not remain pointing forward through the transonic region (the projectile starts to exhibit an unwanted precession or coning motion called limit cycle yaw that, if not damped out, can eventually end in uncontrollable tumbling along the length axis). However, even if the projectile has sufficient stability (static and dynamic) to be able to fly through the transonic region and stays pointing forward, it is still affected. The erratic and sudden CP shift and (temporary) decrease of dynamic stability can cause significant dispersion (and hence significant accuracy decay), even if the projectile's flight becomes well behaved again when it enters the subsonic region. This makes accurately predicting the ballistic behavior of projectiles in the transonic region very difficult. Because of this, marksmen normally restrict themselves to engaging targets within the supersonic range of the projectile used. In 2015 the American ballistician Bryan Litz introduced the "Extended Long Range" concept to define rifle shooting at ranges where supersonic fired (rifle) bullets enter the transonic region. According to Litz, "Extended Long Range starts whenever the bullet slows to its transonic range. As the bullet slows down to approach Mach 1, it starts to encounter transonic effects, which are more complex and difficult to account for, compared to the supersonic range where the bullet is relatively well-behaved."

The ambient air density has a significant effect on dynamic stability during transonic transition. Though the ambient air density is a variable environmental factor, adverse transonic transition effects can be negated better by a projectile traveling through less dense air, than when traveling through denser air. Projectile or bullet length also affects limit cycle yaw. Longer projectiles experience more limit cycle yaw than shorter projectiles of the same diameter. Another feature of projectile design that has been identified as having an effect on the unwanted limit cycle yaw motion is the chamfer at the base of the projectile. At the very base, or heel of a projectile or bullet, there is a 0.25 to 0.50 mm (0.01 to 0.02 in) chamfer, or radius. The presence of this radius causes the projectile to fly with greater limit cycle yaw angles. Rifling can also have a subtle effect on limit cycle yaw. In general faster spinning projectiles experience less limit cycle yaw.

Research into Guided Projectiles

To circumvent the transonic problems encountered by spin-stabilized projectiles, projectiles can theoretically be guided during flight. The Sandia National Laboratories announced in January 2012 it has researched and test-fired 4-inch (102 mm) long prototype dart-like, self-guided bullets for small-caliber, smooth-bore firearms that could hit laser-designated targets at distances of more than a mile (about 1,610 meters or 1760 yards). These projectiles are not spin stabilized and the flight path can be course adjusted with an electromagnetic actuator 30 times per second. The researchers also claim they have video of the bullet radically pitching as it exits the barrel and pitching less as it flies down range, a disputed phenomenon known to long-range firearms experts as "going to sleep". Because the bullet's motions settle the longer it is in flight, accuracy improves at longer ranges, Sandia researcher Red Jones said. "Nobody had ever seen that, but we've got high-speed video photography that shows that it's true," he said. Since Sandia is seeking a private company partner to complete testing of the prototype and bring a guided bullet to the marketplace, the future of this technology remains uncertain.

Testing the Predictive Qualities of Software

Due to the practical inability to know in advance and compensate for all the variables of flight, no software simulation, however advanced, will yield predictions that will always perfectly match real world trajectories. It is however possible to obtain predictions that are very close to actual flight behavior.

Empirical Measurement Method

Ballistic prediction computer programs intended for (extreme) long ranges can be evaluated by conducting field tests at the supersonic to subsonic transition range (the last 10 to 20% of the supersonic range of the rifle/cartridge/bullet combination). For a typical .338 Lapua Magnum rifle for example, shooting standard 16.2 gram (250 gr) Lapua Scenar GB488 bullets at 905 m/s (2969 ft/s) muzzle velocity, field testing of the software should be done at ≈ 1200–1300 meters (1312 - 1422 yd) under International Standard Atmosphere sea level conditions (air density ρ = 1.225 kg/m³). To check how well the software predicts the trajectory at shorter to medium range, field tests at 20, 40 and 60% of the supersonic range have to be conducted. At those shorter to medium ranges, transonic problems and hence unbehaved bullet flight should not occur, and the BC is less likely to be transient. Testing the predictive qualities of software at (extreme) long ranges is expensive because it consumes ammunition; the actual muzzle velocity of all shots fired must be measured to be able to make statistically dependable statements. Sample groups of less than 24 shots may not obtain the desired statistically significant confidence interval.

Doppler Radar Measurement Method

Governments, professional ballisticians, defence forces and a few ammunition manufacturers use Doppler radars and/or telemetry probes fitted to larger projectiles to obtain precise real world data regarding the flight behavior of the specific projectiles of their interest and thereupon compare the gathered real world data against the predictions calculated by ballistic computer programs. The normal shooting or aerodynamics enthusiast, however, has no access to such expensive professional measurement devices. Authorities and projectile manufacturers are generally reluctant to share the results of Doppler radar tests and the test derived drag coefficients (C_d) of projectiles with the general public.

In January 2009 the Sacndinavian ammunition manufacturer Nammo/Lapua published Doppler radar test-derived drag coefficient data for most of their rifle projectiles. In 2015 the US ammunition manufacturer Berger Bullets announced the use of Doppler radar in unison with PRODAS 6 DoF software to generate trajectory solutions. In 2016 US ammunition manufacturer Hornady announced the use of Doppler radar derived drag data in software utilizing a modified point mass model to generate trajectory solutions. With the measurement derived C_d data engineers can create algorithms that utilize both known mathematical ballistic models as well as test specific, tabular data in unison. When used by predictive software like QuickTARGET Unlimited, Lapua Edition, Lapua Ballistics or Hornady 4DOF the Doppler radar test-derived drag coefficient data can be used for more accurate external ballistic predictions.

Some of the Lapua-provided drag coefficient data shows drastic increases in the measured drag around or below the Mach 1 flight velocity region. This behavior was observed for most of the measured small calibre bullets, and not so much for the larger calibre bullets. This implies some (mostly smaller calibre) rifle bullets exhibited more limit cycle yaw (coning and/or tumbling) in the transonic/subsonic flight velocity regime. The information regarding unfavourable transonic/subsonic flight behavior for some of the tested projectiles is important. This is a limiting factor for extended range shooting use, because the effects of limit cycle yaw are not easily predictable and potentially catastrophic for the best ballistic prediction models and software.

Presented C_d data can not be simply used for every gun-ammunition combination, since it was measured for the barrels, rotational (spin) velocities and ammunition lots the Lapua testers used during their test firings. Variables like differences in rifling (number of grooves, depth, width and other dimensional properties), twist rates and/or muzzle velocities impart different rotational (spin) velocities and rifling marks on projectiles. Changes in such variables and projectile production lot variations can yield different downrange interaction with the air the projectile passes through that can result in (minor) changes in flight behavior. This particular field of external ballistics is currently (2009) not elaborately studied nor well understood.

Predictions of Several Drag Resistance Modelling and Measuring methods

The method employed to model and predict external ballistic behavior can yield differing results with increasing range and time of flight. To illustrate this several external ballistic behavior prediction methods for the Lapua Scenar GB528 19.44 g (300 gr) 8.59 mm (0.338 in) calibre very-low-drag rifle bullet with a manufacturer stated G1 ballistic coefficient (BC) of 0.785 fired at 830 m/s (2723 ft/s) muzzle velocity under International Standard Atmosphere sea level conditions (air density ρ = 1.225 kg/m³), Mach 1 = 340.3 m/s, Mach 1.2 = 408.4 m/s), predicted this for the projectile velocity and time of flight from 0 to 3,000 m (0 to 3,281 yd):

Range (m)	0	300	600	900	1,200	1,500	1,800	2,100	2,400	2,700	3,000
Radar test derived drag coefficients method V (m/s)	830	711	604	507	422	349	311	288	267	247	227
Time of flight (s)	0.0000	0.3918	0.8507	1.3937	2.0435	2.8276	3.7480	4.7522	5.8354	7.0095	8.2909
Total drop (m)	0.000	0.715	3.203	8.146	16.571	30.035	50.715	80.529	121.023	173.998	241.735
G1 drag model method V (m/s)	830	718	615	522	440	374	328	299	278	261	248

Time of flight (s)	0.0000	0.3897	0.8423	1.3732	2.0009	2.7427	3.6029	4.5642	5.6086	6.7276	7.9183
Total drop (m)	0.000	0.710	3.157	7.971	16.073	28.779	47.810	75.205	112.136	160.739	222.430
Pejsa drag model method V (m/s)	830	712	603	504	413	339	297	270	247	227	208
Time of flight (s)	0.0000	0.3902	0.8479	1.3921	2.0501	2.8556	3.8057	4.8682	6.0294	7.2958	8.6769
Total drop (m)	0.000	0.719	3.198	8.129	16.580	30.271	51.582	82.873	126.870	185.318	260.968
G7 drag model method V (m/s)	830	713	606	508	418	339	303	283	265	249	235
Time of flight (s)	0.0000	0.3912	0.8487	1.3901	2.0415	2.8404	3.7850	4.8110	5.9099	7.0838	8.3369
Total drop (m)	0.000	0.714	3.191	8.109	16.503	30.039	51.165	81.863	123.639	178.082	246.968
6 DoF modeling method V (m/s)	830	711	605	508	422	349	311	288	267	245	222

The table shows that the traditional Siacci/Mayevski G1 drag curve model prediction method generally yields more optimistic results compared to the modern Doppler radar test derived drag coefficients (C_d) prediction method. At 300 m (328 yd) range the differences will be hardly noticeable, but at 600 m (656 yd) and beyond the differences grow over 10 m/s (32.8 ft/s) projectile velocity and gradually become significant. At 1,500 m (1,640 yd) range the projectile velocity predictions deviate 25 m/s (82.0 ft/s), which equates to a predicted total drop difference of 125.6 cm (49.4 in) or 0.83 mrad (2.87 MOA) at 50° latitude.

The Pejsa drag model closed-form solution prediction method, without slope constant factor fine tuning, yields very similar results in the supersonic flight regime compared to the Doppler radar test derived drag coefficients (C_d) prediction method. At 1,500 m (1,640 yd) range the projectile velocity predictions deviate 10 m/s (32.8 ft/s), which equates to a predicted total drop difference of 23.6 cm (9.3 in) or 0.16 mrad (0.54 MOA) at 50° latitude.

The G7 drag curve model prediction method (recommended by some manufacturers for very-low-drag shaped rifle bullets) when using a G7 ballistic coefficient (BC) of 0.377 yields very similar results in the supersonic flight regime compared to the Doppler radar test derived drag coefficients (C_d) prediction method. At 1,500 m (1,640 yd) range the projectile velocity predictions have their maximum deviation of 10 m/s (32.8 ft/s). The predicted total drop difference at 1,500 m (1,640 yd) is 0.4 cm (0.16 in) at 50° latitude. The predicted total drop difference at 1,800 m (1,969 yd) is 45.0 cm (17.7 in), which equates to 0.25 mrad (0.86 MOA).

The Lapua Ballistics 6 DoF modeling can currently (2016) not provide a completely comparable data set to add to the table. The 6 DoF velocity predictions are however identical up to 2,400 m (2,625 yd) and slightly deviate at 2,700 m (2,953 yd) and 3,000 m (3,281 yd) compared to the Doppler radar test derived drag coefficients (C_d) prediction method. The Lapua Ballistics 6 DoF modeling estimates bullet stability ((S_d) and (S_g)) that gravitates to over-stabilization for ranges over 2,400 m (2,625 yd) for this bullet.

Decent prediction models are expected to yield similar results in the supersonic flight regime. The five example models down to 1,200 m (1,312 yd) all predict supersonic Mach 1.2⁺ projectile velocities and total drop differences within a 51 cm (20.1 in) bandwidth. In the transonic flight regime at 1,500 m (1,640 yd) the models predict projectile velocities around Mach 1.0 to Mach 1.1 and total drop differences within a much larger 150 cm (59 in) bandwidth.

External Factors

Wind

Wind has a range of effects, the first being the effect of making the projectile deviate to the side (horizontal deflection). From a scientific perspective, the "wind pushing on the side of the projectile" is not what causes horizontal wind drift. What causes wind drift is drag. Drag makes the projectile turn into the wind, much like a weather vane, keeping the centre of air pressure on its nose. This causes the nose to be cocked (from your perspective) into the wind, the base is cocked (from your perspective) "downwind." So, (again from your perspective), the drag is pushing the projectile downwind in a nose to tail direction.

Wind also causes aerodynamic jump which is the vertical component of cross wind deflection caused by lateral (wind) impulses activated during free flight of a projectile or at or very near the muzzle leading to dynamic imbalance. The amount of aerodynamic jump is dependent on cross wind speed, the gyroscopic stability of the bullet at the muzzle and if the barrel twist is clockwise or anti-clockwise. Like the wind direction reversing the twist direction will reverse the aerodynamic jump direction.

A somewhat less obvious effect is caused by head or tailwinds. A headwind will slightly increase the relative velocity of the projectile, and increase drag and the corresponding drop. A tailwind will reduce the drag and the projectile/bullet drop. In the real world, pure head or tailwinds are rare, since wind is seldomly constant in force and direction and normally interacts with the terrain it is blowing over. This often makes ultra long range shooting in head or tailwind conditions difficult.

Vertical Angles

The vertical angle (or elevation) of a shot will also affect the trajectory of the shot. Ballistic tables for small calibre projectiles (fired from pistols or rifles) assume a horizontal line of sight between the shooter and target with gravity acting perpendicular to the earth. Therefore, if the shooter-to-target angle is up or down, (the direction of the gravity component does not change with slope direction), then the trajectory curving acceleration due to gravity will actually be less, in proportion to the cosine of the slant angle. As a result, a projectile fired upward or downward, on a so-called "slant range," will over-shoot the same target distance on flat ground. The effect is of sufficient magnitude that hunters must adjust their target hold off accordingly in mountainous terrain. A well known formula for slant range adjustment to horizontal range hold off is known as the Rifleman's rule. The Rifleman's rule and the slightly more complex and less well known Improved Rifleman's rule models produce sufficiently accurate predictions for many small arms applications. Simple prediction models however ignore minor gravity effects when shooting uphill or downhill. The only practical way to compensate for this is to use a ballistic computer program. Besides gravity at very steep angles over long distances, the effect of air density changes the projectile encounters during flight become problematic. The mathematical prediction models available for inclined fire scenarios, depending on the amount and direction (uphill or downhill) of the inclination angle and range, yield varying accuracy expectation levels. Less advanced ballistic computer programs predict the same trajectory for uphill and downhill shots at the same vertical angle and range. The more advanced programs factor in the small effect of gravity on uphill and on downhill shots resulting in slightly differing trajectories at the same vertical angle and range. No publicly available ballistic computer program currently (2017) accounts for the complicated phenomena of differing air densities the projectile encounters during flight.

Ambient Air Density

Air pressure, temperature, and humidity variations make up the ambient air density. Humidity has a counter intuitive impact. Since water vapor has a density of 0.8 grams per litre, while dry air averages about 1.225 grams per litre, higher humidity actually decreases the air density, and therefore decreases the drag. (An easy way to remember that water vapor reduces air density is to observe that clouds float.)

Long Range Factors

Gyroscopic Drift (Spin Drift)

Gyroscopic drift is an interaction of the bullet's mass and aerodynamics with the atmosphere that it is flying in. Even in completely calm air, with no sideways air movement at all, a spin-stabilized projectile will experience a spin-induced sideways component, due to a gyroscopic phenomenon known as "yaw of repose." For a right hand (clockwise) direction of rotation this component will always be to the right. For a left hand (counterclockwise) direction of rotation this component will always be to the left. This is because the projectile's longitudinal axis (its axis of rotation) and the direction of the velocity vector of the center of gravity (CG) deviate by a small angle, which is said to be the equilibrium yaw or the yaw of repose. The magnitude of the yaw of repose angle is typically less than 0.5 degree. Since rotating objects react with an angular velocity vector 90 degrees from the applied torque vector, the bullet's axis of symmetry moves with a component in the vertical plane and a component in the horizontal plane; for right-handed (clockwise) spinning bullets, the bullet's axis of symmetry deflects to the right and a little bit upward with respect to the direction of the velocity vector, as the projectile moves along its ballistic arc. As the result of this small inclination, there is a continuous air stream, which tends to deflect the bullet to the right. Thus the occurrence of the yaw of repose is the reason for the bullet drifting to the right (for right-handed spin) or to the left (for left-handed spin). This means that the bullet is "skidding" sideways at any given moment, and thus experiencing a sideways component.

The following variables affect the magnitude of gyroscopic drift:

- Projectile or bullet length: longer projectiles experience more gyroscopic drift because they produce more lateral "lift" for a given yaw angle.

- Spin rate: faster spin rates will produce more gyroscopic drift because the nose ends up pointing farther to the side.

- Range, time of flight and trajectory height: gyroscopic drift increases with all of these variables.

- density of the atmosphere: denser air will increase gyroscopic drift.

Doppler radar measurement results for the gyroscopic drift of several US military and other very-low-drag bullets at 1000 yards (914.4 m) look like this:

Bullet type	US military M193 Ball (5.56×45mm NATO)	US military M118 Special Ball (7.62×51mm NATO)	Palma Sierra MatchKing	LRBT J40 Match	Sierra MatchKing	Sierra MatchKing	LRBT J40 Match	LRBT J40 Match
Projectile mass (in grains and g)	55 grains (3.56 g)	173 grains (11.21 g)	155 grains (10.04 g)	190 grains (12.31 g)	220 grains (14.26 g)	300 grains (19.44 g)	350 grains (22.68 g)	419 grains (27.15 g)
Projectile diameter (in inches and mm)	.224 inches (5.69 mm)	.308 inches (7.82 mm)	.308 inches (7.82 mm)	.308 inches (7.82 mm)	.308 inches (7.82 mm)	.338 inches (8.59 mm)	.375 inches (9.53 mm)	.408 inches (10.36 mm)
Gyroscopic drift (in inches and mm)	23.00 inches (584.20 mm)	11.50 inches (292.10 mm)	12.75 inches (323.85 mm)	3.00 inches (76.20 mm)	7.75 inches (196.85 mm)	6.50 inches (165.10 mm)	0.87 inches (22.10 mm)	1.90 inches (48.26 mm)

The table shows that the gyroscopic drift cannot be predicted on weight and diameter alone. In order to make accurate predictions on gyroscopic drift several details about both the external and internal ballistics must be considered. Factors such as the twist rate of the barrel, the velocity of the projectile as it exits the muzzle, barrel harmonics, and atmospheric conditions, all contribute to the path of a projectile.

Magnus Effect

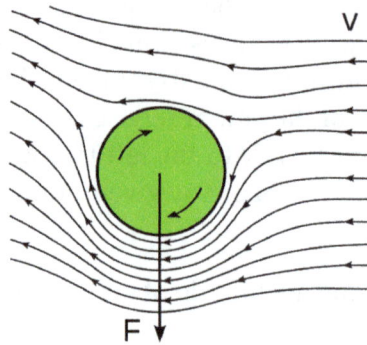

The Magnus effect. *V* represents the wind, the arrow *F* is the resulting
Magnus force towards the side of lower pressure.

Spin stabilized projectiles are affected by the Magnus effect, whereby the spin of the bullet creates a force acting either up or down, perpendicular to the sideways vector of the wind. In the simple case of horizontal wind, and a right hand (clockwise) direction of rotation, the Magnus effect induced pressure differences around the bullet cause a downward (wind from the right) or upward (wind from the left) force viewed from the point of firing to act on the projectile, affecting its point of impact. The vertical deflection value tends to be small in comparison with the horizontal wind induced deflection component, but it may nevertheless be significant in winds that exceed 4 m/s (14.4 km/h or 9 mph).

Magnus Effect and Bullet Stability

The Magnus effect has a significant role in bullet stability because the Magnus force does not act upon the bullet's center of gravity, but the center of pressure affecting the *yaw* of the bullet. The Magnus effect will act as a *destabilizing* force on any bullet with a center of pressure located *ahead* of the center of gravity, while conversely acting as a *stabilizing* force on any bullet with the center of pressure located *behind* the center of gravity. The location of the center of pressure depends on the flow field structure, in other words, depending on whether the bullet is in supersonic, transonic or subsonic flight. What this means in practice depends on the shape and other attributes of the bullet, in any case the Magnus force greatly affects stability because it tries to "twist" the bullet along its flight path.

Paradoxically, very-low-drag bullets due to their length have a tendency to exhibit greater Magnus destabilizing errors because they have a greater surface area to present to the oncoming air they are travelling through, thereby reducing their aerodynamic efficiency. This subtle effect is one of the reasons why a calculated C_d or BC based on shape and sectional density is of limited use.

Poisson Effect

Another minor cause of drift, which depends on the nose of the projectile being above the trajectory, is the Poisson Effect. This, if it occurs at all, acts in the same direction as the gyroscopic drift and is even less important than the Magnus effect. It supposes that the uptilted nose of the projectile causes an air cushion to build up underneath it. It further supposes that there is an increase of friction between this cushion and the projectile so that the latter, with its spin, will tend to roll off the cushion and move sideways.

This simple explanation is quite popular. There is, however, no evidence to show that increased pressure means increased friction and unless this is so, there can be no effect. Even if it does exist it must be quite insignificant compared with the gyroscopic and Coriolis drifts.

Both the Poisson and Magnus Effects will reverse their directions of drift if the nose falls below the trajectory. When the nose is off to one side, as in equilibrium yaw, these effects will make minute alterations in range.

Coriolis Drift

The Coriolis effect causes Coriolis drift, both horizontally and vertically. The deflection is to the right of the trajectory in the northern hemisphere, to the left in the southern hemisphere, upward for eastward shots, and downward for westward shots. The vertical Coriolis deflection is also known as the Eötvös effect. Coriolis drift is not an aerodynamic effect; it is a consequence of the rotation of the Earth.

The magnitude of the Coriolis effect is small. For small arms, the magnitude of the Coriolis effect is generally insignificant (for high powered rifles in the order of about 10 cm (3.9 in) at 1,000 m (1,094 yd)), but for ballistic projectiles with long flight times, such as extreme long-range rifle projectiles, artillery, and rockets like intercontinental ballistic missiles, it is a significant factor in calculating the trajectory. The magnitude of the drift depends on the firing and target location, azimuth of firing, projectile velocity and time of flight.

Horizontal Effect

Viewed from a non-rotating reference frame (i.e. not one rotating with the Earth) and ignoring the forces of gravity and air resistance, a projectile moves in a straight line. When viewed from a reference frame fixed with respect to the Earth, that straight trajectory appears to curve sideways. The direction of this horizontal curvature is to the right in the northern hemisphere and to the left in the southern hemisphere, and does not depend on the azimuth of the shot. The curvature is largest at the poles and zero at the equator.

Vertical (Eötvös) Effect

The Eötvös effect changes the perceived gravitational pull on a moving object based on the relationship between the direction and velocity of movement and the direction of the Earth's rotation. It causes subtle counterintuitive trajectory variations during flight.

The Eötvös effect is largest at the equator and decreases to zero at the poles. It causes eastward-traveling projectiles to deflect upward, and westward-traveling projectiles to deflect downward. The effect is less pronounced for trajectories in other directions, and is zero for trajectories aimed due north or south. In the case of large changes of momentum, such as a spacecraft being launched into Earth orbit, the effect becomes significant. It contributes to the fastest and most fuel-efficient path to orbit: a launch from the equator that curves to a directly eastward heading.

Equipment Factors

Though not forces acting on projectile trajectories there are some equipment related factors that influence trajectories. Since these factors can cause otherwise unexplainable external ballistic flight behavior they have to be briefly mentioned.

Lateral Jump

Lateral jump is caused by a slight lateral and rotational movement of a gun barrel at the instant of firing. It has the effect of a small error in bearing. The effect is ignored, since it is small and varies from round to round.

Lateral Throw-off

Lateral throw-off is caused by mass imbalance in applied spin stabilized projectiles or pressure imbalances during the transitional flight phase when a projectile leaves a gun barrel off axis leading to static imbalance. If present it causes dispersion. The effect is unpredictable, since it is generally small and varies from projectile to projectile, round to round and/or gun barrel to gun barrel.

Maximum Effective Small Arms Range

The maximum practical range of all small arms and especially high-powered sniper rifles depends mainly on the aerodynamic or ballistic efficiency of the spin stabilised projectiles used. Long-range shooters must also collect relevant information to calculate elevation and windage corrections to be able to achieve first shot strikes at point targets. The data to calculate these fire control corrections has a long list of variables including:

- ballistic coefficient or test derived drag coefficients (Cd)/behavior of the bullets used

- height of the sighting components above the rifle bore axis

- the zero range at which the sighting components and rifle combination were sighted in

- bullet mass

- actual muzzle velocity (powder temperature affects muzzle velocity, primer ignition is also temperature dependent)

- range to target

- supersonic range of the employed gun, cartridge and bullet combination

- inclination angle in case of uphill/downhill firing

- target speed and direction

- wind speed and direction (main cause for horizontal projectile deflection and generally the hardest ballistic variable to measure and judge correctly. Wind effects can also cause vertical deflection.)

- air pressure, temperature, altitude and humidity variations (these make up the ambient air density)

- Earth's gravity (changes slightly with latitude and altitude)

- gyroscopic drift (horizontal and vertical plane gyroscopic effect — often known as spin drift - induced by the barrel's twist direction and twist rate)

- Coriolis effect drift (latitude, direction of fire and northern or southern hemisphere data dictate this effect)

- Eötvös effect (interrelated with the Coriolis effect, latitude and direction of fire dictate this effect)

- aerodynamic jump (the vertical component of cross wind deflection caused by lateral (wind) impulses activated during free flight or at or very near the muzzle leading to dynamic imbalance)

- lateral throw-off (dispersion that is caused by mass imbalance in the applied projectile or it leaving the barrel off axis leading to static imbalance)

- the inherent potential accuracy and adjustment range of the sighting components

- the inherent potential accuracy of the rifle

- the inherent potential accuracy of the ammunition

- the inherent potential accuracy of the computer program and other firing control components used to calculate the trajectory

The ambient air density is at its maximum at Arctic sea level conditions. Cold gunpowder also produces lower pressures and hence lower muzzle velocities than warm powder. This means that the maximum practical range of rifles will be at it shortest at Arctic sea level conditions.

The ability to hit a point target at great range has a lot to do with the ability to tackle environmental and meteorological factors and a good understanding of exterior ballistics and the limitations of equipment. Without (computer) support and highly accurate laser rangefinders and meteorological measuring equipment as aids to determine ballistic solutions, long-range shooting beyond 1000 m (1100 yd) at unknown ranges becomes guesswork for even the most expert long-range marksmen.

Using Ballistics Data

Here is an example of a ballistic table for a .30 calibre Speer 169 grain (11 g) pointed boat tail match bullet, with a BC of 0.480. It assumes sights 1.5 inches (38 mm) above the bore line, and sights adjusted to result in point of aim and point of impact matching 200 yards (183 m) and 300 yards (274 m) respectively.

Range		0	100 yd 91 m	200 yd 183 m	300 yd 274 m	400 yd 366 m	500 yd 457 m
Velocity	(ft/s)	2,700	2,512	2,331	2,158	1,992	1,834
	(m/s)	823	766	710	658	607	559
Zeroed for 200 yards/184 m							
Height	(in)	−1.5	2.0	0	−8.4	−24.3	−49.0
	(mm)	−38	51	0	−213	−617	−1245
Zeroed for 300 yards/274 m							
Height	(in)	−1.5	4.8	5.6	0	−13.1	−35.0
	(mm)	−38	122	142	0	−333	−889

This table demonstrates that, even with a fairly aerodynamic bullet fired at high velocity, the "bullet drop" or change in the point of impact is significant. This change in point of impact has two important implications. Firstly, estimating the distance to the target is critical at longer ranges, because the difference in the point of impact between 400 and 500 yd (460 m) is 25–32 in (depending on zero), in other words if the shooter estimates that the target is 400 yd away when it is in fact 500 yd away the shot will impact 25–32 in (635–813 mm) below where it was aimed, possibly missing the target completely. Secondly, the rifle should be zeroed to a distance appropriate to the typical range of targets, because the shooter might have to aim so far above the target to compensate for a large bullet drop that he may lose sight of the target completely (for instance being outside the field of view of a telescopic sight). In the example of the rifle zeroed at 200 yd (180 m), the shooter would have to aim 49 in or more than 4 ft (1.2 m) above the point of impact for a target at 500 yd.

Freeware Small Arms External Ballistics Software

- Ballistic_XLR. (MS Excel spreadsheet)] - A substantial enhancement & modification of the Pejsa spreadsheet.

- GNU Exterior Ballistics Computer (GEBC) - An open source 3DOF ballistics computer for Windows, Linux, and Mac - Supports the G1, G2, G5, G6, G7, and G8 drag models. Created and maintained by Derek Yates.

- 6mmbr.com ballistics section links to / hosts 4 freeware external ballistics computer programs.

- 2DOF & 3DOF R.L. McCoy - Gavre exterior ballistics (zip file) - Supports the G1, G2, G5, G6, G7, G8, GS, GL, GI, GB and RA4 drag models

- PointBlank Ballistics (zip file) - Siacci/Mayevski G1 drag model.

- Remington Shoot! A ballistic calculator for Remington factory ammunition (based on Pinsoft's Shoot! software). - Siacci/Mayevski G1 drag model.

- JBM's small-arms ballistics calculators Online trajectory calculators - Supports the G1, G2, G5, G6, G7 (for some projectiles experimentally measured G7 ballistic coefficients), G8, GI, GL and for some projectiles doppler radar-test derived (C_d) drag models.

- Pejsa Ballistics (MS Excel spreadsheet) - Pejsa model.

- Sharpshooter Friend (Palm PDA software) - Pejsa model.

- Quick Target Unlimited, Lapua Edition - A version of QuickTARGET Unlimited ballistic software (requires free registration to download) - Supports the G1, G2, G5, G6, G7, G8, GL, GS Spherical 9/16"SAAMI, GS Spherical Don Miller, RA4, Soviet 1943, British 1909 Hatches Notebook and for some Lapua projectiles doppler radar-test derived (Cd) drag models.

- Lapua Ballistics Exterior ballistic software for Java or Android mobile phones. Based on doppler radar-test derived (Cd) drag models for Lapua projectiles and cartridges.

- Lapua Ballistics App 6 DoF model limited to Lapua bullets for Android and iOS.

- BfX - Ballistics for Excel Set of MS Excel add-ins functions - Supports the G1, G2, G5, G6, G7 G8 and RA4 and Pejsa drag models as well as one for air rifle pellets. Able to handle user supplied models, e.g. Lapua projectiles doppler radar-test derived (Cd) ones.

- GunSim "GunSim" free browser-based ballistics simulator program for Windows and Mac.

- BallisticSimulator "Ballistic Simulator" free ballistics simulator program for Windows.

- ChairGun Pro free ballistics for rim fire and pellet guns.

- 5H0T Free online web-based ballistics calculator, with data export capability and charting.

Propellant

A propellant or propellent is a chemical substance used in the production of energy or pressurized gas that is subsequently used to create movement of a fluid or to generate propulsion of a vehicle, projectile, or other object. Common propellants are energetic materials and consist of a fuel like gasoline, jet fuel, rocket fuel, and an oxidizer. Propellants are burned or otherwise decomposed to produce the propellant gas. Other propellants are simply liquids that can readily be vaporized.

In rockets and aircraft, propellants are used to produce a gas that can be directed through a nozzle, thereby producing thrust. In rockets, rocket propellant produces an exhaust, and the exhausted material is usually expelled under pressure through a nozzle. The pressure may be from a compressed gas, or a gas produced by a chemical reaction. The exhaust material may be a gas, liquid, plasma, or, before the chemical reaction, a solid, liquid, or gel. In aircraft, the propellant is usually a fuel and is combusted with the air.

In firearm ballistics, propellants fill the interior of an ammunition cartridge or the chamber of a gun or cannon, leading to the expulsion of a bullet or shell (gunpowder, smokeless powder, and large gun propellants). Explosives can be placed in a sealed tube and act as a deflagrant low explosive charge in mining and demolition, to produce a low velocity heave effect (gas pressure blasting).

Cold gas propellants may be used to fill an expansible bag or membrane, such as an automotive airbag (gas generator propellants) or in pressurised dispensing systems, such as aerosol sprays, to force a material through a nozzle. Examples of can propellants include nitrous oxide that is dissolved in canned whipped cream, and the dimethyl ether or low-boiling alkane used in hair spray. Rocket propellant may be expelled through an expansion nozzle as a cold gas, that is, without energetic mixing and combustion, to provide small changes in velocity to spacecraft by the use of cold gas thrusters.

Aerosol Sprays

In aerosol spray cans, the propellant is simply a pressurized gas in equilibrium with its liquid (at its saturated vapour pressure). As some gas escapes to expel the payload, more liquid evaporates, maintaining an even pressure.

Propellant used for Propulsion

Technically, the word propellant is the general name for chemicals used to create thrust. For vehicles, the term propellant refers only to chemicals that are stored within the vehicle prior to use, and excludes atmospheric gas or other material that may be collected in operation.

Among the English-speaking layperson, used to having fuels propel vehicles on Earth, the word fuel is inappropriately used. In Germany, the word *Treibstoff*—literally "drive-stuff"—is used; in France, the word *ergols* is used; it has the same Greek roots as hypergolic, a term used in English for propellants that combine spontaneously and do not have to be set ablaze by auxiliary ignition system.

To attain a useful density for storage, most propellants are either solid or liquid.

Solid Propellant

In ballistics and pyrotechnics, a propellant is a generic name for chemicals used for propelling projectiles from guns and other firearms.

Solid propellants are usually made from low-explosive materials, but may include high-explosive chemical ingredients that are diluted and burned in a controlled way (deflagration) rather than detonation. The controlled burning of the propellant composition usually produces thrust by gas pressure and can accelerate a projectile, rocket, or other vehicle. In this sense, common or well-known propellants include, for firearms, artillery, and solid-propellant rockets:

- Gun propellants, such as:

    o Gunpowder (black powder)

    o Nitrocellulose-based powders

    o Cordite

    o Ballistite

    o Smokeless powders

- Composite propellants made from a solid oxidizer such as ammonium perchlorate or ammonium nitrate, a synthetic rubber such as HTPB, PBAN, or Polyurethane (or energetic polymers such as polyglycidyl nitrate or polyvinyl nitrate for extra energy), optional high-explosive fuels (again, for extra energy) such as RDX or nitroglycerin, and usually a powdered metal fuel such as aluminum.

- Some amateur propellants use potassium nitrate, combined with sugar, epoxy, or other fuels and binder compounds.

- Potassium perchlorate has been used as an oxidizer, paired with asphalt, epoxy, and other binders.

Propellants that explode in operation are of little practical use currently, although there have been experiments with Pulse Detonation Engines. Also the newly synthesized bishomocubane based compounds are under consideration in the research stage as both solid and liquid propellants of the future.

Grain

Solid propellants are used in forms called grains. A grain is any individual particle of propellant regardless of the size or shape. The shape and size of a propellant grain determines the burn time, amount of gas, and rate produced from the burning propellant and, as a consequence, thrust vs time profile.

There are three types of burns that can be achieved with different grains.

Progressive Burn

Usually a grain with multiple perforations or a star cut in the center providing a lot of surface area.

Degressive Burn

Usually a solid grain in the shape of a cylinder or sphere.

Neutral Burn

Usually a single perforation; as outside surface decreases the inside surface increases at the same rate.

Composition

There are four different types of solid propellant compositions:

Single-Based Propellant

A single based propellant has nitrocellulose as its chief explosives ingredient. Stabilizers and other additives are used to control the chemical stability and enhance the propellant's properties.

Double-Based Propellant

Double-based propellants consist of nitrocellulose with nitroglycerin or other liquid organic nitrate explosives added. Stabilizers and other additives are also used. Nitroglycerin reduces smoke and increases the energy output. Double-based propellants are used in small arms, cannons, mortars and rockets.

Triple-Based Propellant

Triple-based propellants consist of nitrocellulose, nitroguanidine, nitroglycerin or other liquid organic nitrate explosives. Triple-based propellants are used in cannons.

Composite

Composites contain no nitrocellulose, nitroglycerin, nitroquanidine or any other organic nitrate. Composites usually consist of a fuel such as metallic aluminum, a combustible binder such as synthetic rubber or HTPB, and an oxidizer such as ammonium perchlorate. Composite propellants are used in large rocket motors.

Liquid Propellant

In rockets, three main liquid bipropellant combinations are used: cryogenic oxygen and hydrogen, cryogenic oxygen and a hydrocarbon, and storable propellants.

- cryogenic oxygen-hydrogen combination system - Used in upper stages and sometimes in booster stages of space launch systems.This is a nontoxic combination. This gives high specific impulse and is ideal for high-velocity missions

- cryogenic oxygen-hydrocarbon propellant system - Used for many booster stages of space launch vehicles as well as a smaller number of second stages. This combination of fuel/oxidizer has high density and hence allows for a more compact booster design.

- storable propellant combinations - used in almost all bipropellant low-thrust, auxiliary or reaction control rocket engines, as well as in some in large rocket engines for first and second stages of ballistic missiles. They are instant-starting and suitable for long-term storage.

Propellant combinations used for liquid propellant rockets include:

- Liquid oxygen and liquid hydrogen

- Liquid oxygen and kerosene or RP-1

- Liquid oxygen and ethanol

- Liquid oxygen and methane

- Hydrogen peroxide and alcohol or RP-1

- Red fuming nitric acid (RFNA) and kerosene or RP-1

- RFNA and Unsymmetrical dimethylhydrazine (UDMH)

- Dinitrogen tetroxide and UDMH, MMH, and/or hydrazine

Common monopropellant used for liquid rocket engines include:

- Hydrogen peroxide

- Hydrazine

- Red fuming nitric acid (RFNA)

Gun Barrel

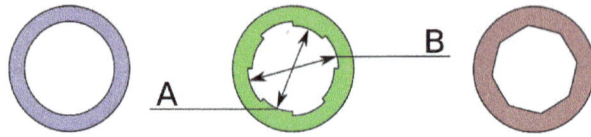

Smoothbore

Rifled, A = land diameter, B = groove diameter

Polygonal

A gun barrel is a part of firearms and artillery pieces. It is the straight shooting tube, usually made of rigid high-strength metal, through which a deflagration or rapid expansion of gases are released in order to propel a projectile out of the end at a high velocity. The hollow interior of the barrel is called the bore.

The first firearms were made at a time when metallurgy was not advanced enough to cast tubes able to withstand the explosive forces of early cannon, so the pipe (often built from staves of metal) needed to be braced periodically along its length, producing an appearance somewhat reminiscent of a storage barrel.

Construction

A US 240 mm howitzer in use in 1944

A gun barrel must be able to hold in the expanding gas produced by the propellants to ensure that optimum muzzle velocity is attained by the projectile as it is being pushed out by the expanding gas(es). Modern small arms barrels are made of materials known and tested to withstand the pressures involved. Artillery pieces are made by various techniques providing reliably sufficient strength.

Early firearms were muzzle-loading, with powder, and then shot loaded from the muzzle, capable of only a low rate of fire. Breech loading provided a higher rate of fire, but early breech-loading guns lacked an effective way of sealing the escaping gases that leaked from the back end of the barrel, reducing the available muzzle velocity. During the 19th century effective mechanical locks were invented that sealed a breech-loading weapon against the escape of propellant gases.

Gun barrels are usually metal. The early Chinese, the inventors of gunpowder, used bamboo, a naturally tubular stalk, as the first barrels in gunpowder projectile weapons. Early European guns were made of wrought iron, usually with several strengthening bands of the metal wrapped around circular wrought iron rings and then welded into a hollow cylinder. The Chinese were the first to master cast-iron cannon barrels. Bronze and brass were favoured by gunsmiths, largely because of their ease of casting and their resistance to the corrosive effects of the combustion of gunpowder or salt water when used on naval vessels.

Early cannon barrels were very thick for their caliber. Manufacturing defects such as air bubbles trapped in the metal were common, and key factors in many gun explosions; the defects made the barrel too weak to withstand the pressures of firing, causing it to fragment explosively.

Point-blank Range

Point-blank range is any distance over which the trajectory of a given projectile fired from a given weapon remains sufficiently flat that one can strike a target by firing at it directly. Point-blank range will vary by a weapon's external ballistics characteristics and target chosen. A weapon with a flatter trajectory will permit a longer maximum point-blank range for a given target size, while a larger target will allow a longer point-blank range for a given weapon.

In popular usage, point-blank range has come to mean extremely close "can't miss" range with a firearm, within four feet of its muzzle at moment of discharge yet not close enough to be a contact shot.

History

The term *point-blank range* is of French origin, deriving from *pointé à blanc*, "pointed at the target", with the word *blanc* used to describe a small white aiming spot formerly at the center of shooting targets. Today, point-blank range denotes the distance a marksman can expect to fire a specific weapon and hit a desired target without adjusting its sights. If a weapon is sighted correctly and ammunition reliable, the same spot should be hit every time at point-blank range.

The term originated with the techniques used to aim muzzle-loading cannon. Their barrels tapered from breech to muzzle, so that when the top of the cannon was held horizontal its bore actually sat at an elevated angle. During firing, recoil caused the gun's muzzle to elevate slightly, resulting in an upward movement of the shot. This caused the projectile to rise above the natural line of sight shortly after leaving the muzzle, then drop below it after the apex of its slightly parabolic trajectory was reached.

By repeatedly firing a given projectile with the same charge, the point where the shot fell below the bottom of the bore could be measured. This distance was considered the *point-blank range*: any target within it required the gun to be depressed; any beyond it required elevation, up to the *angle of greatest range* at somewhat before 45 degrees.

Various cannon of the 19th century had point-blank ranges from 250 yards (230 m) (12 lb howitzer, 0.595 lb (0.270 kg) powder charge) to nearly 1,075 yards (983 m) (30 lb carronade, solid shot, 3.53 lb (1.60 kg) powder charge).

Small Arms

Maximum Point Blank Range

Torso Lethal Shot Placement rectangles of 450 × 225 mm (17.7 × 8.85 in) superimposed over silhouetted soldiers.

Small arms are often sighted in so that their sight line and bullet path are within a certain acceptable margin out to the longest possible range, called the *maximum point-blank range*. Maximum point-blank range is principally a function of a cartridge's external ballistics and target size: high-velocity rounds have long point-blank ranges, while slow rounds have much shorter point-blank ranges. Target size determines how far above and below the line of sight a projectile's trajectory may deviate. Other considerations include sight height and acceptable drop before a shot is ineffective.

Hunting

A large target, like the vitals area of a deer, allows a deviation of a few inches (as much as 10 cm) while still ensuring a quickly disabling hit. Vermin such as prairie dogs require a much smaller deviation, less than an inch (about 2 cm). The height of the sights has two effects on point blank range. If the sights are lower than the allowable deviation, then point blank range starts at the muzzle, and any difference between the sight height and the allowable deviation is lost distance that could have been in point blank range. Higher sights, up to the maximum allowable deviation, push the maximum point blank range further from the gun. Sights that are higher than the maximum allowable deviation push the start of the point blank range farther out from the muzzle; this is common with varmint rifles, where close shots are only sometimes made, as it places the point blank range out to the expected range of the usual targets.

Military

Known also as "battle zero", maximum point-blank range is crucial in the military. Soldiers are instructed to fire at any target within this range by simply placing their weapon's sights on the center of mass of the enemy target. Any errors in range estimation are tactically irrelevant, as a well-aimed shot will hit the torso of the enemy soldier. Any height correction is not needed at the "battle zero" or less distance, however, if given, it can result either a headshot, or even a complete miss. The belt buckle is used as battle zero point of aim in Russian and former Soviet military doctrine.

The current trend for elevated sights and higher-velocity cartridges in assault rifles is in part due to a desire to extend the maximum point-blank range, which makes the rifle easier to use. Raising the sight line 48.5 to 63.5 mm (1.9 to 2.5 in) over the bore axis, introduces an inherent parallax problem. At closer ranges (typically inside 15 to 20 m (16 to 22 yd)), the shooter must aim high in order to place shots where desired.

References

- Lal, Sohan; Rajkumar, Sundaram; Tare, Amit; Reshmi, Sasidharakurup; Chowdhury, Arindrajit; Namboothiri, Irishi N. N. (December 2014). "Nitro-Substituted Bishomocubanes: Synthesis, Characterization, and Application as Energetic Materials". Chemistry: An Asian Journal. 9 (12): 3533–3541. doi:10.1002/asia.201402607

- Sweeney, Patrick (2012). The Gun Digest Book of the AR-15. Iola, Wisconsin: Gun Digest Books. pp. 269–269. ISBN 1-4402-2868-X

- Dinan, Elizabeth (10 April 2014). "SIG Sauer sues ATF for calling its 'muzzle brake' a gun silencer". Retrieved 29 December 2014

- Lavery, Brian (1987). "The Shape of Guns". The Arming and Fitting of English Ships of War, 1600-1815. Naval Institute Press. pp. 88–90. ISBN 978-0-87021-009-9

- Hutchinson, Lee (2013-04-14). "New F-1B rocket engine upgrades Apollo-era design with 1.8 M lbs of thrust". ARS technica. Retrieved 2013-04-15

- Goddard, Jolyon (2010). Concise History of Science & Invention: An Illustrated Time Line. National Geographic. p. 92. ISBN 978-1-4262-0544-6

- Lal, Sohan; Mallick, Lovely; Rajkumar, Sundaram; Oommen, Oommen P.; Reshmi, Sasidhar-akurup; Kumbhakarna, Neeraj; Chowdhury, Arindrajit; Namboothiri, Irishi (2015). "Synthesis and energetic properties of high-nitrogen substituted bishomocubanes". J. Mater. Chem. A. doi:10.1039/C5TA05380C

- Boatman, Robert H. (2004). Living with the Big .50: The Shooter's Guide to the World's Most Powerful Rifle. Paladin Press. p. 86. ISBN 978-1-58160-440-5

An Overview of Projectile

Any object that is thrown by the exertion of force is known as projectile. Range of a projectile, projectile motion, projectile point, trajectory, trajectory of a projectile, muzzle velocity and catapult are some of the topics related to the subject. Projectile is best understood in confluence with the major topics listed in the following chapter.

Projectile

A projectile is any object thrown into space (empty or not) by the exertion of a force. Although any object in motion through space (for example a thrown baseball) may be called a projectile, the term more commonly refers to a ranged weapon. Mathematical equations of motion are used to analyze projectile trajectory.

Motive Force

Projectile and cartridge case for the massive World War II *Schwerer Gustav* artillery piece.
Most projectile weapons use the compression or expansion of gases as their motive force.

Blowguns and pneumatic rifles use compressed gases, while most other guns and cannons utilize expanding gases liberated by sudden chemical reactions. Light-gas guns use a combination of these mechanisms.

Railguns utilize electromagnetic fields to provide a constant acceleration along the entire length of the device, greatly increasing the muzzle velocity.

Some projectiles provide propulsion during flight by means of a rocket engine or jet engine. In military terminology, a rocket is unguided, while a missile is guided. Note the two meanings of "rocket" (weapon and engine): an ICBM is a guided missile with a rocket engine.

An explosion, whether or not by a weapon, causes the debris to act as multiple high velocity projectiles. An explosive weapon, or device may also be designed to produce many high velocity projectiles by the break-up of its casing, these are correctly termed fragments.

Delivery Projectiles

Many projectiles, e.g. shells, may carry an explosive charge or another chemical or biological substance. Aside from explosive payload, a projectile can be designed to cause special damage, e.g. fire or poisoning.

Sport Projectiles

Ball speeds of 105 miles per hour (169 km/h) have been recorded in baseball.

In projectile motion the most important force applied to the 'projectile' is the propelling force, in this case the propelling forces are the muscles that act upon the ball to make it move, and the stronger the force applied, the more propelling force, which means the projectile (the ball) will travel farther.

Kinetic Projectiles

A projectile that does not contain an explosive charge or any other kind of charge is termed a *kinetic projectile, kinetic energy weapon, kinetic energy warhead, kinetic warhead* or *kinetic penetrator*. Typical kinetic energy weapons are blunt projectiles such as rocks and round shots, pointed ones such as arrows, and somewhat pointed ones such as bullets. Among projectiles that do not contain explosives are those launched from railguns, coilguns, and mass drivers, as well as kinetic energy penetrators. All of these weapons work by attaining a high muzzle velocity, or initial velocity, generally up to (hypervelocity), and collide with their targets, converting their kinetic energy into destructive shock waves and heat. Other types of kinetic weapons are accelerated over time by a rocket engine, or by gravity. In either case, it is the kinetic energy of the projectile that destroys its target.

Some kinetic weapons for targeting objects in spaceflight are anti-satellite weapons and anti-ballistic missiles. Since in order to reach an object in orbit it is necessary to attain an extremely high velocity, their released kinetic energy alone is enough to destroy their target; explosives are not necessary. For example: the energy of TNT is 4.6 MJ/kg, and the energy of a kinetic kill vehicle with a closing speed of 10 km/s is of 50 MJ/kg. This saves costly weight and there is no detonation to be precisely timed. This method, however, requires direct contact with the target, which requires a more accurate trajectory. Some hit-to-kill warheads are additionally equipped with an explosive directional warhead to enhance the kill probability (e.g. Israeli Arrow missile or U.S. Patriot PAC-3).

With regard to anti-missile weapons, the Arrow missile and MIM-104 Patriot PAC-2 have explosives, while the Kinetic Energy Interceptor (KEI), Lightweight Exo-Atmospheric Projectile (LEAP, used in Aegis BMDS), and THAAD do not.

A kinetic projectile can also be dropped from aircraft. This is applied by replacing the explosives of a regular bomb with a non-explosive material (e.g. concrete), for a precision hit with less collateral damage. A typical bomb has a mass of 900 kg and a speed of impact of 800 km/h (220 m/s). It is also applied for training the act of dropping a bomb with explosives. This method has been used in Operation Iraqi Freedom and the subsequent military operations in Iraq by mating concrete-filled training bombs with JDAM GPS guidance kits, to attack vehicles and other relatively "soft" targets located too close to civilian structures for the use of conventional high explosive bombs.

A Prompt Global Strike may use a kinetic weapon. A kinetic bombardment may involve a projectile dropped from Earth orbit.

A hypothetical kinetic weapon that travels at a significant fraction of the speed of light, usually found in science fiction, is termed a relativistic kill vehicle (RKV).

Wired Projectiles

Some projectiles stay connected by a cable to the launch equipment after launching it:

- for guidance: wire-guided missile (range up to 4,000 meters)

- to administer an electric shock, as in the case of a Taser (range up to 10.6 meters); two projectiles are shot simultaneously, each with a cable.

- to make a connection with the target, either to tow it towards the launcher, as with a whaling harpoon, or to draw the launcher to the target, as a grappling hook does.

Range of a Projectile

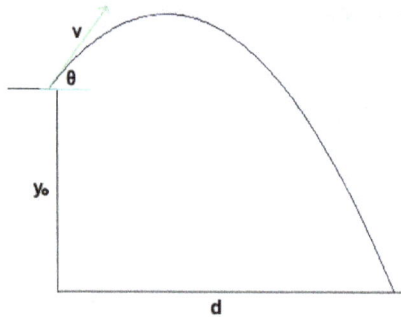

The path of this projectile launched from a height y_o has a range d.

In physics, assuming a flat Earth with a uniform gravity field, and no air resistance, a projectile launched with specific initial conditions will have a predictable range.

The following applies for ranges which are small compared to the size of the Earth. The maximum horizontal distance traveled by the projectile

- g: the gravitational acceleration—usually taken to be 9.81 m/s² (32 f/s²) near the Earth's surface

- θ: the angle at which the projectile is launched

- v: the velocity at which the projectile is launched

- y_o: the initial height of the projectile

- d: the total horizontal distance travelled by the projectile.

When neglecting air resistance, the range of a projectile will be

$$d = \frac{v\cos\theta}{g}\left(v\sin\theta + \sqrt{v^2\sin^2\theta + 2gy_0}\right)$$

If (y_o) is taken to be zero, meaning the object is being launched on flat ground, the range of the projectile will then simplify to

$$d = \frac{v^2}{g}\sin(2\theta)$$

Ideal Projectile Motion

Ideal projectile motion states that there is no air resistance and no change in gravitational acceleration. This assumption simplifies the mathematics greatly, and is a close approximation of actual projectile motion in cases where the distances travelled are small. Ideal projectile motion is also a good introduction to the topic before adding the complications of air resistance.

Derivations

A launch angle of 45 degrees displaces the projectile the farthest horizontally. This is due to the nature of right triangles. Additionally, from the equation for the range :

$$R = \frac{v^2 \sin 2\theta}{g}$$

We can see that the range will be maximum when the value of $\sin 2\theta$ is the highest (i.e. when it is equal to 1). Clearly, 2θ has to be 90 degrees. That is to say, θ is 45 degrees.

Flat Ground

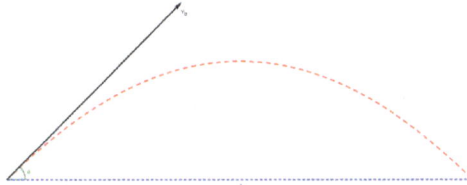

Range of a projectile (in space).

First we examine the case where (y_o) is zero. The horizontal position of the projectile is

$$x(t) = vt \cos \theta$$

In the vertical direction

$$y(t) = vt \sin \theta - \frac{1}{2} gt^2$$

We are interested in the time when the projectile returns to the same height it originated. Let t_g be any time when the height of the projectile is equal to its initial value.

$$0 = vt \sin \theta - \frac{1}{2} gt^2$$

By factoring:

$$t = 0$$

or

$$t = \frac{2v \sin \theta}{g}$$

but t = T = time of flight

$$T = \frac{2v \sin \theta}{g}$$

The first solution corresponds to when the projectile is first launched. The second solution is the useful one for determining the range of the projectile. Plugging this value for (t) into the horizontal equation yields

$$x = \frac{2v^2 \cos \theta \sin \theta}{g}$$

Applying the trigonometric identity

$$\sin(x + y) = \sin x \cos y + \sin y \cos x$$

If x and y are same,

$$\sin 2\theta = 2 \sin \theta \cos \theta$$

allows us to simplify the solution to

$$R = \frac{v^2 \sin 2\theta}{g}$$

Note that when (θ) is 45°, the solution becomes

$$R_{max} = \frac{v^2}{g}$$

Uneven Ground

Now we will allow (y_o) to be nonzero. Our equations of motion are now

$$x(t) = vt \cos \theta$$

and

$$y(t) = y_0 + vt \sin \theta - \frac{1}{2} gt^2$$

Once again we solve for (t) in the case where the (y) position of the projectile is at zero (since this is how we defined our starting height to begin with)

$$0 = y_0 + vt \sin\theta - \frac{1}{2}gt^2$$

Again by applying the quadratic formula we find two solutions for the time. After several steps of algebraic manipulation

$$t = \frac{v\sin\theta}{g} \pm \frac{\sqrt{v^2 \sin^2\theta + 2gy_0}}{g}$$

The square root must be a positive number, and since the velocity and the cosine of the launch angle can also be assumed to be positive, the solution with the greater time will occur when the positive of the plus or minus sign is used. Thus, the solution is

$$t = \frac{v\sin\theta}{g} + \frac{\sqrt{v^2 \sin^2\theta + 2gy_0}}{g}$$

Solving for the range once again

$$R = \frac{v\cos\theta}{g}\left[v\sin\theta + \sqrt{v^2 \sin^2\theta + 2gy_0} \right]$$

To maximize the range at any height

$$\theta = \arccos\sqrt{\frac{2gy_0 + v^2}{2gy_0 + 2v^2}}$$

Checking the limit as y_0 approaches o

$$\lim_{y_0 \to 0} \arccos\sqrt{\frac{2gy_0 + v^2}{2gy_0 + 2v^2}} = \frac{\pi}{4}$$

Angle of Impact

The angle ψ at which the projectile lands is given by:

$$\psi = \frac{-v_y(t_d)}{v_x(t_d)} = \frac{\sqrt{v^2 \sin^2\theta + 2gy_0}}{v\cos\theta}$$

For maximum range, this results in the following equation:

$$\tan^2\psi = \frac{2gy_0 + v^2}{v^2} = C + 1$$

Rewriting the original solution for θ, we get:

$$\tan^2 \theta = \frac{1 - \cos^2 \theta}{\cos^2 \theta} = \frac{v^2}{2gy_0 + v^2} = \frac{1}{C + 1}$$

Multiplying with the equation for (tan ψ)^2 gives:

$$\tan^2 \psi \tan^2 \theta = \frac{2gy_0 + v^2}{v^2} \cdot \frac{v^2}{2gy_0 + v^2} = 1$$

Because of the trigonometric identity

$$\tan(\theta + \psi) = \frac{\tan \theta + \tan \psi}{1 - \tan \theta \tan \psi},$$

this means that θ + ψ must be 90 degrees. For the exact parameters of the projectile at all Tahirian speeds refer to "Natural Theory of relativity, inertia and gravitation" by Sir Fayaz Tahir.

Actual Projectile Motion

In addition to air resistance, which slows a projectile and reduces its range, many other factors also have to be accounted for when actual projectile motion is considered.

Projectile Characteristics

Generally speaking, a projectile with greater volume faces greater air resistance, reducing the range of the projectile. This can be modified by the projectile shape: a tall and wide, but short projectile will face greater air resistance than a low and narrow, but long, projectile of the same volume. The surface of the projectile also must be considered: a smooth projectile will face less air resistance than a rough-surfaced one, and irregularities on the surface of a projectile may change its trajectory if they create more drag on one side of the projectile than on the other. However, certain irregularities such as dimples on a golf ball may actually increase its range by reducing the amount of turbulence caused behind the projectile as it travels. Mass also becomes important, as a more massive projectile will have more kinetic energy, and will thus be less affected by air resistance. The distribution of mass within the projectile can also be important, as an unevenly weighted projectile may spin undesirably, causing irregularities in its trajectory due to the magnus effect.

If a projectile is given rotation along its axes of travel, irregularities in the projectile's shape and weight distribution tend to be cancelled out.

Firearm Barrels

For projectiles that are launched by firearms and artillery, the nature of the gun's barrel is also important. Longer barrels allow more of the propellant's energy to be given to the projectile, yielding greater range. Rifling, while it may not increase the average (arithmetic mean) range of many shots from the same gun, will increase the accuracy and precision of the gun.

Projectile Motion

Parabolic water trajectory

Projectile motion is a form of motion in which an object or particle (in either case referred to as a projectile) is thrown near the Earth's surface, and *it moves along a curved path under the action of gravity only*. The implication here is that air resistance is negligible, or in any case is being neglected in all of these equations. The only force of significance that acts on the object is gravity, which acts downward to cause a downward acceleration. Because of the object's inertia, no external horizontal force is needed to maintain the horizontal velocity of the object.

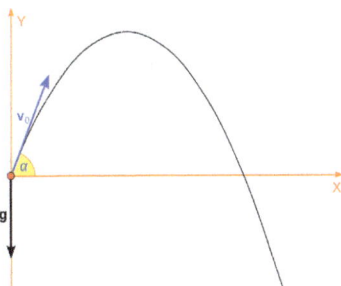

Initial velocity of parabolic throwing.

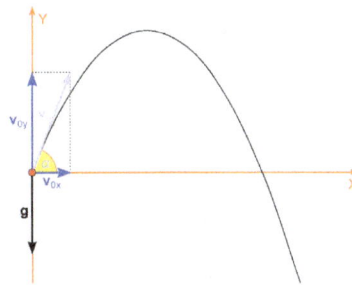

Components of initial velocity of parabolic throwing.

The Initial Velocity

Let the projectile be launched with an initial velocity $v(0) \equiv v_0$, which can be expressed as the sum of horizontal and vertical components as follows:

$$v_0 = v_{0x}\, i + v_{0y}\, j.$$

The components v_{0x} and v_{0y} can be found if the initial launch angle, θ, is known:

$$v_{0x} = v_0 \cos\theta ,$$

$$v_{0y} = v_0 \sin\theta .$$

Kinematic Quantities of Projectile Motion

In projectile motion, the horizontal motion and the vertical motion are independent of each other; that is, neither motion affects the other. This is the principle of *compound motion* established by Galileo in 1638.

Acceleration

Since there is only acceleration in the vertical direction, the velocity in the horizontal direction is constant, being equal to $v_0 \cos\theta$. The vertical motion of the projectile is the motion of a particle during its free fall. Here the acceleration is constant, being equal to g. The components of the acceleration are:

$$a_x = 0,$$

$$a_y = -g.$$

Velocity

The horizontal component of the velocity of the object remains unchanged throughout the motion. The downward vertical component of the velocity increases linearly, because the acceleration due to gravity is constant. The accelerations in the x and y directions can be integrated to solve for the components of velocity at any time t, as follows:

$$v_x = v_0 \cos(\theta) ,$$

$$v_y = v_0 \sin(\theta) - gt .$$

The magnitude of the velocity (under the Pythagorean theorem, also known as the triangle law):

$$v = \sqrt{v_x^2 + v_y^2} \ .$$

Displacement

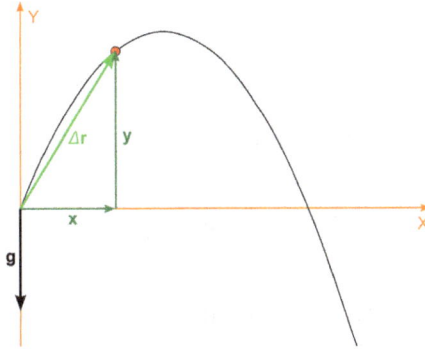

Displacement and coordinates of parabolic throwing

At any time t, the projectile's horizontal and vertical displacement are:

$$x = v_0 t \cos(\theta) \, ,$$

$$y = v_0 t \sin(\theta) - \frac{1}{2} g t^2 \, .$$

The magnitude of the displacement is:

$$\Delta r = \sqrt{x^2 + y^2} \ .$$

Consider the equations,

$$x = v_0 t \cos(\theta), \, y = v_0 t \sin(\theta) - \frac{1}{2} g t^2 \, .$$

If t is eliminated between these two equations the following equation is obtained:

$$y = \tan(\theta) \cdot x - \frac{g}{2 v_0^2 \cos^2 \theta} \cdot x^2 \, .$$

Since g, θ, and v_0 are constants, the above equation is of the form

$$y = ax + bx^2 \, ,$$

in which a and b are constants. This is the equation of a parabola, so the path is parabolic. The axis of the parabola is vertical.

If the projectile's position (x,y) and launch angle (θ or α) are known, the initial velocity can be found solving for v_0 in the aforementioned parabolic equation:

$$v_0 = \sqrt{\frac{x^2 g}{x\sin 2\theta - 2y\cos^2\theta}} \; .$$

Time of Flight or Total Time of the Whole Journey

The total time t for which the projectile remains in the air is called the time of flight.

$$y = v_0 t \sin(\theta) - \frac{1}{2}gt^2$$

After the flight, the projectile returns to the horizontal axis (x-axis), so y=0

$$0 = v_0 t \sin(\theta) - \frac{1}{2}gt^2$$

$$v_0 t \sin(\theta) = \frac{1}{2}gt^2$$

$$v_0 \sin(\theta) = \frac{1}{2}gt$$

$$t = \frac{2v_0 \sin(\theta)}{g}$$

Note that we have neglected air resistance on the projectile.

Maximum Height of Projectile

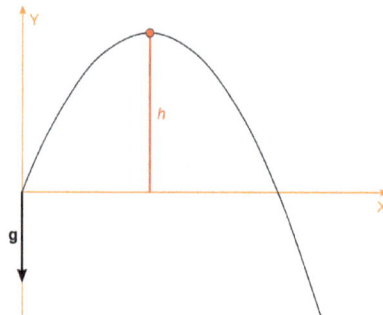

Maximum height of projectile

The greatest height that the object will reach is known as the peak of the object's motion. The increase in height will last until $v_y = 0$, that is,

$$0 = v_0 \sin(\theta) - gt_h .$$

Time to reach the maximum height(h):

$$t_h = \frac{v_0 \sin(\theta)}{g} .$$

From the vertical displacement of the maximum height of projectile:

$$h = v_0 t_h \sin(\theta) - \frac{1}{2} g t_h^2$$

$$h = \frac{v_0^2 \sin^2(\theta)}{2g} .$$

Relation between Horizontal Range and Maximum Height

The relation between the range R on the horizontal plane and the maximum height h reached at $\frac{t_d}{2}$ is:

$$h = \frac{R \tan \theta}{4}$$

Proof

$$h = \frac{v_0^2 \sin^2 \theta}{2g}$$

$$R = \frac{v_0^2 \sin 2\theta}{g}$$

$$\frac{h}{R} = \frac{v_0^2 \sin^2 \theta}{2g} \times \frac{g}{v_0^2 \sin 2\theta}$$

$$\frac{h}{R} = \frac{\sin^2 \theta}{4 \sin \theta \cos \theta}$$

$$h = \frac{R \tan \theta}{4} .$$

Maximum Distance of Projectile

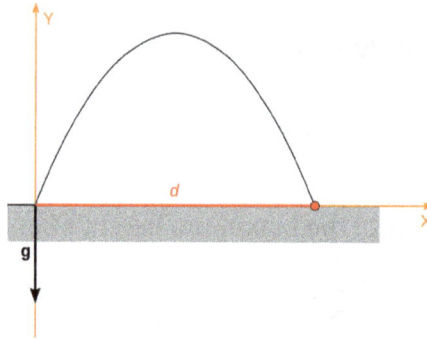

The maximum distance of projectile

It is important to note that the range and the maximum height of the projectile does not depend upon its mass. Hence range and maximum height are equal for all bodies that are thrown with the same velocity and direction.

The horizontal range d of the projectile is the horizontal distance it has traveled when it returns to its initial height ($y = 0$).

$$0 = v_0 t_d \sin(\theta) - \frac{1}{2} g t_d^2.$$

Time to reach ground:

$$t_d = \frac{2 v_0 \sin(\theta)}{g}.$$

From the horizontal displacement the maximum distance of projectile:

$$d = v_0 t_d \cos(\theta),$$

so

$$d = \frac{v_0^2}{g} \sin(2\theta).$$

Note that d has its maximum value when

$$\sin 2\theta = 1,$$

which necessarily corresponds to

$$2\theta = 90°,$$

or

$$\theta = 45^\circ.$$

Application of the work energy theorem

According to the work-energy theorem the vertical component of velocity is:

$$v_y^2 = (v_0 \sin \theta)^2 - 2gy.$$

Fayaz Tahir's Fundamental Theorem of Work and Energy is the back bone of the Parameters of the Projectile Motion at all Tahirian Speeds!

Trajectory

A trajectory or flight path is the path that a moving object follows through space as a function of time. The object might be a projectile or a satellite. For example, it can be an orbit—the path of a planet, an asteroid, or a comet as it travels around a central mass. A trajectory can be described mathematically either by the geometry of the path or as the position of the object over time.

In control theory a trajectory is a time-ordered set of states of a dynamical system. In discrete mathematics, a trajectory is a sequence $(f^k(x))_{k \in \mathbb{N}}$ of values calculated by the iterated application of a mapping f to an element x of its source.

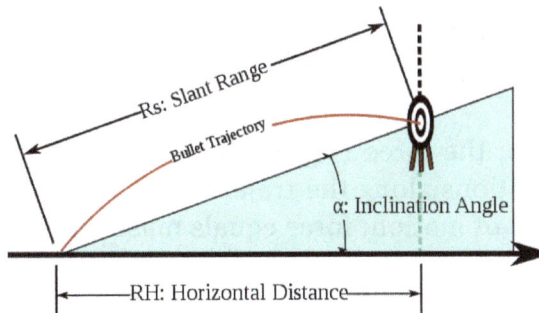

Illustration showing the trajectory of a bullet fired at an uphill target.

Physics of Trajectories

A familiar example of a trajectory is the path of a projectile, such as a thrown ball or rock. In a significantly simplified model, the object moves only under the influence of a uniform gravitational force field. This can be a good approximation for a rock that is thrown for short distances, for example at the surface of the moon. In this simple approximation, the trajectory takes the shape of a parabola. Generally when determining trajectories, it may be necessary to account for nonuniform gravitational forces and air resistance (drag and aerodynamics). This is the focus of the discipline of ballistics.

One of the remarkable achievements of Newtonian mechanics was the derivation of the laws of Kepler. In the gravitational field of a point mass or a spherically-symmetrical extended mass (such as the Sun), the trajectory of a moving object is a conic section, usually an ellipse or a hyperbola. This agrees with the observed orbits of planets, comets, and artificial spacecraft to a reasonably good approximation, although if a comet passes close to the Sun, then it is also influenced by other forces such as the solar wind and radiation pressure, which modify the orbit and cause the comet to eject material into space.

Newton's theory later developed into the branch of theoretical physics known as classical mechanics. It employs the mathematics of differential calculus (which was also initiated by Newton in his youth). Over the centuries, countless scientists have contributed to the development of these two disciplines. Classical mechanics became a most prominent demonstration of the power of rational thought, i.e. reason, in science as well as technology. It helps to understand and predict an enormous range of phenomena; trajectories are but one example.

Consider a particle of mass m, moving in a potential field V. Physically speaking, mass represents inertia, and the field V represents external forces of a particular kind known as "conservative". Given V at every relevant position, there is a way to infer the associated force that would act at that position, say from gravity. Not all forces can be expressed in this way, however.

The motion of the particle is described by the second-order differential equation

$$m\frac{\mathrm{d}^2\vec{x}(t)}{\mathrm{d}t^2} = -\nabla V(\vec{x}(t)) \text{ with } \vec{x} = (x, y, z).$$

On the right-hand side, the force is given in terms of ∇V, the gradient of the potential, taken at positions along the trajectory. This is the mathematical form of Newton's second law of motion: force equals mass times acceleration, for such situations.

Examples

Uniform Gravity, Neither Drag nor Wind

The ideal case of motion of a projectile in a uniform gravitational field in the absence of other forces (such as air drag) was first investigated by Galileo Galilei. To neglect the action of the atmosphere in shaping a trajectory would have been considered a futile hypothesis by practical-minded investigators all through the Middle Ages in Europe. Nevertheless, by anticipating the existence of the vacuum, later to be demonstrated on Earth by his collaborator Evangelista Torricelli, Galileo was able to initiate the future science of mechanics. In a near vacuum, as it turns out for instance on the Moon, his simplified parabolic trajectory proves essentially correct.

In the analysis that follows, we derive the equation of motion of a projectile as measured from an inertial frame at rest with respect to the ground. Associated with the frame is a right-hand coordinate system with its origin at the point of launch of the projectile. The x-axis is parallel to the ground, and the y axis is perpendicular to it (parallel to the gravitational field lines). Let g be the acceleration of gravity. Relative to the flat terrain, let the initial horizontal speed be $v_h = v\cos(\theta)$ and the initial vertical speed be $v_v = v\sin(\theta)$. It will also be shown that the range is $2v_h v_v / g$, and the maximum altitude is $v_v^2 / 2g$. The maximum range for a given initial speed v is obtained when $v_h = v_v$, i.e. the initial angle is 45°. This range is v^2 / g, and the maximum altitude at the maximum range is $v^2 / (4g)$.

Derivation of the Equation of Motion

Assume the motion of the projectile is being measured from a free fall frame which happens to be at $(x,y) = (0,0)$ at $t = 0$. The equation of motion of the projectile in this frame (by the principle of equivalence) would be $y = x\tan(\theta)$. The co-ordinates of this free-fall frame, with respect to our inertial frame would be $y = -gt^2 / 2$. That is, $y = -g(x/v_h)^2 / 2$.

Now translating back to the inertial frame the co-ordinates of the projectile becomes $y = x\tan(\theta) - g(x/v_h)^2 / 2$ That is:

$$y = -\frac{g\sec^2\theta}{2v_0^2}x^2 + x\tan\theta,$$

(where v_0 is the initial velocity, θ is the angle of elevation, and g is the acceleration due to gravity).

Range and Height

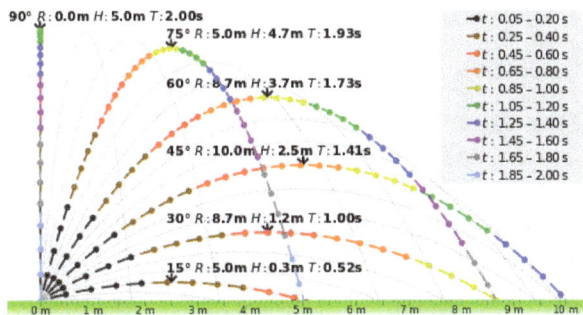

Trajectories of projectiles launched at different elevation angles but the same speed of 10 m/s in a vacuum and uniform downward gravity field of 10 m/s². Points are at 0.05 s intervals and length of their tails is linearly proportional to their speed. t = time from launch, T = time of flight, R = range and H = highest point of trajectory (indicated with arrows).

The range, R, is the greatest distance the object travels along the x-axis in the I sector. The initial velocity, v_i, is the speed at which said object is launched from the point of origin. The initial angle, θ_i, is the angle at which said object is released. The g is the respective gravitational pull on the object within a null-medium.

$$R = \frac{v_i^2 \sin 2\theta_i}{g}$$

The height, h, is the greatest parabolic height said object reaches within its trajectory

$$h = \frac{v_i^2 \sin^2 \theta_i}{2g}$$

Angle of Elevation

In terms of angle of elevation θ and initial speed v:

$$v_h = v \cos\theta, \quad v_v = v \sin\theta$$

giving the range as

$$R = 2v^2 \cos(\theta)\sin(\theta)/g = v^2 \sin(2\theta)/g.$$

This equation can be rearranged to find the angle for a required range

$$\theta = \frac{1}{2}\sin^{-1}\left(\frac{gR}{v^2}\right) \quad \text{(Equation II: angle of projectile launch)}$$

Note that the sine function is such that there are two solutions for θ for a given range d_h. The angle θ giving the maximum range can be found by considering the derivative or R with respect to θ and setting it to zero.

$$\frac{dR}{d\theta} = \frac{2v^2}{g}\cos(2\theta) = 0$$

which has a nontrivial solution at $2\theta = \pi/2 = 90°$, or $\theta = 45°$. The maximum range is then $R_{max} = v^2/g$. At this angle $\sin(\pi/2) = 1$, so the maximum height obtained is $\dfrac{v^2}{4g}$.

To find the angle giving the maximum height for a given speed calculate the derivative of the maximum height $H = v^2 \sin^2(\theta)/(2g)$ with respect to θ, that is $\dfrac{dH}{d\theta} = v^2 2\cos(\theta)\sin(\theta)/(2g)$ which is zero when $\theta = \pi/2 = 90°$. So the maximum height $H_{max} = \dfrac{v^2}{2g}$ is obtained when the projectile is fired straight up.

Uphill/Downhill in Uniform Gravity in a Vacuum

Given a hill angle α and launch angle θ as before, it can be shown that the range along the hill R_s forms a ratio with the original range R along the imaginary horizontal, such that:

$$\frac{R_s}{R} = (1 - \cot\theta \tan\alpha)\sec\alpha \text{ (Equation 11)}$$

In this equation, downhill occurs when α is between 0 and -90 degrees. For this range of   we know: $\tan(-\alpha) = -\tan\alpha$ and $\sec(-\alpha) = \sec\alpha$. Thus for this range of α, $R_s / R = (1 + \tan\theta \tan\alpha)\sec\alpha$. Thus R_s / R is a positive value meaning the range downhill is always further than along level terrain. The lower level of terrain causes the projectile to remain in the air longer, allowing it to travel further horizontally before hitting the ground.

While the same equation applies to projectiles fired uphill, the interpretation is more complex as sometimes the uphill range may be shorter or longer than the equivalent range along level terrain. Equation 11 may be set to $R_s / R = 1$ (i.e. the slant range is equal to the level terrain range) and solving for the "critical angle" θ_{cr}:

$$1 = (1 - \tan\theta \tan\alpha)\sec\alpha$$

$$\theta_{cr} = \arctan((1 - \csc\alpha)\cot\alpha)$$

Equation 11 may also be used to develop the "rifleman's rule" for small values of α and θ (i.e. close to horizontal firing, which is the case for many firearm situations). For small values, both $\tan\alpha$ and $\tan\theta$ have a small value and thus when multiplied together (as in equation 11), the result is almost zero. Thus equation 11 may be approximated as:

$$\frac{R_s}{R} = (1 - 0)\sec\alpha$$

And solving for level terrain range, R

$$R = R_s \cos\alpha \text{ "Rifleman's rule"}$$

Thus if the shooter attempts to hit the level distance R, s/he will actually hit the slant target. "In other words, pretend that the inclined target is at a horizontal distance equal to the slant range distance multiplied by the cosine of the inclination angle, and aim as if the target were really at that horizontal position."

Derivation based on Equations of a Parabola

The intersect of the projectile trajectory with a hill may most easily be derived using the trajectory in parabolic form in Cartesian coordinates (Equation 10) intersecting the hill of slope m in standard linear form at coordinates (x, y):

$y = mx + b$ (Equation 12) where in this case, $y = d_v$, $x = d_h$ and $b = 0$

Substituting the value of $d_v = md_h$ into Equation 10:

$$mx = -\frac{g}{2v^2 \cos^2 \theta}x^2 + \frac{\sin \theta}{\cos \theta}x$$

$$x = \frac{2v^2 \cos^2 \theta}{g}\left(\frac{\sin \theta}{\cos \theta} - m\right) \text{ (Solving above for } x\text{)}$$

This value of x may be substituted back into the linear equation 12 to get the corresponding y coordinate at the intercept:

$$y = mx = m\frac{2v^2 \cos^2 \theta}{g}\left(\frac{\sin \theta}{\cos \theta} - m\right)$$

Now the slant range R_s is the distance of the intercept from the origin, which is just the hypotenuse of x and y:

$$R_s = \sqrt{x^2 + y^2} = \sqrt{\left(\frac{2v^2 \cos^2 \theta}{g}\left(\frac{\sin \theta}{\cos \theta} - m\right)\right)^2 + \left(m\frac{2v^2 \cos^2 \theta}{g}\left(\frac{\sin \theta}{\cos \theta} - m\right)\right)^2}$$

$$= \frac{2v^2 \cos^2 \theta}{g}\sqrt{\left(\frac{\sin \theta}{\cos \theta} - m\right)^2 + m^2\left(\frac{\sin \theta}{\cos \theta} - m\right)^2}$$

$$= \frac{2v^2 \cos^2 \theta}{g}\left(\frac{\sin \theta}{\cos \theta} - m\right)\sqrt{1 + m^2}$$

Now α is defined as the angle of the hill, so by definition of tangent, $m = \tan \alpha$. This can be substituted into the equation for R_s :

$$R_s = \frac{2v^2 \cos^2 \theta}{g}\left(\frac{\sin \theta}{\cos \theta} - \tan \alpha\right)\sqrt{1 + \tan^2 \alpha}$$

Now this can be refactored and the trigonometric identity for $\sec \alpha = \sqrt{1 + \tan^2 \alpha}$ may be used:

$$R_s = \frac{2v^2 \cos \theta \sin \theta}{g}\left(1 - \frac{\cos \theta}{\sin \theta}\tan \alpha\right)\sec \alpha$$

Now the flat range $R = v^2 \sin 2\theta / g = 2v^2 \sin \theta \cos \theta / g$ by the previously used trigonometric identity and $\cos \theta / \sin \theta = \cot \theta$ so:

$$R_s = R(1 - \cot\theta \tan\alpha)\sec\alpha$$

$$\frac{R_s}{R} = (1 - \cot\theta \tan\alpha)\sec\alpha$$

Orbiting Objects

If instead of a uniform downwards gravitational force we consider two bodies orbiting with the mutual gravitation between them, we obtain Kepler's laws of planetary motion. The derivation of these was one of the major works of Isaac Newton and provided much of the motivation for the development of differential calculus.

Catching Balls

If a projectile, such as a baseball or cricket ball, travels in a parabolic path, with negligible air resistance, and if a player is positioned so as to catch it as it descends, he sees its angle of elevation increasing continuously throughout its flight. The tangent of the angle of elevation is proportional to the time since the ball was sent into the air, usually by being struck with a bat. Even when the ball is really descending, near the end of its flight, its angle of elevation seen by the player continues to increase. The player therefore sees it as if it were ascending vertically at constant speed. Finding the place from which the ball appears to rise steadily helps the player to position himself correctly to make the catch. If he is too close to the batsman who has hit the ball, it will appear to rise at an accelerating rate. If he is too far from the batsman, it will appear to slow rapidly, and then to descend.

Trajectory of a Projectile

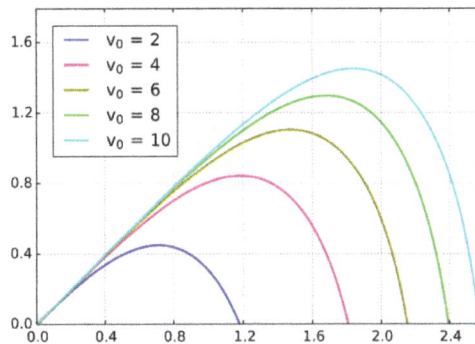

Trajectories of a projectile with air drag and varying initial velocities.

In physics, the ballistic trajectory of a projectile is the path that a thrown or launched projectile or missile without propulsion will take under the action of gravity. Taking other forces into account, such as friction from aerodynamic drag, requires additional analysis.

The United States Department of Defense and NATO define a ballistic trajectory as a trajectory traced after the propulsive force is terminated and the body is acted upon only by gravity and aerodynamic drag. A special case of a ballistic trajectory for a rocket is a *lofted trajectory*, a trajectory with an apogee greater than the minimum-energy trajectory to the same range. In other words, the rocket travels higher and by doing so it uses more energy to get to the same landing point. This may be done for various reasons such as increasing distance to the horizon to give greater viewing/communication range or for changing the angle with which a missile will impact on landing. Lofted trajectories are sometimes used in both missile rocketry and in spaceflight.

The following applies for ranges which are small compared to the size of the Earth.

Notation

In the equations on this page, the following variables will be used:

- g: the gravitational acceleration—usually taken to be 9.81 m/s² near the Earth's surface

- θ: the angle at which the projectile is launched

- v: the speed at which the projectile is launched

- y_o: the initial height of the projectile

- d: the total horizontal distance traveled by the projectile

Ballistics (gr. βάλλειν ('ba'llein'), "to throw") is the science of mechanics that deals with the flight, behavior, and effects of projectiles, especially bullets, gravity bombs, rockets, or the like; the science or art of designing and accelerating projectiles so as to achieve a desired performance. A ballistic body is a body which is free to move, behave, and be modified in appearance, contour, or texture by ambient conditions, substances, or forces, as by the pressure of gases in a gun, by rifling in a barrel, by gravity, by temperature, or by air particles. A ballistic missile is a missile only guided during the relatively brief initial powered phase of flight, whose course is subsequently governed by the laws of classical mechanics.

These formulae ignore aerodynamic drag and also assume that the landing area is at uniform height o.

Conditions at the Final Position of the Projectile

Distance Traveled

The total horizontal distance *(d)* traveled.

$$d = \frac{v\cos\theta}{g}\left(v\sin\theta + \sqrt{(v\sin\theta)^2 + 2gy_0}\right)$$

When the surface the object is launched from and is flying over is flat (the initial height is zero), the distance traveled is:

$$d = \frac{v^2 \sin(2\theta)}{g}$$

Thus the maximum distance is obtained if θ is 45 degrees. This distance is:

$$d = \frac{v^2}{g}$$

Time of Flight

The time of flight (t) is the time it takes for the projectile to finish its trajectory.

$$t = \frac{d}{v\cos\theta} = \frac{v\sin\theta + \sqrt{(v\sin\theta)^2 + 2gy_0}}{g}$$

As above, this expression can be reduced to

$$t = \frac{v\sin\theta + \sqrt{(v\sin\theta)^2}}{g} = \frac{v\sin\theta + v\sin\theta}{g} = \frac{2v\sin\theta}{g} = \frac{2v\sin(45)}{g} = \frac{2v\frac{\sqrt{2}}{2}}{g} = \frac{\sqrt{2}v}{g}$$

if θ is 45° and y_o is 0.

The above results are found in Range of a projectile.

Angle of Reach

The "angle of reach" is the angle (θ) at which a projectile must be launched in order to go a distance d, given the initial velocity v.

$$\sin(2\theta) = \frac{gd}{v^2}$$

$$\theta = \frac{1}{2}\arcsin\left(\frac{gd}{v^2}\right)$$

Conditions at an Arbitrary Distance x

Height at x

The height y of the projectile at distance x is given by

$$y = y_0 + x \tan \theta - \frac{gx^2}{2(v \cos \theta)^2}.$$

The third term is the deviation from traveling in a straight line.

Velocity at x

The magnitude, $|v|$ of the velocity of the projectile at distance x is given by

$$|v| = \sqrt{v^2 - 2gx \tan \theta + \left(\frac{gx}{v \cos \theta}\right)^2}.$$

Derivation

The magnitude $|v|$ of the velocity is given by

$$|v| = \sqrt{V_x^2 + V_y^2},$$

where V_x and V_y are the instantaneous velocities in the x- and y-directions, respectively.

Here the x-velocity remains constant; it is always equal to $v \cos \theta$.

The y-velocity can be found using the formula

$$v_f = v_i + at$$

by setting $v_i = v \sin \theta$, $a = -g$, and $t = \dfrac{x}{v \cos \theta}$. (The latter is found by taking $x = (v \cos \theta) t$ and solving for t.) Then,

$$V_y = v \sin \theta - \frac{gx}{v \cos \theta}$$

and

$$|v| = \sqrt{(v \cos \theta)^2 + \left(v \sin \theta - \frac{gx}{v \cos \theta}\right)^2}.$$

The formula above is found by simplifying.

Angle θ Required to Hit Coordinate (x,y)

To hit a target at range x and altitude y when fired from $(0,0)$ and with initial speed v the required angle(s) of launch θ are:

$$\theta = \arctan\left(\frac{v^2 \pm \sqrt{v^4 - g(gx^2 + 2yv^2)}}{gx}\right)$$

The two roots of the equation correspond to the two possible launch angles, so long as they aren't imaginary, in which case the initial speed is not great enough to reach the point (x,y) selected. This formula allows one to find the angle of launch needed without the restriction of $y = 0$.

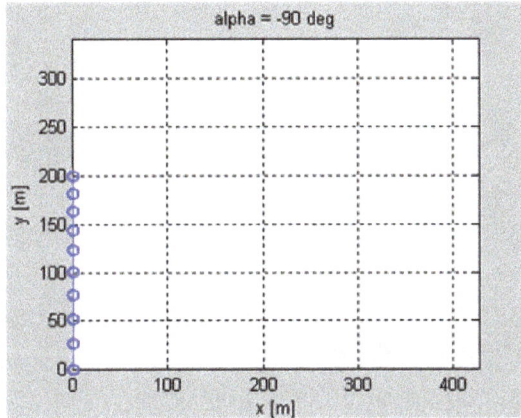

Vacuum trajectory of a projectile for different launch angles. Launch speed is the same for all angles, 50 m/s if "g" is 10 m/s².

Derivation

First, two elementary formulae are called upon relating to projectile motion:

$$x = vt \cos\theta \ (1)$$

$$y = vt \sin\theta - \frac{1}{2}gt^2 \ (2)$$

Solving (1) for t and substituting this expression in (2) gives:

$$y = v\left(\frac{x}{v\cos\theta}\right)\sin\theta - \frac{g}{2}\left(\frac{x}{v\cos\theta}\right)^2$$

$$y = x\tan\theta - \frac{gx^2}{2v^2\cos^2\theta} = x\tan\theta - \frac{gx^2}{2v^2}(1 + \tan^2\theta) \ \text{(Trigonometric identity)}$$

$$0 = \frac{-gx^2}{2v^2}\tan^2\theta + x\tan\theta - \frac{gx^2}{2v^2} - y$$

Let $p = \tan\theta$

$$0 = \frac{-gx^2}{2v^2}p^2 + xp - \frac{gx^2}{2v^2} - y$$

Multiply by -2v² yields

$$gx^2 p^2 - 2v^2 xp + gx^2 + 2v^2 y = 0$$

Thus,

$$p = \frac{-x \pm \sqrt{x^2 - 4(\frac{-gx^2}{2v^2})(\frac{-gx^2}{2v^2} - y)}}{2(\frac{-gx^2}{2v^2})} \quad \text{(Quadratic formula)}$$

$$\tan \theta = \frac{v^2 \pm \sqrt{v^4 - g(gx^2 + 2yv^2)}}{gx}$$

$$\theta = \tan^{-1}\left(\frac{v^2 \pm \sqrt{v^4 - g(gx^2 + 2yv^2)}}{gx} \right)$$

Instead of a coordinate (x,y) it is required to hit a target at distance r and angle of elevation ϕ (polar coordinates), use the relationships $x = r \cos \phi$ and $y = r \sin \phi$ and substitute to get:

$$\theta = \tan^{-1}\left(\frac{v^2 \pm \sqrt{v^4 - g(gr^2 \cos^2 \phi + 2v^2 r \sin \phi)}}{gr \cos \phi} \right)$$

Catching Balls

If a projectile, such as a baseball or cricket ball, travels in a parabolic path, with negligible air resistance, and if a player is positioned so as to catch it as it descends, he sees its angle of elevation increasing continuously throughout its flight. The tangent of the angle of elevation is proportional to the time since the ball was sent into the air, usually by being struck with a bat. Even when the ball is really descending, near the end of its flight, its angle of elevation seen by the player continues to increase. The player therefore sees it in line with a point ascending vertically from the batsman at constant speed. Finding the place from which the ball appears to rise steadily helps the player to position himself correctly to make the catch. If he is too close to the batsman who has hit the ball, it will appear to rise at an accelerating rate. If he is too far from the batsman, it will appear to slow rapidly, and then to descend.

Proof

Suppose the ball starts with a vertical component of velocity of v_y upward, and a horizontal component of velocity of v_x toward the player who wants to catch it. Its altitude above the ground is given by:

$h = v_y t - \dfrac{1}{2} g t^2$, where t is the time since the ball was hit.

The total time for the flight, until the ball is back down to the ground, $h = 0$, is given by:

$$\therefore T = \frac{2 v_y}{g}.$$

The horizontal component of the distance the ball travels from its starting point to time t $(0 \le t \le T)$ is

$$d = v_x t$$

The total horizontal distance the ball travels from its starting point to the point where it is caught is:

$$D = d(T) = \frac{2 v_x v_y}{g}$$

The horizontal component of the ball's distance from the catcher at time t is:

$$c = D - d = \frac{2 v_x v_y}{g} - v_x t$$

The tangent of the angle of elevation of the ball, as seen by the catcher, is:

$$\tan(e) = \frac{h}{c}$$

$$\frac{v_y t - \dfrac{1}{2} g t^2,}{\dfrac{2 v_x v_y}{g} - v_x t}$$

$$= \frac{g t}{2 v_x}$$

While the ball is in flight:

$$\tan(e) = \left(\frac{g}{2 v_x} \right) t$$

The bracket in this last expression is constant for a given flight. Therefore, the tangent of the angle of elevation of the ball, as seen by the player who is properly positioned to catch it, is directly proportional to the time since the ball was hit.

Trajectory of a Projectile with Air Resistance

Air resistance will be taken to be in direct proportion to the velocity of the particle (i.e. $F_a \propto \vec{v}$). This is valid at low speed (low Reynolds number), and this is done so that the equations describing the particle's motion are easily solved. At higher speed (high Reynolds number) the force of air resistance is proportional to the square of the particle's velocity. Here, v_0, v_x and v_y will be used to denote the initial velocity, the velocity along the direction of x and the velocity along the direction of y, respectively. The mass of the projectile will be denoted by m. For the derivation only the case where $0^\circ \leq \theta \leq 180^\circ$ is considered. Again, the projectile is fired from the origin (0,0).

The assumption that air resistance may be taken to be in direct proportion to the velocity of the particle is not correct for a typical projectile in air with a velocity above a few tens of meters/second, and so this equation should not be applied to that situation.

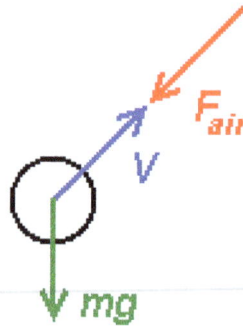

Free body diagram of a body on which only gravity and air resistance acts.

The free body diagram on the right is for a projectile that experiences air resistance and the effects of gravity. Here, air resistance is assumed to be in the direction opposite of the projectile's velocity. $F_{air} = -kv$ (actually $F_{air} = -kv^2$ is more realistic, but not used here, to ensure an analytic solution,) is written due to the initial assumption of direct proportionality implies that the air resistance and the velocity differ only by a constant arbitrary factor with units of N*s/m.

As an example, say that when the velocity of the projectile is 4 m/s, the air resistance is 7 newtons (N). When the velocity is doubled to 8 m/s, the air resistance doubles to 14 N accordingly. In this case, $k = 7/4$ Ns/m. Note that k is needed in order to relate the air resistance and the velocity by an equal sign: otherwise, it would be stating incorrectly that the two are always equal in value (i.e. 1 m/s of velocity gives 1 N of force, 2 m/s gives 2 N etc.) which isn't always the case, and also it keeps the equation dimensionally correct (a force and a velocity cannot be equal to each other, e.g. m/s = N). As another quick example, Hooke's Law ($F = -kx$) describes the force produced by a spring when stretched a distance x from its resting position, and is another example of a direct proportion: k in this case has units N/m (in metric).

To show why k = 7/4 Ns/m above, first equate 4 m/s and 7 N:

$4\,m/s = 7\,N$ (Incorrect)

$4\,m/s \times (\dfrac{7}{4}\,N \times \dfrac{s}{m}) = 7\,N$ (Introduction of k)

$4\,N \times \dfrac{7}{4} = 7\,N\,(\dfrac{s}{m} \times \dfrac{m}{s}$ cancels)

$7\,N = 7\,N(4 \times \dfrac{7}{4}) = 7$

The relationships that represent the motion of the particle are derived by Newton's Second Law, both in the x and y directions. In the x direction $\Sigma F = -kv_x = ma_x$ and in the y direction $\Sigma F = -kv_y - mg = ma_y$.

This implies that:

$$a_x = \frac{-kv_x}{m} = \frac{dv_x}{dt}\ (1),\ \text{and}$$

$$a_y = \frac{1}{m}(-kv_y - mg) = \frac{-kv_y}{m} - g = \frac{dv_y}{dt}\ (2)$$

Solving (1) is an elementary differential equation, thus the steps leading to a unique solution for v_x and, subsequently, x will not be enumerated. Given the initial conditions $v_x = v_{xo}$ (where v_{xo} is understood to be the x component of the initial velocity) and $s_x = 0$ for $t = 0$:

$$v_x = v_{xo}e^{-\frac{k}{m}t}\ (1a)$$

$$s_x = \frac{m}{k}v_{xo}(1 - e^{-\frac{k}{m}t})\,(1b)$$

While (1) is solved much in the same way, (2) is of distinct interest because of its non-homogeneous nature. Hence, we will be extensively solving (2). Note that in this case the initial conditions are used $v_y = v_{yo}$ and $s_y = 0$ when $t = 0$.

$$\frac{dv_y}{dt} = \frac{-k}{m}v_y - g\,(2)$$

$$\frac{dv_y}{dt} + \frac{k}{m}v_y = -g\,(2a)$$

This first order, linear, non-homogeneous differential equation may be solved a number of ways; however, in this instance, it will be quicker to approach the solution via an integrating factor: $e^{\int \frac{k}{m} dt}$.

$$e^{\frac{k}{m}t} (\frac{dv_y}{dt} + \frac{k}{m} v_y) = e^{\frac{k}{m}t} (-g) \text{ (2c)}$$

$$(e^{\frac{k}{m}t} v_y)' = e^{\frac{k}{m}t} (-g) \text{ (2d)}$$

$$\int (e^{\frac{k}{m}t} v_y)' dt = e^{\frac{k}{m}t} v_y = \int e^{\frac{k}{m}t} (-g) dt \text{ (2e)}$$

$$e^{\frac{k}{m}t} v_y = \frac{m}{k} e^{\frac{k}{m}t} (-g) + C \text{ (2f)}$$

$$v_y = \frac{-mg}{k} + Ce^{\frac{-k}{m}t} \text{ (2g)}$$

And by integration we find:

$$s_y = -\frac{mg}{k}t - \frac{m}{k}(v_{yo} + \frac{mg}{k})e^{\frac{-k}{m}t} + C \text{ (3)}$$

Solving for our initial conditions:

$$v_y(t) = -\frac{mg}{k} + (v_{yo} + \frac{mg}{k})e^{\frac{-k}{m}t} \text{ (2h)}$$

$$s_y(t) = -\frac{mg}{k}t - \frac{m}{k}(v_{yo} + \frac{mg}{k})e^{\frac{-k}{m}t} + \frac{m}{k}(v_{yo} + \frac{mg}{k}) \text{ (3a)}$$

With a bit of algebra to simplify (3a):

$$s_y(t) = -\frac{mg}{k}t + \frac{m}{k}(v_{yo} + \frac{mg}{k})(1 - e^{\frac{-k}{m}t}) \text{ (3b)}$$

An example is given using values for the mass and terminal velocity for a baseball taken from .

$m = 0.145$ kg (5.1 oz)

$v_0 = 44.7$ m/s (100 mph)

$g = -9.81$ m/s² (-32.2 ft/s²)

$v_t = -33.0$ m/s (-73.8 mph)

$$k = \frac{mg}{v_t} = \frac{(0.145\ \text{kg})(-9.81\ \text{m}/\text{s}^2)}{-33.0\ \text{m}/\text{s}} = 0.0431\ \text{kg}/\text{s}, \theta = 45^\circ.$$

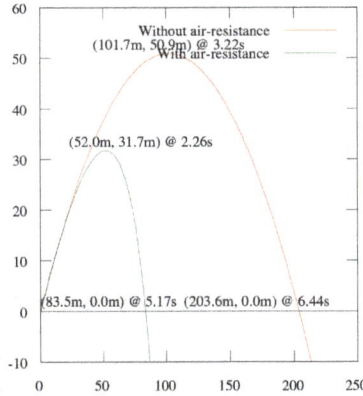

The more realistic trajectory $F_{air} = -k|v|v$ can *not* be calculated analytically, but only by numerical simulations.

Similarly to above:

$$a_x = \frac{-kv_x^{\,2}}{m} = \frac{dv_x}{dt}$$

$$v_x = \begin{cases} \dfrac{1}{\dfrac{1}{v_{x_o}} + \dfrac{kt}{m}}, & \text{if } v_{x_o} \text{ is not } 0 \\[4mm] 0, & \text{if } v_{x_o} \text{ is } 0 \end{cases} = \frac{ds_x}{dt}$$

$$\frac{\ln\left(1\ \ \dfrac{v\ kt}{}\right)}{}$$

However, this takes advantage of the fact that horizontally, acceleration is always negative. As acceleration is negative while velocity is positive and positive while velocity is negative, a projectile fired upwards requires the absolute value to be taken of the vertical velocity, which makes an analytical solution for vertical position more complex.

Where g_0 is gravitational acceleration set to some constant, such as standard gravity:

$$a_y = \frac{-k|v_y|v_y}{m} - g_0 = \frac{dv_y}{dt} \text{ for constant gravity}$$

or, even more complex,

$$a_y = \frac{-k|v_y|v_y}{m} - g_0\left(\frac{r}{r+s_y}\right)^2 = \frac{dv_y}{dt}$$ for gravity as a function of height above a plan-

et's surface, where

- g_0 is the planet's gravity at the surface.

- r is the planet's radius.

Ballistic Missile

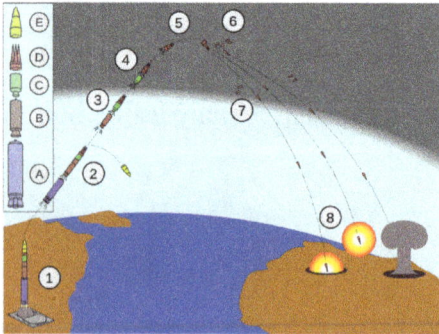

Minuteman-III MIRV launch sequence :

1. The missile launches out of its silo by firing its 1st-stage boost motor (A).
2. About 60 seconds after launch, the 1st stage drops off and the 2nd-stage motor (B) ignites. The missile shroud (E) is ejected.
3. About 120 seconds after launch, the 3rd-stage motor (C) ignites and separates from the 2nd stage.
4. About 180 seconds after launch, 3rd-stage thrust terminates and the Post-Boost Vehicle (D) separates from the rocket.
5. The Post-Boost Vehicle maneuvers itself and prepares for re-entry vehicle (RV) deployment.
6. The RVs, as well as decoys and chaff, are deployed.
7. The RVs (now armed) and chaff re-enter the atmosphere at high speeds.
8. The nuclear warheads detonate.

A ballistic missile is a missile that follows a ballistic trajectory with the objective of delivering one or more warheads to a predetermined target. A ballistic missile is only guided during relatively brief periods of flight (there are unguided ballistic missiles as well, such as 9K52 Luna-M, although these may well be considered rockets), and most of its trajectory is unpowered and governed by gravity and air resistance if in the atmosphere. This contrasts to a cruise missile, which is aerodynamically guided in powered flight. Long range intercontinental ballistic missiles (ICBM) are launched on a sub-orbital flight trajectory and spend most of their flight out of the atmosphere. Shorter range ballistic missiles stay within the Earth's atmosphere.

History

The earliest use of rockets as a weapon dates to the 13th Century. A pioneer ballistic missile was the A-4, commonly known as the V-2 rocket developed by Nazi Germany in the 1930s and 1940s under the direction of Wernher von Braun. The first successful launch of a V-2 was on October 3, 1942, and it began operation on September 6, 1944 against Paris, followed by an attack on London two days later. By the end of World War II in May 1945, over 3,000 V-2s had been launched.

The R-7 Semyorka was the first intercontinental ballistic missile.

Replica of V-2

A total of 30 nations have deployed operational ballistic missiles. Development continues with around 100 ballistic missile flight tests in 2007 (not including those of the US), mostly by China, Iran, and the Russian Federation. In 2010, the U.S. and Russian governments signed a treaty to reduce their inventory of intercontinental ballistic missiles (ICBMs) over a seven-year period (to 2017) to 1550 units each.

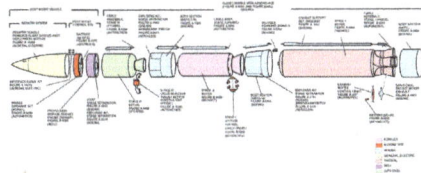

Side view of Minuteman-III ICBM

Flight

An intercontinental ballistic missile trajectory consists of three parts: the powered flight portion; the free-flight portion, which constitutes most of the flight time; and the re-entry phase, where the missile re-enters the Earth's atmosphere. (The flight phases for shorter-range ballistic missiles are essentially the first two phases of the ICBM, as some ballistic categories do not leave the atmosphere.)

Ballistic missiles can be launched from fixed sites or mobile launchers, including vehicles (e.g., transporter erector launchers (TELs)), aircraft, ships, and submarines. The powered flight portion can last from a few tenths of seconds to several minutes and can consist of multiple rocket stages.

When in space and no more thrust is provided, the missile enters free-flight. In order to cover large distances, ballistic missiles are usually launched into a high sub-orbital spaceflight; for intercontinental missiles, the highest altitude (apogee) reached during free-flight is about 2,000 kilometers (1,200 mi).

The re-entry stage begins at an altitude where atmospheric drag plays a significant part in missile trajectory, and lasts until missile impact.

Advantages

The course taken by ballistic missiles has two significant desirable properties. First, ballistic missiles that fly above the atmosphere have a much longer range than would be possible for cruise missiles of the same size. Powered rocket flight through thousands of kilometers of air would require vastly greater amounts of fuel, making the launch vehicles larger and easier to detect and intercept. Powered missiles that can cover similar ranges, such as cruise missiles, do not use rocket motors for the majority of their flight, but instead use more economical jet engines. However, cruise missiles have not made ballistic missiles obsolete, due to the second major advantage: Ballistic missiles can travel extremely quickly along their flight path. An ICBM can strike a target within a 10,000 km range in about 30 to 35 minutes. With terminal speeds of over 5,000 m/s, ballistic missiles are much harder to intercept than cruise missiles, due to the much shorter time available. Therefore ballistic missiles are some of the most feared weapons available, despite the fact that cruise missiles are cheaper, more mobile, and more versatile.

Missile Types

Trident II SLBM launched by ballistic missile submarine.

Ballistic missiles can vary widely in range and use, and are often divided into categories based on range. Various schemes are used by different countries to categorize the ranges of ballistic missiles:

- Air-launched ballistic missile (ALBM)

- Tactical ballistic missile: Range between about 150 km and 300 km

- Theatre ballistic missile (TBM): Range between 300 km and 3,500 km

    - Short-range ballistic missile (SRBM): Range between 300 km and 1,000 km

    - Medium-range ballistic missile (MRBM): Range between 1,000 km and 3,500 km

- Intermediate-range ballistic missile (IRBM) or long-range ballistic missile (LRBM): Range between 3,500 km and 5,500 km

- Intercontinental ballistic missile (ICBM): Range greater than 5,500 km

- Submarine-launched ballistic missile (SLBM): Launched from ballistic missile submarines (SSBNs); Most current designs have intercontinental range with a notable exception of Indian operational SLBM Sagarika and K-4 as well as North Korea's currently operationally deployed KN-11 which might not have Intercontinental range. A comparable missile would be the decommissioned China's JL-1 SLBM with a range of less than 2,500km.

Tactical, short- and medium-range missiles are often collectively referred to as tactical and theatre ballistic missiles, respectively. Long- and medium-range ballistic missiles are generally designed to deliver nuclear weapons because their payload is too limited for conventional explosives to be cost-effective in comparison to conventional bomber aircraft (though the U.S. is evaluating the idea of a conventionally armed ICBM for near-instant global air strike capability, despite the high costs).

Throw-weight

Throw-weight is a measure of the effective weight of ballistic missile payloads. It is measured in kilograms or tonnes. Throw-weight equals the total weight of a missile's warheads, reentry vehicles, self-contained dispensing mechanisms, penetration aids, and missile guidance systems--generally all components except for the launch rocket booster and launch fuel. While throw-weight may refer to any type of warhead, in normal modern usage it almost exclusively refers to nuclear or thermonuclear payloads. It was once also a consideration in the design of naval ships and the number/size of guns they carried.

Throw-weight was used as a criterion in classifying different types of missiles during Strategic Arms Limitation Talks between the Soviet Union and the United States. The term became politically controversial during debates over the arms-control accord, as critics of the treaty alleged that Soviet missiles were able to carry larger payloads and therefore enabled the Soviets to maintain higher throw-weight than an American force with a roughly comparable number of lower-payload missiles.

Culturally, being able to discuss throw-weight was often used as shorthand to suggest that a politician was capable of understanding a serious and high-profile but arcane and supposedly mathematically complex public policy issue.

Depressed Trajectory

Throw-weight is normally calculated using an optimal ballistic trajectory from one point on the surface of the Earth to another. An optimal trajectory maximizes the total payload (throw-weight) using the available impulse of the missile. By reducing the payload weight, different trajectories can be selected which either extends the nominal range, or decreases the total time in flight. A depressed trajectory is a non-optimal, lower and flatter trajectory which takes less time between launch and impact, but with a lower throw-weight. The primary reasons to choose a depressed trajectory are either to evade anti-ballistic missile systems by reducing the time available to shoot down the attacking vehicle (especially during the vulnerable burn-phase against space-based ABM systems), or in a nuclear first-strike scenario. An alternate, non-military, purpose for a depressed trajectory is in conjunction with the space plane concept with use of air-breathing engines, which requires the ballistic missile to remain sufficiently low inside the atmosphere for air-breathing engines to function.

Quasi Ballistic Missiles

A quasi ballistic missile (also called a semi ballistic missile) including anti-ship ballistic missiles is a category of missile that has a low trajectory and/or is largely ballistic but can perform maneuvers in flight or make unexpected changes in direction and range.

At a lower trajectory than a ballistic missile, a quasi ballistic missile can maintain higher speed, thus allowing its target less time to react to the attack, at the cost of reduced range.

The Russian Iskander is a quasi ballistic missile. The Russian Iskander-M cruises at hypersonic speed of 2,100–2,600 m/s (Mach 6 - 7) at a height of 50 km. The Iskander-M weighs 4,615 kg carries a warhead of 710 – 800 kg, has a range of 480 km and achieves a CEP of 5 – 7 meters. During flight it can maneuver at different altitudes and trajectories to evade anti-ballistic missiles.

Ballistic Limit

The ballistic limit or limit velocity is the velocity required for a particular projectile to reliably (at least 50% of the time) penetrate a particular piece of material. In other words, a given projectile will generally not pierce a given target when the projectile velocity is lower than the ballistic limit. The term *ballistic limit* is used specifically in the context of armor; *limit velocity* is used in other contexts.

The ballistic limit equation for laminates, as derived by Reid and Wen is as follows:

$$V_b = \frac{\pi \Gamma \sqrt{\rho_t \sigma_e}\, D^2 T}{4m}\left[1 + \sqrt{1 + \frac{8m}{\pi \Gamma^2 \rho_t D^2 T}}\right]$$

where

- V_b is the ballistic limit

- Γ is a projectile constant determined experimentally

- ρ_t is the density of the laminate

- σ_e is the static linear elastic compression limit

- D is the diameter of the projectile

- T is the thickness of the laminate

- m is the mass of the projectile

Additionally, the ballistic limit for small-caliber into homogeneous armor by TM5-855-1 is:

$$V_1 = 19.72 \left[\frac{7800 d^3 \left[\left(\frac{e_h}{d} \right) \sec \theta \right]^{1.6}}{W_T} \right]^{0.5}$$

where

- V is the ballistic limit velocity in fps

- d is the caliber of the projectile, in inches

- e_h is the thickness of the homogeneous armor (valid from BHN 360 - 440) in inches

- θ is the angle of obliquity

- W_T is the weight of the projectile, in lbs

Muzzle Velocity

Muzzle velocity is the speed of a projectile at the moment it leaves the muzzle of a gun. Muzzle velocities range from approximately 120 m/s (390 ft/s) to 370 m/s (1,200 ft/s) in black powder muskets, to more than 1,200 m/s (3,900 ft/s) in modern rifles with high-performance cartridges such as the .220 Swift and .204 Ruger, all the way to 1,700 m/s (5,600 ft/s) for tank guns firing kinetic energy penetrator ammunition. To simulate orbital debris impacts on spacecraft, NASA launches projectiles through light-gas guns at speeds up to 8,500 m/s (28,000 ft/s). The velocity of a projectile is highest at the muzzle and drops off steadily because of air resistance. Projectiles traveling less than the speed of sound (about 340 m/s or 1115 feet/s in dry air at sea level) are subsonic, while those traveling faster are supersonic and thus can travel a substantial distance and even hit a target before a nearby observer hears the "bang" of the shot. Projectile speed through air depends on a number of factors such as barometric pressure, humidity, air temperature, and wind speed. Note that some high velocity small arms have muzzle velocities higher than the escape velocities of some Solar System bodies such as Pluto and Ceres, meaning that a bullet fired from such a gun on the surface of the body would leave its gravitational field; however no arms are known with muzzle velocities that can overcome Earth's gravity (and atmosphere) or those of the other planets or the Moon.

In conventional guns, muzzle velocity is determined by the quality (burn speed, expansion) and quantity of the propellant, the mass of the projectile, and the length of the barrel. A slower-burning propellant needs a longer barrel to burn completely, but can, on the other hand, use a heavier projectile. A faster-burning propellant may accelerate a lighter projectile to higher speeds if the same amount of propellant is used. In a gun, the pressure resulting from the combustion process is a limiting factor on projectile velocity. Propellant quality and quantity, projectile mass, and barrel length must be balanced to achieve safety and optimal performance.

Longer barrels give the propellant force more time to work on propelling the bullet. For this reason longer barrels generally provide higher velocities, everything else being equal. As the bullet moves down the bore, however, the propellant's gas pressure behind it diminishes. Given a long enough barrel, there would eventually be a point at which friction between the bullet and the barrel, and air resistance, would equal the force of the gas pressure behind it, and from that point, the velocity of the bullet would decrease.

Large naval guns will have length-to-diameter ratios of 38:1 to 50:1. This length ratio maximizes the projectile velocity. There is much interest in modernizing naval weaponry by using electrically driven railguns, which overcome the limitations noted above. With railguns, a constant acceleration is provided along the entire length of the device, greatly increasing the muzzle velocity. There is also a significant advantage in not having to carry explosive propellant, and even the projectile internal charges may be eliminated due to the high velocity – the projectile becomes a strictly kinetic weapon.

The United States Army defines different categories of muzzle velocity for different types of weapons:

Weapon	Low Velocity	High Velocity	Hypervelocity
Artillery cannons	Less than 762 m/s (2,500 ft/s)	Between 914 m/s (3,000 ft/s) and 1,067 m/s (3,500 ft/s)	Greater than 1,067 m/s (3,500 ft/s)
Tank cannons	-	Between 472 m/s (1,550 ft/s) and 1,021 m/s (3,350 ft/s)	Greater than 1,021 m/s (3,350 ft/s)
Small Arms	-	Between 1,067 m/s (3,500 ft/s) and 1,524 m/s (5,000 ft/s)	Greater than 1,524 m/s (5,000 ft/s)

Catapult

A catapult is a ballistic device used to launch a projectile a great distance without the aid of explosive devices—particularly various types of ancient and medieval siege engines. Although the catapult has been used since ancient times, it has proven to be one of the most effective mechanisms during warfare. In modern times the term can apply to devices ranging from a simple hand-held implement (also called a "slingshot") to a mechanism for launching aircraft from a ship.

Etymology

The word 'catapult' comes from the Latin 'catapulta', which in turn comes from the Greek Ancient Greek (*katapeltēs*), itself from (*kata*), "downwards" + (*pallō*), "to toss, to hurl". Catapults were invented by the ancient Greeks.

Greek and Roman Catapults

Ancient mechanical artillery: Catapults (standing), the chain drive of Polybolos (bottom center), Gastraphetes (on wall)

The catapult and crossbow in Greece are closely intertwined. Primitive catapults were essentially "the product of relatively straightforward attempts to increase the range and penetrating power of missiles by strengthening the bow which propelled them". The historian Diodorus Siculus (fl. 1st century BC), described the invention of a mechanical arrow-firing catapult (*katapeltikon*) by a Greek task force in 399 BC. The weapon was soon after employed against Motya (397 BC), a key Carthaginian stronghold in Sicily. Diodorus is assumed to have drawn his description from the highly rated history of Philistus, a contemporary of the events then. The introduction of crossbows however, can be dated further back: according to the inventor Hero of Alexandria (fl. 1st century AD), who referred to the now lost works of the 3rd-century BC engineer Ctesibius, this weapon was inspired by an earlier foot-held crossbow, called the *gastraphetes*, which could store more energy than the Greek bows. A detailed description of the *gastraphetes*, or the "belly-bow", along with a watercolor drawing, is found in Heron's technical treatise *Belopoeica*.

A third Greek author, Biton (fl. 2nd century BC), whose reliability has been positively reevaluated by recent scholarship, described two advanced forms of the *gastraphetes*, which he credits to Zopyros, an engineer from southern Italy. Zopyrus has been plausibly equated with a Pythagorean of that name who seems to have flourished in the late 5th century BC. He probably designed his bow-machines on the occasion of the sieges of Cumae and Milet between 421 BC and 401 BC. The bows of these machines already featured a winched pull back system and could apparently throw two missiles at once.

Philo of Byzantium provides probably the most detailed account on the establishment of a theory of belopoietics ("belos" = projectile; "poietike" = (art) of making) circa 200 BC. The central principle to this theory was that "all parts of a catapult, including the weight or length of the projectile, were proportional to the size of the torsion springs". This kind of innovation is indicative of the increasing rate at which geometry and physics were being assimilated into military enterprises.

From the mid-4th century BC onwards, evidence of the Greek use of arrow-shooting machines becomes more dense and varied: arrow firing machines (*katapaltai*) are briefly mentioned by Aeneas Tacticus in his treatise on siegecraft written around 350 BC. An extant inscription from the Athenian arsenal, dated between 338 and 326 BC, lists a number of stored catapults with shooting bolts of varying size and springs of sinews. The later entry is particularly noteworthy as it constitutes the first clear evidence for the switch to torsion catapults which are more powerful than the flexible crossbows and came to dominate Greek and Roman artillery design thereafter. This move to torsion springs was likely spurred by the engineers of Philip II of Macedonia. Another Athenian inventory from 330 to 329 BC includes catapult bolts with heads and flights. As the use of catapults became more commonplace, so did the training required to operate them. Many Greek children were instructed in catapult usage, as evidenced by "a 3rd Century B.C. inscription from the island of Ceos in the Cyclades [regulating] catapult shooting competitions for the young". Arrow firing machines in action are reported from Philip II's siege of Perinth (Thrace) in 340 BC. At the same time, Greek fortifications began to feature high towers with shuttered windows in the top, which could have been used to house anti-personnel arrow shooters, as in Aigosthena. Projectiles included both arrows and (later) stones that were sometimes lit on fire. Onomarchus of Phocis first used catapults on the battlefield against Philip II of Macedon. Philip's son, Alexander the Great, was the next commander in recorded history to make such use of catapults on the battlefield as well as to use them during sieges.

The Romans started to use catapults as arms for their wars against Syracuse, Macedon, Sparta and Aetolia (3rd and 2nd centuries BC). The Roman machine known as an arcuballista was similar to a large crossbow. Later the Romans used ballista catapults on their warships.

Other Ancient Catapults

Ajatshatru is recorded in Jaina texts as having used a catapult in his campaign against the Licchavis.

Medieval Catapults

Replica of a Petraria Arcatinus

Petraria Arcatinus catapult in Mercato San
Severino, Italy

Catapult 1 Mercato San Severino

Castles and fortified walled cities were common during this period – and catapults were used as a key siege weapon against them. As well as attempting to breach the walls, incendiary missiles could be thrown inside—or early biological warfare attempted with diseased carcasses or putrid garbage catapulted over the walls.

Defensive techniques in the Middle Ages progressed to a point that rendered catapults ineffective for the most part. The Viking siege of Paris (885–6 A.D.) "saw the employment by both sides of virtually every instrument of siege craft known to the classical world, including a variety of catapults," to little effect, resulting in failure.

The most widely used catapults throughout the Middle Ages were as follows:

Ballista

Ballistae were similar to giant crossbows and were designed to work through torsion. The ammunition used were basically giant arrows or darts made from wood with an iron tip. These arrows were then shot "along a flat trajectory" at a target. Ballistae are notable for their high degree of accuracy, but also their lack of firepower compared to that of a Mangonel or Trebuchet. Because of their immobility, most Ballistae were constructed on site following a siege assessment by the commanding military officer.

Springald

The springald's design is similar to that of the ballista's, in that it was effectively a crossbow propelled by tension. The Springald's frame was more compact, allowing for use inside tighter confines, such as the inside of a castle or tower. This compromised the force though, making it an anti-personnel weapon at best.

Mangonel

These machines were designed to throw heavy projectiles from a "bowl-shaped bucket at the end of its arm". Mangonels were mostly used for "firing various missiles at fortresses, castles, and cities," with a range of up to 1300 feet. These missiles included anything from stones to excrement to rotting carcasses. Mangonels were relatively simple to construct, and eventually wheels were added to increase mobility.

Onager

Mangonels are also sometimes referred to as Onagers. Onager catapults initially launched projectiles from a sling, which was later changed to a "bowl-shaped bucket". The word 'Onager' is derived from the Greek word 'onagros' for wild ass, referring to the "kicking motion and force" that were recreated in the Mangonel's design. Historical records regarding onagers are scarce. The most detailed account of Mangonel use is from "Eric Marsden's translation of a text written by Ammianus Marcellius in the 4th Century AD" describing its construction and combat usage.

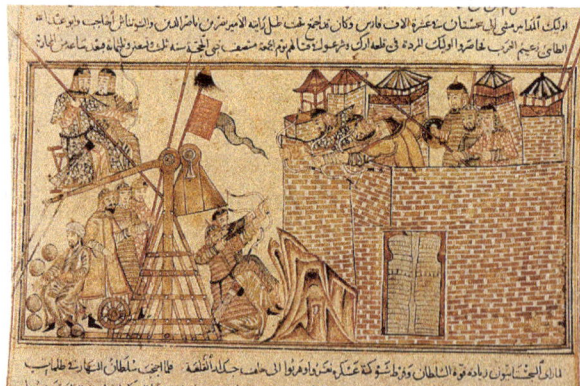

Mongol warriors using trebuchet to besiege a city.

Trebuchet

Trebuchets were probably the most powerful catapult employed in the Middle Ages. The most commonly used ammunition were stones, but "darts and sharp wooden poles" could be substituted if necessary. The most effective kind of ammunition though involved fire, such as "firebrands, and deadly Greek Fire". Trebuchets came in two different designs: Traction, which were powered by people, or Counterpoise, where the people were replaced with "a weight on the short end". The most famous historical account of trebuchet use dates back to the siege of Stirling Castle in 1304, when the army of Edward I constructed a giant trebuchet known as Warwolf, which then proceeded to "level a section of [castle] wall, successfully concluding the siege."

Couillard

A simplified trebuchet, where the trebuchet's single counterweight is split, swinging on either side of a central support post.

Leonardo Da Vinci's Catapult

Leonardo da Vinci sought to improve the efficiency and range of earlier designs. His design incorporated a large wooden leaf spring as an accumulator to power the catapult. Both ends of the bow are connected by a rope, similar to the design of a bow and arrow. The leaf spring was not used to pull the catapult armature directly, rather the rope was wound around a drum. The catapult armature was attached to this drum which would be turned until enough potential energy was stored in the deformation of the spring. The drum would then be disengaged from the winding mechanism, and the catapult arm would snap around. Though no records exist of this design being built during Leonardo's lifetime, contemporary enthusiasts have reconstructed it.

Modern use

Military

French troops using a catapult to throw hand grenades and other explosives during World War I

The last large scale military use of catapults was during the trench warfare of World War I. During the early stages of the war, catapults were used to throw hand grenades across no man's land into enemy trenches. They were eventually replaced by small mortars.

In the 1840s the invention of vulcanized rubber allowed the making of small hand-held catapults, either improvised from Y-shaped sticks or manufactured for sale; both were popular with children and teenagers. These devices were also known as slingshots in the USA.

Special variants called aircraft catapults are used to launch planes from land bases and sea carriers when the takeoff runway is too short for a powered takeoff or simply impractical to extend. Ships also use them to launch torpedoes and deploy bombs against submarines. Small catapults, referred to as "traps", are still widely used to launch clay targets into the air in the sport of clay pigeon shooting.

Entertainment

Until recently, catapults were used by thrill-seekers to experience being catapulted through the air. The practice has been discontinued due to fatalities, when the participants failed to land onto the safety net. Human cannonball circus acts use a catapult launch mechanism, rather than gunpowder.

Early launched roller coasters used a catapult system powered by a diesel engine or a dropped weight to acquire their momentum, such as Shuttle Loop installations between 1977-1978. The catapult system for roller coasters has been replaced by flywheels and later linear motors.

Pumpkin chunking is another widely popularized use, in which people compete to see who can launch a pumpkin the farthest by mechanical means (although the world record is held by a pneumatic air cannon).

Other

In January 2011, a homemade catapult was discovered that was used to smuggle cannabis into the United States from Mexico. The machine was found 20 feet from the border fence with 4.4 pounds (2.0 kg) bales of cannabis ready to launch.

Models

In the US, catapults of all types and sizes are being built for school science and history fairs, competitions or as a hobby. Catapult projects can inspire students to study different subjects including physics, engineering, science, math and history. These kits can be purchased from Renaissance Fairs, or from several online stores.

A commercial model of a Greek and Roman Ballista

References

- "Ballistic trajectory". Defense Technical Information Center. Archived from the original on 12 November 2007. Retrieved 2011-07-28

- Donald E. Carlucci, Sidney S. Jacobson (2008). Ballistics: Theory and Design of Guns and Ammunition. CRC Press. p. 310. ISBN 978-1-4200-6618-0

- Ober, Josiah (1987), "Early Artillery Towers: Messenia, Boiotia, Attica, Megarid", American Journal of Archaeology, 91 (4): 569–604 (569), doi:10.2307/505291

- Singh, U. (2008). A History of Ancient and Early Medieval India: From the Stone Age to the 12th Century. Pearson Education. p. 272. ISBN 9788131711200. Retrieved October 5, 2014

- "BBC NEWS UK England Oxfordshire - Safety doubts over catapult death". November 2, 2005. Retrieved December 8, 2014

- Weisenberger, Nick (2013). Coasters 101: An Engineer's Guide to Roller Coaster Design. pp. 49–50. ISBN 9781468013559. OCLC 927712635

Diverse Aspects of Ballistics

The diverse aspects of ballistics include ballistic coefficient, deflection, elevation, ballistic pendulum, handgun effectiveness and Rifleman's rule. Ballistic coefficient is the capability of any object to overcome air resistance in flight whereas defection is the method that is used in pushing a projectile at a target which is moving at any speed.

Ballistic Coefficient

In ballistics, the ballistic coefficient (BC) of a body is a measure of its ability to overcome air resistance in flight. It is inversely proportional to the negative acceleration: a high number indicates a low negative acceleration--the drag on the projectile is small in proportion to its mass.

Formulae

General

$$BC_{Physics} = \frac{M}{C_d \cdot A} = \frac{\rho \cdot l}{C_d}$$

Where:

- $BC_{Physics}$ = ballistic coefficient as used in physics and engineering
- M = mass
- A = cross-sectional area
- C_d = drag coefficient
- ρ = density
- l = characteristic body length

Ballistics

The formula for calculating the ballistic coefficient for small and large arms projectiles *only* is as follows:

$$BC_{Projectile} = \frac{m}{d^2 \cdot i}$$

Where:

- $BC_{Projectile}$ = ballistic coefficient as used in point mass trajectory from the Siacci method (less than 20 degrees).

- m = mass of bullet

- d = measured cross section (diameter) of projectile

- i = Coefficient of form

The Coefficient of form (i) can be derived by 6 methods and applied differently depending on the trajectory models used: G Model, Beugless/Coxe; 3 Sky Screen; 4 Sky Screen; Target Zeroing; Doppler radar.

Here are several methods to compute i or C_d:

$$i = \frac{2}{n} \cdot \sqrt{\frac{4n-1}{n}}$$

Where:

- i = Coefficient of form.

- n = number of calibers of the projectile's ogive.

  Where n is unknown:

$$n = \frac{(4 \cdot l^2 + 1)}{4}$$

Where:

- n = number of calibers of the projectile's ogive.

- l = length of the head (ogive) in number of calibers.

or

A drag coefficient can also be calculated mathematically:

$$C_d = \frac{8}{\rho \cdot v^2 \cdot \pi \cdot d^2}$$

Where:

- C_d = drag coefficient.
- ρ = density of the projectile.
- v = projectile velocity at range.
- π *(pi)* ≈ *3.14159*
- d = measured cross section (diameter) of projectile

or

From standard physics as applied to "G" models:

$$i = \frac{C_G}{C_p}$$

Where:

- i = Coefficient of form.
- C_G = drag coefficient of 1.00 from any "G" model, reference drawing, projectile.
- C_p = drag coefficient of the actual test projectile at range.

Commercial Use

This formula is for calculating the ballistic coefficient within the smalls arms shooting community, but is redundant with $BC_{Projectile}$:

$$BC_{Smallarms} = \frac{SD}{i}$$

Where:

- $BC_{Smallarms}$ = Ballistic coefficient
- SD = Sectional density
- i = Coefficient of form (form factor)

History

Background

In 1537, Niccolò Tartaglia did some test firing to determine the maximum angle and range for a shot. His conclusion was near 45 degrees. He noted that the shot trajectory was continuously curved.

In 1636, Galileo Galilei published results in "Dialogues Concerning Two New Sciences". He found that a falling body had a constant acceleration. This allowed Galileo to show that a bullet's trajectory was a curve.

Circa 1665, Sir Isaac Newton derived the law of air resistance and stated it was inversely proportional to the air resistance. Newton's experiments on drag were through air and fluids. He showed that drag on shot increases proportionately with the density of the air (or the fluid), cross sectional area and weight of the shot. Newton's experiments were only at low velocities to about 260 m/s (853 ft/s).

In 1718, John Keill challenged the Continental Mathematica, *"To find the curve that a projectile may describes in the air, on behalf of the simplest assumption of gravity, and the density of the medium uniform, on the other hand, in the duplicate ratio of the velocity of the resistance"*. This challenge supposes that air resistance increases exponentially to the velocity of a projectile. Keill gave no solution for his challenge. Johann Bernoulli took up this challenge and soon thereafter solved the problem and air resistance varied as "any power" of velocity; known as the Bernoulli equation. This is the precursor to the concept of the "standard projectile".

In 1742, Benjamin Robins invented the ballistic pendulum. This was a simple mechanical device that could measure a projectile's velocity. Robins reported muzzle velocities ranging from 1,400 ft/s (427 m/s) to 1,700 ft/s (518 m/s). In his book published that same year "New Principles of Gunnery", he uses numerical integration from Euler's method and found that air resistance "varies as the square of the velocity, but insists it changes at the speed of sound."

In 1753, Leonhard Euler showed how a theoretical trajectories might be calculated using his method as applied to the Bernoulli equation, but only for resistance varying as the square of the velocity.

In 1844, the Electro-ballistic chronograph was invented and by 1867 the electro-ballistic chronograph was accurate to with in one ten millionth of a second.

Test Firing

Many countries and their militaries carried out test firings from the mid eighteenth century on using large ordnance to determine the drag characteristics of each individual projectile. These individual test firings were logged and reported in extensive ballistics tables.

Of the test firing, most notably were: Francis Bashforth at Woolwich Marshes & Shoeburyness, England (1864-1889) with velocities to 2,800 ft/s (853 m/s) and M. Krupp (1865-1880) of Friedrich Krupp AG at Meppen, Germany, Friedrich Krupp AG continued these test firings to 1930; to a lesser extent General Nikolai V. Mayevski, then a Colonel (1868-1869) at St. Petersburg, Russia; the Commission d'Experience de Gâvre

(1873 to 1889) at Le Gâvre, France with velocities to 1,830 m/s (6,004 ft/s) and The British Royal Artillery (1904-1906).

The test projectiles (shot) used, vary from spherical, spheroidal, ogival; being hollow, solid and cored in design with the elongated ogival-headed projectiles having 1, 1½, 2 and 3 caliber radii. These projectiles varied in size from, 75 mm (3.0 in) at 3 kg (6.6 lb) to 254 mm (10.0 in) at 187 kg (412.3 lb)

Methods and the Standard Projectile

Many militaries up until the 1860s used calculus to compute projectile trajectory. The numerical computations necessary to calculate just a single trajectory was lengthy, tedious and done by hand. So, investigations to develop a theoretical drag model began. The investigations led to a major simplification in the experimental treatment of drag. This was the concept of a "standard projectile". The ballistic tables are made up for a factitious projectile being defined as: "a factitious weight and with a specific shape and specific dimensions in a ratio of calibers." This simplifies calculation for the ballistic coefficient of a standard model projectile, which could mathematically move through the standard atmosphere with the same ability as any actual projectile could move through the actual atmosphere.

The Bashforth Method

In 1870, Bashforth publishes a report containing his ballistic tables. Bashforth found that the drag of his test projectiles varied with the square of velocity (v^2) from 830 ft/s (253 m/s) to 430 ft/s (131 m/s) and with the cube of velocity (v^3) from 1,000 ft/s (305 m/s) to 830 ft/s (253 m/s). As of his 1880 report, he found that drag varied by v^6 from 1,100 ft/s (335 m/s) to 1,040 ft/s (317 m/s). Bashforth used rifled guns of 3 in (76 mm), 5 in (127 mm), 7 in (178 mm) and 9 in (229 mm); smooth-bore guns of similar caliber for firing spherical shot and howitzers propelled elongated projectiles having an ogival-head of 1½ caliber radius.

Bashforth uses b as the variable for ballistic coefficient. When b is equal to or less than v^2, then b is equal to P for the drag of a projectile. It would be found that air does not deflect off the front of a projectile in the same direction, when there are of differing shapes. This prompted the introduction of a second factor to b, the coefficient of form (i). This is particularly true at high velocities, greater than 830 ft/s (253 m/s). Hence, Bashforth introduced the "undetermined multiplier" of any power called the k factor that compensate for this unknown effects of drag above 830 ft/s (253 m/s); $k > i$. Bashforth then integrated k and i as K_v.

Although Bashforth did not conceive the "restricted zone", he showed mathematically there were 5 restricted zones. Bashforth did not propose a standard projectile, but was well aware of the concept.

The Mayevski/Siacci Method

In 1872, General Mayevski published his report *Trité Balistique Extérieure*, which included the Mayevski model. Using his ballistic tables along with Bashforth's tables from the 1870 report, Mayevski created an analytical math formula that calculated the air resistances of a projectile in terms of log A and the value *n*. Although Mayevski's math used a differing approach than Bashforth, the resulting calculations of air resistance was the same. Mayevski proposed the restricted zone concept and found there to be 6 restricted zones for projectiles.

Circa 1886, General Mayevski published the results from a discussion of experiments made by M. Krupp (1880). Though the ogival-headed projectiles used varied greatly in caliber, they had essentially the same proportions as the standard projectile, being mostly 3 caliber in length, with an ogive of 2 calibers radius. Giving the standard projectile dimensionally as 10 cm (3.9 in) and 1 kg (2.2 lb).

In 1880, Colonel Francesco Siacci published his work "Balistica". Siacci found as did those who came before him that the resistance and density of the air becomes greater and greater as a projectile displaced the air at higher and higher velocities.

Siacci's method was for flat-fire trajectories with angles of departure of less than 20 degrees. He found that the angle of departure is sufficiently small to allow for air density to remain the same and was able to reduce the ballistics tables to easily tabulated quadrants giving distance, time, inclination and altitude of the projectile. Using Bashforth's *k* and Mayevski's tables, Siacci created a 4 zone model. Siacci used Mayevski's standard projectile. From this method and standard projectile, Siacci formulated a short cut.

Siacci found that within a low velocity restricted zone, projectiles of similar shape, and velocity in the same air density behave similar; $\frac{\delta w}{d^2}$ or $\frac{\delta}{C}$. Siacci used the variable C for ballistic coefficient. Meaning, air density is the generally the same for flat-fire trajectories, thus sectional density is equal to the ballistic coefficient and air density can be dropped. Then as the velocity rises to Bashforth's k for high velocity when C requires the introduction of i. Following within today's currently used ballistic trajectory tables for an average ballistic coeficient: $\frac{m}{d^2} \cdot \frac{P_0}{P}$ would equal $\frac{m}{d^2 i}$ equals $\frac{SD}{i}$ as BC.

Siacci wrote that within any restricted zone, C being the same for two or more projectiles, the trajectories differences will be minor. Therefore, C agrees with an average curve, and this average curve applies for all projectiles. Therefore, a single trajectory can be computed for the standard projectile without having to resort to tedious calculus methods, and then a trajectory for any actual bullet with known C can be computed from the standard trajectory with just simple algebra.

The Ballistic Tables

The aforementioned ballistics tables are generally: functions, air density, projectile time at range, range, degree of projectile departure, weight and diameter to facilitate the calculation of ballistic formulae. These formulae produce the projectile velocity at range, drag and trajectories. The modern day commercially published ballistic tables or software computed ballistics tables for small arms, sporting ammunition are exterior ballistic, trajectory tables.

The 1870 Bashforth tables were to 2,800 ft/s (853 m/s). Mayevski, using his tables, supplemented by the Bashforth tables (to 6 restricted zones) and the Krupp tables. Mayevski conceived a 7th restricted zone and extended the Bashforth tables to 1,100 m/s (3,609 ft/s). Mayevski converted Bashforth's data from Imperial units of measure to metric units of measure (now in SI units of measure). In 1884, James Ingalls published his tables in the U.S. Army Artillery Circular M using the Mayevski tables. Ingalls extended Mayevski's ballistics tables to 5,000 ft/s (1,524 m/s) within an 8th restricted zone, but still with the same n value (1.55) as Mayevski's 7th restricted zone. Ingalls, converted Mayevski's results back to Imperial units. The British Royal Artillery results were very similar to those of Mayevski's and extended their tables to 5,000 ft/s (1,524 m/s) within the 8th restricted zone changing the n value from 1.55 to 1.67. These ballistic tables were published in 1909 and almost identical to those of Ingalls. In 1971 the Sierra Bullet company calculated their ballistic tables to 9 restricted zones but only within 4,400 ft/s (1,341 m/s).

The G Model

In 1881, the Commission d'Experience de Gâvre did a comprehensive survey of data available from their tests as well as other countries. After adopting a standard atmospheric condition for the drag data the Gavre drag function was adopted. This drag function was known as the Gavre function and the standard projectile adopted was the Type 1 projectile. Thereafter, the Type 1 standard projectile was renamed by Ballistics Section of Aberdeen Proving Grounds in Maryland, USA as G_1 after the Commission d'Experience de Gâvre. For practical purposes the subscript 1 in G_1 is generally written in normal font size as G1.

The general form for the calculations of trajectory adopted for the G model is the Siacci method. The standard model projectile is a "fictitious projectile" used as mathematical basis for the calculation of actual projectile's trajectory when an initial velocity is known. The G1 model projectile adopted is in dimensionless measures of 2 caliber radius ogival-head and 3.28 caliber in length. By calculation this leaves the body length 1.96 caliber and head, 1.32 caliber long.

Over the years there has been some confusion as to adopted size, weight and radius ogival-head of the G1 standard projectile. This misconception may be explained by Colonel Ingalls in the 1886 publication, Exterior Ballistics in the Plan Fire; page 15, *In the following tables the first and second columns give the velocities and corresponding resistance, in pounds, to an elongated one inch in diameter and having an ogival head of one and a half calibers. They were deduced from Bashforth's experiments by Professor A. G. Greenhill, and are taken from his papers published in the Proceedings of the Royal Artillery Institution, No 2, Vol. XIII.* Further it is discussed that said projectile's weight was one pound.

For the purposes of mathematical convenience for any standard projectile (G) the *BC* is 1.00. Where as the projectile's sectional density (SD) is dimensionless with a mass of 1 divided by the square of the diameter of 1 caliber equaling an SD of 1. Then the standard projectile is assigned a coefficient of form of 1. Following that $BC = \frac{SD}{i} = \frac{1}{1} = 1.00$. *BC*, as a general rule, within flat-fire trajectory, is carried out to 2 decimal points. *BC* is commonly found within commercial publications to be carried out to 3 decimal points as few sporting, small arms projectiles rise to the level of 1.00 for a ballistic coefficient.

When using the Siacci method for different G models, the formula used to compute the trajectories is the same. What differs is retardation factors found through testing of actual projectiles that are similar in shape to the standard project reference. This creates slightly different set of retardation factors between differing G models. When the correct G model retardation factors are applied within the Siacci mathematical formula for the same G model *BC*, a corrected trajectory can be calculated for any G model.

Another method of determining trajectory and ballistic coefficient was developed and published by Wallace H. Coxe and Edgar Beugless of DuPont in 1936. This method is by shape comparison an logarithmic scale as drawn on 10 charts. The method estimates the ballistic coefficient related to the drag model of the Ingalls tables. When matching an actual projectile against the drawn caliber radii of Chart No. 1, it will provide *i* and by using Chart No. 2, *C* can be quickly calculated. Coxe and Beugless used the variable *C* for ballistic coefficient.

The Siacci method was abandoned by the end of the World War I for artillery fire. But the U.S. Army Ordnance Corps continued using the Siacci method into the middle of the 20th century for direct (flat-fire) tank gunnery. The development of the electromechanical analog computer contributed to the calculation of aerial bombing trajectories during World War II. After World War II the advent of the silicon semiconductor based digital computer made it possible to create trajectories for the guided missiles/bombs, intercontinental ballistic missiles and space vehicles.

Between World War I and II the U.S. Army Ballistics research laboratories at Aberdeen Proving Grounds, Maryland, USA developed the standard models for G2, G5, G6. In 1965, Winchester Western published a set of ballistics tables for G1, G5, G6 and GL. In 1971 Sierra Bullet Company retested all their bullets and concluded that the G5 model was not the best model for their boat tail bullets and started using the G1 model. This was fortunate, as the entire commercial sporting and firearms industries had based their calculations on the G1 model. The G1 model and Mayevski/Siacci Method continue to be the industry standard today. This benefit allows for comparison of all ballistic tables for trajectory within the commercial sporting and firearms industry.

In recent years there have been vast advancements in the calculation of flat-fire trajectories with the advent of Doppler radar and the personal computer and handheld computing devices. Also, the newer methodology proposed by Dr. Arthur Pejsa and the use of the G7 model used by Mr. Brian Litz, ballistic engineer for Berger Bullets, LLC for calculating boat tailed spitzer rifle bullet trajectories and 6 Dof model based software have improved the prediction of flat-fire trajectories.

Differing Mathematical Models and Bullet Ballistic Coefficients

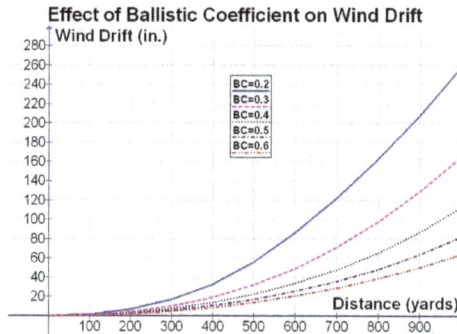

Effect of Ballistic Coefficient on Wind Drift

Wind drift calculations for rifle bullets of differing G1 BCs fired with a muzzle |velocity of 2,950 ft/s (900 m/s) in a 10 mph (16 km/h) crosswind.

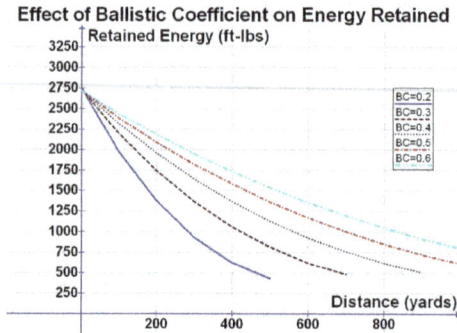

Effect of Ballistic Coefficient on Energy Retained

Energy calculations for 9.1 grams (140 gr) rifle bullets of differing G1 BCs fired with a muzzle velocity of 2,950 feet per second (900 m/s).

Most ballistic mathematical models and hence tables or software take for granted that one specific drag function correctly describes the drag and hence the flight characteristics of a bullet related to its ballistic coefficient. Those models do not differentiate between wadcutter, flat-based, spitzer, boat-tail, very-low-drag, etc. bullet types or shapes. They assume one invariable drag function as indicated by the published BC. Several different drag curve models optimized for several standard projectile shapes are available, however.

The resulting drag curve models for several standard projectile shapes or types are referred to as:

- G1 or Ingalls (flatbase with 2 caliber (blunt) nose ogive - by far the most popular)

- G2 (Aberdeen J projectile)

- G5 (short 7.5° boat-tail, 6.19 calibers long tangent ogive)

- G6 (flatbase, 6 calibers long secant ogive)

- G7 (long 7.5° boat-tail, 10 calibers tangent ogive, preferred by some manufacturers for very-low-drag bullets)

- G8 (flatbase, 10 calibers long secant ogive)

- GL (blunt lead nose)

Since these standard projectile shapes differ significantly the Gx BC will also differ significantly from the Gy BC for an identical bullet. To illustrate this the bullet manufacturer Berger has published the G1 and G7 BCs for most of their target, tactical, varmint and hunting bullets. Other bullet manufacturers like Lapua and Nosler also published the G1 and G7 BCs for most of their target bullets. How much a projectile deviates from the applied reference projectile is mathematically expressed by the form factor (i). The applied reference projectile shape always has a form factor (i) of exactly 1. When a particular projectile has a sub 1 form factor (i) this indicates that the particular projectile exhibits lower drag than the applied reference projectile shape. A form factor (i) greater than 1 indicates the particular projectile exhibits more drag than the applied reference projectile shape. In general the G1 model yields comparatively high BC values and is often used by the sporting ammunition industry.

The Transient Nature of Bullet Ballistic Coefficients

Variations in BC claims for exactly the same projectiles can be explained by differences in the ambient air density used to compute specific values or differing range-speed measurements on which the stated G1 BC averages are based. Also, the BC changes during a projectile's flight, and stated BCs are always averages for particular range-speed regimes.

For the precise establishment of BCs (or perhaps the scientifically better expressed drag coefficients), Doppler radar-measurements are required. The normal shooting or aerodynamics enthusiast, however, has no access to such expensive professional measurement devices. Weibel 1000e or Infinition BR-1001 Doppler radars are used by governments, professional ballisticians, defense forces, and a few ammunition manufacturers to obtain exact real world data on the flight behavior of projectiles of interest.

Doppler radar measurement results for a lathe turned monolithic solid .50 BMG very-low-drag bullet (Lost River J40 13.0 millimetres (0.510 in), 50.1 grams (773 gr) monolithic solid bullet / twist rate 1:380 millimetres (15 in)) look like this:

Range (m)	500	600	700	800	900	1000	1100	1200
Ballistic coefficient (G1)	1.040	1.051	1.057	1.063	1.064	1.067	1.068	1.068

Range (m)	1300	1400	1500	1600	1700	1800	1900	2000
Ballistic coefficient (G1)	1.068	1.066	1.064	1.060	1.056	1.050	1.042	1.032

The initial rise in the BC value is attributed to a projectile's always present yaw and precession out of the bore. The test results were obtained from many shots, not just a single shot. The bullet was assigned 1.062 for its BC number by the bullet's manufacturer, Lost River Ballistic Technologies.

Measurements on other bullets can give totally different results. How different speed regimes affect several 8.6 mm (.338 in calibre) rifle bullets made by the Finnish ammunition manufacturer Lapua can be seen in the .338 Lapua Magnum product brochure which states Doppler radar established BC data.

General Trends

Sporting bullets, with a calibre d ranging from 4.4 to 12.7 millimetres (0.172 to 0.50 in), have BCs in the range 0.12 to slightly over 1.00. Those bullets with the higher BCs are the most aerodynamic, and those with low BCs are the least. Very-low-drag bullets with BCs ≥ 1.10 can be designed and produced on CNC precision lathes out of mono-metal rods, but they often have to be fired from custom made full bore rifles with special barrels.

Ammunition makers often offer several bullet weights and types for a given cartridge. Heavy-for-caliber pointed (spitzer) bullets with a boattail design have BCs at the higher end of the normal range, whereas lighter bullets with square tails and blunt noses have lower BCs. The 6 mm and 6.5 mm cartridges are probably the most well known for having high BCs and are often used in long range target matches of 300 m (328 yd) – 1,000 m (1,094 yd). The 6 and 6.5 have relatively light recoil compared to high BC bullets of greater caliber and tend to be shot by the winner in matches where accuracy is key. Examples include the 6mm PPC, 6mm Norma BR, 6x47mm SM, 6.5×55mm Swedish Mauser, 6.5×47mm Lapua, 6.5 Creedmoor, 6.5 Grendel, .260 Remington, and the 6.5-284. The 6.5 mm is also a popular hunting caliber in Europe.

In the United States, hunting cartridges such as the .25-06 Remington (a 6.35 mm caliber), the .270 Winchester (a 6.8 mm caliber), and the .284 Winchester (a 7 mm caliber) are used when high BCs and moderate recoil are desired. The .30-06 Springfield and .308 Winchester cartridges also offer several high-BC loads, although the bullet weights are on the heavy side.

In the larger caliber category, the .338 Lapua Magnum and the .50 BMG are popular with very high BC bullets for shooting beyond 1,000 meters. Newer chamberings in the larger caliber category are the .375 and .408 Cheyenne Tactical and the .416 Barrett.

Information Sources

For many years, bullet manufacturers were the main source of ballistic coefficients for use in trajectory calculations. However, in the past decade or so, it has been shown that ballistic coefficient measurements by independent parties can often be more accurate than manufacturer specifications. Since ballistic coefficients depend on the specific firearm and other conditions that vary, it is notable that methods have been developed for individual users to measure their own ballistic coefficients.

Satellites and Reentry Vehicles

Satellites in Low Earth Orbit (LEO) with high ballistic coefficients experience smaller perturbations to their orbits due to atmospheric drag.

The ballistic coefficient of an atmospheric reentry vehicle has a significant effect on its behavior. A very high ballistic coefficient vehicle would lose velocity very slowly and would impact the Earth's surface at higher speeds. In contrast a low ballistic coefficient would reach subsonic speeds before reaching the ground.

In general, reentry vehicles that carry human beings back to Earth from space have high drag and a correspondingly low ballistic coefficient. Vehicles that carry nuclear weapons launched by an intercontinental ballistic missile (ICBM), by contrast, have a high ballistic coefficient, which enables them to travel rapidly from space to a target on land. That makes the weapon less affected by crosswinds or other weather phenomena, and harder to track, intercept, or otherwise defend against.

Deflection (Ballistics)

Deflection is a technique used for effectively propelling a projectile at a moving target, also known as "leading the target", i.e. shooting ahead of a moving target so that the target and projectile will collide. This technique is only necessary when the target will have moved a sufficient distance to fully displace its position during the time the projectile would take to reach the target's range. This can become the case over long distances (e.g. a distant target for a skilled sniper), due to fast moving targets (e.g. an opposing aircraft in an aerial dogfight), or while using relatively slow projectiles (e.g. a crossbow bolt or a basketball thrown to a running teammate). During World War II, U.S. Navy pilots were taught explicitly about the concept in order to capitalize on the advantages of the F4F Wildcat.

Modern day fighter aircraft have automated deflection sights, where a computer calculates lead and projects the solution onto a heads up display (HUD). The visual assistance with targeting the gun is offset by the enormous speed and agility of modern aircraft, compared to the days when targeting was less advanced.

In artillery, deflection is also used against fixed targets to compensate for windage and range. Due to the Earth's rotation, surface points have different velocities and curved motion, leading to apparent Coriolis drift of a long-range target.

Leading targets is the practice of aiming one's weapon ahead of the target so that the projectile will hit its mark. Over reasonably short ranges, leading is typically unnecessary when using firearms, but it is still relevant for sniping where the bullet may take a second or more to reach its target, as well as for weapons such as bows that use lower-velocity projectiles. It is generally unnecessary for guided projectiles, although the autonomous guiding mechanism may be designed to calculate a flight path to lead its targets on its own to ensure an interception.

Elevation (Ballistics)

In ballistics, the elevation is the angle between the horizontal plane and the direction of the barrel of a gun, mortar or heavy artillery. Originally, elevation was a *linear* measure of how high the gunners had to physically lift the muzzle of a gun up from the gun carriage to hit targets at a certain distance.

Pre-WWI and WWI

Though early 20th-century firearms were relatively easy to fire, artillery was not. Before and during World War I, the only way to effectively fire artillery was plotting points on a plane.

Most artillery units seldom employed their cannons in small numbers. Instead of using pin-point artillery firing they used old means of "fire for effect" using artillery en masse. This tactic was employed successfully by past armies.

But changes have been made since past wars and in World War I, artillery was more accurate than before, although not as accurate as modern artillery guns. The tactics of artillery from previous wars were carried on, and still had similar success. Warships and battleships also carried large caliber guns that needed to be elevated to certain degrees to accurately hit targets, and they also had the similar drawbacks of land artillery.

WWII and Beyond

As time passed on, more accurate artillery guns were made, and they came in different varieties. Small artillery pieces were used as mortars, medium sized artillery guns became tank guns, and the largest artillery guns became long range land batteries and battleship armaments.

With the introduction of better tanks in World War II, elevation was once again a problem for tank gunners, which had to aim through the Gunner's Auxiliary Sights (GAS) or even through iron sights. Though the problem was not that evident as tanks fired rounds at a higher velocity than normal artillery, making aiming less of a hassle.

As with World War I, World War II artillery was almost like its old counterpart. But in the war came the introduction of the FCS or the fire-control system, which made firing artillery accurately easier.

With the advancements in the 21st century, it has become easy to determine how much elevation a gun needed to hit a target. The laser rangefinder is a modern component of FCS, and can accurately determine the range of the target, thereby calculating how much elevation the gun needs, making modern day guns highly accurate.

Ballistic Pendulum

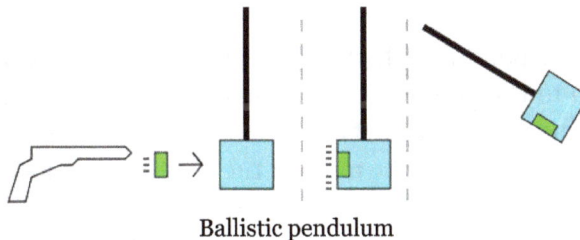

Ballistic pendulum

A ballistic pendulum is a device for measuring a bullet's momentum, from which it is possible to calculate the velocity and kinetic energy. Ballistic pendulums have been largely rendered obsolete by modern chronographs, which allow direct measurement of the projectile velocity.

Although the ballistic pendulum is considered obsolete, it remained in use for a significant length of time and led to great advances in the science of ballistics. The ballistic pendulum is still found in physics classrooms today, because of its simplicity and usefulness in demonstrating properties of momentum and energy. Unlike other methods of measuring the speed of a bullet, the basic calculations for a ballistic pendulum do not require any measurement of time, but rely only on measures of mass and distance.

In addition its primary uses of measuring the velocity of a projectile or the recoil of a gun, the ballistic pendulum can be used to measure any transfer of momentum. For example, a ballistic pendulum was used by physicist C. V. Boys to measure the elasticity of golf balls, and by physicist Peter Guthrie Tait to measure the effect that spin had on the distance a golf ball traveled.

History

The ballistic pendulum was invented in 1742 by English mathematician Benjamin Robins (1707–1751), and published in his book *New Principles of Gunnery*, which revolutionized the science of ballistics, as it provided the first way to accurately measure the velocity of a bullet.

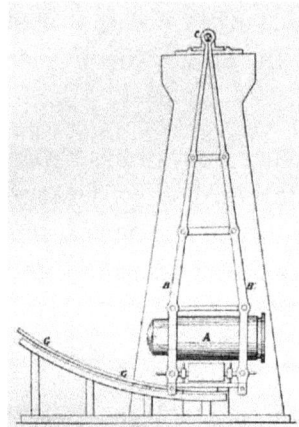

Ballistic pendulum (1911)

Robins used the ballistic pendulum to measure projectile velocity in two ways. The first was to attach the gun to the pendulum, and measure the recoil. Since the momentum of the gun is equal to the momentum of the ejecta, and since the projectile was (in those experiments) the large majority of the mass of the ejecta, the velocity of the bullet could be approximated. The second, and more accurate method, was to directly measure the bullet momentum by firing it into the pendulum. Robins experimented with musket balls of around one ounce in mass (30 g), while other contemporaries used his methods with cannon shot of one to three pounds (0.5 to 1.4 kg).

Robins' original work used a heavy iron pendulum, faced with wood, to catch the bullet. Modern reproductions, used as demonstrations in physics classes, generally use a heavy weight suspended by a very fine, lightweight arm, and ignore the mass of the pendulum's arm. Robins' heavy iron pendulum did not allow this, and Robins' mathematical approach was slightly more complex. He used the period of oscillation and mass of the pendulum (both measured with the bullet included) to calculate the rotational inertia of the pendulum, which was then used in the calculations. Robins also used a length of ribbon, loosely gripped in a clamp, to measure the travel of the pendulum. The pendulum would draw out a length of ribbon equal to the chord of pendulum's travel.

The first system to supplant ballistic pendulums with direct measures of projectile speed was invented in 1808, during the Napoleonic Wars and used a rapidly rotating shaft of known speed with two paper disks on it; the bullet was fired through the disks, parallel to the shaft, and the angular difference in the points of impact provided an elapsed time over the distance between the disks. A direct electromechanical clockwork measure appeared in 1840, with a spring-driven clock started and stopped by electromagnets, whose current was interrupted by the bullet passing through two meshes of fine wires, again providing the time to traverse the given distance.

Mathematical Derivations

Most physics textbooks provide a simplified method of calculation of the bullet's velocity that uses the mass of the bullet and pendulum and the height of the pendulum's travel to calculate the amount of energy and momentum in the pendulum and bullet system. Robins' calculations were significantly more involved, and used a measure of the period of oscillation to determine the rotational inertia of the system.

Simple Derivation

We begin with the motion of the bullet-pendulum system from the instant the pendulum is struck by the bullet.

Given g, the acceleration due to gravity, and h, the final height of the pendulum, it is possible to calculate the initial velocity of the bullet-pendulum system using conservation of mechanical energy (kinetic energy + potential energy). Let this initial velocity be denoted by v_1. Suppose the masses of the bullet and pendulum are m_b and m_p respectively.

The initial kinetic energy of the system $K_{initial} = \dfrac{1}{2}(m_b + m_p) \cdot v_1^2$

Taking the initial height of the pendulum as the potential energy reference ($U_{initial} = 0$), the final potential energy when the bullet-pendulum system comes to a stop ($K_{final} = 0$) is given by $U_{final} = (m_b + m_p) \cdot g \cdot h$

So, by the conservation of mechanical energy, we have:

$$K_{initial} = U_{final}$$

$$\frac{1}{2}(m_b + m_p) \cdot v_1^2 = (m_b + m_p) \cdot g \cdot h$$

Solve for velocity to obtain: $v_1 = \sqrt{2 \cdot g \cdot h}$

We can now use momentum conservation for the bullet-pendulum system to get the speed of the bullet, v_0, before it struck the pendulum. Equating the momentum of the bullet before it is fired to that of the bullet-pendulum system as soon as the bullet strikes the pendulum (and using $v_1 = \sqrt{2 \cdot g \cdot h}$ from above), we get:

$$m_b \cdot v_0 = (m_b + m_p) \cdot \sqrt{2 \cdot g \cdot h}$$

Solving for v_0:

$$v_0 = \frac{(m_b + m_p) \cdot \sqrt{2 \cdot g \cdot h}}{m_b} = (1 + \frac{m_p}{m_b}) \cdot \sqrt{2 \cdot g \cdot h}$$

Robins' Formula

Robins' original book had some omitted assumptions in the formula; for example, it did not include a correction to account for a bullet impact that did not match the center of mass of the pendulum. An updated formula, with this omission corrected, was published in the Philosophical Transactions of the Royal Society the following year. Swiss mathematician Leonhard Euler, unaware of this correction, independently corrected this omission in his annotated German translation of the book. The corrected formula, appearing in a 1786 edition of the book, was:

$$v = 614.58 gc \cdot \frac{p+b}{birn}$$

where:

- v is the velocity of the ball in units per second

- b is the mass of the ball

- p is the mass of the pendulum

- g is the distance from pivot to the center of gravity

- i is the distance from pivot to the point of the ball's impact

- c is the chord, as measured by the ribbon described in Robins' apparatus

- r is the radius, or distance from the pivot the attachment of the ribbon

- n is the number of oscillations made by the pendulum in one minute

Robins used feet for length and ounces for mass, though other units, such as inches or pounds, may be substituted as long as consistency is maintained.

Poisson's Formula

A rotational inertia based formula similar to Robins' was derived by French mathematician Siméon Denis Poisson and published in *The Mécanique Physique*, for measuring the bullet velocity by using the recoil of the gun:

$$mvcf = Mbk'\sqrt{gh}$$

where:

- m is the mass of the bullet

- v is the velocity of the bullet

- c is the distance from pivot to the ribbon

- f is the distance from bore axis to pivot point

- M is the combined mass of gun and pendulum

- b is the chord measured by the ribbon

- k' is the radius from pivot to the center of mass of gun and pendulum (measured by oscillation, as per Robins)

- g is gravitational acceleration

- h is the distance from the center of mass of the pendulum to the pivot

k' can be calculated with the equation:

$$T = \pi \sqrt{\frac{k'^2}{gh}}$$

Where T is half the period of oscillation.

Ackley's Ballistic Pendulum

P. O. Ackley, a gunsmith, author, and researcher, described how to construct and use a ballistic pendulum in his 1962 Handbook for Shooters & Reloaders, Volume I. Ackley's pendulum used a parallelogram linkage, with a standardized size that allowed a simplified means of calculating the velocity.

Ackley's pendulum used pendulum arms of exactly 66.25 inches (168.3 cm) in length, from bearing surface to bearing surface, and used turnbuckles located in the middle of the arms to provide a means of setting the arm length precisely. Ackley recommends masses for the body of the pendulum for various calibers as well; 50 pounds (22.7 kg) for rimfire up through the .22 Hornet, 90 pounds (40.9 kg) for .222 Remington through .35 Whelen, and 150 pounds (68.2 kg) for magnum rifle calibers. The pendulum is made of heavy metal pipe, welded shut at one end, and packed with paper and sand to stop the bullet. The open end of the pendulum was covered in a sheet of rubber, to allow the bullet to enter and prevent material from leaking out.

To use the pendulum, it is set up with a device to measure the horizontal distance of the pendulum swing, such as a light rod that would be pushed backwards by the rear of the pendulum as it moved. The shooter is seated at least 15 feet (5 m) back from the pendulum (reducing the effects of muzzle blast on the pendulum) and a bullet is fired into the pendulum. To calculate the velocity of the bullet given the horizontal swing, the following formula is used:

$$V = \frac{Mp}{Mb} 0.2018D$$

where:

- V is the velocity of the bullet, in feet per second

- Mp is the mass of the pendulum, in grains

- Mb is the mass of the bullet, in grains

- D is the horizontal travel of the pendulum, in inches

For more accurate calculations, a number of changes are made, both to the construction and the use of the pendulum. The construction changes involve the addition of a small box on top of the pendulum. Before weighing the pendulum, the box is filled with a number of bullets of the type being measured. For each shot made, a bullet can be removed from the box, thus keeping the mass of the pendulum constant. The measurement change involves measuring the period of the pendulum. The pendulum is swung, and the number of complete oscillations is measured over a long period of time, five to ten minutes. The time is divided by the number of oscillations to obtain the period. Once this is done, the formula $C = \frac{pi}{T12}$ generates a more precise constant to replace the value 0.2018 in the above equation. Just like above, the velocity of the bullet is calculated using the formula:

$$V = \frac{Mp}{Mb} CD$$

Handgun Effectiveness

Handgun effectiveness is a measure of the stopping power of a handgun: its ability to incapacitate a hostile target as quickly and efficiently as possible.

Overview

Most handgun projectiles have significantly lower energy than centerfire rifles. What they lack in power, they make up for it in light weight, small size, concealability and practicality. The lack of power they possess, and caliber/bullet effectiveness, are widely debated topics with growing experimental research among civilians, law enforcement, ammunition companies, and the military. Factors that can influence handgun effectiveness include handgun design, bullet types and bullet capabilities (e.g. wound mechanisms, penetration, velocity).

Factors

Cavitation

Most handgun projectiles wound primarily through the size of the hole they produce. This hole is known as a permanent cavity. For comparison, rifles wound through temporary cavitation as well as permanent cavitation. A temporary cavity is also known as a stretch cavity. This is because it acts to stretch the permanent cavity, increasing the wounding potential. The potential for wounding via temporary cavity depends on the elasticity of the tissue, bullet fragmentation, and the rate of energy transfer. Many handgun bullets do not create significant wounding via temporary cavitation, but the potential is there if the bullet fragments, strikes inelastic tissue (liver, spleen, kidneys, CNS), or if the bullet transfers over 500 ft·lbf (680 J) of energy per foot of penetration. These phenomena are unrelated to low-pressure cavitation in liquids.

Penetration

There are many factors used to measure a handgun's effectiveness. One of them is penetration. The FBI's requirement for all service rounds is 12 inches (30 cm) penetration or greater in calibrated ballistic gelatin. This generally ensures a bullet will reach the vital organs from most angles. Penetration is arguably the most important factor in handgun wounding potential, because the vital areas must be destroyed or damaged to incapacitate.

Ballistic Pressure Wave/Hydrostatic Shock

There is a significant body of evidence that Hydrostatic shock (more precisely known as the ballistic pressure wave) can contribute to handgun bullet effectiveness.

Recent work published by scientists M Courtney and A Courtney provides compelling support for the role of a ballistic pressure wave in incapacitation and injury. This work builds upon the earlier works of Suneson et al. where the researchers implanted high-speed pressure transducers into the brain of pigs and demonstrated that a significant pressure wave reaches the brain of pigs shot in the thigh. These scientists observed neural damage in the brain caused by the distant effects of the ballistic pressure wave originating in the thigh.

The results of Suneson et al. were confirmed and expanded upon by a later experiment in dogs which "confirmed that distant effect exists in the central nervous system after a high-energy missile impact to an extremity. A high-frequency oscillating pressure wave with large amplitude and short duration was found in the brain after the extremity impact of a high-energy missile . . ." Wang et al. observed significant damage in both the hypothalamus and hippocampus regions of the brain due to remote effects of the ballistic pressure wave.

Damage to other nerves than those in the brain may lead to partial incapacitation (inability to walk or use the arms).

Caliber

Another factor is expansion and caliber. Many civilians and practically all law enforcement agencies use jacketed hollow point or some form of expanding ammunition. This increases the chance of a handgun bullet striking a vital organ, and increases blood loss. Because of this, two different calibers could theoretically produce almost identical incapacitation results, provided the two penetrate the same area, and the small caliber expands to the size of the larger.

It is generally agreed that most intermediate handgun calibers will perform similarly, since their wounding principles are the same.

One-shot Stops

The only scientifically proven and biologically possible way to guarantee instant incapacitation is through destruction of the central nervous system or brain. This will usually cease all motor-related and voluntary actions. If the central nervous system is not damaged or destroyed, there will be no immediate incapacitation. To allow room for error, since a central nervous system hit is very unlikely, most people use expanding ammunition. This will increase the odds of striking a part of the central nervous system, and cause faster blood loss.

For example, a popular caliber in the United States is the .45 ACP. It is the largest practical handgun caliber in use, featuring a bullet that is .452 inches in diameter. With well made expanding ammunition, a .452 bullet often expands to .70 caliber or larger. With a 9 mm Luger bullet, for example, its normal .355 diameter might be hoped to expand to .50 caliber or larger. This could give a preference for larger caliber bullets, as they do not rely on expansion as much as smaller caliber bullets do to provide incapacitation. However, multiple tests and actual data from shootings have not found this to be true.

The most popular round in the United States is the 9 mm Luger. The variety of handguns and ammunition available for this round is much higher than any other caliber. The amount of energy delivered to a body is dependent on both the weight of the bullet and its velocity as basic physics dictates. Thus NATO prefers the 9 mm Luger parabellum to a larger but slower round such as the .45 ACP

Bullet expansion in handguns is desirable not solely for incapacitation, but also so the bullet will not exit the target. An expanding bullet will stop in the target and "dump" all its energy there, rather than overpenetrating and possibly endangering people behind the target. Since all handgun rounds are marginal at best, the one with the most energy and which expends all that energy in a target is the one that is most effective.

Rifleman's Rule

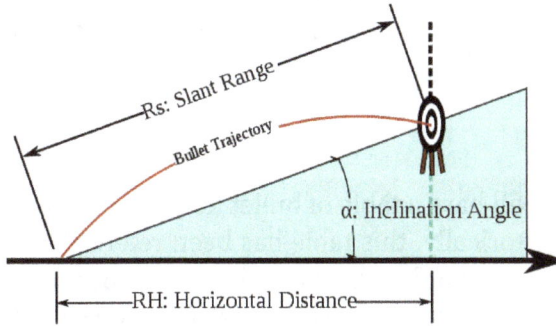

Illustration of the Shooting Scenario.

Rifleman's rule is a "rule of thumb" that allows a rifleman to accurately fire a rifle that has been calibrated for horizontal targets at uphill or downhill targets. The rule provides an equivalent horizontal range setting for engaging a target at a known uphill or downhill distance from the rifle (called the slant range). For a bullet to strike a target at a slant range of R_S and an incline of α, the rifle sight must be adjusted as if the shooter were aiming at a horizontal target at a range of $R_H = R_S \cos(\alpha)$. Figure above illustrates the shooting scenario. The rule holds for inclined and declined shooting (all angles measured with respect to horizontal). Very precise computer modeling and empirical evidence suggests that the rule does appear to work with reasonable accuracy in air and with both bullets and arrows.

Background

Definitions

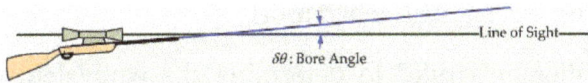

Illustration of a Rifle Showing Line of Sight and Bore Angle.

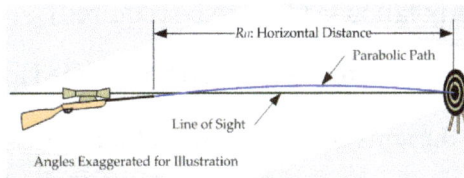

Illustration of a Rifle Showing the LOS and Bore Angle.

There is a device that is mounted on the rifle called a sight. While there are many forms of rifle sight, they all permit the shooter to set the angle between the bore of the rifle and the line of sight (LOS) to the target. Figure illustrates the relationship between the LOS and bore angle.

This relationship between the LOS to the target and the bore angle is determined through a process called "zeroing." The bore angle is set to ensure that a bullet on a parabolic trajectory will intersect the LOS to the target at a specific range. A properly adjusted rifle barrel and sight are said to be "zeroed." Figure illustrates how the LOS, bullet trajectory, and range (R_H) are related.

Procedure

In general, the shooter will have a table of bullet heights with respect to the LOS versus horizontal distance. Historically, this table has been referred to as a "drop table." The drop table can be generated empirically using data taken by the shooter at a rifle range; calculated using a ballistic simulator; or is provided by the rifle/cartridge manufacturer. The drop values are measured or calculated assuming the rifle has been zeroed at a specific range. The bullet will have a drop value of zero at the zero range. Table gives a typical example of a drop table for a rifle zeroed at 100 meters.

Table: Example Bullet Drop Table

Range (meters)	0	100	200	300	400	500
Bullet Height (cm)	-1.50	0.0	-2.9	-11.0	-25.2	-46.4

If the shooter is engaging a target on an incline and has a properly zeroed rifle, the shooter goes through the following procedure:

1. Determine the slant range to the target (measurement can be performed using various forms of range finders, e.g. laser rangefinder)

2. Determine the elevation angle of the target (measurement can be made using various devices, e.g. sight attached unit)

3. Apply the "rifleman's rule" to determine the equivalent horizontal range ($R_H = R_S \cos(\alpha)$)

4. Use the bullet drop table to determine the bullet drop over that equivalent horizontal range (interpolation is likely to be required)

5. Compute the bore angle correction that is to be applied to the sight. The correction is computed using the equation angle correction $= -\dfrac{\text{bullet drop}}{R_H}$ (in radians).

6. Adjust the bore angle by the angle correction.

Example

Assume a rifle is being fired that shoots with the bullet drop table given in the table above. This means that the rifle sight setting for any range from 0 to 500 meters is available. The sight adjustment procedure can be followed step-by-step.

1. Determine the slant range to the target.

Assume that a range finder is available that determines that the target is exactly 300 meters distance.

2. Determine the elevation angle of the target.

Assume that an angle measurement tool is used that measures the target to be at an angle of $20°$ with respect to horizontal.

3. Apply the rifleman's rule to determine the equivalent horizontal range.

$$R_H = 300 \text{ meters} \cos(20°) = 282 \text{ meters}$$

4. Use the bullet drop table to determine the bullet drop over that equivalent horizontal range.

Linear interpolation can be used to estimate the bullet drop as follows:

$$\text{bullet drop} = \frac{-11.0 \cdot (282 - 200) + -2.9 \cdot (300 - 282)}{300 - 200} = -9.5 \text{ cm}$$

5. Compute the bore angle correction that is to be applied to the sight.

$$\text{angle correction} = -\frac{-9.5 \text{ cm}}{282 \text{ meters}} = 0.00094 \text{ radians} = 3.2' \text{ (minutes of angle)}$$

6. Adjust the bore angle by the angle correction.

The gun sight is adjusted up by 3.2' in order to compensate for the bullet drop. The gunsights are usually adjustable in unit of minutes, half minutes, or quarter minutes of angle.

Analysis

Zeroing the Rifle

Let $\delta\theta$ be the bore angle required to compensate for the bullet drop caused by gravity. Standard practice is for the shooter to zero their rifle at a standard range, such as 100 or 200 meters. Once the rifle is zeroed, adjustments to $\delta\theta$ are made for other ranges relative to this zero setting. One can calculate $\delta\theta$ using standard Newtonian dynamics as follows.

Two equations can be set up that describe the bullet's flight in a vacuum, (presented for computational simplicity compared to solving equations describing trajectories in an atmosphere).

$$x(t) = v_{bullet} \cos(\delta\theta)t \quad \text{(Equation 1)}$$

$$y(t) = v_{bullet} \sin(\delta\theta)t - \frac{1}{2}gt^2 \quad \text{(Equation 2)}$$

Solving Equation 1 for t yields Equation 3.

$$t = \frac{x}{v_{bullet} \cos(\delta\theta)} \quad \text{(Equation 3)}$$

Equation 3 can be substituted in Equation 2. The resulting equation can then be solved for x assuming that $y = 0$ and $t \neq 0$, which produces Equation 4.

$$y(t) = 0 = \left(v_{bullet} \sin(\delta\theta) - \frac{1}{2}gt \right)t$$

$$0 = v_{bullet} \sin(\delta\theta)t - \frac{1}{2}gt^2$$

$$v_{bullet} \sin(\delta\theta) = \frac{1}{2}gt = \frac{1}{2}g\frac{x}{v_{bullet} \cos(\delta\theta)}$$

$$x = \frac{v_{bullet}^2 2\sin(\delta\theta)\cos(\delta\theta)}{g} \quad \text{(Equation 4)}$$

where v_{bullet} is the speed of the bullet, x is the horizontal distance, y is the vertical distance, g is the Earth's gravitational acceleration, and t is time.

When the bullet hits the target (i.e. crosses the LOS), $x = R_H$ and $y = 0$. Equation 4 can be simplified assuming $x = R_H$ to obtain Equation 5.

$$R_H = \frac{v_{bullet}^2 2\sin(\delta\theta)\cos(\delta\theta)}{g} = \frac{v_{bullet}^2 \sin(2\delta\theta)}{g} \quad \text{(Equation 5)}$$

The zero range, R_H, is important because corrections due to elevation differences will be expressed in terms of changes to the horizontal zero range.

For most rifles, $\delta\theta$ is quite small. For example, the standard 7.62 mm (0.308 in) NATO bullet is fired with a muzzle velocity of 853 m/s (2800 ft/s). For a rifle zeroed at 100 meters, this means that $\delta\theta = 0.039°$.

While this definition of $\delta\theta$ is useful in theoretical discussions, in practice $\delta\theta$ must also account for the fact that the rifle sight is actually mounted above the barrel by several centimeters. This fact is important in practice, but is not required to understand the rifleman's rule.

Inclined Trajectory Analysis

The situation of shooting on an incline is illustrated below.

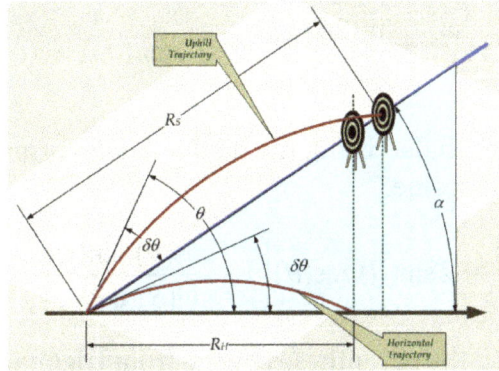

Illustration of Shooting on an Incline.

Figure illustrates both the horizontal shooting situation and the inclined shooting situation. When shooting on an incline with a rifle that has been zeroed at R_H, the bullet will impact along the incline as if it were zeroed at a longer range R_S. Observe that if the rifleman does not make a range adjustment, his rifle will appear to hit above its intended aim point. In fact, riflemen often report their rifle "shoots high" when they engage a target on an incline and they have not applied the rifleman's rule.

Equation 6 is the exact form of the rifleman's equation. It is derived from Equation 11 in Trajectory.

$$R_S = R_H \left(1 - \tan(\delta\theta)\tan(\alpha)\right)\sec(\alpha) \quad \text{(Equation 6)}$$

The complete derivation of Equation 6 is given below. Equation 6 is valid for all $\delta\theta$, α, and R_H. For small $\delta\theta$ and α, we can say that $1 - \tan(\delta\theta)\tan(\alpha) \approx 1$. This means we can approximate R_S as shown in Equation 7.

$$R_S \approx R_H \sec(\alpha) \quad \text{(Equation 7)}$$

Since the $\sec(\alpha) \geq 1$, we can see that a bullet fired up an incline with a rifle that was zeroed at R_H will impact the incline at a distance $R_S > R_H$. If the rifleman wishes to adjust his rifle to strike a target at a distance R_H instead of R_S along an incline, he needs to adjust the bore angle of his rifle so that the bullet will strike the target at R_H. This requires adjusting the rifle to a horizontal zero distance setting of $R_{Zero} = R_H \cos(\alpha)$. Equation 8 demonstrates the correctness of this assertion.

$$R_S = R_{Zero} \sec(\alpha) = \left(R_H \cos(\alpha)\right)\sec(\alpha) = R_H \quad \text{(Equation 8)}$$

This completes the demonstration of the rifleman's rule that is seen in routine practice. Slight variations in the rule do exist.

Derivation

Equation 6 can be obtained from the following equation

$$R_S = \frac{v_{Bullet}^2 \sin(2\theta)}{g}(1 - \cot(\theta)\tan(\alpha))\sec(\alpha)$$

This expression can be expanded using the double-angle formula for the sine and the definitions of tangent and cosine.

$$R_S = \frac{v_{Bullet}^2}{g}2\sin(\theta)\cos(\theta)\left(1 - \frac{\cos(\theta)}{\sin(\theta)}\frac{\sin(\alpha)}{\cos(\alpha)}\right)\sec(\alpha)$$

Multiply the expression in the parentheses by the front trigonometric term.

$$R_S = \frac{v_{Bullet}^2}{g}2\left(\sin(\theta)\cos(\theta) - \frac{\cos(\theta)^2\sin(\alpha)}{\cos(\alpha)}\right)\sec(\alpha)$$

Extract the factor $\cos(\theta)/\cos(\alpha)$ from the expression in parentheses.

$$R_S = \frac{v_{Bullet}^2}{g}2\frac{\cos(\theta)}{\cos(\alpha)}\left(\sin(\theta)\cos(\alpha) - \sin(\alpha)\cos(\theta)\right)\sec(\alpha)$$

The expression inside the parentheses is in the form of a sine difference formula. Also, multiply the resulting expression by the factor $\cos(\theta-\alpha)/\cos(\theta-\alpha)$.

$$R_S = \frac{v_{Bullet}^2}{g}\left(2\sin(\theta-\alpha)\cos(\theta)-\alpha)\right)\frac{\cos(\theta)}{\cos(\alpha)\cos(\theta-\alpha)}\sec(\alpha)$$

Factor the expression $\sin(2(\theta-\alpha))$ from the expression inside the parentheses. In addition, add and subtract the expression $\cos(\alpha)\cos(\theta-\alpha)$ inside the parentheses.

$$R_S = \frac{v_{Bullet}^2}{g}\sin(2(\theta-\alpha))\left(\frac{\cos(\alpha)\cos(\theta-\alpha)+\cos(\theta)-\cos(\alpha)\cos(\theta-\alpha)}{\cos(\alpha)\cos(\theta-\alpha)}\right)\sec(\alpha)$$

Let $\delta\theta = \theta-\alpha$.

$$R_S = \frac{v_{Bullet}^2}{g}\sin(2\delta\theta)\left(\frac{\cos(\alpha)\cos(\theta-\alpha)+\cos(\theta)-\cos(\alpha)\cos(\theta-\alpha)}{\cos(\alpha)\cos(\theta-\alpha)}\right)\sec(\alpha)$$

Let $R_H = \frac{v_{Bullet}^2 \sin(2\delta\theta)}{g}$ (see Equation 1) and simplify the expression in parentheses.

$$R_S = R_H\left(1+\frac{\cos(\theta)-\cos(\alpha)\cos(\theta-\alpha)}{\cos(\alpha)\cos(\theta-\alpha)}\right)\sec(\alpha)$$

Expand $\cos(\theta - \alpha) = \cos(\theta)\cos(\alpha) + \sin(\theta)\sin(\alpha)$.

$$R_S = R_H \left(1 + \frac{\cos(\theta) - \cos(\alpha)\big(\cos(\alpha)\cos(\theta) + \sin(\alpha)\sin(\theta)\big)}{\cos(\alpha)\cos(\theta - \alpha)} \right) \sec(\alpha)$$

Distribute the factor $\cos(\alpha)$ through the expression.

$$R_S = R_H \left(1 + \frac{\cos(\theta) - \cos(\alpha)^2 \cos(\theta) - \cos(\alpha)\sin(\theta)\sin(\alpha)}{\cos(\alpha)\cos(\theta - \alpha)} \right) \sec(\alpha)$$

Factor out the $\cos(\alpha)$ and substitute $\sin(\alpha)^2 = 1 - \cos(\alpha)^2$.

$$R_S = R_H \left(1 + \frac{\cos(\theta)\sin(\alpha)^2 - \cos(\alpha)\sin(\theta)\sin(\alpha)}{\cos(\alpha)\cos(\theta - \alpha)} \right) \sec(\alpha)$$

Factor out $\sin(\alpha)$.

$$R_S = R_H \left(1 + \frac{\sin(\alpha)\big(\cos(\theta)\sin(\alpha) - \cos(\alpha)\sin(\theta)\big)}{\cos(\alpha)\cos(\theta - \alpha)} \right) \sec(\alpha)$$

Substitute $-\sin(\theta - \alpha) = \cos(\theta)\sin(\alpha) - \sin(\theta)\cos(\alpha)$ into the equation.

$$R_S = R_H \left(1 - \frac{\sin(\alpha)\sin(\theta - \alpha)}{\cos(\alpha)\cos(\theta - \alpha)} \right) \sec(\alpha)$$

Substitute the definitions of $\tan(\delta\theta)$, $\tan(\alpha)$, and $\delta\theta = \theta - \alpha$ into the equation.

$$R_S = R_H \big(1 - \tan(\alpha)\tan(\delta\theta) \big) \sec(\alpha)$$

This completes the derivation of the exact form of the rifleman's rule.

References

- Zubrin, Robert (1996). The case for Mars: the plan to settle the red planet and why we must. New York: Free Press. ISBN 978-0-684-83550-1

- Jeremy J. Hollerman; Martin L. Fackler; Douglas M. Coldwell; Yoram Ben-Menachem (October 1990). "Gunshot Wounds: 1. Bullets, Ballistics, and Mechanisms of Injury" (PDF). American Journal of Roentgenology vol.155 no. 4: 685–690

- Moss, Leeming and Farrar (1995). Brassey's Land Warfare Series: Military Ballistics. Royal Military College of Science, Shrivenham, UK. p. 86. ISBN 978-1857530841

- Wheeler, Robin (2009). "Apollo lunar landing launch window: The controlling factors and constraints". NASA. Retrieved 2009-10-27

- Wang Q, Wang Z, Zhu P, Jiang J: Alterations of the Myelin Basic Protein and Ultrastructure in the Limbic System and the Early Stage of Trauma-Related Stress Disorder in Dogs. The Journal of Trauma. 56(3):604-610; 2004

- Rinker, Robert A. (1999). Understanding Firearm Ballistics; 3rd Edition. Mulberry House Publishing. p. 176. ISBN 978-0964559844

- Suneson A, Hansson HA, Seeman T: Pressure Wave Injuries to the Nervous System Caused by High Energy Missile Extremity Impact: Part I. Local and Distant Effects on the Peripheral Nervous System. A Light and Electron Microscopic Study on Pigs. The Journal of Trauma. 30(3):281-294; 1990

- Pejsa, Arthur, New Exact Small Arms Ballistics: The Source Book for Riflemen, 2008; pg29, Kenwood Publishing ISBN 978-0974990262

Permissions

All chapters in this book are published with permission under the Creative Commons Attribution Share Alike License or equivalent. Every chapter published in this book has been scrutinized by our experts. Their significance has been extensively debated. The topics covered herein carry significant information for a comprehensive understanding. They may even be implemented as practical applications or may be referred to as a beginning point for further studies.

We would like to thank the editorial team for lending their expertise to make the book truly unique. They have played a crucial role in the development of this book. Without their invaluable contributions this book wouldn't have been possible. They have made vital efforts to compile up to date information on the varied aspects of this subject to make this book a valuable addition to the collection of many professionals and students.

This book was conceptualized with the vision of imparting up-to-date and integrated information in this field. To ensure the same, a matchless editorial board was set up. Every individual on the board went through rigorous rounds of assessment to prove their worth. After which they invested a large part of their time researching and compiling the most relevant data for our readers.

The editorial board has been involved in producing this book since its inception. They have spent rigorous hours researching and exploring the diverse topics which have resulted in the successful publishing of this book. They have passed on their knowledge of decades through this book. To expedite this challenging task, the publisher supported the team at every step. A small team of assistant editors was also appointed to further simplify the editing procedure and attain best results for the readers.

Apart from the editorial board, the designing team has also invested a significant amount of their time in understanding the subject and creating the most relevant covers. They scrutinized every image to scout for the most suitable representation of the subject and create an appropriate cover for the book.

The publishing team has been an ardent support to the editorial, designing and production team. Their endless efforts to recruit the best for this project, has resulted in the accomplishment of this book. They are a veteran in the field of academics and their pool of knowledge is as vast as their experience in printing. Their expertise and guidance has proved useful at every step. Their uncompromising quality standards have made this book an exceptional effort. Their encouragement from time to time has been an inspiration for everyone.

The publisher and the editorial board hope that this book will prove to be a valuable piece of knowledge for students, practitioners and scholars across the globe.

Index

A

Aerodynamic Projectiles, 3
Aerosol Sprays, 122
Ambient Air Density, 109, 114, 119, 185
Ammunition Cartridge Collectors, 67
Artillery Software, 106
Astrodynamics, 7

B

Baffles, 70, 73-75, 87, 89-90
Ballistic Coefficient, 14, 80, 100, 104-105, 107-108, 112, 118, 176-178, 180-181, 183-187
Ballistic Conduction, 17, 34-35, 37-38
Ballistic Missile, 1, 162-166, 187
Ballistic Pendulum, 179, 189-190, 193
Ballistic Tables, 98, 101, 113, 180-183
Bashforth Method, 180
Battle Zero, 99, 129
Black Powder, 19-20, 22, 39, 48, 51-53, 55-59, 64-65, 123, 167
Bore Characteristics, 28
Bore Diameter, 27, 29, 100
Bullet Drop, 96-97, 113, 120, 198-199
Bullet Path, 97-98, 128

C

Carbon Nanotubes, 37-38
Caseless Ammunition, 20-21
Combustible Ingredients, 54
Controlled Penetration, 11
Coriolis Drift, 117, 188
Cruise Missile, 1, 162

D

Depleted Uranium Alloys, 16
Double-base Propellants, 19
Drag Curve Models, 101-102, 184
Drag Resistance, 102

E

Energy Transfer, 27, 195
Expanding Bullets, 11, 13

External Ballistics, 6-7, 65, 68-69, 95-96, 100, 111, 120-121, 127-128

F

Firearm Noise Anatomy, 72
Flash Suppressor, 71, 77, 84-87
Flat Point, 12, 99, 108
Forensic Ballistics, 7
Fragmentation Bomblets, 15
Fragmenting Bullets, 12-13
Frangible Bullets, 11, 13

G

G Model, 177, 182-183
Graphene Nanoribbon, 36, 38
Guided Projectiles, 109, 188
Gunpowder, 17, 19, 23, 39-60, 65, 67, 119, 122-123, 127, 174
Gunshot Residue, 17, 60-62, 67
Gyroscopic Drift, 104, 115, 117, 119

H

Handgun Effectiveness, 176, 194-195

I

Improvised Silencers, 78
Initial Velocity Calculation, 69
Internal Ballistics, 5-6, 17-18, 48, 68, 115

K

Kinetic Friction, 23
Kinetic Projectiles, 132

L

Landauer-büttiker Formalism, 37
Lateral Jump, 118
Lateral Throw-off, 118
Liquid Propellant, 125
Load Density, 21

M

Magnus Effect, 116-117, 138
Manges Model, 105

Mathematical Models, 99, 184
Maximum Penetration, 10
Medieval Catapults, 171
Motive Force, 131
Muzzle Brake, 26-28, 70, 85, 88-92, 94, 129
Muzzle Energy, 32-34, 80
Muzzle Pressure Concerns, 26

N
Nitrocellulose, 19, 26, 56-57, 123-124

O
Optical Analogies, 37
Orbiting Objects, 151

P
Packing Material, 75
Pejsa Model, 102-105, 121
Penetrator Material, 16
Point-blank Range, 99, 127-129
Powered Flight, 162-163
Pressure-velocity Relationships, 24
Projectile Mass, 27, 115, 168
Propellant Burnout, 26
Propellant Charge, 21
Propellant Design, 25
Propulsive Nozzle, 1

R
Reentry Vehicles, 187
Rimfire, 17, 29-30, 62-67, 70, 82, 193

S
Scattering Mechanisms, 35-36
Serpentine, 47, 54
Shaped Charge, 15-16
Si Nanowires, 38
Smokeless Powder, 21-24, 52, 60, 64, 122
Solid Propellants, 20, 123-124
Spacers, 74-75
Sport Projectiles, 132
Static Friction, 23
Stoneware Bombs, 40
Subsonic Ammunition, 79-81
Sulfur-free Gunpowder, 57
Suppressor, 22, 70-79, 81-87

T
Target Shooting, 9-10, 39, 66, 96
Terminal Ballistics, 6-8, 15
Test Firing, 103, 178-179
Throw-weight, 165-166
Transitional Ballistics, 6, 68-69
Transonic Problem, 108

U
Unguided Bombs, 1

V
Venting Direction, 90
Vertical Angles, 113

W
Wired Projectiles, 133

Part 1: Introduction to Business Law

Chapter 1: The Importance of Understanding Business Law
Chapter 2: Basic Legal Concepts and Principles
Chapter 3: Sources of Law
Chapter 4: Types of Legal Systems

Part 2: Business Organizations and Corporate Governance

Chapter 5: Types of Business Organizations
Chapter 6: Corporate Governance Principles and Practices
Chapter 7: Board of Directors and Executive Compensation
Chapter 8: Securities Law and Corporate Finance

Part 3: Contract Law and Legal Issues in Business Transactions

Chapter 9: Formation and Enforcement of Contracts
Chapter 10: Breach of Contract and Remedies
Chapter 11: Legal Issues in Business Transactions
Chapter 12: International Business Law
Chapter 13: Conclusion

Part 4: Securities Laws and Regulations

Chapter 14: Introduction to Securities Law
Chapter 15: Securities Act of 1933 and 1934
Chapter 16: Securities Exchange Act of 1934
Chapter 17: Regulation A and D Exemptions
Chapter 18: Regulation A+

Part 5: Securities Regulation Agencies and Enforcement

Chapter 19: Securities and Exchange Commission (SEC)
Chapter 20: Financial Industry Regulatory Authority (FINRA)
Chapter 21: State Securities Regulators

Part 6: Disclosure Requirements and Liability

Chapter 22: Disclosure Requirements for Public Companies
Chapter 23: Liability for Securities Fraud
Chapter 24: Insider Trading and Other Securities Violations
Chapter 25: Conclusion

Part 7: Public Offering Process and Requirements

Chapter 26: Pre-IPO Planning and Preparation

Chapter 27: Financial and Legal Due Diligence
Chapter 28: IPO Alternatives and Considerations
Chapter 29: Preparing for Public Scrutiny

Part 10: Financial Statement Analysis and Reporting

Chapter 30: Introduction to Financial Statement Analysis
Chapter 31: Analysis of Income Statements, Balance Sheets, and Cash Flow Statements
Chapter 32: Ratio Analysis and Interpretation

Part 11: Financial Reporting Standards and Regulations

Chapter 33: Generally Accepted Accounting Principles (GAAP)
Chapter 34: International Financial Reporting Standards (IFRS)
Chapter 35: Sarbanes-Oxley Act of 2002

Part 12: International Accounting Standards

Chapter 36: Global Financial Reporting Standards
Chapter 37: Overview of IFRS and its Adoption Around the World
Chapter 38: Comparison with US GAAP
Chapter 39: Challenges and Opportunities in Global Financial Reporting

Part 13: How Web 3 and Bitcoin are Changing the Financial Landscape and Laws as it Relates to Business:

Chapter 40: Explanation of what Web 3 and Bitcoin are
Chapter 41 The Emergence of Web 3 and Bitcoin
Chapter 43: Impacts on the Financial Landscape
Chapter 44: Impacts on Business and Commerce
Chapter 45: Legal and Regulatory Issues

PART 1:
INTRODUCTION TO
BUSINESS LAW

Business law is a crucial field of study for anyone interested in the world of commerce. This section aims to provide students with a comprehensive introduction to business law and its significance in the business world. By studying business law, students will gain an understanding of the basic concepts and principles that underlie legal systems around the world, as well as how these principles are applied to business operations. In this section, we will cover a wide range of topics, from the sources of law to dispute resolution and emerging trends in business law.

Basic Legal Concepts and Principles:
In this section, we will explore the fundamental concepts and principles of business law. We will discuss the difference between civil and criminal law, the role of precedent in legal decisions, and the importance of contracts in business transactions. Students will learn how to read and interpret legal texts and how to apply legal principles to real-world scenarios.

Sources of Law:
The legal system is complex and has many different sources of law. In this section, students will learn about the primary sources of law, including common law, statutory law, and administrative law. We will explore how these sources of law interact and shape the legal landscape, as well as the roles they play in the business world.

Types of Legal Systems:
Different countries have different legal systems, each with its own unique characteristics. In this section, we will examine the most common types of legal systems, including civil law, common law, and Sharia law. Students will learn about the history and development of these legal systems and how they are applied in practice.

Business Ethics and Corporate Social Responsibility:

Business law is closely tied to business ethics and corporate social responsibility. In this section, we will examine the ethical and social responsibilities that businesses have, including their impact on the environment, their treatment of employees, and their role in society. Students will learn how legal and ethical principles intersect and how businesses can operate in a socially responsible manner.

The Role of Law in Business:
The legal system plays a crucial role in the world of business. In this section, we will examine the ways in which business law affects business operations, including the legal requirements for forming a business, the role of contracts in business transactions, and the legal protections afforded to businesses and their owners.

The Importance of Legal Research and Analysis:
Legal research and analysis are critical skills for anyone interested in business law. In this section, students will learn about the process of legal research and analysis, including how to identify and interpret legal sources, how to analyze legal texts, and how to apply legal principles to real-world scenarios. This section will also provide practical guidance on conducting legal research and analysis, including tips for using legal databases and other resources.

Business Law and Dispute Resolution:
Disputes are a common occurrence in the world of business. In this section, we will explore the various methods of dispute resolution available to businesses, including litigation, arbitration, and mediation. Students will learn about the advantages and disadvantages of each method and how to choose the best approach for a given situation.

International Business Law:
In today's globalized world, businesses must navigate an increasingly complex legal landscape. In this section, we will examine the key legal issues that arise in international business, including cross-border transactions, intellectual property protection, and compliance with international trade agreements. Students will learn about the legal frameworks that govern international business and how to navigate the challenges of doing business across borders.

Emerging Trends in Business Law:
The field of business law is constantly evolving. In this section, we will explore some of the emerging trends in business law, including the impact of technology on the legal landscape, the rise of alternative forms of business organization, and the increasing importance of social responsibility and sustainability in business operations. Students will learn about the challenges and opportunities presented by these trends and how they may shape the future of business law.

One of the most significant trends in business law is the impact of technology on the legal landscape. With the rise of new technologies such as artificial intelligence and blockchain, the legal system must adapt to keep up with the changing business environment. For example, blockchain technology is being used to create smart contracts that automatically execute contractual obligations, reducing the need for intermediaries and potentially changing the way contracts are enforced.

Another emerging trend in business law is the rise of alternative forms of business organization, such as social enterprises and benefit corporations. These types of businesses prioritize social and environmental goals alongside profit, and often require unique legal structures to operate effectively. Business law must adapt to accommodate these new business models and provide legal frameworks that support their growth and sustainability.

The importance of social responsibility and sustainability in business operations is also increasing, with more companies recognizing the need to prioritize environmental and social issues alongside financial performance. Business law can play a crucial role in supporting these efforts, by providing legal structures and incentives that encourage companies to operate in a socially responsible and sustainable manner.

Overall, the future of business law is likely to be shaped by these and other emerging trends, as the legal system adapts to the changing needs of the business world. Students of business law will need to stay up-to-date with these trends and understand how they impact the legal landscape, in order to succeed in the dynamic and evolving field of business law. In the following chapters, we will explore these emerging trends in more detail, and examine their implications for the practice of business law.

CHAPTER 1: THE IMPORTANCE OF UNDERSTANDING BUSINESS LAW

Business law plays a crucial role in shaping the modern business environment. Understanding the fundamentals of business law is essential for students seeking a bachelor's degree in business, as well as for professionals in various fields such as Business Owners/Entrepreneurs, Corporate Executives, In-house Counsel/Legal Departments, Compliance Officers, Financial Analysts, Investment Bankers, Accountants, Regulators. In this chapter, we will explore the importance of understanding business law and the key concepts and principles that underpin the legal framework for businesses.

Importance of Understanding Business Law

Business law governs the interactions between businesses, individuals, and government entities. It provides a framework for conducting business transactions, resolving disputes, and protecting intellectual property. Business law also regulates the behavior of businesses, ensuring that they operate in a socially responsible and sustainable manner.

One of the key benefits of understanding business law is that it can help individuals and businesses to make informed decisions. For example, a business owner who understands contract law can negotiate and draft contracts that are beneficial to their company. Similarly, an investor who understands securities law can make informed investment decisions.

Furthermore, understanding business law can help individuals and businesses to avoid legal issues and liabilities. Non-compliance with legal regulations can result in costly fines, legal disputes, and reputational damage. By understanding the legal landscape, individuals and businesses can mitigate these risks and ensure that they are operating in a legal and ethical manner.

Key Concepts and Principles

Business law encompasses a wide range of legal concepts and principles. Some of the key concepts and principles that underpin business law include:

Contracts: Contracts are legally binding agreements between two or more parties. They are used in a variety of business transactions, including employment contracts, lease agreements, and purchase contracts.

Tort law: Tort law is a branch of civil law that deals with harm caused by one party to another. It includes areas such as negligence, defamation, and product liability.

Intellectual property law: Intellectual property law protects the rights of creators and owners of intellectual property, including patents, trademarks, and copyrights.

Corporate law: Corporate law governs the formation, operation, and dissolution of corporations.

Securities law: Securities law regulates the issuance and trading of securities, such as stocks and bonds.

Antitrust law: Antitrust law is designed to promote competition and prevent monopolies.

Labor law: Labor law regulates the relationship between employers and employees, including issues such as wages, hours, and working conditions.

Environmental law: Environmental law regulates the impact of businesses on the environment, including issues such as pollution and conservation.

By understanding these key concepts and principles, individuals and businesses can navigate the legal landscape with greater confidence and make informed decisions that are in compliance with legal regulations.

Conclusion

In conclusion, understanding business law is essential for anyone seeking to succeed in the modern business environment. By understanding the key concepts and principles that underpin business law, individuals and businesses can make informed decisions, avoid legal issues and liabilities, and operate in a socially responsible and sustainable manner. This chapter provides an overview of the importance of understanding business law, and subsequent chapters will delve into greater detail on specific areas of business law. Throughout the chapter, we will provide real-world examples from a variety of fields, and provide math equations, problems, and exercises to illustrate key concepts.

The role of law in business

Law plays a critical role in regulating business activities and ensuring that companies operate in a fair and transparent manner. In this section, we will explore the different aspects of the role of law in business and how it helps to create a level playing field for all stakeholders.

Providing a Framework for Business Operations

The law provides a framework for businesses to operate within, and it sets the rules of the game that all players must follow. This framework encompasses a wide range of legal areas, including contract law, tort law, property law, employment law, and intellectual property law, among others.

One of the primary benefits of having a legal framework is that it helps to create certainty and predictability for businesses. Companies can plan and execute their operations with confidence, knowing that the law provides a set of rules that they must follow. This also makes it easier for businesses to attract investment and financing, as investors and lenders are more likely to be confident in a company that operates within a stable legal environment.

Protecting Business Interests

Another critical role that the law plays in business is to protect the interests of companies and their stakeholders. This protection comes in many forms, including:

Protecting intellectual property rights, such as patents, trademarks, and copyrights
Regulating competition to prevent monopolies and promote fair market competition
Ensuring that contracts are enforceable and that parties can rely on the law to hold up their end of the bargain
Providing legal remedies for parties who have suffered harm due to the actions of others
In addition to protecting businesses, the law also provides safeguards for consumers and other stakeholders, such as employees and the general public. For

example, laws governing product safety and environmental protection help to ensure that companies operate in a socially responsible manner and do not harm the public or the environment.

Resolving Disputes

Despite the best efforts of businesses and their legal teams, disputes are inevitable in the course of business operations. The law provides a means for parties to resolve disputes in a fair and impartial manner, through a system of courts and other dispute resolution mechanisms.

The legal system allows parties to present their cases to an independent decision-maker, such as a judge or arbitrator, who can make a ruling based on the evidence presented. This helps to ensure that disputes are resolved in a fair and equitable manner, without resorting to violence or other forms of coercion.

Promoting Social Responsibility

In recent years, there has been a growing emphasis on the role of business in promoting social responsibility and sustainability. The law has played an important role in driving this trend, by creating legal obligations for companies to act in a socially responsible manner.

For example, laws governing environmental protection and labor rights require companies to take steps to minimize their impact on the environment and to treat their employees fairly. Similarly, laws governing corporate governance require companies to act in the best interests of their stakeholders, including shareholders, employees, and the broader community.

Conclusion:

The role of law in business is multifaceted and critical to the success of companies and the broader economy. By providing a framework for operations, protecting business interests, resolving disputes, and promoting social responsibility, the law helps to create a level playing field for all stakeholders and promote a more equitable and sustainable business environment.

The consequences of ignoring legal requirements

In the world of business, there are numerous legal requirements that companies and individuals must comply with. Ignoring these legal requirements can have severe consequences, including legal liabilities, reputational damage, and financial losses. In this section, we will explore the consequences of ignoring legal requirements and why it is essential to comply with legal requirements.

Legal Liabilities:

One of the most significant consequences of ignoring legal requirements is the potential for legal liabilities. Companies that fail to comply with legal requirements can face lawsuits and other legal actions, resulting in significant financial losses. For example, if a company violates labor laws and fails to pay overtime to its employees, it may face a lawsuit and be required to pay back wages, penalties, and legal fees.

Moreover, ignoring legal requirements can lead to criminal liabilities. Companies and individuals who engage in illegal activities, such as fraud or embezzlement, can face criminal charges and imprisonment. For instance, Bernard Madoff was sentenced to 150 years in prison for running a Ponzi scheme that defrauded investors of billions of dollars.

Reputational Damage:

Ignoring legal requirements can also result in reputational damage. When companies engage in unethical or illegal behavior, it can damage their brand and reputation, leading to a loss of customer trust and loyalty. For example, in 2010, BP's Deepwater Horizon oil spill in the Gulf of Mexico caused significant reputational damage, resulting in a loss of customers and revenues.

Moreover, reputational damage can also impact a company's ability to attract and retain employees. In today's world, employees value ethical behavior and social responsibility in their employers. Companies that ignore legal requirements may struggle to attract and retain top talent, resulting in reduced productivity and increased turnover.

Financial Losses:

Ignoring legal requirements can also result in significant financial losses. Companies that fail to comply with legal requirements may face fines, penalties, and legal fees, resulting in increased costs. Moreover, the cost of defending lawsuits and legal actions can be significant, leading to financial losses and decreased profitability.

For example, in 2017, Wells Fargo was fined $185 million for opening unauthorized accounts and charging customers fees. The company also faced lawsuits from customers and shareholders, resulting in significant legal fees and reputational damage.

Conclusion:

In conclusion, ignoring legal requirements can have severe consequences for companies and individuals. Legal liabilities, reputational damage, and financial losses can all result from failing to comply with legal requirements. Therefore, it is essential for businesses and individuals to understand the legal requirements that apply to them and take the necessary steps to comply with them. By doing so, companies and individuals can protect themselves from potential legal and financial risks and maintain their reputation and financial stability.

The benefits of compliance

In the world of business, compliance refers to the adherence of companies to legal regulations and industry standards. Compliance is essential for businesses to operate legally and ethically, and failing to comply with these requirements can have severe consequences for businesses and their stakeholders. However, compliance is not just about avoiding negative consequences. In this section, we will explore the benefits of compliance for businesses, including improved reputation, increased profitability, and reduced legal risks.

Improved Reputation

One of the most significant benefits of compliance for businesses is an improved reputation. Compliance demonstrates a commitment to ethics and integrity, which can lead to increased trust and loyalty from customers, employees, and other stakeholders. Compliance can also help businesses differentiate themselves from their competitors by demonstrating their commitment to ethical behavior and social responsibility.

For example, the pharmaceutical company Novo Nordisk has made compliance and ethical behavior a core part of its business strategy. The company's commitment to compliance has led to a strong reputation for ethical behavior and has helped it attract top talent and customers who value ethical business practices. In contrast, companies that have a history of non-compliance, such as Enron, have suffered significant reputational damage that has taken years to repair.

Increased Profitability

Compliance can also have a positive impact on a business's profitability. By avoiding legal penalties and fines associated with non-compliance, businesses can save money and reinvest those savings into other areas of the business. Compliance can also help businesses avoid costly lawsuits and other legal disputes, which can be time-consuming and expensive.

In addition, compliance can help businesses identify areas for improvement and optimize their operations. For example, complying with environmental regulations can lead to the implementation of more efficient and sustainable practices, which can reduce costs and improve profitability over the long term.

Reduced Legal Risks

Compliance is essential for businesses to reduce legal risks. Non-compliance can result in a range of legal consequences, including fines, lawsuits, and even criminal charges. These consequences can be devastating for businesses and their stakeholders, leading to significant financial losses and reputational damage.

In contrast, compliance can help businesses avoid legal risks and ensure they are operating within legal boundaries. Compliance can also help businesses navigate complex legal requirements, such as tax laws or data protection regulations, reducing the risk of unintentional non-compliance.

Conclusion

Compliance is not just a legal requirement, but it is also a critical aspect of business success. Compliance can lead to improved reputation, increased profitability, and reduced legal risks. By investing in compliance, businesses can demonstrate their commitment to ethics and integrity, and create a sustainable competitive advantage in their industry.

To reinforce these concepts, let's consider the following example: ABC Corporation, a publicly traded company, recently implemented a compliance program that focused on improving environmental sustainability. The program resulted in significant cost savings through the implementation of more energy-efficient practices and reducing waste. Additionally, ABC Corporation's commitment to sustainability helped it attract environmentally conscious customers and investors, improving its reputation and overall profitability. In contrast, XYZ Corporation, a competitor of ABC Corporation, failed to comply with environmental regulations, resulting in legal penalties and reputational damage. As a result, XYZ Corporation suffered financial losses and struggled to regain the trust of its stakeholders.

CHAPTER 2: BASIC LEGAL CONCEPTS AND PRINCIPLES

In the world of business, understanding the basic legal concepts and principles is essential for success. Business owners, executives, legal departments, and compliance officers must be knowledgeable about the law to avoid legal risks, make informed decisions, and comply with legal requirements. This chapter will provide an introduction to the basic legal concepts and principles that are relevant to business, including contract law, tort law, property law, and intellectual property law.

To begin with, contract law is a crucial aspect of business law as it governs the creation and enforcement of agreements between parties. Business owners and executives must understand the essential elements of a contract, such as offer, acceptance, consideration, and mutual assent, to create valid and enforceable contracts. This chapter will delve into the various types of contracts and their enforceability, the legal remedies available for breach of contract, and the legal implications of contract interpretation.

Another critical area of business law is tort law, which involves civil wrongs that result in harm or injury to another party. In the business context, tort law encompasses a range of issues, including product liability, negligence, and intentional torts. It is important for business owners and executives to understand the legal duties owed to customers, employees, and other parties, and the potential legal consequences of breaching these duties.

Furthermore, property law is a fundamental area of business law that deals with the ownership and use of tangible and intangible assets. Business owners and executives must be aware of the various types of property rights, including real property, personal property, and intellectual property, and the legal protections afforded to them. This chapter will provide an overview of the different forms of property rights and the legal frameworks governing their acquisition, transfer, and protection.

Lastly, intellectual property law is a crucial area of business law that deals with the protection of intangible assets such as patents, trademarks, copyrights, and trade secrets. It is essential for businesses to safeguard their intellectual property to maintain their competitive advantage and prevent others from exploiting their innovations. This chapter will cover the various forms of intellectual property protection available and the legal requirements for obtaining and enforcing them.

In summary, this chapter will provide a comprehensive overview of the basic legal concepts and principles that are essential for understanding business law. By studying these fundamental concepts, students will develop the knowledge and skills necessary to navigate the legal landscape of business and make informed decisions that comply with legal requirements and minimize legal risks.

The structure of the legal system

In order to understand the role of law in business, it is essential to first understand the structure of the legal system. The legal system in any given country is complex and multi-layered, consisting of various courts, statutes, regulations, and agencies. In this section, we will provide an in-depth analysis of the structure of the legal system, including its components and functions.

The Structure of the Legal System:

The legal system can be divided into two main categories: civil law and criminal law. Civil law pertains to disputes between private individuals or organizations, while criminal law deals with offenses against the state or society as a whole. In the United States, the legal system is based on common law, which means that courts interpret laws based on past decisions or precedents.

The following are the main components of the legal system:

✧   Legislative Branch: This branch is responsible for creating laws. In the United States, the legislative branch is comprised of Congress, which is composed of the Senate and the House of Representatives.

✧   Executive Branch: This branch is responsible for enforcing laws. The executive branch in the United States is led by the President.

✧   Judicial Branch: This branch is responsible for interpreting laws and resolving disputes. In the United States, the judicial branch is composed of the Supreme Court and lower federal courts.

✧   Administrative Agencies: These agencies are responsible for implementing laws and regulations. Examples of administrative agencies include the Environmental Protection Agency (EPA), the Federal Communications Commission (FCC), and the Securities and Exchange Commission (SEC).

Functions of the Legal System:

The legal system serves several important functions in society, including:

✧ Maintaining social order: The legal system helps to maintain social order by providing a framework for resolving disputes and punishing offenders.

✧ Protecting individual rights: The legal system protects individual rights by providing a means for individuals to seek redress for grievances and by ensuring that the government does not violate individual rights.

✧ Promoting fairness and equality: The legal system promotes fairness and equality by providing equal protection under the law and ensuring that everyone is held accountable to the same standards.

✧ Encouraging economic development: The legal system encourages economic development by providing a stable and predictable framework for businesses to operate within.

Conclusion:

In conclusion, understanding the structure and functions of the legal system is essential for any business owner, corporate executive, or legal professional. The legal system provides a framework for resolving disputes, protecting individual rights, promoting fairness and equality, and encouraging economic development. By complying with legal requirements and working within the legal system, businesses can operate effectively and responsibly while avoiding the potential consequences of ignoring legal requirements.

The distinction between civil and criminal law

The legal system is an intricate network of laws, regulations, and procedures that govern how society operates. One of the most fundamental distinctions within the legal system is the separation between civil law and criminal law. Understanding this distinction is essential for anyone who wants to navigate the legal system effectively.

This section will explore the differences between civil and criminal law, including their respective purposes, standards of proof, and consequences.

Purpose:

The purpose of civil law is to resolve disputes between individuals or entities. These disputes can involve anything from breach of contract to personal injury claims. Civil law cases are generally brought by one party against another party seeking compensation for damages. The goal of civil law is to provide a resolution to the dispute and to ensure that the injured party is compensated for their losses.

Criminal law, on the other hand, is designed to punish individuals who have committed crimes against society. Criminal law cases are brought by the government and can result in penalties such as fines, probation, and incarceration. The goal of criminal law is to deter criminal behavior and to protect society from dangerous individuals.

Standards of Proof:

The standard of proof required for civil law cases is known as the preponderance of the evidence. This standard requires the plaintiff to prove that it is more likely than not that the defendant caused the harm alleged. In other words, the plaintiff must show that it is at least 51% likely that the defendant is responsible for the damages.

Criminal law cases, on the other hand, require a much higher standard of proof known as beyond a reasonable doubt. This means that the prosecution must prove that the defendant is guilty beyond any reasonable doubt. This is a much higher standard of proof than the preponderance of the evidence, and it is designed to protect individuals from being wrongly convicted.

Consequences:

The consequences of civil law cases are generally monetary in nature. The plaintiff can seek compensation for damages, such as lost wages, medical bills, and property damage. In some cases, the plaintiff may also be awarded punitive damages, which are designed to punish the defendant for their actions.

Criminal law cases, on the other hand, can result in much more severe consequences. These consequences can include fines, probation, and incarceration. In some cases, the defendant may also be required to complete community service or attend counseling. The severity of the consequences depends on the nature of the crime committed and the defendant's criminal history.

Examples:

To illustrate the differences between civil and criminal law, let's consider a few examples. Suppose that a driver runs a red light and causes a car accident that results in injury to another driver. The injured driver could file a civil lawsuit against the driver who ran the red light seeking compensation for medical bills, lost wages, and other damages.

Now suppose that the driver who ran the red light was intoxicated at the time of the accident. In this case, the government could bring criminal charges against the driver for driving under the influence. If convicted, the driver could face fines, probation, or even incarceration.

Conclusion:

In conclusion, the distinction between civil and criminal law is an essential concept to understand for anyone who wants to navigate the legal system effectively. While civil law is designed to resolve disputes between individuals or entities, criminal law is designed to punish individuals who have committed crimes against society. The standards of proof and consequences also differ significantly between civil and criminal law cases. Understanding these differences can help individuals make informed decisions about how to proceed when faced with legal issues.

The role of precedent

In the legal system, precedent refers to the decisions made by higher courts that lower courts are bound to follow in similar cases. This concept of stare decisis, meaning "let the decision stand," is a fundamental principle of common law systems such as that in the United States. Precedent plays a critical role in the legal system as it provides a sense of stability and consistency, but it also raises important questions about the role of judges and the limits of their power. In this section, we will explore the role of precedent in the legal system, how it is established, and the impact it has on legal decisions.

Establishment of Precedent:

Precedent is established through the court system, beginning with a case that is appealed to a higher court. The decision made by the higher court creates a precedent that lower courts must follow when presented with a similar case. The precedent set by a court can be binding or persuasive. A binding precedent is one that must be followed by lower courts within the same jurisdiction, while a persuasive precedent is one that is not binding but can be influential in a decision.

The impact of precedent is not limited to the jurisdiction where it was established. Precedent can have an impact in other jurisdictions, and even in other countries, depending on the strength and persuasiveness of the precedent. For example, a U.S. Supreme Court decision on a constitutional issue can have implications for courts in other countries that have similar constitutional provisions.

The Role of Precedent:

Precedent serves several important functions in the legal system. First, it provides consistency and predictability in the law, allowing individuals and organizations to plan their actions with a degree of certainty about the legal consequences. This consistency is particularly important in areas such as contract law and property law, where parties rely on the stability of legal rules to make informed decisions.

Second, precedent promotes fairness and equality in the legal system. The principle of stare decisis ensures that similar cases are treated similarly, regardless of the individual characteristics of the parties involved. This principle is essential in maintaining the public's trust in the legal system and ensuring that justice is served.

Third, precedent helps to ensure that the law is developed through a democratic process, rather than through the whims of individual judges. By following precedent, judges are bound by the decisions of their predecessors and must make decisions based on established legal principles and rules.

However, the reliance on precedent has also been criticized for limiting judicial discretion and inhibiting the development of the law in response to changing social and technological circumstances. Critics argue that the rigid application of precedent can lead to unjust results, particularly in cases where the precedent was established in a different time or context.

Conclusion:

In conclusion, precedent is a fundamental principle of the legal system that provides stability, consistency, and fairness. Precedent ensures that similar cases are treated similarly and promotes the development of the law through established legal principles and rules. However, the reliance on precedent also raises important questions about the role of judges and the limits of their power. As the legal system continues to evolve, it is essential that judges and legal scholars continue to evaluate the role of precedent in the law and work to ensure that it is used in a way that promotes justice and fairness for all.

CHAPTER 3: SOURCES OF LAW

In the previous chapters, we discussed the basic legal concepts and principles as well as the structure of the legal system. In this chapter, we will explore the sources of law. Understanding the sources of law is essential for anyone who seeks to comprehend and navigate the legal system.

In general, the sources of law can be divided into two broad categories: primary sources and secondary sources. Primary sources are the actual laws that govern our society. They are the authoritative statements of the legal rules and regulations that we must follow. Primary sources of law include constitutions, statutes, administrative regulations, and case law.

Constitutions are fundamental documents that establish the basic framework of government and individual rights. In the United States, the Constitution is the supreme law of the land, and all other laws must conform to it. Statutes are written laws passed by legislative bodies, such as the U.S. Congress or state legislatures. Administrative regulations are rules and regulations created by administrative agencies, such as the Environmental Protection Agency or the Federal Communications Commission, to carry out their functions. Case law, also known as common law, is the body of law created by judges when they decide cases. Case law is based on precedent, which is the principle that judges should follow the decisions of earlier cases that involve similar legal issues.

Secondary sources of law are materials that interpret, explain, or analyze the primary sources of law. Secondary sources are not themselves legal rules or regulations, but they are valuable tools for understanding and applying the law. Secondary sources include legal treatises, law review articles, and legal encyclopedias.

In this chapter, we will examine each of these sources of law in more detail, discussing their characteristics, strengths, and weaknesses. We will also explore the role of these sources in different areas of law, including business law, criminal law, and administrative law. Finally, we will analyze the interplay between these sources of law and how they shape the legal system as a whole.

By the end of this chapter, you will have a comprehensive understanding of the various sources of law and their importance in our legal system. This knowledge will be essential for anyone seeking to work in legal fields such as business, government, or law enforcement. Let us begin our exploration of the sources of law.

The Constitution and its amendments

The Constitution of the United States is the supreme law of the land, and its amendments provide additional protections to citizens. The Constitution lays out the framework for our government, including the roles and powers of the three branches of government, the rights of citizens, and the relationship between the federal government and the states. Understanding the Constitution and its amendments is essential for anyone who wants to understand the legal system in the United States.

This section will provide an overview of the Constitution and its amendments, including the history of their creation and their significance in American law. We will also discuss some of the key concepts and principles that are found in the Constitution, such as federalism, separation of powers, and individual rights.

The Creation of the Constitution

The Constitution was created in 1787 at the Constitutional Convention in Philadelphia. The delegates who attended the convention were tasked with creating a new system of government that would replace the Articles of Confederation, which had proven to be ineffective in governing the newly formed United States.

The delegates to the convention included some of the most prominent figures in American history, such as George Washington, Benjamin Franklin, James Madison, and Alexander Hamilton. The Constitution they created was a product of compromise, as the delegates had differing opinions on how the government should be structured.

The Constitution is divided into seven articles, each of which outlines different aspects of the government. The first three articles describe the structure and powers of the three branches of government: the legislative, executive, and judicial branches. The fourth article outlines the relationship between the federal government and the states, while the fifth article provides procedures for amending the Constitution. The sixth article establishes the Constitution as the supreme law of the land, and the seventh article sets out the procedures for ratifying the Constitution.

The Bill of Rights

One of the most important aspects of the Constitution is the Bill of Rights, which consists of the first ten amendments. The Bill of Rights was added to the

Constitution in 1791, and it provides additional protections for individual rights and limits the powers of the federal government.

The Bill of Rights includes provisions such as freedom of speech, religion, and the press, the right to bear arms, and protection against unreasonable searches and seizures. These rights are essential to the functioning of a free and democratic society, and they have been the subject of many court cases over the years.

Other Amendments

In addition to the Bill of Rights, there have been 17 other amendments to the Constitution. Some of the most significant amendments include the 13th Amendment, which abolished slavery, the 14th Amendment, which guarantees equal protection under the law, and the 19th Amendment, which gave women the right to vote.

The Constitution and the Legal System

The Constitution is the foundation of the legal system in the United States. It establishes the framework for our government, outlines the powers and responsibilities of each branch, and provides protections for individual rights. The Constitution also establishes the relationship between the federal government and the states, and it provides procedures for amending the Constitution.

Understanding the Constitution and its amendments is essential for anyone who wants to work in the legal system, whether as a lawyer, judge, or law enforcement officer. The Constitution and its amendments provide the basis for many of the laws and court cases that shape our legal system today.

Conclusion

The Constitution of the United States and its amendments are the foundation of our legal system. They provide the framework for our government, establish the relationship between the federal government and the states, and provide protections for individual rights. Understanding the Constitution and its amendments is essential for anyone who wants to understand the legal system in the United States, and it is an important part of American history and culture.

Statutory law

Statutory law refers to laws that are enacted by a legislative body. Statutes are written laws that set out specific rules and regulations to be followed by individuals and businesses within a given jurisdiction. They are an essential source of law and form the backbone of legal systems in many countries. This section will provide an in-depth analysis of statutory law, exploring its definition, sources, interpretation, and application. It will also examine the advantages and disadvantages of statutory law and provide examples of statutory law from various fields.

Definition of Statutory Law:

Statutory law refers to written laws that are enacted by a legislative body such as a parliament or congress. These laws are typically codified, meaning that they are organized into a single document or set of documents that contain all the rules and regulations that govern a particular area of law. Statutory law can cover a wide range of legal issues, from criminal law to commercial law, and can be enforced by government agencies or through private litigation.

Sources of Statutory Law:

The primary source of statutory law is the legislative body, which can be either a national or regional parliament, congress, or other legislative assembly. These bodies are responsible for creating and enacting laws that are binding on all individuals and businesses within their jurisdiction. Statutes may also be created by administrative agencies that are authorized by the legislature to create rules and regulations to implement and enforce the statutes.

Interpretation of Statutory Law:

The interpretation of statutory law is the process of determining the meaning and scope of the language used in the statute. The interpretation of statutory law can be a complex process, as the language used in statutes is often technical and precise. Courts play a critical role in interpreting statutory law, and their decisions can have a significant impact on the application of the law in practice.

Application of Statutory Law:

Statutory law applies to all individuals and businesses within the jurisdiction of the legislative body that enacted the statute. It is enforceable through both criminal and civil courts, and violations of statutory law can result in fines, imprisonment, or other penalties. Statutory law can also create rights and obligations for individuals and businesses, such as the obligation to pay taxes or the right to receive government benefits.

Advantages and Disadvantages of Statutory Law:

Statutory law has several advantages over other sources of law. It is often clear and specific, providing individuals and businesses with a clear understanding of their rights and obligations. It is also flexible, as legislative bodies can amend and update statutes as needed to reflect changing social and economic conditions.

However, statutory law also has several disadvantages. It can be rigid, as it is often difficult to change or amend statutes once they are enacted. It can also be complex and difficult to understand, particularly for individuals and businesses without legal expertise. In addition, statutory law can be subject to political influences, as legislative bodies may be influenced by special interest groups or lobbyists.

Examples of Statutory Law:

Statutory law can be found in many fields, including criminal law, commercial law, and employment law. In criminal law, statutes define specific crimes and the penalties for committing them. In commercial law, statutes regulate the formation and operation of businesses, including the sale of goods and services. In employment law, statutes set out the rights and obligations of employers and employees, including minimum wage requirements and workplace safety standards.

Mathematical Applications:

One example of a mathematical application in statutory law is the calculation of damages in a civil lawsuit. When a plaintiff is awarded damages in a civil case, the amount is often determined by a formula that takes into account the actual harm suffered by the plaintiff, such as medical expenses, lost wages, and pain and suffering. For example, the formula for calculating damages in a personal injury case might be:

Damages = Medical Expenses + Lost Wages + Pain and Suffering

In this formula, each of the factors represents a specific amount that can be quantified. Medical expenses refer to the costs of any medical treatment or therapy that the plaintiff received as a result of their injury. Lost wages refer to the income that the plaintiff was unable to earn as a result of the injury. Pain and suffering is a more subjective factor that takes into account the emotional and psychological impact of the injury on the plaintiff's life.

Another example of a mathematical application in statutory law is the use of statistical analysis in antitrust cases. Antitrust laws are designed to promote competition and prevent monopolies from forming in the market. When there is a suspicion that a company has engaged in anticompetitive behavior, statistical analysis can be used to demonstrate the effects of that behavior on the market.

For example, suppose that a company is accused of engaging in price-fixing with its competitors. One way to analyze the effects of this behavior on the market is to use regression analysis to determine the relationship between the price of the company's products and other variables, such as the price of competing products or the overall demand for the product. If the analysis shows a strong correlation between the company's prices and the prices of its competitors, this could be evidence of price-fixing.

In addition to these examples, mathematical applications can also be found in areas such as tax law, securities regulation, and intellectual property law. For example, tax law often involves calculations of income, deductions, and credits, which require a basic understanding of arithmetic and algebra. Securities regulation involves analyzing financial statements and using statistical models to predict the performance of investments. Intellectual property law may involve calculations of damages in patent infringement cases.

Overall, the use of mathematics in statutory law demonstrates the importance of quantitative reasoning and analysis in the legal profession. As laws become more complex and the stakes become higher, the ability to understand and apply mathematical concepts is increasingly important for lawyers and legal professionals.

Common law

Common law is an important source of law in many countries, including the United States, the United Kingdom, Canada, and Australia. It is a system of law based on judicial decisions, rather than on written laws or statutes. Common law is derived from the principles and precedents established by previous court decisions, and is developed and refined over time by judges and legal scholars.

Historically, common law developed in England following the Norman Conquest in 1066. Prior to the Conquest, England had a decentralized legal system in which different regions had their own customs and laws. However, after the Conquest, the King's courts became the primary forum for resolving disputes, and judges began to rely more heavily on earlier decisions when making rulings. Over time, these decisions became the basis for the common law system that we know today.

One of the key features of common law is the doctrine of stare decisis, which means "to stand by things decided." This doctrine holds that courts are bound by the decisions of higher courts in similar cases, and that they should follow established legal precedent when deciding new cases. As a result, common law is characterized by its emphasis on precedent and the importance of previous decisions in shaping the law.

One area in which common law is particularly important is in the development of tort law. Tort law is the branch of law that deals with civil wrongs, such as negligence or intentional harm, that result in harm to another person. In common law systems, tort law has evolved over time through the accumulation of judicial decisions and the establishment of legal precedent. For example, the famous case of Donoghue v Stevenson in 1932 established the principle of negligence, which has since become a cornerstone of tort law in many common law countries.

Another important aspect of common law is its flexibility and adaptability. Because common law is based on judicial decisions rather than on written laws or statutes, it can be adapted and refined over time to reflect changing social norms and values. This allows common law to evolve and develop in response to new and emerging legal issues, such as those related to emerging technologies like social media or artificial intelligence.

Mathematical applications in common law are less prevalent than in statutory law, as common law is largely based on precedent and the interpretation of legal principles rather than on specific numerical calculations. However, there are some areas in which mathematical concepts and calculations are important in common law, such as in cases involving economic damages or in cases where statistical evidence is used to prove a claim.

One example of a mathematical application in common law is the use of statistical analysis in discrimination cases. In employment discrimination cases, statistical evidence can be used to establish patterns of discrimination or to demonstrate that a particular employment practice has a disparate impact on certain groups. For example, statistical evidence might be used to show that a company's hiring practices consistently result in fewer women or minorities being hired than would be expected based on their qualifications and the available pool of applicants.

In addition to statistical analysis, mathematical concepts such as probability and game theory can also be used in common law. For example, game theory can be used to analyze legal disputes involving multiple parties or to predict the outcome of a particular legal strategy. Probability theory can also be used to assess the likelihood of a particular legal outcome or to evaluate the strength of a particular legal argument.

Overall, common law is an important source of law in many countries, and plays a vital role in shaping legal systems and ensuring that justice is served. While mathematical applications in common law may be less common than in other areas of law, they nevertheless have an important role to play in shaping legal precedent and ensuring that the law is applied fairly and consistently.

In conclusion, the sources of law are complex and multifaceted. From constitutional law to statutory law to common law, each source has its own unique characteristics, historical development, and current applications. Understanding the sources of law is essential for lawyers, judges, and legal scholars, as well as for individuals and businesses who must navigate the legal system.

One important aspect of the sources of law is the role of precedent. Precedent plays a critical role in the common law system, where judges rely on previous court decisions to interpret and apply the law. Precedent allows for consistency and predictability in the law, while also allowing for flexibility and adaptation to changing circumstances.

Moreover, mathematical applications in the law, including the use of probability and statistics, have become increasingly important in recent years. These tools can help lawyers and judges make more informed and accurate decisions in a variety of legal contexts, from criminal trials to civil lawsuits to regulatory compliance.

In sum, the sources of law are a fascinating and complex area of study, with implications for a wide range of legal and non-legal fields. Whether you are a lawyer, a business owner, or simply an interested citizen, an understanding of the sources of law can help you navigate the legal system and make informed decisions.

CHAPTER 4: TYPES OF LEGAL SYSTEMS

Legal systems exist in every country around the world to provide order, protect individuals and businesses, and enforce laws. However, legal systems vary widely from country to country, and understanding the differences between them is critical for anyone working in a globalized economy. In this chapter, we will explore the different types of legal systems and their characteristics, including common law, civil law, religious law, and mixed legal systems.

The study of legal systems can be complex and nuanced, and requires a deep understanding of both the legal and cultural contexts in which they operate. Additionally, the different types of legal systems can interact in complex ways, particularly in the context of international business transactions or legal disputes. As such, it is essential for professionals in a variety of fields, including business owners and entrepreneurs, corporate executives, in-house counsel and legal departments, compliance officers, financial analysts, investment bankers, accountants, and regulators to have a solid understanding of the different types of legal systems and how they function.

In this chapter, we will provide a comprehensive overview of each type of legal system, including its history, key features, and examples of countries where it is in use. We will also explore the advantages and disadvantages of each system, as well as the potential challenges that may arise when operating across different legal systems.

Finally, we will consider the implications of the rise of globalization on legal systems around the world, and the ways in which legal systems are evolving to meet the demands of an increasingly interconnected world. We will conclude the chapter by reflecting on the importance of understanding legal systems for anyone working in a global context, and the ways in which such understanding can inform better decision-making and facilitate smoother cross-border transactions.

Common law systems

The legal systems of different countries can be broadly classified into two categories: civil law and common law. Civil law systems trace their origins to ancient Rome and rely heavily on written codes and statutes. In contrast, common law systems evolved in England and rely more on judicial decisions and precedents.

Common law systems are based on the principle of stare decisis, which means that judges are bound to follow the precedents set by higher courts in earlier cases. This system creates a body of case law that serves as a source of law in itself. Unlike civil law systems, common law systems place a greater emphasis on the role of judges in interpreting and applying the law.

The common law system has been adopted by many countries around the world, including the United States, Canada, Australia, New Zealand, and India. These countries have adapted the common law system to their own legal and cultural contexts, resulting in some variations in the way the system operates.

This section will provide an overview of common law systems, including their history, key characteristics, and examples of how they are used in different countries. We will also explore the advantages and disadvantages of the common law system, and how it compares to other legal systems.

History of Common Law Systems

The origins of the common law system can be traced back to medieval England, where judges began to develop a system of law based on their own decisions and interpretations of earlier cases. Over time, a body of common law developed that applied throughout the country.

The common law system became more established in England during the reign of King Henry II in the 12th century. Henry II created a system of royal courts that were responsible for enforcing the law throughout the kingdom. These courts were staffed by judges who were trained in the law and who were expected to apply the principles of common law in their decisions.

In the centuries that followed, the common law system continued to evolve in England. Judges began to rely more on precedent and less on traditional customs and practices, and the law became more complex and specialized.

When English colonists settled in North America, they brought the common law system with them. This system became the basis for the legal systems of the United States and Canada, as well as other countries that were once part of the British Empire.

Key Characteristics of Common Law Systems

One of the key characteristics of the common law system is the importance of precedent. In common law systems, judges are bound to follow the precedents set by higher courts in earlier cases. This creates a body of case law that serves as a source of law in itself.

Another characteristic of the common law system is the role of judges in interpreting and applying the law. In common law systems, judges have a greater degree of discretion in deciding cases than in civil law systems. Judges are expected to apply the principles of common law to the specific facts of each case, and to consider the impact of their decisions on future cases.

Common law systems also tend to rely on adversarial proceedings, in which opposing parties present their cases to a judge or jury. This is in contrast to civil law systems, which often rely on inquisitorial proceedings, in which a judge investigates the facts of a case and reaches a decision based on the evidence presented.

Examples of Common Law Systems

Common law systems are used in many countries around the world, including the United States, Canada, Australia, New Zealand, and India. While these countries share a common legal heritage, there are some differences in the way the common law system is used in each country.

For example, in the United States, the common law system is used in the federal courts and in most state courts. However, each state has its own body of case law, and there may be variations in the way the common law is interpreted and applied in different states. In addition, the United States legal system is a dual system, which means that there are both federal and state laws that govern different aspects of life. This can make the application of the common law more complex, as there may be different interpretations of the same legal principles in different jurisdictions.

Similarly, in Canada, the common law system is used in the federal courts and in most provincial and territorial courts. However, the legal system in Quebec is based on civil law, which is a different legal system altogether. As a result, there are differences in the way legal cases are interpreted and decided in Quebec compared to the rest of Canada.

In Australia, the common law system is used in both federal and state courts. However, the legal system in the state of Victoria has been influenced by the civil law system used in continental Europe. This has led to some unique features in the way legal cases are decided in Victoria, such as the use of juries in civil cases.

In New Zealand, the common law system is used in the higher courts, while the lower courts use a mix of common law and statute law. The legal system in New

Zealand is also unique in that the country has a treaty with the indigenous Maori people, known as the Treaty of Waitangi. This treaty has had an impact on the way the legal system is interpreted and applied in cases involving Maori rights and interests.

Finally, in India, the common law system is used in the courts, but it is heavily influenced by the country's religious and cultural traditions. For example, Indian law recognizes the concept of dharma, which is a Hindu concept of morality and duty, and this concept can sometimes be used to interpret legal cases.

Overall, while common law systems share many similarities, there are also important differences in the way they are used in different countries. These differences are shaped by a variety of factors, including historical traditions, cultural values, and political institutions. Understanding these differences is essential for anyone seeking to navigate the legal systems of different countries and to engage in international business and commerce.

Civil law systems

Civil law systems are one of the major legal systems used around the world. They are based on written laws and codes, which are used to resolve disputes between individuals or entities. Unlike common law systems, civil law systems do not rely heavily on precedent, and judges are not bound by previous decisions.

Civil law systems are used in many countries, including France, Germany, Japan, and most countries in Latin America. In this section, we will explore the key features of civil law systems, including their history, sources of law, legal codes, and procedures.

History of Civil Law Systems:

The roots of civil law systems can be traced back to ancient Rome. The Roman legal system was based on written laws and codes, and it heavily influenced the development of modern civil law systems.

During the medieval period, the Roman legal system was largely forgotten in Europe, and laws were based on custom and tradition. However, in the 11th century, a revival of Roman law began, and it was later codified in the Corpus Juris Civilis, a collection of laws compiled by the Byzantine emperor Justinian I.

In the following centuries, civil law systems spread throughout Europe, and they were further developed and refined in countries such as France and Germany. Today, civil law systems are used in many countries around the world, and they continue to evolve and adapt to changing social and economic conditions.

Sources of Law in Civil Law Systems:

In civil law systems, the primary sources of law are legislative acts, such as codes, statutes, and regulations. These laws are written and published by government bodies, and they are binding on all individuals and entities within the jurisdiction.

In addition to legislative acts, civil law systems also recognize the importance of judicial decisions, legal doctrine, and academic scholarship. However, these sources of law are not as influential in civil law systems as they are in common law systems.

Legal Codes in Civil Law Systems:

One of the most distinctive features of civil law systems is the use of legal codes. These codes are comprehensive and systematic compilations of laws and regulations that cover a wide range of legal issues, such as contracts, property, and torts.

Legal codes are designed to be clear, concise, and accessible to all individuals and entities. They are intended to provide a complete and consistent set of rules that can be applied to all legal disputes.

Civil law systems also recognize the importance of judicial interpretation of legal codes. However, judges are not bound by precedent, and they have more discretion to interpret and apply the law than in common law systems.

Procedures in Civil Law Systems:

Civil law systems have a distinctive set of procedures for resolving legal disputes. These procedures are designed to be efficient, affordable, and accessible to all individuals and entities.

In civil law systems, legal disputes are typically resolved in specialized courts, such as civil courts or commercial courts. These courts have jurisdiction over specific types of legal disputes, and they are staffed by judges who are trained in the relevant areas of law.

In addition to specialized courts, civil law systems also have a range of alternative dispute resolution mechanisms, such as mediation and arbitration. These mechanisms are designed to be faster and less costly than traditional litigation, and they are increasingly popular in many civil law jurisdictions.

Examples of Civil Law Systems:

Civil law systems are used in many countries around the world, including France, Germany, Japan, and most countries in Latin America. In each of these countries, civil law is used to resolve disputes between individuals and entities.

For example, in France, civil law is used to resolve a wide range of legal disputes, including contract disputes, property disputes, and family law disputes. The French legal system is based on a comprehensive set of legal codes, which provide a complete and consistent set of rules for resolving legal disputes.

Conclusion:

In conclusion, civil law systems have their own unique characteristics and play a crucial role in shaping legal systems across the globe. These systems are primarily based on written legal codes and statutes, which provide comprehensive and consistent guidelines for resolving legal disputes.

Civil law systems are used in many countries around the world, including France, Germany, Spain, Italy, and Japan. In these countries, civil law is used to resolve a wide range of legal disputes, including contract disputes, property disputes, and family law disputes.

One of the key features of civil law systems is the emphasis on legal codes and statutes, which are designed to provide clarity and predictability in legal decision-making. The use of legal codes also means that civil law systems tend to rely less on judicial precedent than common law systems.

However, this does not mean that civil law systems lack flexibility or adaptability. In fact, many civil law systems are designed to be updated and revised over time to reflect changes in society and to ensure that the law remains relevant and effective.

Overall, civil law systems are an important component of the global legal landscape, and they provide a valuable alternative to common law systems. By understanding the key features and characteristics of civil law systems, legal professionals can better navigate the complexities of global legal systems and provide effective representation to their clients.

Religious law systems

Religious law systems, also known as theocratic legal systems, are legal systems that are based on religious texts, teachings, and traditions. In these legal systems, religious laws are considered to be the supreme source of authority, and the judiciary and legal institutions are tasked with interpreting and enforcing these laws. Religious law systems have been in existence for thousands of years, and they continue to be used in many parts of the world today. This section will explore the history, principles, and examples of religious law systems.

History of Religious Law Systems:

Religious law systems have been in existence for thousands of years, and they have played an important role in shaping the legal systems of many countries. The earliest religious law systems can be traced back to ancient civilizations such as Egypt, Mesopotamia, and Greece. These civilizations had complex legal systems that were based on religious beliefs and traditions.

One of the most well-known religious law systems is Islamic law, also known as Sharia law. Islamic law is based on the teachings of the Quran, which is the holy book of Islam, and the Sunnah, which are the sayings and actions of the Prophet Muhammad. Islamic law has been used in many countries throughout history, including Iran, Pakistan, and Saudi Arabia.

Another important religious law system is Jewish law, also known as Halakha. Jewish law is based on the Torah, which is the holy book of Judaism, and the Talmud, which is a collection of Jewish oral traditions. Jewish law has been used in many countries throughout history, including Israel and some parts of Europe.

Principles of Religious Law Systems:

Religious law systems are based on a set of fundamental principles that are rooted in religious beliefs and traditions. One of the key principles of religious law systems is that the law is divine in origin and is therefore absolute and unchanging. This means that the law is not subject to human interpretation or amendment, and it cannot be altered to suit changing circumstances or social norms.

Another principle of religious law systems is that the law is comprehensive and all-encompassing. This means that the law covers all aspects of human life, including religious practices, family law, criminal law, and commercial law. Religious law systems are designed to provide a complete and consistent set of rules for governing all aspects of human behavior.

Examples of Religious Law Systems:

There are many examples of religious law systems in the world today. One of the most well-known examples is Islamic law, which is used in many countries in the Middle East and Asia. Islamic law covers a wide range of legal topics, including criminal law, family law, and commercial law. Islamic law is based on the principles of the Quran and the Sunnah, and it is interpreted and applied by Islamic scholars and judges.

Another example of a religious law system is Jewish law, which is used in Israel and some parts of Europe. Jewish law covers a wide range of legal topics, including family law, criminal law, and commercial law. Jewish law is based on the principles of the Torah and the Talmud, and it is interpreted and applied by Jewish scholars and judges.

Conclusion:

Religious law systems are an important part of the legal landscape in many parts of the world. These legal systems are based on religious texts, teachings, and traditions, and they provide a comprehensive and consistent set of rules for governing human behavior. While religious law systems have been criticized for being inflexible and lacking in human rights protections, they continue to play an important role in shaping the legal systems of many countries. Understanding the principles and examples of religious law systems is essential for students seeking a comprehensive understanding of legal systems around the world.

PART 2: BUSINESS ORGANIZATIONS AND CORPORATE GOVERNANCE

Business organizations play a critical role in the global economy, providing goods and services to consumers, creating jobs, and generating wealth for investors. The legal framework that governs these organizations has evolved over time, as societies have sought to balance the interests of investors, managers, employees, and other stakeholders.

Part 2 of this work focuses on business organizations and corporate governance, exploring the legal and ethical issues that arise in the formation, operation, and regulation of corporations and other business entities. This part provides a comprehensive overview of the different types of business organizations, the legal and regulatory framework that governs their formation and operation, and the principles and practices of corporate governance.

The part begins with an overview of the different types of business organizations, including sole proprietorships, partnerships, limited liability companies, and corporations. It explains the advantages and disadvantages of each type of entity, and provides guidance on choosing the right business structure for a given venture.

Next, the part explores the legal and regulatory framework that governs the formation and operation of corporations, including corporate governance, securities regulation, and antitrust law. It examines the role of the board of directors, the duties and responsibilities of corporate officers and directors, and the mechanisms for shareholder participation and control.

The part also delves into the principles and practices of corporate governance, exploring the ethical and legal standards that govern the behavior of corporate managers, directors, and shareholders. It examines the key principles of corporate social responsibility, including the duty to respect human rights, protect the environment, and promote sustainable development.

Overall, Part 2 provides a comprehensive and in-depth exploration of the legal and ethical issues that arise in the formation, operation, and regulation of business organizations. It provides a solid foundation for anyone seeking to understand the complex legal and regulatory environment in which modern corporations operate, and the principles and practices of effective corporate governance.

CHAPTER 5: TYPES OF BUSINESS ORGANIZATIONS

The decision to start a business is not an easy one, and it involves several key considerations, including the type of business organization to choose. The type of business organization selected has important implications for the business owners in terms of legal and tax obligations, as well as liability for the business's debts and obligations. Therefore, it is crucial for business owners, entrepreneurs, corporate executives, in-house counsel/legal departments, compliance officers, financial analysts, investment bankers, accountants, and regulators to have a thorough understanding of the different types of business organizations.

In this chapter, we will explore the different types of business organizations, including sole proprietorships, partnerships, limited liability companies (LLCs), corporations, and nonprofit organizations. We will provide a detailed analysis of each type of organization, including the legal and tax implications, advantages, disadvantages, and practical considerations.

First, we will examine sole proprietorships, which are the simplest form of business organization. Sole proprietorships are owned and operated by a single individual, and they are relatively easy to establish and maintain. However, sole proprietors are personally liable for all of the business's debts and obligations, which can be a significant risk.

Next, we will look at partnerships, which are similar to sole proprietorships but involve two or more owners. Partnerships can be either general partnerships, in which all partners are personally liable for the business's debts and obligations, or limited partnerships, in which one or more partners have limited liability. Partnerships can be advantageous because they allow for shared ownership and management responsibilities, but they can also be complex and require a formal partnership agreement.

We will then move on to limited liability companies (LLCs), which are a popular choice for many small businesses. LLCs provide limited liability protection to their owners while still allowing for flexibility in management and taxation. However, LLCs can be subject to complex tax rules, and they require formal legal documentation to establish.

Next, we will explore corporations, which are separate legal entities that are owned by shareholders. Corporations provide limited liability protection to their shareholders and have formal management structures. However, corporations are subject to complex regulations, and they can be expensive to establish and maintain.

Finally, we will examine nonprofit organizations, which are designed to serve a specific charitable or social purpose. Nonprofit organizations can take several forms, including charities, foundations, and social enterprises. Nonprofits are typically exempt from certain taxes and have special legal obligations related to their charitable purpose.

Overall, the type of business organization selected can have significant implications for a business's success and sustainability. By understanding the advantages and disadvantages of each type of organization, business owners can make informed decisions about which type of organization is best suited for their needs. In the following sections, we will provide a detailed analysis of each type of business organization, including practical examples and exercises to reinforce understanding.

Sole proprietorship

A sole proprietorship is a type of business organization that is owned and operated by a single person. It is one of the simplest and most common forms of business ownership, particularly for small businesses. In this section, we will discuss the characteristics of a sole proprietorship, the advantages and disadvantages of this form of business organization, and the legal requirements for establishing and operating a sole proprietorship.

Characteristics of a Sole Proprietorship

A sole proprietorship is a business that is owned and operated by a single individual. In this form of business organization, the owner and the business are considered to be one and the same, meaning that the owner is personally responsible for all debts and obligations of the business. The owner is also entitled to all profits and has complete control over all business decisions.

Sole proprietorships are typically small businesses that do not require a lot of capital to start. The owner may use their personal savings or obtain a loan to fund the business. The business may operate under the owner's name or a fictitious business name, also known as a "doing business as" (DBA) name.

Advantages of a Sole Proprietorship

One of the main advantages of a sole proprietorship is that it is easy and inexpensive to set up and operate. The owner has complete control over the business and can make decisions quickly without consulting with others. The business is also easy to dissolve if the owner decides to close it.

Another advantage is that the owner is entitled to all profits of the business. Unlike other forms of business organization, there are no shareholders or partners to share profits with. Additionally, the owner may be able to take advantage of certain tax deductions and credits that are available to small businesses.

Disadvantages of a Sole Proprietorship

One of the main disadvantages of a sole proprietorship is that the owner is personally responsible for all debts and obligations of the business. This means that if the business is unable to pay its debts, the owner's personal assets may be at risk.

Another disadvantage is that a sole proprietorship may have limited access to capital. Since the owner is solely responsible for the business, it may be difficult to obtain financing from banks or investors.

Additionally, a sole proprietorship may have limited growth potential. The owner may not have the resources or expertise to expand the business beyond a certain point.

Legal Requirements for Establishing and Operating a Sole Proprietorship

In most cases, there are no specific legal requirements for establishing a sole proprietorship. However, depending on the type of business, the owner may need to obtain certain licenses or permits to operate.

The owner should also register the business name with the appropriate state agency and obtain any necessary tax identification numbers. Additionally, the owner should keep accurate records of all business transactions and file the appropriate tax returns.

Conclusion

In conclusion, a sole proprietorship is a simple and common form of business organization that is owned and operated by a single individual. It is easy and inexpensive to set up and operate, and the owner is entitled to all profits. However, the owner is personally responsible for all debts and obligations of the business, and the business may have limited access to capital and growth potential. Business owners and entrepreneurs should carefully consider the advantages and disadvantages of a sole proprietorship before deciding to establish this type of business organization.

Partnership

Partnerships are a type of business organization that is owned and managed by two or more people. Unlike sole proprietorships, partnerships allow for shared ownership, responsibility, and liability among the partners. In a partnership, each partner contributes to the business in terms of capital, labor, or expertise, and shares in the profits and losses of the business.

Partnerships can be either general partnerships or limited partnerships. In a general partnership, all partners are responsible for the management of the business and share in the profits and losses equally. In a limited partnership, there are both general partners who manage the business and limited partners who provide capital but do not participate in the management of the business.

This section will provide a detailed analysis of partnerships, including their advantages and disadvantages, formation, management, dissolution, and taxation. We will also discuss the legal requirements for partnerships, as well as some common challenges that partnerships face.

Advantages of Partnerships

Partnerships offer several advantages for entrepreneurs and small business owners. One of the most significant advantages of partnerships is that they allow for shared responsibility and decision-making among the partners. This can be especially beneficial for businesses that require diverse skills and expertise.

Partnerships also allow for shared financial resources, which can make it easier for partners to secure financing and manage cash flow. Partnerships can also offer tax advantages, as profits and losses are reported on the partners' individual tax returns rather than the business tax return.

Another advantage of partnerships is that they can be relatively easy and inexpensive to form. Partnerships can be formed with a simple written agreement between the partners, and there are no legal requirements for registering a partnership with the state.

Disadvantages of Partnerships

Despite their advantages, partnerships also have some disadvantages that should be considered. One of the main disadvantages of partnerships is that all partners share unlimited liability for the business's debts and obligations. This means that if the business incurs debts or is sued, each partner can be held personally liable for the entire amount owed.

Another potential disadvantage of partnerships is that they can be difficult to manage when partners have conflicting goals or visions for the business. Without clear communication and a strong partnership agreement, disputes between partners can arise, potentially leading to the dissolution of the partnership.

Formation of Partnerships

Forming a partnership requires a written agreement between the partners that outlines the terms of the partnership. The agreement should include details such as the name and purpose of the partnership, the contributions of each partner, the division of profits and losses, and the responsibilities of each partner.

Partnership agreements should also address potential issues that may arise, such as how decisions will be made, how disputes will be resolved, and what happens if a partner wants to leave the partnership. It is important for partners to consult with a lawyer to ensure that their partnership agreement is legally binding and protects their interests.

Management of Partnerships

Partnerships are managed by the partners, who are responsible for making decisions and managing the day-to-day operations of the business. In a general partnership, all partners have equal management rights and responsibilities. In a limited partnership, the general partners are responsible for managing the business, while the limited partners have no management authority.

Partnership management requires clear communication and collaboration among the partners. It is important for partners to establish a system for making decisions, such as through regular meetings or by designating one partner as the managing partner.

Dissolution of Partnerships

Partnerships can be dissolved for a variety of reasons, such as when a partner wants to leave the partnership or when the partners no longer wish to continue the business. The process for dissolution should be outlined in the partnership agreement, and may involve a vote by the partners or the appointment of a third-party mediator to help resolve disputes.

When a partnership is dissolved, the partners must wind up the affairs of the partnership, including paying off any debts and dividing up the remaining assets. If the partnership agreement does not specify how assets are to be divided, the default rule is that assets are divided equally among the partners.

One important consideration in the dissolution of a partnership is the treatment of partnership liabilities. Partnerships are not separate legal entities, so each partner is personally liable for the partnership's debts. This means that when a partnership is dissolved, partners may still be liable for any outstanding debts of the partnership, even if they did not personally incur those debts.

To avoid this risk, partners may consider creating a limited liability partnership (LLP) or a limited liability company (LLC), which provide some protection from personal liability for the debts of the business.

In addition to voluntary dissolution, partnerships can also be terminated involuntarily by court order. This may occur in cases of fraud, mismanagement, or other serious breaches of the partnership agreement or applicable law.

Partnerships can also be converted into other forms of business organizations, such as a limited liability company (LLC) or a corporation. This can be done for a variety of reasons, such as to limit personal liability for business debts or to raise capital by selling shares of stock to investors.

In summary, partnerships are a popular form of business organization that offer many benefits, including shared management and liability, flexible tax treatment, and the ability to raise capital through contributions from multiple partners. However, partnerships also have some drawbacks, such as the potential for personal liability for business debts and the risk of disputes among partners. Careful consideration of the advantages and disadvantages of partnerships is essential when choosing a form of business organization, and seeking the advice of legal and financial professionals is highly recommended.

Limited liability company

A Limited Liability Company (LLC) is a popular business structure that combines the benefits of both partnerships and corporations. Like a partnership, an LLC provides its owners with flexibility and pass-through taxation. At the same time, an LLC offers limited liability protection to its owners, similar to that of a corporation. As a result, LLCs are a common choice for small business owners, entrepreneurs, and investors.

In this section, we will discuss the key features of LLCs, their advantages and disadvantages, and the process for forming and operating an LLC. We will also examine the taxation of LLCs and compare them to other business structures, such as sole proprietorships, partnerships, and corporations.

Features of LLCs

LLCs are characterized by several key features, including limited liability protection, pass-through taxation, and flexible management structure. Let's take a closer look at each of these features.

Limited Liability Protection

One of the main advantages of an LLC is the limited liability protection it offers to its owners. This means that the personal assets of LLC owners are protected from business liabilities and debts. In other words, if the LLC is sued or goes bankrupt, the owners are not personally responsible for paying the debts and liabilities of the business.

Pass-through Taxation

LLCs are also popular because they are pass-through entities for tax purposes. This means that the profits and losses of the business pass through to the owners' personal tax returns. This allows LLC owners to avoid the double taxation that is common with corporations.

Flexible Management Structure

LLCs offer a flexible management structure that allows owners to run the business as they see fit. Unlike corporations, which have a more rigid management structure with a board of directors and officers, LLCs can be managed by their owners or by a designated manager. This flexibility allows LLCs to adapt to the needs and preferences of their owners.

Advantages of LLCs

LLCs offer several advantages over other business structures, including:

Limited liability protection: As mentioned earlier, LLCs provide limited liability protection to their owners, which can help protect personal assets in the event of business-related lawsuits or debts.

Pass-through taxation: LLCs are pass-through entities, which means that business income is taxed only once, at the owner's personal income tax rate. This can help LLC owners avoid double taxation, which is common with corporations.

Flexible management structure: LLCs offer a flexible management structure that allows owners to manage the business as they see fit. This can be especially helpful for small businesses that require more hands-on management.

Easy to form and maintain: LLCs are relatively easy to form and maintain, with fewer formalities and less paperwork than corporations.

Disadvantages of LLCs

Despite their advantages, LLCs also have some disadvantages, including:

Limited life: LLCs have a limited life span, which means that they may dissolve upon the death, retirement, or departure of one or more of the owners.

Self-employment taxes: LLC owners may be subject to self-employment taxes, which can be higher than payroll taxes for employees.

State-specific regulations: LLCs are subject to state-specific regulations, which can vary widely from state to state. This can make it difficult for businesses operating in multiple states.

Forming an LLC

The process for forming an LLC varies from state to state, but typically involves the following steps:

Choose a name for your LLC: The name must be unique and not already in use by another business.

File articles of organization: This is a legal document that establishes the LLC and its operating agreement. This document must be filed with the state in which the LLC is formed.

Obtain any necessary licenses and permits: Depending on the nature of the business, there may be specific licenses and permits required to operate legally. These requirements also vary by state and industry.

Create an operating agreement: This document outlines the ownership structure and management of the LLC, as well as how profits and losses will be distributed among the members.

Issue membership interests: LLCs are owned by their members, who hold a percentage of ownership known as membership interests. These interests can be assigned or sold, and the ownership structure can change over time.

Obtain an employer identification number (EIN): This is a unique identifier assigned by the Internal Revenue Service (IRS) to businesses for tax purposes.

Once these steps are completed, the LLC is considered legally formed and can begin operating as a separate entity from its owners. It is important to note that LLCs are not taxed as separate entities, but rather as pass-through entities, meaning that the income and losses of the business are reported on the individual tax returns of the members.

One of the primary advantages of forming an LLC is the limited liability protection it offers to its members. This means that members are not personally responsible for the debts and obligations of the business, and their personal assets are protected from lawsuits or other legal claims against the LLC.

Another advantage of an LLC is its flexibility in terms of management and taxation. LLCs can be managed by the members themselves, or by a designated manager or board of managers. They also have the option to be taxed as a partnership, an S corporation, or even as a C corporation, depending on their specific needs and goals.

However, there are also some potential disadvantages to consider when forming an LLC. One of these is the cost of formation, as the process can involve legal fees and other expenses. Additionally, LLCs may be subject to certain taxes and fees in some states, such as franchise taxes or annual report fees.

Furthermore, while LLCs offer limited liability protection, this protection may not be absolute. In some cases, courts may "pierce the corporate veil" and hold members personally liable for the debts and obligations of the LLC if they have engaged in fraudulent or illegal activities or have not followed proper procedures for maintaining the separation between their personal and business finances.

Overall, an LLC can be a highly advantageous structure for small businesses and entrepreneurs looking to protect their personal assets while maintaining flexibility and control over their business operations. However, it is important to carefully consider the specific needs and goals of the business before choosing this form of organization, and to seek the advice of legal and financial professionals as needed.

Corporation

Corporations are one of the most common forms of business organization, and they have been around for centuries. They are characterized by their separate legal identity, limited liability, and ability to raise capital through the sale of stock. This section will provide an in-depth analysis of corporations, including their advantages and disadvantages, the process for forming a corporation, and the various types of corporations.

Advantages of Corporations:

One of the main advantages of a corporation is that it has a separate legal identity from its owners, which means that the corporation can enter into contracts, sue and be sued, and own property in its own name. This provides a level of protection for the owners, who are not personally liable for the debts and obligations of the corporation. Limited liability is particularly important for corporations engaged in high-risk activities, such as manufacturing or construction, where the potential for lawsuits and other liabilities is high.

Another advantage of corporations is their ability to raise capital through the sale of stock. This allows corporations to grow and expand their operations, and it also allows investors to participate in the growth and success of the corporation. Corporations can issue different classes of stock with different voting and dividend rights, which allows them to raise capital from a variety of sources.

Corporations are also more durable than other forms of business organization. Unlike sole proprietorships and partnerships, corporations have perpetual existence, which means that they can continue to operate even after the death or retirement of their owners. This provides a level of stability and continuity that is important for the long-term success of the business.

Disadvantages of Corporations:

One of the main disadvantages of corporations is their complexity and expense. Corporations are subject to a variety of legal and regulatory requirements, including state and federal tax laws, securities laws, and corporate governance laws. This can make it more difficult and expensive to set up and operate a corporation compared to other forms of business organization.

Corporations are also subject to double taxation. This means that the corporation itself is taxed on its profits, and then the shareholders are taxed on any dividends they receive. This can reduce the amount of money available for distribution to shareholders and can make corporations less attractive to investors.

Another disadvantage of corporations is that they are subject to more public scrutiny than other forms of business organization. Because corporations issue stock and are subject to securities laws, they must disclose certain information to the public, including financial statements and other business information. This can make it more difficult for corporations to maintain their privacy and to protect their trade secrets and other confidential information.

Forming a Corporation:

The process for forming a corporation involves several steps. The first step is to choose a name for the corporation, which must be unique and not already in use by another business. Once a name has been chosen, the next step is to file articles of incorporation with the state in which the corporation will be formed.

The articles of incorporation must include certain information, such as the name and address of the corporation, the purpose of the corporation, and the number and type of shares of stock that will be authorized. Once the articles of incorporation have been filed and approved, the corporation is considered to be a legal entity and can begin to conduct business.

Types of Corporations:

There are several different types of corporations, each with its own advantages and disadvantages. The most common types of corporations are C corporations and S corporations.

C corporations are the most common type of corporation, and they are subject to double taxation. C corporations are also subject to more regulations and legal requirements than other types of corporations, which can make them more expensive to operate.

S corporations are a special type of corporation that allows the corporation to avoid double taxation. Instead, the income of the corporation is passed through to the shareholders, who are then taxed on their individual tax returns. S corporations are limited in terms of ownership and are only allowed to have up to 100 shareholders who must be US citizens or residents. Additionally, S corporations are required to follow strict eligibility criteria, including that they must be a domestic corporation, have only one class of stock, and have no more than 100 shareholders.

Another type of corporation is a nonprofit corporation, which is formed for charitable, religious, educational, or other charitable purposes. Nonprofit corporations are not subject to federal income tax, but they are still required to file annual reports with the IRS and maintain accurate financial records.

In addition to these types of corporations, there are also professional corporations, which are typically used by licensed professionals such as doctors, lawyers, and accountants. These corporations are subject to special regulations and restrictions, but they offer certain benefits such as liability protection and the ability to take advantage of certain tax benefits.

Overall, the type of corporation that is most appropriate for a particular business will depend on a variety of factors, including the business's goals, ownership structure, and tax considerations. It is important for business owners to consult with legal and financial professionals when deciding on the appropriate type of corporation for their business.

Math exercise:
Assume that a C corporation earns $500,000 in revenue and has $200,000 in deductible expenses. The corporate tax rate is 21%. What is the corporation's taxable income and tax liability?

Solution:
Revenue - Deductible expenses = Taxable income
$500,000 - $200,000 = $300,000

Taxable income x Corporate tax rate = Tax liability
$300,000 x 0.21 = $63,000

Therefore, the corporation's taxable income is $300,000 and its tax liability is $63,000.

CHAPTER 6: CORPORATE GOVERNANCE PRINCIPLES AND PRACTICES

Corporate governance is the set of processes, principles, and practices by which a company is directed and controlled. It encompasses the relationships among a company's management, board of directors, shareholders, and other stakeholders, and it is critical to the long-term success of a business. Effective corporate governance helps to ensure that a company operates with integrity, accountability, and transparency, and that it manages risks appropriately.

The importance of corporate governance has become increasingly apparent in recent years, as numerous high-profile corporate scandals and failures have highlighted the consequences of poor governance. The collapse of Enron, for example, was largely attributed to a failure of corporate governance, as the company's board of directors failed to exercise appropriate oversight of management's actions. Similarly, the financial crisis of 2008 was in part the result of a lack of effective governance and risk management within the banking sector.

In response to these failures, regulators around the world have increased their focus on corporate governance, and companies themselves have become more aware of the need to adopt good governance practices. This has led to the development of a range of corporate governance principles and guidelines, such as those issued by the Organisation for Economic Co-operation and Development (OECD), the International Corporate Governance Network (ICGN), and the United Nations Global Compact.

This chapter provides an overview of the principles and practices of corporate governance, including the roles and responsibilities of boards of directors, the relationship between management and the board, the importance of effective risk management, and the impact of governance on stakeholders. It also examines the challenges and opportunities associated with corporate governance, including the need for companies to balance the interests of different stakeholders and to adapt to changing market conditions.

Overall, the chapter aims to provide a comprehensive understanding of the importance of corporate governance and the key principles and practices that companies can adopt to ensure that they operate with integrity, accountability, and transparency, and that they deliver long-term value to their stakeholders.

The role of the board of directors

The board of directors is a critical component of corporate governance. The board is responsible for overseeing the management of the company and ensuring that the company operates in the best interests of its shareholders. The board plays a crucial role in setting the strategic direction of the company and ensuring that management executes that strategy. In this section, we will explore the role of the board of directors, including its composition, responsibilities, and best practices.

Composition of the Board:

The composition of the board of directors is crucial to its effectiveness. A well-composed board brings together individuals with diverse skills, experiences, and backgrounds. The board should include both insiders, such as the CEO and other top executives, as well as outsiders who can bring an objective perspective to the board's deliberations.

Board members should be chosen based on their expertise, experience, and ability to contribute to the company's strategic direction. The ideal board should include individuals with experience in finance, marketing, operations, and other critical areas of the business. Additionally, board members should have a track record of success in their respective fields and be able to provide meaningful insights and guidance to management.

Responsibilities of the Board:

The primary responsibility of the board of directors is to oversee the management of the company. The board sets the strategic direction of the company, approves major business decisions, and monitors the performance of management. The board is responsible for ensuring that the company operates in compliance with all applicable laws and regulations and that it maintains a strong ethical culture.

The board is also responsible for ensuring that the company has adequate resources to pursue its strategic objectives. This includes approving the company's budget and monitoring its financial performance. Additionally, the board is responsible for ensuring that the company has a sound risk management program in place to identify and mitigate potential risks.

Best Practices for the Board:

There are several best practices that the board of directors can follow to ensure its effectiveness. These include:

Regularly reviewing the company's strategy and ensuring that management is executing that strategy effectively.

Ensuring that the company has a strong ethical culture and that it operates in compliance with all applicable laws and regulations.

Ensuring that the company has adequate resources to pursue its strategic objectives.

Ensuring that the company has a sound risk management program in place.

Regularly evaluating the performance of management and making changes as necessary.

Ensuring that the board is composed of individuals with diverse skills, experiences, and backgrounds.

Providing meaningful oversight and guidance to management.

Regularly reviewing the company's financial performance and making adjustments as necessary.

Conclusion:

In conclusion, the board of directors plays a critical role in corporate governance. A well-composed and effective board can help ensure that the company operates in the best interests of its shareholders and that it achieves its strategic objectives. By following best practices, the board can provide meaningful oversight and guidance to management, and ensure that the company operates in compliance with all applicable laws and regulations.

Shareholder rights

In a corporate structure, shareholders are considered the owners of the company. As such, they have certain rights that they can exercise to influence and monitor the actions of the board of directors and the management team. These rights include the right to vote on major corporate decisions, the right to access information about the company, and the right to receive dividends. This section will provide a detailed analysis of shareholder rights and their importance in corporate governance.

Voting Rights

Shareholders have the right to vote on major corporate decisions, such as the election of directors, mergers and acquisitions, and amendments to the company's bylaws. This right is usually exercised at the annual general meeting (AGM) of shareholders, where the board of directors presents the company's financial statements and proposes new resolutions.

One of the key ways that shareholders can exercise their voting rights is by casting their votes for the election of directors. The board of directors is responsible for overseeing the management of the company and making important strategic decisions. Shareholders can use their voting rights to elect directors who they believe will act in their best interests and hold the management team accountable.

Shareholders can also use their voting rights to block major corporate actions that they disagree with, such as mergers or acquisitions. For example, if shareholders believe that a proposed acquisition is not in their best interests, they can vote against it and prevent it from being approved.

Information Rights

Another important right that shareholders have is the right to access information about the company. This includes access to financial statements, audit reports, and other important documents. Shareholders can use this information to monitor the company's performance, assess the risks that the company faces, and make informed decisions about whether to buy or sell shares.

In addition to accessing information, shareholders also have the right to request additional information from the company. For example, if shareholders believe that the company's financial statements are incomplete or misleading, they can request additional information to help them better understand the company's financial position.

Dividend Rights

Shareholders also have the right to receive dividends. Dividends are payments made by the company to its shareholders, usually in the form of cash or stock. Dividends are typically paid out of the company's profits and are a way for shareholders to share in the company's financial success.

The amount of the dividend that shareholders receive is usually determined by the board of directors. Shareholders can use their voting rights to influence the board's decision on whether to pay a dividend and how much to pay.

Limitations on Shareholder Rights

While shareholder rights are an important aspect of corporate governance, they are not absolute. There are several limitations on shareholder rights that are designed to protect the company and its stakeholders.

One limitation on shareholder rights is the board of directors' power to make decisions that are in the best interests of the company as a whole. While shareholders have the right to vote on major corporate decisions, the board of directors is ultimately responsible for making these decisions and can override the wishes of shareholders if it believes that doing so is necessary to protect the company's interests.

Another limitation on shareholder rights is the principle of majority rule. In most cases, decisions are made based on the majority of votes cast. This means that minority shareholders may not have as much influence as majority shareholders, even if they disagree with the decisions being made.

Conclusion

Shareholder rights are an important aspect of corporate governance. They provide shareholders with a way to influence and monitor the actions of the board of directors and the management team. By exercising their voting rights, accessing information about the company, and receiving dividends, shareholders can hold the company accountable and make informed decisions about their investments. While there are limitations on shareholder rights, these are designed to protect the company and its stakeholders and ensure that decisions are made in the best interests of all parties involved.

It is important for shareholders to understand their rights and how to exercise them effectively. This requires education and engagement with the company, as well as the willingness to take action when necessary. Shareholders should also be aware of the potential risks and benefits of investing in a particular company, including the company's corporate governance structure and the strength of its shareholder rights.

In conclusion, shareholder rights play a crucial role in corporate governance, providing a mechanism for shareholders to hold the board of directors and management team accountable and make informed decisions about their investments.

While there are limitations on these rights, they are designed to balance the interests of shareholders with those of the company and its stakeholders. By understanding their rights and how to exercise them effectively, shareholders can play an important role in promoting good corporate governance and contributing to the long-term success of the company.

Ethical considerations in corporate governance

Corporate governance is not only about compliance with legal regulations and maximizing shareholder value, but also about ethical considerations. Ethical considerations play an important role in shaping the culture of the organization and the decisions made by the board of directors and the management team. Ethical behavior in corporate governance is crucial for the sustainability and success of the company, as well as the trust and confidence of stakeholders. In this section, we will discuss the importance of ethics in corporate governance, the ethical principles that should guide decision-making, and the challenges and risks associated with ethical lapses in corporate governance.

Importance of Ethics in Corporate Governance

Ethics refers to a set of moral principles that guide behavior and decision-making. In the context of corporate governance, ethics refers to the principles and values that should guide the behavior and decision-making of the board of directors and the management team. Ethical behavior is important for several reasons:

1.   Reputation: The reputation of a company is built on its ethical behavior. A company that is perceived to be ethical and responsible is more likely to attract customers, investors, and talented employees.

2.   Stakeholder Trust: Ethical behavior is essential for building trust and confidence with stakeholders, such as customers, investors, employees, and the public. Without trust, it is difficult to establish long-term relationships and to maintain a social license to operate.

3.   Risk Management: Ethical behavior is also important for managing risk. Unethical behavior can result in legal and financial risks, as well as reputational risks that can harm the company and its stakeholders.

4.   Social Responsibility: Ethical behavior is a social responsibility of corporations. Companies have a responsibility to act in the best interests of society, to respect human rights, and to minimize negative impacts on the environment.

Principles of Ethical Corporate Governance

The principles of ethical corporate governance can be divided into two categories: principles of conduct and principles of responsibility.

Principles of Conduct

1.   Honesty: Honesty is essential in corporate governance. The board of directors and the management team should be honest and transparent in their communication with stakeholders, including customers, investors, and employees.

2.   Integrity: Integrity refers to the adherence to moral and ethical principles. The board of directors and the management team should demonstrate integrity in their behavior and decision-making.

3.   Fairness: Fairness refers to the impartial treatment of stakeholders. The board of directors and the management team should treat all stakeholders fairly and without bias.

4.   Responsibility: Responsibility refers to the obligation to act in the best interests of the company and its stakeholders. The board of directors and the management team should take responsibility for their decisions and actions.

Principles of Responsibility

1.   Sustainability: Sustainability refers to the responsibility to consider the long-term impact of the company's decisions and actions on society and the environment.

2.   Accountability: Accountability refers to the responsibility to answer for one's decisions and actions. The board of directors and the management team should be accountable to stakeholders for their decisions and actions.

3.   Ethical Leadership: Ethical leadership refers to the responsibility to set an example of ethical behavior and to promote ethical behavior throughout the organization.

Challenges and Risks Associated with Ethical Lapses in Corporate Governance

While ethical behavior is essential for corporate governance, there are many challenges and risks associated with ethical lapses. Some of the common challenges and risks include:

1. Conflicts of Interest: Conflicts of interest can arise when the interests of the board of directors or the management team conflict with the interests of the company or its stakeholders.

2. Corruption: Corruption refers to the abuse of power for personal gain. Corruption can take many forms, including bribery, kickbacks, and embezzlement.

3. Fraud: Fraud refers to the deliberate misrepresentation of information for personal gain. Examples of fraud include falsifying financial statements, insider trading, and Ponzi schemes.

4. Lack of Transparency: Lack of transparency refers to a lack of openness and accountability in a company's operations. When a company fails to provide clear and accurate information about its finances, operations, or business practices, it can erode trust and confidence in the company.

5. Whistleblower Retaliation: Whistleblowers are employees who report illegal or unethical behavior within a company. Retaliation against whistleblowers can take many forms, including termination, demotion, or harassment.

These challenges and risks can have serious consequences for companies and their stakeholders. For example, ethical lapses can lead to reputational damage, legal liability, and financial losses. In some cases, unethical behavior can even lead to the downfall of a company.

To mitigate these risks, companies need to have strong systems of corporate governance in place. This includes implementing ethical standards, providing training and education to employees, and creating a culture of transparency and accountability.

Additionally, regulators and other oversight bodies play an important role in ensuring ethical behavior in corporate governance. These bodies may conduct investigations, impose fines and penalties, or even revoke a company's license to operate.

In conclusion, ethical considerations are an essential aspect of corporate governance. Companies that prioritize ethics and transparency are more likely to earn the trust and loyalty of their stakeholders and achieve long-term success. However, ethical lapses can have serious consequences, and companies must be vigilant in mitigating the risks associated with unethical behavior.

CHAPTER 7: BOARD OF DIRECTORS AND EXECUTIVE COMPENSATION

In recent years, there has been increased attention on the role of boards of directors in setting executive compensation. The compensation of top executives has risen significantly, while average worker pay has remained relatively stagnant. This has led to concerns about income inequality and the fairness of executive pay.

Boards of directors play a critical role in determining executive compensation, but their decisions are often influenced by a variety of factors, including market trends, peer comparisons, and individual performance. Additionally, there may be conflicts of interest among board members, who may have personal or professional relationships with the executives they are responsible for compensating.

In this chapter, we will explore the role of boards of directors in setting executive compensation. We will examine the various factors that influence their decision-making, including market forces, individual performance, and peer comparisons. We will also discuss the potential conflicts of interest that can arise and how these can be mitigated.

Through our analysis, we will seek to provide a comprehensive understanding of the complexities of executive compensation and the role that boards of directors play in setting it. We will examine the various perspectives on this issue, including those of business owners/entrepreneurs, corporate executives, in-house counsel/legal departments, compliance officers, financial analysts, investment bankers, accountants, regulators, and other stakeholders. Additionally, we will provide exercises and problems to help students understand the key concepts and apply them in real-world situations.

The composition of the board of directors

The board of directors is a key element of corporate governance, responsible for overseeing the management of the company and representing the interests of its stakeholders. The composition of the board is crucial for its effectiveness and the overall success of the company. This section will explore the various factors that determine the composition of the board of directors, including its size, diversity, independence, and expertise. We will also discuss the benefits and challenges of having a diverse board and the role of shareholders in shaping the composition of the board.

Board Size:

The size of the board of directors varies depending on the size and complexity of the company. Small companies may have a board of only a few members, while larger companies may have a board of up to 20 members. However, research has shown that the optimal size of a board is between 7 to 11 members. Having a smaller board allows for more effective communication and decision-making, while a larger board can lead to communication and coordination challenges.

Board Diversity:

Board diversity refers to the variety of backgrounds, experiences, and perspectives of the board members. A diverse board can bring a range of perspectives and ideas to the decision-making process, which can lead to better outcomes. Diversity can also help the board better reflect the interests of its stakeholders and the broader community.

However, achieving diversity on the board can be challenging. Historically, many boards have been dominated by white, male, and older members. While progress has been made in recent years, there is still a long way to go. Some companies have implemented diversity initiatives to promote greater diversity on the board, including setting diversity targets, using diverse search firms, and providing training on unconscious bias.

Board Independence:

Board independence refers to the extent to which the board members are free from any conflicts of interest and have the ability to make decisions that are in the best interests of the company and its stakeholders. Independent directors are not employees of the company and do not have any significant business relationships with the company, which can help ensure that their decisions are not influenced by personal interests.

The presence of independent directors on the board is crucial for effective corporate governance. Independent directors can act as a check on the power of the

CEO and the management team, providing an objective perspective on strategic decisions and ensuring that the company is being run in the best interests of all stakeholders.

Board Expertise:

Board expertise refers to the skills and knowledge that the board members bring to the company. A board with diverse expertise can help the company navigate complex business challenges and make informed decisions. Expertise can come from a variety of fields, including finance, law, marketing, and technology.

It is important for the board to have a mix of both industry-specific and general business expertise. Industry-specific expertise can help the board better understand the competitive landscape and industry trends, while general business expertise can help the board make more strategic decisions about the company as a whole.

Role of Shareholders:

While the composition of the board is ultimately determined by the company's management team and board, shareholders can play a significant role in shaping the board's composition. Shareholders can use their voting power to elect board members and can also propose resolutions that impact the board's composition, such as proposals to increase diversity or set term limits for directors.

Shareholders can also engage with the company's management team and board to express their views on the composition of the board and other corporate governance matters. Engaging with the board can help shareholders better understand the company's strategy and decision-making process and can also help the board better understand the concerns and interests of its shareholders.

Conclusion:

The composition of the board of directors is a crucial element of corporate governance. A well-composed board can help ensure that the company is being run in the best interests of all stakeholders and can help the company navigate complex business challenges. Factors such as board size, diversity, and expertise are all important considerations when selecting board members. However, there is no one-size-fits-all approach to board composition, as the needs of each company can vary depending on industry, size, and other factors.

It is important for companies to regularly evaluate and update their board composition to ensure that it remains effective and responsive to changing business environments. This may involve bringing on new board members with different backgrounds and skillsets, or making changes to the board's committees and leadership structure.

Overall, the composition of the board of directors plays a critical role in shaping the direction and success of a company. By carefully selecting and cultivating a diverse and knowledgeable board, companies can ensure that they are well-equipped to navigate the complex challenges of modern business and provide long-term value for all stakeholders.

Duties of the board of directors

The board of directors is a vital component of corporate governance. Its primary function is to oversee the management team and ensure that the company is being run in the best interests of all stakeholders. In this section, we will discuss the duties of the board of directors in detail, including their fiduciary responsibilities, oversight responsibilities, and strategic responsibilities.

Fiduciary Responsibilities:

The board of directors has several fiduciary responsibilities to the company and its stakeholders. These responsibilities include:

1.   Duty of Care: The duty of care requires board members to exercise reasonable care when making decisions on behalf of the company. This includes conducting due diligence, seeking expert advice when necessary, and making informed decisions based on all available information.

2.   Duty of Loyalty: The duty of loyalty requires board members to act in the best interests of the company and its stakeholders. Board members must avoid conflicts of interest and disclose any conflicts that do arise.

3.   Duty of Obedience: The duty of obedience requires board members to ensure that the company complies with all applicable laws and regulations.

Oversight Responsibilities:

In addition to their fiduciary responsibilities, the board of directors also has several oversight responsibilities. These responsibilities include:

1.   Oversight of Management: The board of directors is responsible for overseeing the management team and ensuring that they are running the company in a responsible and ethical manner.

2.   Financial Oversight: The board of directors is responsible for overseeing the financial health of the company. This includes reviewing financial reports, setting budgets, and approving major financial transactions.

3.   Risk Oversight: The board of directors is responsible for overseeing the company's risk management policies and procedures. This includes identifying potential risks, developing risk mitigation strategies, and monitoring the effectiveness of these strategies.

Strategic Responsibilities:

Finally, the board of directors also has several strategic responsibilities. These responsibilities include:

1.   Setting Strategy: The board of directors is responsible for setting the overall strategy and direction of the company.

2.   Succession Planning: The board of directors is responsible for planning for the succession of key executives and ensuring that the company has a strong leadership team in place.

3.   Stakeholder Engagement: The board of directors is responsible for engaging with stakeholders, including shareholders, employees, customers, and the community. This includes communicating with stakeholders, listening to their concerns, and addressing any issues that arise.

Conclusion:

In conclusion, the board of directors has several important duties in corporate governance. These include their fiduciary responsibilities, oversight responsibilities, and strategic responsibilities. By fulfilling these duties, the board of directors can help ensure that the company is being run in the best interests of all stakeholders and can help the company navigate complex business challenges.

Executive compensation and its role in corporate governance

Executive compensation is a critical element of corporate governance that involves the use of various forms of compensation to attract, retain, and motivate top-level executives. The issue of executive compensation has gained significant attention in recent years due to the increasing size and complexity of compensation packages, which often include large amounts of stock options, bonuses, and other benefits. While executive compensation can help align the interests of executives with those of the company's shareholders and other stakeholders, it can also create conflicts of interest and raise questions about the overall fairness and effectiveness of corporate governance.

This section will provide an in-depth analysis of executive compensation and its role in corporate governance. We will explore the various forms of executive compensation, the challenges associated with executive compensation, and the potential benefits and drawbacks of different compensation structures. We will also examine the role of executive compensation in aligning the interests of executives with those of the company's stakeholders and how it can be used to promote corporate governance.

Forms of Executive Compensation:
Executive compensation can take many forms, including salaries, bonuses, stock options, restricted stock units (RSUs), stock awards, and other benefits. Each form of compensation has its unique benefits and drawbacks, and different companies may use different forms of compensation based on their specific needs and goals.

1.   Salaries: Salaries are a fixed form of compensation paid to executives and are typically based on their position, experience, and other factors. Salaries are considered a relatively stable form of compensation that helps provide executives with a steady income stream.

2.   Bonuses: Bonuses are typically paid to executives based on their performance or the company's performance. Bonuses can be used to incentivize executives to achieve specific goals or to reward them for achieving exceptional results. However, bonuses can also create conflicts of interest if they are not tied to the long-term interests of the company.

3.   Stock Options: Stock options give executives the right to purchase company stock at a predetermined price. Stock options can provide executives with a significant financial incentive to improve the company's performance, as the value of their options will increase as the company's stock price rises. However, stock options can also create conflicts of interest if executives prioritize short-term gains over long-term value creation.

Restricted Stock Units (RSUs): RSUs are similar to stock options but do not provide executives with the right to purchase stock. Instead, RSUs represent a promise to deliver a certain number of shares of company stock at a future date. RSUs can be used to incentivize executives to stay with the company over the long term, as the value of their RSUs will only be realized if they remain with the company.

Stock Awards: Stock awards are a form of compensation in which executives are given company stock outright. Stock awards can be used to incentivize executives to focus on the long-term interests of the company, as the value of their stock will increase as the company's performance improves.

Other Benefits: Executives may also receive other benefits, such as retirement plans, health insurance, and other perks. These benefits can help attract and retain top talent but can also create conflicts of interest if they are not aligned with the long-term interests of the company.

Challenges Associated with Executive Compensation:
Executive compensation can create several challenges for corporate governance. One of the most significant challenges is ensuring that executive compensation aligns with the long-term interests of the company and its stakeholders. Executives may prioritize short-term gains over long-term value creation, which can result in a misalignment of interests.

Another challenge associated with executive compensation is the potential for conflicts of interest. Executives may prioritize their own interests over those of the company's shareholders and other stakeholders, which can lead to unethical behavior and poor decision-making.

Finally, executive compensation can create a perception of unfairness among employees and other stakeholders. When executives receive large compensation packages, it can lead to resentment and a feeling of inequality among lower-level employees, who may feel that their contributions to the company are undervalued in comparison.

In order to address these challenges, it is important for companies to design executive compensation plans that align with the company's long-term interests and its stakeholders. One way to do this is to tie executive compensation to the company's performance over a longer period of time, such as five or ten years. This can help ensure that executives are incentivized to create long-term value for the company, rather than just short-term gains.

Another approach is to ensure that executive compensation is tied to specific performance metrics that are aligned with the company's goals and values. For example, a company that values sustainability might tie executive compensation to the company's carbon footprint or other sustainability metrics.

In addition, it is important for companies to be transparent about their executive compensation practices and to engage in regular communication with their stakeholders. This can help build trust and reduce the perception of unfairness associated with executive compensation.

Despite these challenges, executive compensation can play an important role in corporate governance. When properly designed and aligned with the company's interests, it can incentivize executives to create long-term value for the company and its stakeholders. However, it is important for companies to be mindful of the potential challenges associated with executive compensation and to take steps to address them in order to ensure that executive compensation is serving its intended purpose.

As future business leaders, it is important for students to understand the complexities associated with executive compensation and its role in corporate governance. By learning about these challenges and exploring potential solutions, students can develop the skills and knowledge necessary to design effective executive compensation plans and to promote ethical and sustainable corporate governance practices.

To reinforce this learning, students can engage in exercises such as analyzing real-world case studies of executive compensation plans and proposing alternative compensation plans that align with the company's goals and values. Students can also explore the ethical implications of different executive compensation practices and engage in debates and discussions about the role of executive compensation in corporate governance. By doing so, students can develop a deeper understanding of this important topic and build the skills necessary to make informed and ethical decisions as future business leaders.

CHAPTER 8: SECURITIES LAW AND CORPORATE FINANCE

Securities laws are a vital component of corporate finance, governing the issuance and trading of securities in public markets. These laws regulate the process by which companies raise capital from investors and the disclosures that companies must make to investors. Securities laws aim to protect investors from fraud and ensure that investors have access to accurate and timely information about the securities they are buying.

Corporate finance is the area of finance that deals with the financial decisions made by corporations. It includes the analysis of how corporations raise capital and allocate resources, as well as the management of financial risks. The study of corporate finance is important for understanding how corporations make decisions that affect their profitability and long-term success.

This chapter provides an overview of securities laws and their role in corporate finance. We will discuss the different types of securities, the regulation of securities markets, and the responsibilities of issuers and other market participants. We will also examine the ways in which securities laws affect corporate finance decisions and strategies.

In addition, we will explore some of the current issues and challenges in securities regulation, such as the regulation of emerging technologies and the changing nature of financial markets. Finally, we will provide some practical examples and case studies to illustrate the concepts discussed in this chapter.

By the end of this chapter, readers will have a solid understanding of the fundamentals of securities laws and their role in corporate finance. They will be able to identify the key features of different types of securities, understand the regulatory framework governing securities markets, and appreciate the importance of securities laws in promoting transparency and protecting investors.

The role of the Securities and Exchange Commission (SEC)

The Securities and Exchange Commission (SEC) is an independent federal agency tasked with regulating the securities industry in the United States. Established in 1934 by the Securities Exchange Act, the SEC is responsible for enforcing federal securities laws, protecting investors, and maintaining fair and orderly markets. The SEC's mission is to promote capital formation while ensuring that investors are provided with full and accurate information about investments. In this section, we will discuss the role of the SEC in regulating the securities industry, the structure and organization of the SEC, and the key functions performed by the agency.

The Role of the SEC:

The SEC's role is to regulate the securities industry and protect investors by ensuring that securities offerings are conducted in a fair and transparent manner. The SEC has the power to enforce federal securities laws, investigate potential violations, and bring enforcement actions against individuals and companies that violate these laws. The SEC also has the authority to establish rules and regulations to govern the securities industry.

One of the primary functions of the SEC is to review and approve securities offerings made by public companies. The SEC's review process ensures that investors are provided with accurate and complete information about the securities being offered, including the company's financial condition, management team, and risk factors. The SEC also monitors public companies for compliance with disclosure requirements, ensuring that investors are provided with timely and accurate information about material events that could affect the company's financial performance.

Another important function of the SEC is to regulate the behavior of securities professionals. The SEC sets rules and standards of conduct for brokers, investment advisers, and other securities professionals, and is responsible for enforcing these rules. The SEC also investigates and brings enforcement actions against individuals and firms that engage in fraudulent or manipulative behavior, insider trading, or other violations of securities laws.

The Structure and Organization of the SEC:

The SEC is structured as a five-member commission appointed by the President of the United States, with the advice and consent of the Senate. The commission is led by a chairperson, who is appointed by the President and confirmed by the Senate. The other four commissioners are appointed by the President, with no more than two from the same political party. The commissioners serve staggered five-year terms and can only be removed for cause.

The SEC is divided into several divisions and offices, each with specific responsibilities. These include the Division of Corporation Finance, which reviews securities offerings and monitors public companies for compliance with disclosure requirements; the Division of Enforcement, which investigates potential violations of securities laws and brings enforcement actions against violators; the Office of Compliance Inspections and Examinations, which conducts inspections and examinations of securities firms to ensure compliance with SEC rules and regulations; and the Office of the Chief Accountant, which is responsible for establishing and enforcing accounting and auditing standards for public companies.

Key Functions of the SEC:

The SEC performs several key functions in regulating the securities industry, including:

1.   Protecting Investors: The SEC's primary goal is to protect investors by ensuring that securities offerings are conducted in a fair and transparent manner. The SEC reviews and approves securities offerings, monitors public companies for compliance with disclosure requirements, and investigates and brings enforcement actions against individuals and companies that violate securities laws.

2.   Promoting Capital Formation: The SEC also plays a key role in promoting capital formation by ensuring that companies have access to capital markets. The SEC's review process for securities offerings ensures that investors are provided with accurate and complete information about the securities being offered, which in turn helps to build investor confidence and promote investment in the securities markets.

3.   Regulating Securities Professionals: The SEC sets rules and standards of conduct for securities professionals, including brokers, investment advisers, and other market participants. The SEC is responsible for enforcing these rules and investigating and bringing enforcement actions against individuals and firms that engage in fraudulent or manipulative behavior, insider trading, or other securities violations.

4.   Monitoring Market Activity: The SEC also monitors market activity to detect and prevent market manipulation, insider trading, and other types of securities fraud. The SEC works closely with other regulatory agencies, such as

the Financial Industry Regulatory Authority (FINRA), to ensure that securities professionals are following the rules and to investigate and prosecute violations.

5.	Educating Investors: The SEC provides a wealth of educational resources to help investors make informed decisions about their investments. These resources include investor alerts, guidance on how to avoid investment fraud, and information about the risks and benefits of different types of investments.

In addition to these key functions, the SEC also has a number of other responsibilities, including overseeing the regulation of securities exchanges and self-regulatory organizations, such as FINRA, and participating in international regulatory efforts to promote consistency in securities regulation across different countries.

Overall, the SEC plays a crucial role in promoting fair and transparent securities markets and protecting investors from fraudulent or manipulative behavior. By ensuring that companies have access to capital markets and that investors have access to accurate and complete information, the SEC helps to promote economic growth and stability.

The process of issuing securities

The process of issuing securities is a critical aspect of corporate finance. Companies use securities issuance to raise funds for various purposes, including financing growth, repaying debts, and funding acquisitions. The process involves various stakeholders, including the issuer, underwriters, legal counsel, accountants, and regulatory bodies. This section will provide a detailed analysis of the process of issuing securities, the various types of securities, and the roles of different stakeholders.

Types of Securities:

Securities are financial instruments that represent a legal claim on the issuer's assets or cash flows. There are several types of securities that companies can issue, including equity securities and debt securities.

Equity Securities:

Equity securities represent ownership interests in a company. They give the holder the right to vote on important corporate matters, such as the election of board members and major corporate transactions. The most common type of equity security is common stock, which gives the holder the right to receive dividends and participate in any potential appreciation in the company's stock price. Another type of equity security is preferred stock, which has priority over common stock in terms of dividends and liquidation rights.

Debt Securities:

Debt securities represent the issuer's obligation to repay the holder's investment, usually with interest. The most common type of debt security is bonds, which are issued with a fixed interest rate and a maturity date. Another type of debt security is a debenture, which is an unsecured bond that is not backed by collateral.

The Securities Issuance Process:

The securities issuance process involves several steps and stakeholders, including the issuer, underwriters, legal counsel, accountants, and regulatory bodies. The following is a detailed analysis of the securities issuance process:

Step 1: Determine the Need for Capital

The first step in the securities issuance process is for the company to determine the need for capital. The company must identify the specific purpose for which it requires the funds and the amount of capital it needs.

Step 2: Determine the Type of Security to Issue

Once the company has identified the need for capital, it must determine the type of security to issue. This decision will depend on various factors, including the company's financial condition, the prevailing market conditions, and the investor demand for the particular type of security.

Step 3: Retain Legal Counsel

The company must retain legal counsel to assist with the securities issuance process. The legal counsel will advise the company on the applicable securities laws and regulations and assist in preparing the necessary documents, such as the prospectus.

Step 4: Draft the Prospectus

The prospectus is a legal document that provides detailed information about the issuer, the security being offered, and the terms and conditions of the offering. The prospectus must comply with the securities laws and regulations and must be filed with the regulatory body responsible for overseeing the securities markets, such as the SEC in the United States.

Step 5: Select Underwriters

Underwriters are financial institutions that assist the issuer with the securities issuance process by purchasing the securities from the issuer and then reselling them to the public. The underwriters will also assist with marketing the securities to potential investors.

Step 6: Price the Securities

The underwriters will work with the issuer to determine the price of the securities being offered. The price will depend on various factors, including the prevailing market conditions, the issuer's financial condition, and the demand for the securities.

Step 7: Register the Securities

The securities must be registered with the regulatory body responsible for overseeing the securities markets. The registration process involves filing the prospectus and other required documents with the regulatory body, such as the SEC in the United States.

Step 8: Market and Sell the Securities

Once the securities have been registered with the SEC, the underwriters will begin marketing and selling the securities to potential investors. This is a critical step in the process of issuing securities, as the success of the offering depends on the ability of the underwriters to find buyers for the securities.

The underwriters may use a variety of methods to market the securities, including roadshows, advertisements, and public relations campaigns. They may also work with brokerage firms and other financial intermediaries to reach potential investors.

One important factor in marketing securities is the pricing of the securities. The underwriters will typically set an initial price range for the securities, based on factors such as the company's financial performance, industry trends, and market conditions. They will then work to generate interest in the securities and determine the optimal price for the offering.

Investors who are interested in purchasing the securities will typically place orders through their brokerage firms or other financial intermediaries. The underwriters will then allocate the securities to investors based on a variety of factors, including the investor's size, investment objectives, and relationship with the underwriters.

Once the securities have been sold, the underwriters will typically receive a commission based on the value of the offering. This commission is intended to compensate the underwriters for their efforts in marketing and selling the securities, as well as for the risks they take on in the event that the offering is not successful.

Overall, the process of issuing securities is complex and involves many different parties and steps. However, it is a critical process for companies that are looking to raise capital and grow their businesses. By understanding the key steps involved in the process of issuing securities, companies can better navigate the regulatory requirements and market conditions that are involved in a successful offering.

Math problem:

If a company issues 10 million shares of stock at $25 per share, what is the total value of the offering?

Solution:

The total value of the offering is calculated by multiplying the number of shares by the price per share. In this case, the total value of the offering would be:

10,000,000 shares x $25 per share = $250,000,000

Therefore, the total value of the offering would be $250 million.

Insider trading and securities fraud

Insider trading and securities fraud are serious issues in the securities industry that can have significant consequences for investors, companies, and individuals involved in the illegal activity. In this section, we will discuss what insider trading and securities fraud are, why they are illegal, how they are detected, and the penalties associated with them.

Insider Trading:

Insider trading occurs when individuals who have access to non-public information about a company use that information to make trading decisions. This can give them an unfair advantage over other investors who do not have access to the same information. Insider trading can take many forms, including buying or selling securities, tipping off others to buy or sell securities, or using the information to trade options or other derivatives.

Insider trading is illegal because it undermines the integrity of the securities markets and erodes investor confidence. When individuals engage in insider trading, it can distort the market's perception of a company's true value, leading to inaccurate stock prices and potentially causing significant harm to innocent investors.

There are several ways in which insider trading can be detected. One method is through a review of trading patterns that suggest unusual or suspicious behavior, such as a significant increase in trading volume or a large purchase or sale of securities by an individual who would not normally make such a trade. Another way is through the use of sophisticated surveillance tools, such as data analysis algorithms or insider trading detection software, that can identify patterns of behavior that are indicative of insider trading.

Penalties for insider trading can be severe. In the United States, insider trading is a violation of federal securities laws, and individuals who engage in insider trading can face both civil and criminal penalties. Civil penalties can include fines, disgorgement of profits, and injunctions, while criminal penalties can include fines and imprisonment. In addition to legal penalties, individuals who engage in insider trading can also face significant reputational damage, which can have lasting consequences for their personal and professional lives.

Example:

Let's say that a senior executive at a publicly-traded company learns of a major upcoming announcement that will significantly impact the company's stock price. This executive decides to buy a large amount of the company's stock before the announcement is made, knowing that the stock price will likely increase once the news is public. This would be considered insider trading because the executive used non-public information to make a profit. Meanwhile, other investors who did not have access to this information would not have been able to make the same profit.

Insider trading can also take the form of tipping, in which an insider shares confidential information with someone else who then trades on that information. For example, if the same executive tells a friend or family member about the upcoming announcement and that person buys the stock based on that information, it would still be considered insider trading even though the friend or family member did not have direct access to the information.

Securities Fraud:

Securities fraud occurs when individuals or companies make false or misleading statements or omissions about a company or its securities in order to deceive investors. Securities fraud can take many forms, including false or misleading financial statements, false or misleading press releases, or manipulating the market through schemes such as pump-and-dump or Ponzi schemes.

Securities fraud is illegal because it violates the trust that investors place in the integrity of the securities markets and the information provided by companies. Securities fraud can lead to significant financial losses for investors, as well as damage to the reputation and financial stability of companies that engage in such practices.

There are several ways in which securities fraud can be detected. One method is through a review of financial statements and other disclosures to identify inaccuracies or inconsistencies. Another way is through the use of market surveillance tools that can identify unusual or suspicious trading patterns, such as a sudden increase in trading volume or price movements that are not supported by underlying fundamentals.

Penalties for securities fraud can be severe. In the United States, securities fraud is a violation of federal securities laws, and individuals or companies that engage in securities fraud can face both civil and criminal penalties. Civil penalties can include fines, disgorgement of profits, and injunctions, while criminal penalties can include fines and imprisonment. In addition to legal penalties, individuals or companies that engage in securities fraud can also face significant reputational damage, which can have lasting consequences for their business operations and financial stability.

In conclusion, insider trading and securities fraud are serious issues that can have significant consequences for individuals, companies, and investors. These illegal activities undermine the integrity of the securities markets and erode investor confidence. It is important for investors, companies, and regulatory bodies to be vigilant in detecting and preventing insider trading and securities fraud to ensure the integrity and stability of the securities markets.

Case Study:

Enron was an American energy company that was involved in one of the largest corporate accounting scandals in history.

Enron's management engaged in fraudulent accounting practices to make the company appear more profitable than it actually was. These practices involved the use of off-balance-sheet special purpose entities to hide debt and losses, as well as the manipulation of financial statements and earnings reports.

The fraudulent accounting practices were carried out with the help of the company's auditing firm, Arthur Andersen. In addition, several Enron executives, including CEO Jeffrey Skilling and CFO Andrew Fastow, were found guilty of insider trading, as they sold off their Enron stock before the company's financial troubles became public knowledge.

The scandal came to light in 2001, when Enron's stock price plummeted and the company filed for bankruptcy. The fallout from the scandal was widespread, as many investors lost their life savings, and the public's confidence in the securities markets was severely damaged.

The Enron scandal led to significant changes in the regulatory landscape, including the passage of the Sarbanes-Oxley Act, which established new accounting and reporting requirements for publicly traded companies, as well as stricter penalties for securities fraud.

PART 3: CONTRACT LAW AND LEGAL ISSUES IN BUSINESS TRANSACTIONS

In the world of business, contracts play a vital role in the smooth functioning of transactions. A contract is a legally binding agreement between two or more parties that sets out the terms and conditions of their business relationship. A well-drafted contract helps to establish clarity and certainty in the rights and obligations of each party, and provides a framework for resolving disputes that may arise.

This section will explore the fundamental principles of contract law and their application in business transactions. We will begin by examining the essential elements of a contract and the various types of contracts that exist. We will then move on to discuss the legal issues that arise in the formation, performance, and enforcement of contracts, including issues related to breach of contract, misrepresentation, and duress.

We will also explore the role of contract law in different types of business transactions, such as sales of goods, employment contracts, and intellectual property agreements. We will examine the legal requirements for creating valid contracts in these contexts, and the specific legal issues that may arise in each.

Finally, we will discuss some of the practical considerations that businesses should keep in mind when drafting and negotiating contracts. This will include an examination of common contract clauses and the implications of various contractual provisions.

Overall, this section aims to provide students with a comprehensive understanding of contract law and its application in the context of business transactions. By the end of this section, students should be able to identify the essential elements of a contract, recognize common legal issues that arise in contract formation, performance, and enforcement, and understand the importance of careful drafting and negotiation of contracts in the world of business.

Essential Elements of a Contract:

A contract is a legally binding agreement between two or more parties that creates enforceable rights and obligations. In order for a contract to be valid, it must contain certain essential elements, including:

1.   Offer: An offer is a proposal made by one party to another, indicating a willingness to enter into a contract on certain terms. An offer must be communicated to the offeree and must be sufficiently definite and certain in its terms.

2.   Acceptance: Acceptance is the manifestation of assent by the offeree to the terms of the offer. Acceptance must be communicated to the offeror and must be unconditional and in accordance with the terms of the offer.

3.   Consideration: Consideration is something of value that is exchanged between the parties in return for their promises. Consideration can be in the form of money, goods, services, or a promise to do or refrain from doing something.

4.   Capacity: The parties to a contract must have the legal capacity to enter into a contract. This generally means that they must be of legal age, of sound mind, and not under duress or undue influence.

5.   Legality: The subject matter of the contract must be legal. A contract that involves illegal or immoral activities is void and unenforceable.

Types of Contracts:

There are several types of contracts that may be used in business transactions, including:

1.   Express contracts: Express contracts are contracts in which the terms are explicitly stated, either orally or in writing.

2.   Implied contracts: Implied contracts are contracts that are inferred from the conduct of the parties. These contracts may be created when one party accepts goods or services from another, with the expectation of paying for them.

3.   Unilateral contracts: Unilateral contracts are contracts in which one party makes a promise in exchange for the other party's performance. These contracts are often used in employment agreements, where the employer promises to pay the employee in exchange for the employee's work.

4.   Bilateral contracts: Bilateral contracts are contracts in which both parties make promises to each other. These contracts are often used in sales transactions, where the buyer promises to pay for goods or services, and the seller promises to provide those goods or services.

5.   Executed contracts: Executed contracts are contracts that have been fully performed by both parties.

6.   Executory contracts: Executory contracts are contracts that have not yet been fully performed by one or both parties.

7.   Void contracts: Void contracts are contracts that are not legally enforceable from the beginning. Examples of void contracts include contracts that involve illegal activity or contracts that involve a minor.

8.   Voidable contracts: Voidable contracts are contracts that can be legally avoided or cancelled by one party, due to some type of defect in the contract. For example, a contract may be voidable if one party was coerced into signing it or if there was a mistake or misrepresentation in the terms.

9.   Unenforceable contracts: Unenforceable contracts are contracts that are legally binding, but for some reason cannot be enforced by a court of law. This may be due to a lack of evidence, a statute of limitations, or a violation of a legal requirement such as the statute of frauds.

In order to be valid and enforceable, a contract must meet certain legal requirements, including:

1.   Offer and acceptance: A valid contract requires a clear offer by one party and a clear acceptance of that offer by the other party.

2.   Consideration: Consideration refers to the exchange of something of value between the parties, such as money or services.

3.   Legal purpose: A contract must have a legal purpose in order to be enforceable.

Capacity: The parties to a contract must have the legal capacity to enter into the agreement, meaning they must be of legal age and have the mental capacity to understand the terms of the contract.

In Part 3 of this series, we will explore these legal requirements in more detail, as well as other legal issues that arise in business transactions, including contract interpretation, performance and breach, remedies for breach, and the role of third-party beneficiaries. We will also examine common types of contracts used in specific industries, such as employment contracts and real estate contracts. Understanding contract law is crucial for business owners, executives, and legal professionals in order to ensure that transactions are conducted fairly, efficiently, and in compliance with the law.

CHAPTER 9: FORMATION AND ENFORCEMENT OF CONTRACTS

Contracts are an essential aspect of business transactions. They provide a legal framework for agreements between parties, and are used to govern the rights and obligations of those parties. Contracts can take many forms, ranging from simple agreements to complex legal documents. Understanding the basics of contract law is essential for any businessperson, as it can help to avoid costly legal disputes and ensure that agreements are legally enforceable.

This chapter will explore the formation and enforcement of contracts, including the key elements required for a valid contract, the various types of contracts, and the remedies available in the event of a breach of contract.

Section 1: Elements of a Valid Contract

In order for a contract to be legally binding, it must contain several key elements. These elements are:

✧　　Offer: An offer is a proposal to enter into a contract. It must be made with the intention of creating a legal obligation, and must be communicated to the other party.

✧　　Acceptance: Acceptance is the agreement by the other party to the terms of the offer. It must be communicated to the offeror, and must be unconditional.

✧　　Consideration: Consideration is something of value that is given in exchange for the promise to perform. It can be money, goods, services, or something else of value.

✧　　Capacity: The parties to the contract must have the legal capacity to enter into the contract. This means that they must be of legal age, and must not be under duress or undue influence.

✧　Legality: The contract must be for a lawful purpose. Contracts that are illegal, immoral, or against public policy are not enforceable.

Section 2: Types of Contracts

There are several types of contracts that may be used in business transactions, including:

✧　Express contracts: Express contracts are contracts in which the terms are explicitly stated, either orally or in writing.

✧　Implied contracts: Implied contracts are contracts that are inferred from the conduct of the parties. These contracts may be created when one party accepts goods or services from another, with the expectation of paying for them.

✧　Unilateral contracts: Unilateral contracts are contracts in which one party makes a promise in exchange for the other party's performance. These contracts are often used in employment agreements, where the employer promises to pay the employee in exchange for the employee's work.

✧　Bilateral contracts: Bilateral contracts are contracts in which both parties make promises to each other. These contracts are often used in sales transactions, where the seller promises to deliver goods or services in exchange for payment from the buyer.

Section 3: Remedies for Breach of Contract

When a contract is breached, the non-breaching party may be entitled to various remedies. The most common remedies are:

✧　Damages: Damages are monetary compensation awarded to the non-breaching party to compensate for losses suffered as a result of the breach. There are several types of damages, including compensatory damages, consequential damages, and punitive damages.

✧　Specific performance: Specific performance is a remedy in which the court orders the breaching party to perform the obligations under the contract. This remedy is typically only available in cases involving unique goods or services.

Rescission: Rescission is a remedy in which the contract is cancelled and the parties are restored to their pre-contract positions. This remedy is typically only available in cases involving fraud or misrepresentation.

Conclusion

Contracts are an essential aspect of business transactions, and understanding the basics of contract law is essential for any businessperson. By understanding the key elements of a valid contract, the various types of contracts, and the remedies available in the event of a breach, businesspeople can ensure that their agreements are legally enforceable and that they are protected in the event of a dispute.

It is important to note that contract law can be complex and nuanced, and there are many legal considerations that must be taken into account when negotiating, drafting, and enforcing contracts. Therefore, it is highly recommended that businesses seek the advice of qualified legal counsel when dealing with contractual matters.

In addition, as the business world continues to evolve, new issues and challenges may arise in the context of contract law. For example, the rise of e-commerce and the use of blockchain technology for contracting may present unique legal issues that require further development and clarification.

Overall, the formation and enforcement of contracts is a fundamental aspect of business law, and a solid understanding of this area is essential for anyone seeking to succeed in the world of business.

Elements of a contract

Certainly. In this section, we will discuss the key elements of a contract, including offer, acceptance, consideration, capacity, and legality.

Offer: The first element of a contract is the offer. An offer is a promise or proposal by one party to enter into a contract with another party. The offer must be communicated to the other party and must be clear and definite. It must also indicate the intention to enter into a contract on specific terms.
Example: A seller offers to sell a car to a buyer for $10,000.

Acceptance: The second element of a contract is acceptance. Acceptance is the agreement by the offeree to the terms of the offer. It must be communicated to the offeror and must be unconditional and unequivocal.
Example: The buyer accepts the offer and agrees to purchase the car for $10,000.

Consideration: The third element of a contract is consideration. Consideration is something of value given by both parties in exchange for the promises made in the contract. It must be legally sufficient and can be a promise to do something or refrain from doing something.

Example: The consideration in the car sale contract is $10,000 paid by the buyer to the seller.

Capacity: The fourth element of a contract is capacity. Capacity refers to the legal ability of a party to enter into a contract. Parties must have the mental capacity to understand the terms of the contract, and they must not be under duress or undue influence.

Example: If the buyer is a minor, they may not have the legal capacity to enter into a contract.

Legality: The final element of a contract is legality. A contract must be formed for a legal purpose and must not violate any laws or public policy.

Example: A contract to sell illegal drugs is not enforceable in court.

It is important to note that all of these elements must be present for a contract to be legally enforceable. In addition, the terms of the contract must be specific, definite, and certain. If any of these elements are missing, the contract may be void or unenforceable.

There are also several defenses to the enforcement of a contract, including fraud, misrepresentation, mistake, and duress. These defenses may be used to argue that a contract should not be enforced.

Overall, understanding the elements of a contract is essential for anyone engaging in business transactions. By ensuring that all of these elements are present and that the terms of the contract are clear and specific, parties can avoid disputes and ensure that their agreements are legally enforceable.

Example:

CONTRACT FOR SERVICES

This agreement ("Agreement") is made and entered into on [Date] by and between [Name of Client], with an address at [Address], ("Client") and [Name of Contractor], with an address at [Address], ("Contractor").

Services to be Provided
Contractor agrees to provide the following services to Client:
[List of Services]

Payment
Client agrees to pay Contractor the sum of [Amount] for the services rendered. Payment shall be made in [Payment Terms].

Term of Agreement
This Agreement shall commence on [Start Date] and shall continue until the completion of the services specified in Section 1, unless earlier terminated as provided herein.

Termination
Either party may terminate this Agreement upon [Notice Period] days' written notice to the other party.

Independent Contractor
Contractor is an independent contractor and not an employee of Client. Contractor shall be responsible for payment of all taxes, including self-employment taxes.

Confidentiality
Contractor agrees to keep confidential all information provided by Client or obtained during the performance of the services, and to use such information only for the purposes of performing the services.

Indemnification
Contractor agrees to indemnify and hold harmless Client from any and all claims arising out of the services provided by Contractor.

Governing Law
This Agreement shall be governed by and construed in accordance with the laws of the state of [State], without giving effect to its conflicts of laws principles.

Entire Agreement
This Agreement constitutes the entire agreement between the parties and supersedes all prior agreements and understandings, whether written or oral, relating to the subject matter of this Agreement.

IN WITNESS WHEREOF, the parties have executed this Agreement as of the date first written above.

[Signature of Client] [Signature of Contractor]
Name: _____ Name: _____
Title: _____ Title: _____

Consideration and legality

Consideration and legality are two essential elements of a contract that must be present for the agreement to be valid and enforceable. In this section, we will explore the concept of consideration and legality in detail, and provide examples of how these elements are applied in different types of contracts.

Consideration:
Consideration is the exchange of something of value between the parties to a contract. It is the price that one party pays for the promise of the other. In other words, consideration is what each party gives or receives in return for the agreement. Consideration can take many forms, such as money, goods, services, promises, or forbearance (refraining from doing something). Consideration must be sufficient, but it does not need to be adequate. This means that the value of the consideration does not have to be equivalent to the value of the promise made. However, there must be some value exchanged for the contract to be valid.

For example, imagine that Alice promises to sell her car to Bob for $1. This agreement is not enforceable because there is no consideration. However, if Alice promises to sell her car to Bob for $1,000, and Bob promises to pay Alice the $1,000, then there is consideration on both sides, and the contract is valid.

Consideration must also be given by both parties. If one party promises to do something without receiving anything in return, then there is no consideration, and the agreement is not legally enforceable. This is known as a gift, and it is not a contract because there is no exchange of value.

For example, if Alice promises to give her car to Bob as a gift, and Bob promises nothing in return, then there is no consideration, and the agreement is not a valid contract.

Legality:
The second essential element of a contract is legality. A contract must have a lawful purpose and be in compliance with the law. If the contract involves an illegal activity, then it is not enforceable. This means that a contract for the sale of illegal drugs, for example, would not be enforceable in court. Similarly, contracts that violate public policy, such as contracts to commit a crime, are not enforceable.

For example, imagine that Alice and Bob enter into a contract where Bob promises to commit a burglary for Alice, and Alice promises to pay Bob for his services. This contract is illegal because it involves a criminal act, and it is not enforceable in court.

Moreover, contracts that violate the rights of third parties or violate established legal principles are also considered illegal. For instance, a contract that restricts a person's freedom of speech or freedom of religion would not be enforceable.

In conclusion, consideration and legality are essential elements of a valid and enforceable contract. Consideration requires that each party receives something of value in exchange for their promise, while legality requires that the contract has a lawful purpose and complies with the law. Understanding these elements is crucial for anyone entering into a business transaction or signing a contract.

Examples:
Agreement between John Smith and Jane Doe

This agreement (the "Agreement") is made on April 14, 2023, by and between John Smith ("Smith"), with a mailing address of 123 Main Street, Anytown, USA, and Jane Doe ("Doe"), with a mailing address of 456 Oak Avenue, Anytown, USA.

Recitals:

Smith is the owner of a vintage car, a 1957 Chevrolet Bel Air, which he wishes to sell.

Doe is interested in purchasing the car.

Agreement:

Smith agrees to sell the car to Doe for the sum of $50,000.

Doe agrees to pay the sum of $50,000 to Smith.

The parties agree to execute all documents necessary to effectuate the sale.

This Agreement shall be binding upon and inure to the benefit of the parties hereto and their respective heirs, executors, administrators, and assigns.

The contract appears to have all the elements of a valid contract, including offer, acceptance, and mutual assent. However, it lacks consideration. Neither party is promising to do anything new or different in exchange for the other's promise. Smith is simply promising to sell the car for $50,000, and Doe is promising to pay that amount. There is no additional benefit or detriment to either party, and therefore, the contract lacks consideration and may not be enforceable.

Example :

AGREEMENT FOR THE SALE OF A CAR

This Agreement for the Sale of a Car ("Agreement") is made on [Date], by and between [Seller Name] ("Seller"), with an address at [Address], and [Buyer Name] ("Buyer"), with an address at [Address]. Seller agrees to sell and Buyer agrees to purchase the following described automobile:

Make: Honda
Model: Civic
Year: 2020
Color: Red
License Plate: ABC-1234

Purchase Price: $5,000

Delivery Date: [Date]

TERMS AND CONDITIONS
1.1 Condition of Car. The Seller guarantees that the car is in good condition and has no known mechanical problems.
1.2 Title. The Seller guarantees that the car has a clear title and can be legally sold to the Buyer.
1.3 Payment. The Buyer agrees to pay the purchase price in full upon delivery of the car.
1.4 Delivery. The Seller agrees to deliver the car to the Buyer on the Delivery Date.
1.5 Warranty. The Seller does not provide any warranty for the car.
1.6 Governing Law. This Agreement shall be governed by the laws of the State of [State].

ACKNOWLEDGMENT
2.1 The Buyer acknowledges that he or she has inspected the car and accepts it "as is" without any warranty from the Seller.
2.2 The Buyer acknowledges that he or she is responsible for all costs associated with the registration and licensing of the car.
2.3 The Buyer acknowledges that he or she has read this Agreement and fully understands its terms and conditions.

Although this agreement may appear valid, there are a few reasons why it would not be legally binding. Firstly, the purchase price of $5,000 for a 2020 Honda Civic is significantly below market value and may indicate that the Seller is not acting in good faith. Secondly, the Seller does not provide any warranty for the car, which is a red flag as it suggests that the Seller is aware of some defect in the car that may not be visible to the Buyer. Finally, there is no consideration given by the Buyer in exchange for the car, which is a fundamental requirement of a valid contract. Without consideration, the contract lacks the necessary element of exchange and is therefore unenforceable.

Contract interpretation and enforcement

Contracts are legal agreements that bind parties to certain obligations and responsibilities. When a dispute arises between parties to a contract, the interpretation of the terms of the agreement is critical in determining their respective rights and obligations. Contract interpretation is the process of determining the meaning and intent of the parties to the contract as expressed in the terms of the agreement. This process is crucial to the enforcement of contracts, as the outcome of a legal dispute may depend on the interpretation of the terms of the agreement. In this section, we will discuss the principles of contract interpretation and the methods used to enforce contracts.

Principles of Contract Interpretation:

The goal of contract interpretation is to determine the intent of the parties to the agreement. This is done by examining the terms of the contract and considering the context in which the contract was formed. Courts apply several principles to determine the intent of the parties:

The Plain Meaning Rule: This rule holds that the words of a contract should be given their ordinary and plain meaning. If the language of the contract is clear and unambiguous, the court will enforce the contract according to its plain meaning.

The Parol Evidence Rule: This rule prohibits the introduction of evidence outside the four corners of the contract to interpret the terms of the agreement. If the language of the contract is clear and unambiguous, the court will not consider extrinsic evidence to interpret the terms of the agreement.

The Doctrine of Contra Proferentem: This doctrine holds that ambiguous language in a contract should be construed against the party that drafted the contract.

This is because the drafter of the contract is presumed to have had greater control over the language used in the agreement and should have clarified any ambiguities.

Methods of Contract Enforcement:

Once the terms of the contract have been interpreted, the next step is to enforce the agreement. There are several methods of contract enforcement, including:

Specific Performance: This is a court order requiring a party to perform its obligations under the contract. Specific performance is often used when damages would not adequately compensate the injured party.

Damages: Damages are a monetary award intended to compensate the injured party for the harm caused by the breach of contract. There are several types of damages, including compensatory, consequential, and punitive damages.

Injunction: An injunction is a court order that requires a party to stop or refrain from doing a specific act. Injunctions are often used to prevent a party from breaching a contract.

Conclusion:

In conclusion, contract interpretation and enforcement are critical to the success of any business transaction. By understanding the principles of contract interpretation and the methods used to enforce contracts, business people can protect their interests and ensure that their agreements are legally enforceable. It is important to draft contracts carefully, using clear and unambiguous language, to avoid disputes and ensure that the terms of the agreement are properly enforced.

Examples:

Unclear and Ambiguous Contract for the Sale of a Business:

This agreement is made on this day of __, 20, between the Seller and the Buyer.

The Seller agrees to sell the Business to the Buyer for the agreed-upon price. The Buyer agrees to pay the Seller the agreed-upon price for the Business.

The Seller will transfer all ownership rights of the Business to the Buyer. The Buyer agrees to accept ownership of the Business from the Seller.

The parties agree to sign any additional documents necessary to complete the sale of the Business.

Correct Contract for the Sale of a Business:

This agreement is made on this day of __, 20, between [Seller's Name], with a mailing address of [Seller's Address] (hereinafter referred to as the "Seller"), and [Buyer's Name], with a mailing address of [Buyer's Address] (hereinafter referred to as the "Buyer").

Recitals:

A. The Seller is the owner of all right, title, and interest in and to the business conducted under the name of [Business Name] (the "Business").

B. The Buyer desires to purchase from the Seller, and the Seller desires to sell to the Buyer, all of the assets of the Business, including, without limitation, all tangible and intangible assets of the Business, subject to the liabilities of the Business (the "Assets").

Agreement:

Sale of Assets. Subject to the terms and conditions of this Agreement, the Seller agrees to sell to the Buyer, and the Buyer agrees to purchase from the Seller, the Assets for the purchase price of [Purchase Price] (the "Purchase Price").

Closing Date. The closing of the purchase and sale of the Assets (the "Closing") shall take place on [Closing Date] at a time and place designated by the Seller and the Buyer.

Payment of Purchase Price. The Purchase Price shall be paid by the Buyer to the Seller at the Closing as follows: [Payment Terms].

Representations and Warranties of the Seller. The Seller represents and warrants to the Buyer as follows:

(a) Organization and Authority. The Seller is a legal entity duly organized, validly existing, and in good standing under the laws of the jurisdiction of its organization and has the power and authority to enter into and perform this Agreement.

(b) Title to Assets. The Seller has good and marketable title to the Assets, free and clear of any liens, encumbrances, security interests, or other restrictions on transfer.

(c) Compliance with Laws. The Seller has complied with all applicable federal, state, and local laws, rules, and regulations in the operation of the Business.

(d) No Litigation. There is no litigation or governmental proceeding pending or, to the Seller's knowledge, threatened against the Seller or the Business.

Representations and Warranties of the Buyer. The Buyer represents and warrants to the Seller as follows:
(a) Organization and Authority. The Buyer is a legal entity duly organized, validly existing, and in good standing under the laws of the jurisdiction of its organization and has the power and authority to enter into and perform this Agreement.

(b) Ability to Pay. The Buyer has sufficient funds to pay the Purchase Price and to perform its obligations under this Agreement.

(c) No Litigation. There is no litigation or governmental proceeding pending or, to the Buyer's knowledge, threatened against the Buyer.

Indemnification. The Seller shall indemnify and hold harmless the Buyer from and against any and all claims, damages, liabilities, costs, and expenses (including reasonable attorneys' fees and expenses) arising out of or in connection with any breach of any representation, warranty, covenant, or obligation of the Seller under this Agreement. This indemnification shall survive the closing of the sale.

Dispute Resolution
In the event of any dispute arising out of or relating to this Agreement, the parties shall first attempt to resolve the dispute through mediation, to be conducted by a mediator selected by mutual agreement. If the parties are unable to resolve the dispute through mediation, then the dispute shall be resolved through binding arbitration, in accordance with the rules of the American Arbitration Association. The decision of the arbitrator shall be final and binding on the parties, and judgment may be entered thereon in any court of competent jurisdiction.

Governing Law and Venue
This Agreement shall be governed by and construed in accordance with the laws of the state of [insert state]. Any legal action or proceeding arising out of or relating to this Agreement shall be brought exclusively in the courts of [insert county], [insert

state], and each party hereby consents to the jurisdiction of such courts for the purposes of any such action or proceeding.

Entire Agreement
This Agreement constitutes the entire agreement between the parties with respect to the subject matter hereof and supersedes all prior negotiations, agreements and understandings between the parties, both written and oral. This Agreement may not be amended except in writing signed by both parties.

IN WITNESS WHEREOF, the parties have executed this Agreement as of the date first above written.

Seller:
[Insert Seller's Name]

Buyer:
[Insert Buyer's Name]

CHAPTER 10: BREACH OF CONTRACT AND REMEDIES

Contracts are a cornerstone of modern business and commerce, allowing parties to enter into agreements with one another for the exchange of goods and services. However, as with any human endeavor, mistakes can be made, misunderstandings can arise, and parties may not always perform as expected. When this happens, a contract is said to have been breached.

Breach of contract is a legal term used to describe a situation where one party fails to perform a duty or obligation that they agreed to under a contract. When this happens, the other party may be entitled to remedies for the breach. Remedies are the legal means by which a party can be made whole again, or compensated for the loss suffered as a result of the breach.

In this chapter, we will examine the concept of breach of contract, the types of breaches that can occur, and the remedies available to parties in the event of a breach.

Section 1: Breach of Contract

A breach of contract occurs when one party fails to perform a duty or obligation that they agreed to under a contract. This can take many forms, including failing to deliver goods on time, failing to pay for services rendered, or failing to perform work as agreed.

There are two types of breaches of contract: material breaches and non-material breaches. A material breach is a serious breach that goes to the heart of the contract, such as a failure to deliver goods at all, or delivering goods that are significantly different from what was agreed upon. A non-material breach, on the other hand, is a less serious breach, such as a minor delay in delivery or a small defect in the goods.

Section 2: Remedies for Breach of Contract

When a breach of contract occurs, the innocent party may be entitled to remedies. The most common remedies for breach of contract are damages, specific performance, and cancellation or rescission of the contract.

Damages are a monetary award intended to compensate the innocent party for the loss suffered as a result of the breach. Damages can be either direct or consequential. Direct damages are those that flow directly from the breach, such as the cost of replacing defective goods. Consequential damages, on the other hand, are those that are not a direct result of the breach, but are a foreseeable consequence of the breach, such as lost profits.

Specific performance is a remedy that requires the breaching party to perform the duty or obligation that they agreed to under the contract. This remedy is often used in cases where damages would not be an adequate remedy, such as when the subject matter of the contract is unique.

Cancellation or rescission of the contract is a remedy that allows the innocent party to terminate the contract and be relieved of any further obligations under the contract. This remedy is often used in cases where the breach is so serious that it would be impractical or impossible to continue with the contract.

Section 3: Defenses to Breach of Contract

There are several defenses to breach of contract that a party may raise in response to an allegation of breach. These include impossibility, impracticability, frustration of purpose, and waiver.

Impossibility is a defense that arises when performance under the contract becomes impossible due to unforeseeable events, such as a natural disaster or a change in the law.

Impracticability is a defense that arises when performance under the contract becomes impractical due to unforeseeable events, such as a severe shortage of materials or labor.

Frustration of purpose is a defense that arises when an unforeseeable event makes it impossible for the contract to serve its intended purpose, such as a change in the law that renders the subject matter of the contract illegal.

Waiver is a defense that arises when the innocent party has waived their right to insist on strict performance by the other party. A waiver can be explicit or implied. An explicit waiver occurs when the innocent party expressly agrees to waive their right to insist on strict performance by the other party. An implied waiver occurs when the innocent party acts in a manner that is inconsistent with strict performance by the other party. For example, if the innocent party accepts partial performance by the other party without objection, this may be considered an implied waiver.

Estoppel is a defense that arises when one party has induced the other party to believe that a certain course of conduct is acceptable, and the other party has relied on that belief to their detriment. Estoppel prevents the party that induced the belief from arguing that the other party breached the contract by acting in accordance with that belief. For example, if a landlord tells a tenant that they can sublease the property, and the tenant relies on that representation to sublease the property, the landlord may be estopped from arguing that the tenant breached the lease agreement by subleasing the property.

Remedies for Breach of Contract

When a party breaches a contract, the innocent party may seek a remedy for the breach. The most common remedies for breach of contract are damages, specific performance, and cancellation or rescission of the contract.

Damages are a monetary award designed to compensate the innocent party for the losses they suffered as a result of the breach of contract. There are several types of damages that may be awarded in a breach of contract case, including compensatory damages, consequential damages, and punitive damages.

Compensatory damages are designed to compensate the innocent party for the losses they suffered as a result of the breach of contract. Consequential damages, also known as special or indirect damages, are damages that flow indirectly from the breach of contract, such as lost profits or lost opportunities. Punitive damages are designed to punish the breaching party for their conduct, rather than compensate the innocent party for their losses. However, punitive damages are rarely awarded in breach of contract cases.

Specific performance is an equitable remedy that requires the breaching party to perform their contractual obligations. Specific performance is only available when money damages are not an adequate remedy, such as in cases involving unique or irreplaceable goods or services. Specific performance is not available in cases involving personal services, as it would violate the constitutional prohibition on involuntary servitude.

Cancellation or rescission of the contract is a remedy that allows the innocent party to cancel or rescind the contract and recover any consideration they provided. Cancellation or rescission is only available when the breach of contract is material, or goes to the heart of the contract. A material breach excuses the innocent party from their obligations under the contract, and allows them to recover any consideration they provided.

Conclusion

In conclusion, breach of contract is a serious matter that can have significant consequences for the parties involved. It is important for parties to understand the elements of a contract, the defenses to breach of contract, and the remedies available for breach of contract. By understanding these concepts, parties can avoid disputes and resolve any disputes that do arise in a fair and efficient manner.

Types of breaches

In contract law, a breach of contract refers to a failure to perform one's obligations under the terms of a contract. A breach of contract can be a serious issue for parties involved in a contract and can have legal and financial consequences. Understanding the different types of breaches can help parties determine the best course of action when dealing with a breach of contract.

There are different types of breaches that can occur under a contract. This section will explore the different types of breaches and provide examples of each.

Material Breach:
A material breach occurs when one party fails to perform a major obligation under the contract. This type of breach is often considered the most serious type of breach and can lead to termination of the contract. Examples of a material breach include failure to deliver goods as promised, failure to pay for goods or services, or failure to complete a project on time.

Minor Breach:
A minor breach, also known as a partial breach, occurs when one party fails to perform a minor obligation under the contract. This type of breach does not go to the heart of the contract and usually does not lead to termination of the contract. Examples of a minor breach include a delay in delivery of goods, minor defects in the goods delivered, or failure to meet a minor deadline.

Anticipatory Breach:
An anticipatory breach occurs when one party indicates that they will not be able to perform their obligations under the contract before the performance is due. This type of breach can be problematic because it can make it difficult for the other party to plan and take appropriate action. For example, if a construction company informs the owner of a building that they will not be able to complete the project on time, the owner may need to hire a new construction company and may incur additional costs.

Actual Breach:

An actual breach occurs when one party fails to perform their obligations under the contract at the time of performance. This type of breach is the most common type of breach and can occur in any type of contract. For example, if a supplier fails to deliver goods as promised, they have committed an actual breach.

Fundamental Breach:

A fundamental breach occurs when one party fails to perform an obligation that is essential to the contract. This type of breach can lead to termination of the contract and can have significant legal and financial consequences. For example, if a manufacturer delivers goods that are completely different from what was ordered, they have committed a fundamental breach.

Repudiatory Breach:

A repudiatory breach occurs when one party indicates that they do not intend to perform their obligations under the contract. This type of breach is similar to an anticipatory breach, but it is more explicit. For example, if a buyer informs the seller that they will not be paying for the goods ordered, they have committed a repudiatory breach.

Conclusion:

Understanding the different types of breaches is important for parties involved in a contract. Each type of breach has different legal and financial consequences, and the appropriate course of action will depend on the specific circumstances of the breach. Parties should seek legal advice and carefully consider their options when dealing with a breach of contract.

Examples:

Example of a Material Breach:

Suppose that John hires a construction company to build a house for him. The contract specifies that the construction company must complete the project within six months, and John agrees to pay the company $500,000 upon completion. However, after three months, the construction company has only completed about 10% of the work, and John discovers that the company has been using substandard materials and unskilled laborers, resulting in shoddy workmanship.

In this case, the construction company has committed a material breach of the contract by failing to perform its obligations within the specified time frame and by failing to deliver work of the required quality. This breach is material because it goes to the heart of the contract and undermines the entire purpose of the agreement. As a result, John may be entitled to terminate the contract and seek damages from the construction company for any losses or expenses incurred as a result of the breach.

Example of a Minor Breach:

Suppose you hire a contractor to renovate your kitchen, and the contract specifies that the work must be completed by a certain date. The contractor finishes the work a few days late, but the delay does not cause you any significant harm. This would be considered a minor breach because the contractor did not fulfill the contract exactly as specified, but the breach did not have a major impact on the outcome of the project.

In this scenario, you may be entitled to some compensation for the delay, but it is unlikely to be substantial.

Example of a Anticipatory Breach:

An anticipatory breach occurs when one party to a contract declares, either through words or actions, that they do not intend to fulfill their contractual obligations. Here is an example:

John enters into a contract to sell his car to Sarah for $10,000, with delivery to occur on August 1st. However, on July 25th, John sends Sarah an email stating that he has changed his mind and will not be selling the car after all. This email constitutes an anticipatory breach, as John has indicated that he will not be fulfilling his contractual obligation to deliver the car to Sarah on August 1st. Sarah may now consider the contract to be breached and seek remedies for the breach.

Example of an Actual Breach:

Let's say that Company A and Company B enter into a contract for Company A to provide 1,000 widgets to Company B by a certain date, and in return, Company B will pay Company A $10,000. The contract specifies that the widgets must meet certain specifications, including size, weight, and durability.

However, when Company B receives the shipment of widgets, they discover that many of the widgets are of a different size and weight than specified in the contract, and some are damaged or broken. Company B notifies Company A of the breach, and requests that they either replace the defective widgets or provide a refund for the cost of the defective widgets.

Company A initially refuses to take responsibility for the breach, arguing that the contract did not specify exact sizes and weights for the widgets, and that the damage may have occurred during shipping. However, upon further investigation, it is discovered that Company A did not properly supervise the manufacturing process of the widgets and allowed faulty products to be shipped.

In this scenario, Company A has committed an actual breach of contract by failing to deliver goods that meet the specifications agreed upon in the contract. Company B may be entitled to damages, such as a refund for the cost of the defective widgets, and may choose to terminate the contract.

Example of a Fundamental Breach:

Company A is contracted to deliver 10,000 units of a particular product to Company B by a specific date. The contract specifies that the product must meet certain quality standards, and Company B is relying on this product to fulfill orders from their own customers. Company A fails to deliver the product by the agreed-upon date and when Company B receives the product, they discover that it does not meet the quality standards specified in the contract. The breach by Company A is considered a fundamental breach because it goes to the heart of the contract and renders the entire purpose of the contract frustrated. As a result, Company B is entitled to terminate the contract and seek damages for any losses they have incurred as a result of Company A's breach.

Example of a Repudiatory Breach:

Imagine that Company A enters into a contract with Company B to deliver 1,000 units of a specific product by a certain date. The contract specifies that Company B will pay Company A $100,000 upon delivery of the units. However, a few days before the delivery date, Company A informs Company B that it will not be able to deliver the units on time due to a delay in the manufacturing process. Company A also states that it will not be able to deliver the units at all, as it has decided to focus on a different product line.

In this case, Company A's statement constitutes a repudiatory breach of the contract. By stating that it will not be able to deliver the units on time or at all, Company A has demonstrated a clear intention not to fulfill its contractual obligations. As a result, Company B can treat the contract as terminated and seek damages for any losses incurred as a result of Company A's breach.

Damages and other remedies

In contract law, damages and other remedies are available to compensate the injured party for the harm suffered as a result of a breach of contract. The primary purpose of these remedies is to put the injured party in the same position they would have been in if the contract had been performed as promised. The remedies available for breach of contract can be categorized into two main types: legal remedies and equitable remedies. This section will provide a detailed analysis of damages and other remedies available for breach of contract, including the types of damages, the calculation of damages, and the equitable remedies available.

Types of Damages:

Compensatory Damages:
Compensatory damages are the most common type of damages awarded for breach of contract. These damages are designed to compensate the injured party for the actual loss suffered as a result of the breach. The goal of compensatory damages is to restore the injured party to the same economic position they would have been in if the contract had been fully performed. Compensatory damages are usually awarded for direct losses that are foreseeable and certain. Examples of compensatory damages include:

Expectation damages: These damages are designed to put the injured party in the position they would have been in if the contract had been fully performed. In other words, the damages are meant to compensate the injured party for the benefit they would have received under the contract. The calculation of expectation damages is based on the difference between the contract price and the cost of performance, plus any other damages that may have been incurred as a result of the breach.

Consequential damages: These damages are awarded for losses that were foreseeable at the time the contract was entered into, but that do not flow directly from the breach itself. For example, if a supplier breaches a contract to deliver goods, and the buyer loses profits as a result of the delay, the buyer may be entitled to consequential damages.

Incidental damages: These damages are awarded for expenses incurred as a result of the breach, such as the cost of finding a replacement supplier or the cost of shipping goods that were not delivered as promised.

Punitive Damages:
Punitive damages are awarded to punish the breaching party for their wrongful conduct. These damages are only awarded in exceptional cases, where the breaching

party has engaged in conduct that is particularly egregious or fraudulent. Punitive damages are not available in most contract cases.

Liquidated Damages:

Liquidated damages are a pre-determined amount of damages that are specified in the contract. These damages are designed to compensate the injured party for losses that are difficult to calculate. For example, if a contractor breaches a contract to build a house, the contract may specify that the contractor will pay a certain amount of liquidated damages per day for each day the project is delayed. The amount of liquidated damages must be reasonable and cannot be excessive.

Nominal Damages:

Nominal damages are awarded in cases where there has been a technical breach of contract, but no actual harm has been suffered by the injured party. The purpose of nominal damages is to recognize the breach of contract, but not to compensate the injured party for any actual loss suffered.

Calculation of Damages:

The calculation of damages in a breach of contract case can be complex, as it requires the injured party to prove both the fact of the loss and the amount of the loss suffered. In general, the injured party is entitled to recover the difference between the value of the promised performance and the value of the actual performance received. This calculation can be further broken down into the following steps:

✧ Determine the contract price: The first step in calculating damages is to determine the contract price.

Example: Let's say that Company A enters into a contract with Company B to deliver 1,000 units of a particular product for a total contract price of $100,000. The contract specifies that the delivery will take place in three installments: the first 300 units will be delivered on or before August 1st, the second 300 units on or before August 15th, and the final 400 units on or before August 31st.

If Company B fails to deliver any of the units by the specified dates, they are in breach of the contract. In this case, the contract price is the total amount of $100,000 specified in the contract.

✧ Determine the value of the performance received: The next step is to determine the value of the performance that was actually received. This may involve an assessment of the quality and quantity of the performance, as well as any defects or deficiencies that may have affected its value.

Example: let's say that a company hired a contractor to build a new office building for $1 million. The contract specified that the building would be completed within 12 months. However, due to the contractor's delay, the building was not completed until 15 months later.

To determine the value of the performance received, the company would need to assess the actual completed building and compare it to the building that was promised in the contract. They would need to evaluate the quality of the building, the materials used, and whether it met all the specifications laid out in the contract.

Let's say that upon inspection, the company determined that the completed building was of lower quality than what was promised in the contract, and was only worth $900,000 instead of the contracted price of $1 million. This means that the value of the performance received was $900,000, which is $100,000 less than the contract price.

✧   Determine the value of the promised performance: The injured party must then determine the value of the promised performance, which is the performance that should have been delivered under the terms of the contract.

Example: Let's say that Party A hires Party B to paint a portrait of Party A's family for $1,000. The contract specifies that the portrait will be painted with high-quality materials, be delivered within 30 days of the contract signing, and be a true likeness of the family.

However, when Party A receives the portrait, they find that it was painted with low-quality materials, contains numerous inaccuracies, and was delivered 60 days after the contract was signed.

In this case, the value of the promised performance is the value of a high-quality portrait delivered within 30 days of the contract signing that is a true likeness of Party A's family. This is because the contract specifies that Party A is entitled to this particular performance, and it is the performance that Party A paid for.

The value of the promised performance can be determined by looking at the market value of such a portrait, as well as any additional costs that Party A incurred as a result of the breach, such as the cost of hiring a new artist to create the portrait or the cost of repairing any inaccuracies in the portrait.

Calculate the difference: Once the value of the promised performance and the actual performance received have been determined, the difference between the two can be calculated. This difference represents the amount of damages that the injured party is entitled to.

There are various methods of calculating damages, depending on the circumstances of the case. In some cases, the damages may be calculated based on the cost of replacement or repair of the defective performance. In other cases, the damages may be calculated based on the lost profits or revenue that the injured party suffered as a result of the breach.

In addition to damages, there are other remedies that may be available to the injured party in a breach of contract case. These remedies include:

Example to help illustrate how to calculate the difference between the value of the promised performance and the value of the actual performance received:

Let's say that Company A enters into a contract with Company B to purchase a shipment of 1,000 widgets for $10,000. The contract specifies that the widgets will be delivered within 30 days of the date of the contract.

However, Company B only delivers 800 widgets after 40 days, and the widgets are of a lower quality than what was promised in the contract. Company A had to purchase the remaining 200 widgets from another supplier at a cost of $3,000.

To calculate the damages owed to Company A, the first step is to determine the contract price, which is $10,000.

The next step is to determine the value of the promised performance, which would have been the 1,000 widgets delivered within 30 days at the specified quality. Let's assume that this would have been worth $12,000 to Company A based on their expected resale value.

The third step is to determine the value of the actual performance received. Company A received 800 widgets that were of lower quality than what was promised, which would have been worth $8,000 to Company A based on their resale value.

To calculate the difference, we subtract the value of the actual performance received ($8,000) from the value of the promised performance ($12,000), which gives us $4,000.

Therefore, the damages owed to Company A for the breach of contract would be $4,000.

Specific performance: This remedy requires the breaching party to perform their obligations under the contract as originally agreed. This remedy is typically only available in cases where damages would not adequately compensate the injured party, such as in cases involving unique or irreplaceable goods.

Example: Specific performance is an equitable remedy that involves a court order compelling a party to perform a specific act that was promised under a contract. This remedy is typically used when damages are inadequate or impractical, and the specific act that was promised is unique or irreplaceable.

For example, suppose that a buyer enters into a contract to purchase a rare piece of artwork from a seller. The contract specifies that the artwork is to be delivered on a specific date, and the buyer has already paid the full purchase price. However, on the delivery date, the seller refuses to deliver the artwork, claiming that they have decided to keep it for themselves.

In this situation, the buyer may seek specific performance as a remedy, since the artwork is unique and cannot be easily replaced. The court may order the seller to deliver the artwork to the buyer as promised under the contract, even though this involves compelling the seller to do something they no longer want to do. If the seller fails to comply with the court order, they may face contempt of court charges and other legal consequences.

Injunction: An injunction is a court order that requires the breaching party to stop doing something that is in violation of the contract. This remedy is typically used in cases where the breach involves ongoing conduct, such as the use of trade secrets or confidential information.

Example: Suppose that a company is accused of violating a competitor's intellectual property rights by producing and selling a product that infringes on the competitor's patent. The competitor may seek an injunction to prevent the company from continuing to produce and sell the infringing product. If the court grants the injunction, the company would be prohibited from producing and selling the product, potentially resulting in a significant loss of profits. In this case, the injunction serves as a remedy to prevent further harm to the competitor's intellectual property rights.

Rescission: Rescission is a remedy that allows the injured party to cancel the contract and be returned to their pre-contract position. This remedy is typically used in cases where the breach involves fraud, mistake, or duress.

Example: Suppose that John enters into a contract with Mary to purchase her car for $10,000. After John pays Mary the full amount, he discovers that the car has significant mechanical problems that were not disclosed to him prior to the sale. John can bring a legal action for rescission of the contract, which would allow him to cancel the contract and get his money back.

In this scenario, rescission would be an appropriate remedy because John was induced to enter into the contract by Mary's misrepresentation or failure to disclose material information about the car. Rescission would allow John to undo the contract and return the car to Mary in exchange for a refund of the purchase price.

Mitigation of damages: The injured party has a duty to mitigate their damages by taking reasonable steps to minimize their losses. This may involve finding alternative sources of performance, or taking other actions to limit the extent of their losses.

Example: Suppose that a company hired a contractor to build a new office building. The contract specified that the building would be completed by a certain date and that the total cost would be $10 million. However, due to the contractor's delay, the building was not completed on time, and the final cost ended up being $12 million.

The company could seek damages from the contractor for the $2 million difference in cost. However, in order to mitigate its damages, the company could take steps to reduce the impact of the delay and cost overrun. For example, the company could:

✧ Rent temporary office space to continue its operations while waiting for the building to be completed

✧ Negotiate with suppliers to reduce the costs of furniture and equipment for the new building

✧ Use its own employees to perform some of the work that would have been done by the contractor

By taking these steps, the company can reduce the amount of damages it suffers as a result of the contractor's breach of contract. The company cannot simply sit back and allow the damages to accumulate without taking any action to mitigate them. If the company fails to mitigate its damages, it may be prevented from recovering the full amount of damages from the contractor.

Overall, the choice of remedy will depend on the specific circumstances of the case, including the nature and severity of the breach, the type of contract involved, and the availability of alternative remedies. It is important for parties to carefully consider their options and seek legal advice to determine the best course of action in a breach of contract case.

Exercise 1: Calculation of Damages

What is the first step in calculating damages in a breach of contract case?
a) Determine the value of the performance received
b) Determine the value of the promised performance
c) Determine the contract price
d) Calculate the difference

What does the injured party need to prove in order to calculate damages in a breach of contract case?
a) The fact of the loss
b) The amount of the loss suffered
c) Both a and b
d) None of the above

How is the difference between the promised performance and the actual performance received calculated?
a) By subtracting the contract price from the value of the promised performance
b) By subtracting the value of the performance received from the contract price
c) By adding the value of the promised performance to the value of the performance received
d) By multiplying the value of the performance received by the contract price

Exercise 2: Remedies

What is specific performance?
a) A monetary remedy that compensates the injured party for their losses
b) An order requiring the breaching party to perform their contractual obligations
c) A court order preventing the breaching party from performing their contractual obligations

d) A remedy that cancels the contract and restores the parties to their pre-contractual position

What is an injunction?
a) A monetary remedy that compensates the injured party for their losses
b) An order requiring the breaching party to perform their contractual obligations
c) A court order preventing the breaching party from performing their contractual obligations
d) A remedy that cancels the contract and restores the parties to their pre-contractual position

What is rescission?
a) A monetary remedy that compensates the injured party for their losses
b) An order requiring the breaching party to perform their contractual obligations
c) A court order preventing the breaching party from performing their contractual obligations
d) A remedy that cancels the contract and restores the parties to their pre-contractual position

Exercise 3: Breach of Contract

What is a material breach of contract?
a) A breach that does not affect the essence of the contract
b) A breach that affects the essence of the contract
c) A breach that is trivial or minor
d) A breach that occurs before the contract is formed

What is a repudiatory breach of contract?
a) A breach that does not affect the essence of the contract
b) A breach that affects the essence of the contract
c) A breach that is trivial or minor
d) A breach that occurs before the contract is formed

What is an anticipatory breach of contract?
a) A breach that occurs before the contract is formed
b) A breach that occurs after the contract is formed
c) A breach that is trivial or minor
d) A breach that is threatened or declared before the time for performance

Equitable remedies

In a breach of contract scenario, the aggrieved party may seek both legal and equitable remedies. Legal remedies, such as damages, are monetary in nature and are designed to compensate the injured party for the loss suffered. In contrast, equitable remedies are non-monetary in nature and are designed to put the parties in the position they would have been in had the contract been performed as promised.

Equitable remedies are typically sought when the legal remedy is inadequate, such as when the subject matter of the contract is unique and cannot be easily replaced. This section will discuss the various types of equitable remedies that may be available in a breach of contract scenario, along with the requirements for obtaining them.

Specific Performance:

Specific performance is an equitable remedy that requires the breaching party to perform its contractual obligations as promised. This remedy is typically sought when the subject matter of the contract is unique and cannot be easily replaced. Specific performance is available only when damages would be an inadequate remedy, such as in cases involving the sale of real estate or other unique assets.

To obtain specific performance, the injured party must show that: (1) there is a valid and enforceable contract; (2) the injured party has performed or is willing to perform its obligations under the contract; (3) the breaching party has failed or refuses to perform its obligations; and (4) damages would be an inadequate remedy.

For example, assume that a seller contracts to sell a rare painting to a buyer. The painting is one of a kind, and the buyer cannot find a comparable painting elsewhere. If the seller breaches the contract by refusing to sell the painting, the buyer may seek specific performance to compel the seller to sell the painting.

Injunction:

An injunction is an equitable remedy that requires a party to do or refrain from doing a particular act. In a breach of contract scenario, an injunction may be sought to prevent a party from engaging in conduct that would cause irreparable harm to the other party.

For example, assume that a company contracts with a key employee to work exclusively for the company for a period of three years. If the employee breaches the contract by accepting employment with a competitor, the company may seek an injunction to prevent the employee from working for the competitor during the contractual period.

To obtain an injunction, the injured party must show that: (1) there is a valid and enforceable contract; (2) the injured party has performed or is willing to perform its obligations under the contract; (3) the breaching party has engaged in or is about to engage in conduct that would cause irreparable harm to the injured party; and (4) there is no adequate remedy at law.

Rescission:

Rescission is an equitable remedy that cancels the contract and restores the parties to their pre-contractual positions. This remedy is typically sought when the contract was induced by fraud, misrepresentation, or mistake.

For example, assume that a buyer contracts to purchase a car from a seller. If the seller knowingly misrepresents the condition of the car, the buyer may seek rescission to cancel the contract and obtain a refund of any money paid.

To obtain rescission, the injured party must show that: (1) there was a material misrepresentation, fraud, or mistake; (2) the injured party reasonably relied on the misrepresentation, fraud, or mistake in entering into the contract; and (3) the injured party has not affirmed or ratified the contract.

Accounting:

Accounting is an equitable remedy that requires a party to provide an accounting of profits or losses arising from the breach of contract. This remedy is typically sought when the breaching party has profited from its breach, such as by using confidential information obtained from the injured party.

For example, assume that a former employee of a company starts his own business and uses confidential customer lists and trade secrets he learned while working for the company. The former employer can seek an accounting of the profits earned by the former employee through the use of these confidential resources.

In order to obtain an accounting remedy, the injured party must show that the breaching party has profited from the breach and that the injured party suffered a loss as a result. Once the court has established the amount of the profits earned by the breaching party, the injured party is entitled to recover those profits as damages.

Accounting remedies can be difficult to obtain, as the injured party must show that the breaching party has actually profited from the breach. This can be challenging when the profits earned by the breaching party are difficult to quantify or are a result of a combination of factors.

Reformation

Reformation is an equitable remedy that is available when there has been a mistake or misunderstanding in the terms of a contract. This remedy allows the court to reform, or modify, the contract in order to reflect the true intentions of the parties.

For example, assume that two parties enter into a contract for the sale of a piece of property, but due to a mistake, the contract incorrectly describes the property. The injured party can seek reformation of the contract in order to correct the mistake and reflect the true intent of the parties.

Reformation is not available for all types of contract mistakes, however. In order to obtain reformation, the injured party must show that the mistake was mutual, or that one party knew or should have known about the mistake and failed to correct it. Additionally, the court must be able to determine the true intent of the parties in order to reform the contract.

Restitution

Restitution is an equitable remedy that requires the breaching party to return any property or funds that it obtained as a result of the breach. This remedy is typically sought in cases where the breaching party obtained property or funds through fraud, misrepresentation, or duress.

For example, assume that a seller of a used car falsely represents the condition of the car to the buyer. The buyer can seek restitution of the purchase price of the car in order to be compensated for the fraud committed by the seller.

In order to obtain restitution, the injured party must show that the breaching party obtained property or funds as a result of the breach and that the injured party suffered a loss as a result. Once the court has established the amount of the property or funds obtained by the breaching party, the injured party is entitled to recover those amounts as damages.

Restitution can be a difficult remedy to obtain, as the injured party must show that the breaching party obtained property or funds as a result of the breach. Additionally, the court must be able to determine the amount of the property or funds obtained in order to award restitution as a remedy.

What is the purpose of an equitable remedy?
A) To punish the breaching party
B) To compensate the injured party
C) To prevent future breaches
D) Both B and C

Which of the following is an example of an equitable remedy?
A) Damages
B) Specific performance
C) Rescission
D) All of the above

What is the purpose of an accounting remedy?
A) To punish the breaching party
B) To compensate the injured party
C) To prevent future breaches
D) Both A and B

When is specific performance typically granted?
A) When the contract involves the sale of goods
B) When the subject matter of the contract is unique
C) When the injured party can easily obtain damages
D) Both A and B

Which of the following is an example of an injunction remedy?
A) Ordering the breaching party to perform its contractual obligations
B) Ordering the breaching party to cease a certain activity
C) Ordering the breaching party to pay damages
D) None of the above

CHAPTER 11: LEGAL ISSUES IN BUSINESS TRANSACTIONS

Business transactions are at the heart of modern commerce. From buying and selling goods and services to mergers and acquisitions, businesses engage in a wide range of transactions to achieve their objectives. However, these transactions are not without legal risks. Contracts can be breached, intellectual property can be misappropriated, and disputes can arise between parties. It is essential for business professionals to understand the legal issues that arise in these transactions and how to mitigate these risks.

This chapter will provide an overview of the legal issues that arise in business transactions. We will begin by discussing the various types of business transactions, including mergers and acquisitions, joint ventures, and licensing agreements. We will then explore the legal principles that underpin these transactions, such as contract law, intellectual property law, and antitrust law. Finally, we will discuss how businesses can mitigate legal risks in their transactions through due diligence, risk management strategies, and dispute resolution mechanisms.

Types of Business Transactions:

Business transactions can take many forms, each with its own unique legal issues. The following are some of the most common types of business transactions:

Mergers and Acquisitions: Mergers and acquisitions (M&A) involve the consolidation of two or more companies into a single entity. This type of transaction raises a variety of legal issues, including antitrust law, securities law, and tax law.

Joint Ventures: Joint ventures involve two or more parties coming together to form a new entity for a specific purpose. Joint ventures can take many forms, such as a partnership, limited liability company (LLC), or corporation. Legal issues that arise in joint ventures include contractual obligations, fiduciary duties, and intellectual property ownership.

Licensing Agreements: Licensing agreements involve the transfer of intellectual property rights from one party to another. These agreements can be complex and require careful drafting to ensure that both parties are adequately protected. Legal issues that arise in licensing agreements include ownership of intellectual property, royalties, and breach of contract.

Legal Principles:

Business transactions are governed by a variety of legal principles, including:

Contract Law: Contracts are the foundation of business transactions. A contract is a legally binding agreement between two or more parties that outlines their rights and obligations. Breach of contract is a common legal issue in business transactions.

Intellectual Property Law: Intellectual property (IP) includes patents, trademarks, copyrights, and trade secrets. These rights are essential for businesses to protect their innovations and creations. Legal issues that arise in IP transactions include ownership, infringement, and licensing.

Antitrust Law: Antitrust laws are designed to prevent anti-competitive behavior, such as price-fixing and monopolies. Legal issues that arise in M&A transactions include pre-merger notification, market power, and antitrust enforcement.

Mitigating Legal Risks:

Businesses can mitigate legal risks in their transactions through due diligence, risk management strategies, and dispute resolution mechanisms. Due diligence involves a comprehensive review of a company's legal and financial documents to identify potential legal risks. Risk management strategies include insurance policies, indemnification clauses, and arbitration clauses. Dispute resolution mechanisms include mediation and arbitration, which can be less expensive and time-consuming than traditional litigation.

Conclusion:

Business transactions are an essential part of modern commerce, but they are not without legal risks. Understanding the legal issues that arise in these transactions is essential for businesses to protect their interests and achieve their objectives. By being proactive in identifying and mitigating legal risks, businesses can ensure that their transactions are successful and legally compliant. In the following chapters, we will explore these legal issues in more detail and provide practical guidance for businesses to navigate these complex legal landscapes.

Antitrust laws

Antitrust laws are a set of regulations that are designed to promote competition and protect consumers from anticompetitive behavior in the marketplace. These laws are enforced by government agencies, such as the Federal Trade Commission (FTC) and the Department of Justice (DOJ), and violations can result in significant fines and other penalties.

Antitrust laws aim to prevent companies from engaging in activities that stifle competition and harm consumers. They promote competition by prohibiting anticompetitive practices such as price fixing, bid rigging, market allocation, and monopolization. Antitrust laws are applicable to a wide range of industries, including technology, healthcare, energy, and finance.

History of Antitrust Laws:

The roots of antitrust laws can be traced back to the late 19th century, when several large corporations, such as Standard Oil and the American Tobacco Company, controlled a significant portion of the U.S. economy. These companies used their market power to drive out competitors and raise prices, which harmed consumers and small businesses.

In response to these practices, Congress passed the Sherman Antitrust Act in 1890, which made it illegal for companies to engage in monopolistic practices that restricted trade or commerce. This was followed by the Clayton Antitrust Act in 1914, which further strengthened the provisions of the Sherman Act and prohibited additional anticompetitive practices, such as price discrimination and tying arrangements.

The Federal Trade Commission Act, also passed in 1914, created the FTC, which was charged with enforcing antitrust laws and investigating unfair methods of competition. Since then, antitrust laws have been expanded and strengthened, with additional legislation passed, such as the Robinson-Patman Act in 1936, the Celler-Kefauver Act in 1950, and the Hart-Scott-Rodino Antitrust Improvements Act in 1976.

Antitrust Laws Today:

Today, antitrust laws are enforced by the FTC and the DOJ, which investigate and prosecute companies that engage in anticompetitive behavior. The most common violations of antitrust laws include price fixing, bid rigging, and monopolization.

Price fixing involves companies conspiring to set prices at a certain level, which reduces competition and harms consumers. Bid rigging involves companies conspiring to submit non-competitive bids, which harms consumers and violates procurement laws. Monopolization involves a company acquiring a dominant position in a market and using its market power to exclude competitors and harm consumers.

Antitrust laws also prohibit mergers and acquisitions that would substantially lessen competition in a market. Companies that plan to merge or acquire another company must comply with the Hart-Scott-Rodino Antitrust Improvements Act, which requires them to notify the FTC and the DOJ of their plans and undergo a review process to determine if the merger or acquisition would harm competition.

Conclusion:

Antitrust laws are an essential component of the U.S. economy, as they promote competition and protect consumers from anticompetitive practices. These laws have a long and storied history, and continue to evolve as new forms of anticompetitive behavior emerge. Companies must be vigilant in ensuring that they comply with these laws to avoid significant fines and other penalties, as well as damage to their reputation and brand.

Intellectual property rights

Intellectual property (IP) refers to the intangible property that is the result of human creativity and innovation. The concept of IP rights is based on the recognition that ideas and creative expressions are valuable assets that should be protected by law, just like physical property. The main types of IP rights include patents, trademarks, copyrights, and trade secrets. These rights provide exclusive control over the use, distribution, and commercial exploitation of the protected assets, giving their owners the ability to benefit financially from their creations.

The protection of IP rights is crucial for promoting innovation and creativity, as it encourages individuals and companies to invest time, money, and resources into developing new ideas and products. It also plays an important role in the global economy, as it allows companies to compete fairly and prevents the theft of valuable assets. In this section, we will explore the different types of IP rights and the legal framework that governs their protection.

Types of IP Rights:

Patents:
A patent is a legal right granted to an inventor or a company that provides exclusive control over the production, use, and sale of a new and useful invention for a limited period of time. The purpose of a patent is to encourage innovation by providing inventors with the opportunity to profit from their ideas. In order to be granted a patent, an invention must be novel, non-obvious, and useful. Examples of patentable inventions include new machines, processes, chemicals, and plants.

Patents are a form of intellectual property protection that provide inventors with exclusive rights to their invention for a certain period of time. In the United States, patents are granted by the United States Patent and Trademark Office (USPTO). The process for acquiring a patent can be a lengthy and complicated one, but it provides inventors with the legal protection they need to prevent others from making, using, or selling their invention.

Let's say that John has invented a new type of bicycle that he believes will revolutionize the industry. He decides to seek patent protection to prevent others from copying his invention. The first step in the process is to conduct a patent search to make sure that the invention is new and not already patented. A patent search involves reviewing existing patents, scientific literature, and other resources to determine if there is already a similar invention on the market.

Once John has completed a patent search and determined that his invention is novel, he can begin the process of applying for a patent. The application process can be complex, and it is recommended that inventors seek the advice of a patent attorney to ensure that their application is properly filed and presented.

John's patent application will include a detailed description of the invention, as well as any supporting documents or drawings that are necessary to illustrate how the invention works. In addition, the application will need to include a series of claims that define the scope of the invention and what aspects of the invention are patentable.

After the application is filed with the USPTO, it is reviewed by an examiner who will determine if the invention meets the requirements for patentability. This includes determining if the invention is novel, non-obvious, and useful. If the examiner determines that the invention meets these criteria, the patent will be granted and John will receive a certificate of patent.

Once a patent is granted, the inventor has exclusive rights to the invention for a certain period of time, typically 20 years from the filing date. During this time, the inventor can prevent others from making, using, or selling the invention without their permission. If someone infringes on the patent, the inventor can take legal action to enforce their rights and seek damages for any losses they may have incurred as a result of the infringement.

In conclusion, patents provide inventors with the legal protection they need to prevent others from copying their inventions. The process for acquiring a patent can be complicated and time-consuming, but it is essential for inventors who want to protect their ideas and prevent others from profiting off of their hard work.

Trademarks:

A trademark is a symbol, word, phrase, or design that identifies and distinguishes the goods or services of one company from those of another. The purpose of a trademark is to prevent confusion among consumers and to protect the reputation and goodwill of a company. Examples of trademarks include brand names, logos, and slogans.

Example of Trademarks:

Let's say you're starting a new clothing brand and want to ensure that your brand name and logo are legally protected. You come up with the name "Fashion Forward" and a distinctive logo featuring a stylized "FF" in bold, contrasting colors.

To protect these assets, you decide to register them as trademarks with the U.S. Patent and Trademark Office (USPTO).

The trademark registration process involves several steps:

Conduct a trademark search: Before you begin the application process, it's important to conduct a search to ensure that your proposed trademark is available and not already in use by another party. You can conduct a basic search on the USPTO website for free or hire a trademark attorney to conduct a more comprehensive search.

Submit a trademark application: Once you've confirmed that your trademark is available, you can begin the application process. The USPTO requires detailed information about the trademark, including a description of the goods or services associated with the mark and any relevant dates of use. You will also need to submit a specimen of the mark, such as a photo of the logo on a product or advertising material.

Wait for review: After submitting your application, it will be assigned to an examining attorney who will review it for compliance with trademark laws and regulations. The examination process can take several months to a year or more, depending on the complexity of the application and any objections raised by the examining attorney.

Respond to any objections: If the examining attorney has any objections to your application, they will issue an office action outlining the reasons for the objections. You will have the opportunity to respond to the objections and provide additional evidence or arguments to support your application.

Receive a decision: Once the examination process is complete, the USPTO will issue a decision on your trademark application. If your application is approved, you will receive a Certificate of Registration, which gives you the exclusive right to use the trademark in connection with the goods or services listed in the application.

Maintain your trademark: After your trademark is registered, you will need to maintain it by filing periodic renewal applications and monitoring for potential infringement by other parties.

In our example, after completing the application process, you are able to register "Fashion Forward" and the "FF" logo as trademarks with the USPTO. This gives you exclusive rights to use these marks in connection with clothing and other fashion-related goods and services, and provides legal protection against other parties using similar names or logos that could cause confusion among consumers.

Copyrights:
A copyright is a legal right granted to the creator of an original work of authorship, such as a book, movie, or song. The purpose of a copyright is to protect the creator's expression of an idea, rather than the idea itself. Copyright protection gives the owner exclusive control over the reproduction, distribution, and public performance of the work. In order to be eligible for copyright protection, a work must be original and fixed in a tangible medium of expression.

Example:

Assume that John is an author who has written a book titled "The Art of Gardening." He wishes to obtain a copyright for his book to protect it from unauthorized use and reproduction. The copyright will grant John exclusive rights to reproduce, distribute, display, and perform his book and its contents.

Step 1: Determine if the work is eligible for copyright protection

The first step in obtaining a copyright is to determine if the work is eligible for copyright protection. In general, literary works, including books, are eligible for copyright protection. John's book, "The Art of Gardening," meets this requirement.

Step 2: Complete the copyright application

The next step is to complete the copyright application. The application can be submitted online or by mail, and it must include the following information:

The title of the work

The author's name and contact information

The type of work being registered (in this case, a book)

The date and place of publication

A brief description of the work

A statement confirming the originality of the work and the author's ownership of the copyright

A nonrefundable filing fee (varies depending on the type of work being registered)

John completes the copyright application, paying the appropriate fee, and submits it to the U.S. Copyright Office.

Step 3: Wait for the copyright registration to be processed

After submitting the copyright application, John must wait for it to be processed. The processing time varies depending on the workload of the U.S. Copyright Office. In general, it can take anywhere from several weeks to several months to receive the copyright registration.

Step 4: Receive the copyright registration

Once the copyright registration has been processed, John will receive a certificate of registration from the U.S. Copyright Office. This certificate serves as proof of copyright ownership and should be kept in a safe place.

To ensure that his copyright is fully protected, John should mark his book with the copyright symbol (©), the year of first publication, and his name. This will put potential infringers on notice that the work is protected by copyright.

In summary, obtaining a copyright for a book involves determining eligibility, completing a copyright application, waiting for the registration to be processed, and receiving a certificate of registration. By obtaining a copyright, John has exclusive rights to reproduce, distribute, display, and perform his book and its contents, and can take legal action against anyone who infringes on those rights.

Trade Secrets:

A trade secret is any confidential information that provides a competitive advantage to a company. This can include formulas, processes, designs, and customer lists. The purpose of trade secret protection is to prevent the unauthorized use or disclosure of valuable information that could harm a company's competitiveness. In order to qualify as a trade secret, the information must be kept confidential and must provide a competitive advantage.

Example:

ABC Corporation has developed a proprietary software that analyzes consumer behavior and predicts future trends. The software's algorithms, source code, and data sets are all closely guarded as trade secrets, and are only accessible to a small group of employees with a need to know.

Process for protecting trade secrets:

Identify the trade secret: The first step in protecting a trade secret is to identify what information constitutes a trade secret. This can include technical information, customer lists, financial data, and other confidential business information that provides a competitive advantage.

Limit access: Once the trade secret has been identified, access to the information should be limited to a need-to-know basis. This can be achieved through employee confidentiality agreements, restricted access to the information, and other security measures.

Mark confidential: To help ensure that employees understand the sensitive nature of the information, it should be clearly marked as confidential or proprietary. This can be done through labeling documents, software code, and other materials.

Implement security measures: To prevent unauthorized access to the trade secret, it's important to implement security measures such as password protection, encryption, and physical security measures like locked rooms or file cabinets.

Monitor access: Regularly monitoring access to the trade secret can help to identify any unauthorized access and prevent leaks. This can be achieved through security logs, auditing, and other monitoring tools.

Take legal action: In the event that the trade secret is misappropriated or leaked, legal action can be taken to protect the trade secret. This can include filing a lawsuit for damages, obtaining an injunction to prevent further disclosure, or seeking criminal charges against the individual responsible.

In the case of ABC Corporation, the company has taken steps to protect its proprietary software as a trade secret by limiting access to the information, clearly marking it as confidential, implementing security measures, and monitoring access to the information. If the trade secret were to be misappropriated, the company could take legal action to protect its rights and seek damages against the responsible party.

Legal Framework:

The protection of IP rights is governed by a complex network of national and international laws and treaties. In the United States, IP rights are protected primarily through federal law, including the Patent Act, the Trademark Act, the Copyright Act, and the Economic Espionage Act.

The Patent Act establishes the requirements for patentability and provides the legal framework for obtaining and enforcing patents. The Trademark Act establishes the requirements for trademark registration and provides the legal framework for protecting and enforcing trademarks. The Copyright Act provides the legal framework for protecting and enforcing copyrights. The Economic Espionage Act provides criminal penalties for the theft of trade secrets.

In addition to these federal laws, IP rights are also protected through international treaties, such as the Paris Convention for the Protection of Industrial Property and the Berne Convention for the Protection of Literary and Artistic Works. These treaties provide a framework for the protection of IP rights across national borders and promote international cooperation in the enforcement of IP rights.

Example:

Legal framework refers to the set of laws, regulations, and policies that govern the operation of businesses and individuals in a particular jurisdiction. The legal framework provides the foundation for protecting intellectual property rights, resolving disputes, and promoting fair competition. In this example, we will focus on the United States legal framework.

The process of protecting intellectual property within the legal framework of the United States involves several steps:

Identify the Intellectual Property: The first step is to identify the type of intellectual property that needs protection. This could be a patentable invention, a trademark, a copyright, or a trade secret.

Determine Eligibility: Once the type of intellectual property is identified, the next step is to determine if it is eligible for protection under U.S. law. Each type of intellectual property has specific eligibility criteria.

File for Protection: If the intellectual property is eligible for protection, the next step is to file for protection with the appropriate government agency. For example, patents are filed with the United States Patent and Trademark Office (USPTO), trademarks are filed with the USPTO, and copyrights are registered with the U.S. Copyright Office.

Review and Approval: Once the application is filed, it is reviewed by the appropriate government agency. The review process can take several months to several years, depending on the type of intellectual property and the complexity of the application.

Maintenance: After approval, the owner of the intellectual property must maintain it by paying fees and complying with any other requirements.

Enforcement: In the event of infringement, the owner of the intellectual property can enforce their rights through the court system. This may involve filing a lawsuit and seeking damages for the unauthorized use of the intellectual property.

The legal framework also provides other mechanisms for protecting intellectual property rights, such as trade secret protection and antitrust laws. To protect trade secrets, businesses must take steps to ensure that their confidential information is kept secret and not disclosed to competitors. This may involve the use of non-disclosure agreements (NDAs), limited access to sensitive information, and other measures.

In conclusion, the legal framework provides a comprehensive set of laws and regulations for protecting intellectual property rights. By understanding the eligibility criteria and the process for acquiring protection, individuals and businesses can take steps to safeguard their valuable assets. Additionally, by complying with the legal framework and respecting the intellectual property rights of others, businesses can promote fair competition and innovation.

Conclusion:

The protection of IP rights is essential for promoting innovation, creativity, and economic growth. By providing legal protection for intangible assets, IP rights encourage investment

Employment law and labor relations

Employment law and labor relations are two critical areas of law that govern the relationship between employers and employees. These areas of law are designed to ensure that both employers and employees are treated fairly and that they understand their rights and obligations.

In this section, we will discuss the key elements of employment law and labor relations, including the legal framework that governs the employment relationship, the rights and obligations of employers and employees, and the various legal remedies that are available to protect these rights.

Legal Framework:

The legal framework that governs employment law and labor relations is complex and multifaceted. It includes federal and state statutes, regulations, and case law that have been developed over many years. At the federal level, the primary statutes that govern employment law and labor relations include the Fair Labor Standards Act, the National Labor Relations Act, the Civil Rights Act, and the Americans with Disabilities Act, among others.

At the state level, there are numerous statutes and regulations that govern various aspects of the employment relationship, including minimum wage and overtime requirements, anti-discrimination laws, and workers' compensation laws.

Employment Relationship:

The employment relationship is governed by a complex set of legal rules and regulations that define the rights and obligations of both employers and employees. At a basic level, the employment relationship involves an agreement between an employer and an employee in which the employer agrees to pay the employee a certain wage or salary in exchange for the employee's services.

However, the employment relationship is much more complex than this basic agreement, and it involves a wide range of legal issues, including the following:

1.   Hiring and firing: Employers are generally free to hire and fire employees at will, subject to certain limitations imposed by law. For example, employers cannot discriminate against employees on the basis of race, gender, age, religion, or disability.

Sample Employment Form :

[Company Name]
[Address]
[City, State ZIP]
[Phone Number]
[Email Address]

APPLICATION FOR EMPLOYMENT

Position Applied for: _____

PERSONAL INFORMATION

Full Name: _____

Address: _____

City: _____ State: _____ ZIP: _____

Phone Number: _____ Email Address: _____

Are you legally authorized to work in the United States? Yes/No

Have you ever worked for this company before? Yes/No

If yes, when? _____ Position: _____

How did you hear about this position? _____

AVAILABILITY

What hours and days are you available to work? _____

EDUCATION

High School: _____

College/University: _____

Degree Earned: _____ Major: _____

EMPLOYMENT HISTORY

List your three most recent employers, starting with the most recent one:

Employer: _____ Dates of Employment: _____
Address: _____ Phone Number: _____

Position: _____ Supervisor's Name: _____

Reason for leaving: _____

Employer: _____ Dates of Employment: _____
Address: _____ Phone Number: _____

Position: _____ Supervisor's Name: _____

Reason for leaving: _____

Employer: _____ Dates of Employment: _____
Address: _____ Phone Number: _____

Position: _____ Supervisor's Name: _____

Reason for leaving: _____

REFERENCES

List three professional references:

Name: _____ Relationship: _____
Phone Number: _____ Email: _____

Name: _____ Relationship: _____
Phone Number: _____ Email: _____

Name: _____ Relationship: _____

Phone Number: _____ Email: _____

CERTIFICATION AND AGREEMENT

I certify that the information provided in this application is true and complete to the best of my knowledge. I understand that any false statements or omissions may result in disqualification from employment, or termination if already employed.

I understand that employment with this company is at-will, meaning either the company or I may terminate the employment relationship at any time, for any reason, with or without cause or notice. I also understand that no representative of the company has the authority to make any promises or agreements contrary to the foregoing except in writing, signed by an authorized company representative.

Signature: _____ Date: _____

Sample Termination Form:

[Company Letterhead]

Date: [Date of Termination]

To: [Employee Name]

Dear [Employee Name],

This letter is to inform you that your employment with [Company Name] will be terminated effective [Date of Termination]. The reason for your termination is [Reason for Termination].

As part of the termination process, we request that you return all company property in your possession, including but not limited to keys, access cards, laptops, mobile devices, and any other equipment or materials belonging to the company.

Please be advised that you will receive your final paycheck on your regular pay date, and that this will include any accrued but unused vacation time or other benefits.

We would also like to remind you of your obligations under any confidentiality, non-disclosure, or non-compete agreements that you may have signed during your employment with us. You are required to honor these agreements even after your employment with us has ended.

We wish you the best in your future endeavors.

Sincerely,

[Company Representative]

2.   Wage and hour laws: Employers are required to pay employees at least the minimum wage and to provide overtime pay for employees who work more than 40 hours per week.

3.   Workplace safety: Employers have a legal obligation to provide a safe working environment for their employees and to take steps to prevent workplace accidents and injuries.

4.   Discrimination: Employers cannot discriminate against employees on the basis of race, gender, age, religion, or disability.

5.   Harassment: Employers are required to take steps to prevent and address workplace harassment, including sexual harassment.

Legal Remedies:

There are a variety of legal remedies that are available to employees who believe that their rights have been violated by their employers. These remedies include the following:

Filing a complaint with the Equal Employment Opportunity Commission (EEOC): The EEOC is responsible for enforcing federal anti-discrimination laws, and employees who believe that they have been discriminated against can file a complaint with the EEOC.

The Equal Employment Opportunity Commission (EEOC) provides a number of forms related to employment discrimination complaints that can be accessed on their website. Some of the common forms include:
  ✧   Intake Questionnaire: This form helps the EEOC gather basic information about the discrimination allegations being made.

  ✧   Charge of Discrimination: This form is used to officially file a complaint with the EEOC.

✧ Notice of Charge of Discrimination: This form is sent by the EEOC to the employer to notify them that a complaint has been filed against them.

✧ Request for Information: This form is used by the EEOC to request additional information from the employer during the investigation process.

✧ Notice of Right to Sue: This form is sent by the EEOC to the complainant if they decide not to pursue the case further and want to file a lawsuit instead.

The exact appearance of these forms may vary depending on the specific situation and jurisdiction involved.

Filing a lawsuit: Employees can file a lawsuit against their employers if they believe that their rights have been violated. This can include claims for discrimination, harassment, wage and hour violations, and wrongful termination.

Collective bargaining: Employees who are members of a union can engage in collective bargaining to negotiate better wages, hours, and working conditions.

Conclusion:

Employment law and labor relations are critical areas of law that govern the relationship between employers and employees. These areas of law are designed to ensure that both employers and employees are treated fairly and that they understand their rights and obligations. By understanding the legal framework that governs employment law and labor relations, employees can better protect their rights and ensure that they are treated fairly in the workplace.

CHAPTER 12: INTERNATIONAL BUSINESS LAW

International business law refers to the set of laws, rules, and regulations that govern business transactions between parties from different countries. In today's global economy, businesses often operate in multiple countries, and international business law is critical in ensuring that these operations are conducted in compliance with the laws of all relevant jurisdictions. The scope of international business law is broad and covers a range of areas, including international trade, foreign investment, cross-border transactions, and dispute resolution.

The legal systems of different countries vary significantly, and this can create challenges for businesses that operate across borders. International business law aims to provide a framework that businesses can use to navigate the complexities of doing business in different countries. This includes understanding the legal requirements for establishing a business in a foreign country, complying with local laws and regulations, and resolving disputes that may arise between parties from different jurisdictions.

In this chapter, we will provide an overview of the key principles and concepts of international business law. We will begin by examining the sources of international business law and the various legal systems that businesses may encounter when operating internationally. We will also discuss the role of international organizations, such as the World Trade Organization and the International Chamber of Commerce, in shaping international business law.

Next, we will explore the key legal issues that businesses may face when operating internationally, including international trade agreements, foreign investment laws, intellectual property rights, and dispute resolution mechanisms. We will also examine the different forms of business structures that may be used when operating internationally, such as joint ventures and subsidiaries.

Finally, we will discuss the importance of understanding cultural differences and local customs when doing business in different countries. We will examine how cultural differences can impact business negotiations and how businesses can adapt to different cultural norms to succeed in the international marketplace.

Overall, this chapter will provide a comprehensive overview of the key principles and concepts of international business law. By understanding the legal requirements and challenges of doing business in different countries, businesses can ensure compliance with relevant laws and regulations, minimize legal risks, and maximize their chances of success in the global marketplace.

The role of international law in business

In the age of globalization, international trade and commerce have become commonplace, and the importance of international law has increased significantly. International law plays a crucial role in regulating business practices, protecting the rights of investors, ensuring fair competition, and promoting economic development. This section will explore the role of international law in business, focusing on its importance, sources, and applications.

Importance of International Law in Business:
International law provides a framework for regulating business activities across national borders. It establishes common rules, standards, and norms that govern international trade and investment, and creates a level playing field for businesses of all sizes and types. International law is particularly important for small and medium-sized enterprises (SMEs) that lack the resources and expertise to navigate complex legal regimes in foreign markets. It also protects the rights of investors by ensuring that their investments are safe and secure, and that they are treated fairly by host countries.

Sources of International Law:
International law is derived from a variety of sources, including treaties, customary international law, general principles of law, and the decisions of international tribunals. Treaties are agreements between states that establish binding legal obligations, while customary international law arises from the general practices of states that are accepted as legal norms. General principles of law refer to fundamental principles of justice and fairness that are recognized by the international community, while decisions of international tribunals are considered to be authoritative interpretations of international law.

Applications of International Law in Business:
International law has numerous applications in business, including trade law, investment law, and intellectual property law. Trade law governs the movement of goods and services across national borders, and includes rules on tariffs, quotas, and trade agreements. Investment law protects the rights of investors and regulates the

conduct of host countries, and includes rules on expropriation, fair and equitable treatment, and dispute resolution. Intellectual property law protects the rights of creators and innovators, and includes rules on patents, trademarks, and copyrights.

Conclusion:
In conclusion, the role of international law in business cannot be overstated. It provides a vital framework for regulating business activities across national borders, protects the rights of investors, and promotes economic development. International law is derived from a variety of sources and has numerous applications in business, including trade law, investment law, and intellectual property law. As globalization continues to reshape the business landscape, the importance of international law will only increase, and businesses that understand and comply with international legal regimes will be better positioned to succeed in the global marketplace.

Differences between domestic and international law

Law is the foundation of modern society, regulating behavior and providing a framework for resolving disputes. It is composed of a complex web of statutes, regulations, and judicial decisions that have been developed over centuries. The study of law is essential for anyone who wishes to understand how society functions and how it can be improved.

One of the most significant distinctions in the study of law is the difference between domestic law and international law. While domestic law governs behavior within the borders of a particular nation-state, international law governs behavior that occurs between nations. This section will explore the differences between these two types of law, including their sources, enforcement mechanisms, and cultural context.

Sources of Law:

The sources of domestic law are typically found in the constitution, statutes, and case law of a particular nation-state. In the United States, for example, the Constitution outlines the basic principles of the legal system and provides a framework for the federal government. Statutes are laws passed by the legislative bodies at the federal, state, and local levels, while case law is created by judicial decisions.

In contrast, the sources of international law are more diverse and complex. International law is derived from treaties, custom, and the decisions of international courts and tribunals. Treaties are agreements between nation-states that create legal obligations, while custom refers to the unwritten rules and practices that have developed over time. The decisions of international courts and tribunals help to clarify and interpret these sources of law.

Enforcement Mechanisms:

Domestic law is enforced by the courts and other law enforcement agencies within a particular nation-state. These agencies have the power to arrest, prosecute, and punish individuals who violate domestic laws. The judicial system in a particular nation-state may also have the power to invalidate laws that are found to be unconstitutional.

International law, on the other hand, lacks a centralized enforcement mechanism. While there are international courts and tribunals that can interpret and apply international law, they lack the power to enforce their decisions. Instead, enforcement of international law is typically left to the nation-states themselves. This can result in challenges when nation-states refuse to comply with international legal obligations.

Cultural Context:

Another key difference between domestic and international law is the cultural context in which they operate. Domestic law is shaped by the culture and values of a particular nation-state. For example, laws regarding freedom of speech in the United States reflect the importance of individual rights and the value placed on free expression. Similarly, laws regarding gun ownership reflect the history and culture of the United States.

In contrast, international law is shaped by the diverse cultures and values of the many nation-states that participate in the international legal system. This can result in disagreements and challenges when different nations have different values and beliefs. For example, debates over human rights and environmental protection reflect different cultural values and priorities.

Conclusion:

In conclusion, the differences between domestic and international law are significant and complex. Domestic law governs behavior within a particular nation-state and is enforced by the courts and other law enforcement agencies within that state. International law, on the other hand, governs behavior between nations and lacks a centralized enforcement mechanism. The sources of international law are diverse and include treaties, custom, and the decisions of international courts and tribunals. The cultural context in which these laws operate is also distinct, with domestic law reflecting the culture and values of a particular nation-state and international law reflecting the diverse cultures and values of the many nation-states that participate in the international legal system. Understanding these differences is essential for anyone who wishes to navigate the complex world of modern law.

Trade agreements and their impact on business

In today's globalized world, international trade agreements play a crucial role in shaping business practices and outcomes. A trade agreement is a legal pact between two or more countries that outlines the terms of trade, such as tariffs, quotas, and regulations. The primary objective of trade agreements is to remove barriers to trade and promote economic growth. This section will provide a comprehensive analysis of trade agreements and their impact on businesses.

Overview of Trade Agreements

Trade agreements come in different forms, such as bilateral, regional, and multilateral agreements. Bilateral trade agreements involve two countries and focus on reducing barriers to trade between them. Regional trade agreements involve a group of countries that share geographic proximity or other commonalities, such as culture or language. Multilateral trade agreements involve many countries and are usually negotiated through international organizations such as the World Trade Organization (WTO).

Trade agreements address several issues related to international trade, such as tariffs, non-tariff barriers, intellectual property, investment, and services. These agreements are designed to create a level playing field for all parties involved in international trade, thereby promoting economic growth, reducing poverty, and increasing employment.

The Impact of Trade Agreements on Businesses

Trade agreements have a significant impact on businesses, and their success or failure can affect entire industries and economies. One of the primary benefits of trade agreements for businesses is increased access to foreign markets. When trade barriers are reduced, businesses can expand their customer base and reach new markets, increasing sales and revenue.

Trade agreements can also reduce the cost of doing business. When tariffs and other barriers to trade are lowered, businesses can import and export goods at a lower cost. This can make businesses more competitive and profitable.

However, trade agreements can also have negative impacts on businesses. For example, if a business relies heavily on a particular industry that is negatively affected by a trade agreement, such as tariffs on steel imports, it could suffer significant losses. Additionally, some businesses may not be able to compete with foreign businesses that can produce goods at a lower cost due to lower wages or less stringent regulations.

Examples of Trade Agreements

One of the most well-known trade agreements is the North American Free Trade Agreement (NAFTA), which was signed in 1994 by the United States, Canada, and Mexico. NAFTA eliminated tariffs and other trade barriers between the three countries, creating a massive free trade zone. While NAFTA has been credited with increasing trade and investment between the three countries, it has also been criticized for job losses and environmental damage.

Another example is the Trans-Pacific Partnership (TPP), a regional trade agreement signed in 2016 by 12 countries, including the United States, Japan, and Australia. The TPP aimed to lower trade barriers and promote economic growth in the Asia-Pacific region. However, the agreement was never ratified by the United States and is currently not in force.

The Trans-Pacific Partnership (TPP) was a trade agreement negotiated between 12 Pacific Rim countries, including the United States, Canada, Japan, Australia, and New Zealand. The agreement aimed to promote economic integration and trade liberalization among its signatories, which together accounted for nearly 40% of the world's GDP. The TPP covered a wide range of issues related to trade, including tariffs, intellectual property rights, investment, and labor standards.

However, in 2017, the United States withdrew from the TPP under the Trump administration. There were several reasons for this decision, including concerns about the impact of the agreement on American jobs and the perceived loss of sovereignty to international tribunals.

One of the main criticisms of the TPP was that it would lead to the offshoring of American jobs. Critics argued that the agreement would make it easier for U.S. companies to move their operations to countries with lower labor standards and wages, resulting in job losses in the United States. This concern was particularly acute in the manufacturing sector, where many jobs had already been lost to outsourcing.

Another point of contention was the dispute settlement mechanism included in the agreement. The TPP would have created an Investor-State Dispute Settlement (ISDS) system that allowed foreign investors to sue governments for damages if their investments were harmed by government policies or regulations. Critics of the ISDS system argued that it would undermine the sovereignty of countries and allow multinational corporations to challenge laws and regulations that were intended to protect public health, safety, and the environment.

Overall, the decision to withdraw from the TPP was a controversial one, with supporters and opponents on both sides of the issue. While the TPP was intended to promote trade and economic growth among its signatories, concerns about its impact on American jobs and sovereignty ultimately led to the United States' withdrawal from the agreement.

Conclusion

In conclusion, trade agreements have a significant impact on businesses, both positive and negative. These agreements can provide businesses with increased access to foreign markets, reduce the cost of doing business, and promote economic growth. However, they can also harm businesses that rely heavily on particular industries or are unable to compete with foreign businesses. Understanding the complexities of trade agreements and their impact on businesses is essential for success in today's globalized economy.

CHAPTER 13: CONCLUSION

Part 3 of this book has explored the legal issues that arise in business transactions, with a focus on contract law. We have examined the basics of contract law, including its sources, formation, interpretation, and performance, as well as various types of contracts and their specific legal requirements. We have also delved into the different stages of a business transaction, from negotiation and drafting to execution and enforcement.

Throughout our discussions, we have emphasized the importance of contract law in providing a framework for business relationships and promoting commercial certainty. By setting out clear rights and obligations, contracts allow parties to avoid disputes and ensure that their expectations are met. They also provide a means for parties to allocate risk and protect their interests, whether through provisions addressing warranties, representations, indemnification, or other matters.

However, as we have seen, contract law is not without its challenges and complexities. Parties must navigate a multitude of legal rules and requirements, as well as practical considerations such as negotiation power, bargaining tactics, and market forces. They must also be aware of the risks of contract disputes, which can arise from misunderstandings, ambiguities, changes in circumstances, or breaches of contract.

Moreover, contract law operates in a broader legal and social context, which can shape and influence its application. Business transactions are subject to a range of other legal issues, including intellectual property, securities regulation, antitrust, labor and employment, and international trade. They also operate within a complex economic and political landscape, which can affect the parties' choices and outcomes.

In this final chapter, we reflect on the key themes and takeaways of Part 3, and consider some of the broader implications of contract law and legal issues in business transactions. We will explore the role of contracts in promoting economic growth and innovation, as well as some of the challenges and limitations that arise. We will also consider some of the ethical and social dimensions of contract law, and the ways in which it can contribute to or detract from broader societal goals.

In sum, this book has aimed to provide a comprehensive and accessible introduction to contract law and legal issues in business transactions. We hope that it has equipped readers with the knowledge and tools needed to navigate the complexities of business relationships, and to appreciate the legal and social dimensions of commercial activity.

The importance of understanding the intersection of law and finance

Law and finance are two distinct fields that have a significant impact on businesses. However, the intersection of these two fields is often overlooked, and businesses may not fully appreciate the importance of understanding how they interact. In this section, we will explore the importance of understanding the intersection of law and finance, and how this knowledge can be applied to business transactions.

Importance of Understanding the Intersection of Law and Finance:

The intersection of law and finance is crucial for businesses as it can impact various aspects of their operations, including financing, investment, and risk management. For example, companies seeking financing must comply with legal regulations and contracts, and understanding the legal implications of financing options can help companies make informed decisions. Similarly, investment decisions are influenced by both legal and financial considerations, such as the regulatory environment and the financial viability of potential investments.

Furthermore, businesses must also manage their risk exposure, which requires an understanding of both legal and financial concepts. For example, companies must comply with various regulations that affect their risk exposure, such as labor laws, environmental regulations, and tax laws. Financial risks, such as market fluctuations and credit risk, must also be managed. Understanding the legal and financial aspects of risk management can help companies develop effective risk management strategies.

Legal and Financial Considerations in Business Transactions:

Business transactions, such as mergers and acquisitions, require a deep understanding of both legal and financial concepts. In these transactions, the legal and financial aspects are often intertwined, and it is essential to consider both aspects simultaneously.

For example, when valuing a company, financial analysts must consider legal factors that could impact the company's value, such as legal disputes, intellectual property rights, and regulatory compliance. Similarly, when negotiating the terms of a merger or acquisition, legal considerations such as regulatory compliance, intellectual property rights, and contractual obligations must be taken into account alongside financial considerations such as valuation, financing, and tax implications.

Legal and Financial Expertise in Business Transactions:

Given the importance of legal and financial considerations in business transactions, it is essential for companies to have access to legal and financial expertise. Business owners, corporate executives, in-house counsel, compliance officers, financial analysts, investment bankers, accountants, and regulators all play important roles in ensuring that legal and financial considerations are appropriately addressed in business transactions.

Business owners and corporate executives should have a basic understanding of legal and financial concepts and should be able to identify when legal or financial expertise is required. In-house counsel and compliance officers are responsible for ensuring that businesses comply with legal regulations and contractual obligations, and they should have a deep understanding of legal concepts. Financial analysts, investment bankers, and accountants are responsible for assessing the financial viability of potential investments and should have a deep understanding of financial concepts. Regulators are responsible for overseeing compliance with legal regulations and should have a deep understanding of both legal and financial concepts.

Conclusion:

In conclusion, understanding the intersection of law and finance is critical for businesses. The legal and financial aspects of business transactions are often intertwined, and it is essential to consider both aspects simultaneously. By understanding the legal and financial implications of business transactions, businesses can make informed decisions, manage risk effectively, and maximize their value. Furthermore, legal and financial expertise is essential in ensuring that businesses comply with legal regulations, contractual obligations, and financial considerations. Overall, the intersection of law and finance plays a critical role in business transactions, and businesses must appreciate its importance to succeed in today's global marketplace.

Future developments in business law

Business law is a constantly evolving field that reflects the changing nature of commerce and society. Legal developments have profound implications for businesses, affecting how they operate, their bottom line, and their ability to remain competitive. In this section, we will examine some of the future developments in business law and how they may impact businesses.

Data privacy and cybersecurity

One of the biggest challenges facing businesses today is the protection of data privacy and cybersecurity. With the growing reliance on technology and the increasing use of cloud computing, businesses are more vulnerable than ever to cyber attacks and data breaches. In response to this threat, governments around the world have enacted new regulations to protect personal data and hold companies accountable for breaches. For example, the European Union's General Data Protection Regulation (GDPR) imposes strict requirements on companies that collect and process personal data of EU citizens. Failure to comply can result in fines of up to 4% of the company's global revenue.

As technology continues to advance, businesses will need to stay up-to-date with the latest cybersecurity measures and data privacy regulations. Failure to do so can result in significant reputational damage and legal liability.

Artificial intelligence

Artificial intelligence (AI) is rapidly transforming many industries, including law. AI has the potential to streamline legal processes, reduce costs, and increase efficiency. For example, AI-powered contract review tools can quickly analyze thousands of documents for relevant information, freeing up lawyers to focus on more complex tasks.

However, the use of AI in the legal field also raises significant ethical and legal issues. For example, how can we ensure that AI decisions are fair and unbiased? How do we ensure that AI systems are transparent and explainable? These questions will need to be addressed as AI continues to play a larger role in the legal profession.

Environmental law

The issue of climate change is becoming increasingly urgent, and businesses are under growing pressure to reduce their carbon footprint and adopt sustainable practices. As a result, environmental law is becoming more important than ever. Governments are enacting new regulations to encourage businesses to reduce their environmental impact, and investors are increasingly demanding that companies disclose their sustainability efforts.

In the coming years, we can expect to see more regulations aimed at combating climate change and promoting sustainability. Businesses that fail to take these issues seriously may face reputational damage and legal liability.

International trade

International trade is a critical driver of economic growth, but it is also a complex and constantly changing field. As businesses become more globalized, they will need to navigate an increasingly complex web of international trade regulations, tariffs, and trade agreements.

Recent years have seen a rise in protectionist policies, with many countries imposing tariffs and other barriers to trade. In response, businesses will need to stay up-to-date with the latest developments in international trade and adapt their strategies accordingly.

Conclusion

In conclusion, the field of business law is constantly evolving, and businesses will need to stay up-to-date with the latest legal developments to remain competitive. Data privacy and cybersecurity, artificial intelligence, environmental law, and international trade are just a few of the areas where we can expect to see significant legal developments in the coming years. By understanding and adapting to these developments, businesses can minimize legal risks, protect their bottom line, and stay ahead of the competition.

Recommendations for business leaders.

Business leaders have a significant role to play in shaping the future of their organizations and the business environment as a whole. They must have a deep understanding of the legal, regulatory, and financial landscape to navigate the complex and rapidly changing business environment. In this section, we will provide some recommendations for business leaders to help them achieve success and sustain growth in the long term.

Embrace Diversity and Inclusion:

One of the most critical recommendations for business leaders is to embrace diversity and inclusion in their organizations. Diversity and inclusion are essential for creating a positive work environment that fosters creativity and innovation. A diverse workforce brings a wide range of perspectives, ideas, and experiences to the table, which can help organizations solve complex problems and make better decisions. Additionally, embracing diversity and inclusion can help businesses attract and retain top talent, improve employee morale, and enhance their reputation in the community.

Foster a Culture of Ethics and Compliance:

Business leaders must also foster a culture of ethics and compliance within their organizations. This means setting high ethical standards and ensuring that employees at all levels of the organization adhere to them. Business leaders must also ensure that their organizations comply with all applicable laws and regulations, including those related to labor, environmental protection, and consumer protection. Failure to comply with these laws and regulations can result in significant financial and reputational damage to organizations.

Emphasize Sustainability:

In today's world, sustainability has become an increasingly important consideration for businesses. Business leaders must embrace sustainability and take steps to minimize their organizations' environmental impact. This includes reducing waste, conserving energy, and minimizing their carbon footprint. Emphasizing sustainability can also help businesses attract environmentally conscious consumers and investors, thereby enhancing their reputation and improving their long-term growth prospects.

Invest in Digital Transformation:

Digital transformation has become a necessity for businesses that want to remain competitive in today's rapidly evolving business environment. Business leaders must invest in new technologies, such as artificial intelligence, machine learning, and the Internet of Things, to improve operational efficiency and drive innovation. Additionally, investing in digital transformation can help businesses enhance their customer experience, create new revenue streams, and improve their competitive positioning.

Focus on Talent Development:

Finally, business leaders must focus on talent development to ensure that their organizations have the skills and knowledge they need to succeed in the long term. This means investing in training and development programs to help employees acquire new skills and stay up-to-date with the latest industry trends. Business leaders must also create opportunities for employees to advance within the organization, providing them with a clear path to career progression.

Conclusion:

In conclusion, business leaders must take a proactive approach to navigate the complex and rapidly changing business environment successfully. By embracing diversity and inclusion, fostering a culture of ethics and compliance, emphasizing sustainability, investing in digital transformation, and focusing on talent development, business leaders can position their organizations for long-term success and growth.

These recommendations will enable business leaders to navigate the challenges and opportunities of the business world and create a positive impact in the world.

PART 4: SECURITIES LAWS AND REGULATIONS

Securities laws and regulations are essential for maintaining transparency and integrity in the financial markets. Securities laws and regulations are rules put in place by regulatory bodies such as the Securities and Exchange Commission (SEC) to protect investors from fraudulent activities, market manipulations, and other unethical practices. These laws and regulations cover a wide range of topics including registration requirements for securities offerings, disclosure obligations for companies, insider trading rules, and many others. In this section, we will provide an overview of securities laws and regulations and how they impact businesses, investors, and the broader economy.

Part 4 will begin by introducing the basic principles of securities laws and regulations. We will discuss the objectives and key features of securities laws and regulations and why they are essential for a well-functioning financial system. We will also examine the main players involved in the regulation of securities, including regulatory agencies, self-regulatory organizations, and other industry bodies.

Next, we will delve into the details of securities laws and regulations. We will discuss the various types of securities that are subject to regulation, including stocks, bonds, mutual funds, and exchange-traded funds (ETFs). We will also examine the registration process for securities offerings, the types of disclosures that companies must make to investors, and the rules that govern insider trading.

We will then turn our attention to the role of securities laws and regulations in the broader economy. We will explore the relationship between securities laws and regulations and the efficient allocation of capital. We will examine how securities laws and regulations can help promote investor confidence and support economic growth. We will also discuss the challenges associated with securities laws and regulations, including the potential for overregulation and unintended consequences.

In the final sections of Part 4, we will provide practical guidance for businesses and investors navigating the complex world of securities laws and regulations. We will provide examples of common compliance issues and best practices for avoiding regulatory violations. We will also examine the role of technology in securities regulation and the potential for emerging technologies to transform the regulatory landscape.

Overall, Part 4 will provide a comprehensive overview of securities laws and regulations, their importance in the financial system, and their impact on businesses, investors, and the broader economy. By the end of this section, readers should have a solid understanding of the key principles and practices of securities laws and regulations and be better equipped to navigate the complex regulatory landscape of the financial markets.

CHAPTER 14: INTRODUCTION TO SECURITIES LAW

Securities laws play an important role in regulating the issuance, trading, and exchange of securities. Securities laws are a complex set of regulations designed to ensure that investors receive accurate and complete information about the securities they purchase. This chapter provides an introduction to securities laws, focusing on the key principles and regulations that govern securities transactions.

The securities laws are a combination of federal and state laws. The federal securities laws are primarily governed by the Securities Act of 1933, the Securities Exchange Act of 1934, and the Investment Company Act of 1940. The state securities laws are known as "blue sky" laws and are designed to supplement the federal laws.

The Securities Act of 1933 is the primary federal law that regulates the initial sale of securities. The Act requires that all securities offered for sale to the public must be registered with the Securities and Exchange Commission (SEC) unless they qualify for an exemption. The Act also requires that companies issuing securities provide investors with certain disclosures about the company and the securities being offered.

The Securities Exchange Act of 1934 is the primary federal law that regulates the trading of securities on national stock exchanges. The Act requires companies whose securities are traded on national exchanges to disclose certain information about their business operations and financial condition to the public. The Act also requires that companies disclose information about insider trading and executive compensation.

The Investment Company Act of 1940 is the primary federal law that regulates the organization and operation of investment companies, including mutual funds and exchange-traded funds. The Act requires investment companies to register with the SEC and disclose certain information about their investment objectives, strategies, and risks to investors.

State securities laws, also known as blue sky laws, vary from state to state. These laws generally require that securities offerings be registered with the state before they can be sold to the public. Blue sky laws also provide investors with certain protections, such as anti-fraud provisions, and require brokers and investment advisers to register with the state.

The securities laws serve several important functions. They provide investors with access to accurate and complete information about the securities they purchase, which is essential for making informed investment decisions. They also help prevent fraud and other abusive practices in the securities markets, protecting investors and ensuring the integrity of the markets. In addition, the securities laws help promote transparency and accountability in corporate governance, which is essential for maintaining investor confidence and promoting economic growth.

In this chapter, we will explore the basic principles of securities law, including the types of securities covered by the laws, the key players in the securities markets, and the various regulations governing securities transactions. We will also examine the role of the SEC in enforcing the securities laws, as well as the different types of enforcement actions that can be taken against companies and individuals who violate the laws.

Throughout this chapter, we will use examples and case studies to illustrate key concepts and principles, helping you gain a deeper understanding of the complex world of securities law. We will also provide exercises and problems to help you apply these concepts to real-world situations, preparing you to succeed in a career in finance or law.

Definition of securities and securities law

Securities and securities law are an essential aspect of the modern financial landscape. They play a significant role in capital raising and investing, providing a mechanism for companies to obtain funding and investors to purchase ownership interests in those companies. However, securities transactions also carry risks and can be subject to fraud and other abuses, which have led to the creation of securities laws and regulations. In this section, we will explore the definition of securities and securities law, including the history, purpose, and scope of securities laws and regulations.

What are securities?

Securities are financial instruments that represent ownership or debt in a company, government, or other entity. They are typically bought and sold in capital markets and can take many different forms, such as stocks, bonds, options, and derivatives. Securities can be issued by companies to raise capital, and they can also be traded by investors looking to buy or sell ownership interests or debt obligations in those companies.

The definition of securities is broad and encompasses a wide range of financial instruments. Some securities are straightforward, such as common stocks, which represent ownership interests in a company. Other securities can be more complex, such as options or derivatives, which provide investors with the right to buy or sell an underlying asset at a specified price.

History of securities law

The regulation of securities transactions and markets dates back to the early 20th century in the United States. The Securities Act of 1933 was the first federal securities law, created in response to the stock market crash of 1929 and the subsequent Great Depression. The Securities Act of 1933 requires companies to register their securities offerings with the Securities and Exchange Commission (SEC) and provide detailed information about their business, financial condition, and risk factors to potential investors. The Securities Act of 1934 created the SEC as a regulatory agency to enforce federal securities laws, and it established rules for trading securities on national securities exchanges.

Purpose of securities law

The primary purpose of securities laws and regulations is to protect investors and maintain fair and efficient markets. Securities laws require companies to provide investors with accurate and timely information about their business and financial condition, as well as any risks associated with investing in their securities. Securities laws also establish rules for the conduct of market participants, such as brokers and investment advisors, to ensure that investors are treated fairly and that markets are free from fraud and manipulation.

Scope of securities law

Securities laws and regulations apply to a broad range of securities transactions and market participants. They cover the issuance, sale, and trading of securities by companies, as well as the conduct of market participants, such as brokers, investment advisors, and financial analysts. Securities laws also regulate the activities of stock exchanges and other trading platforms.

Securities laws can be complex and vary by jurisdiction, so it is essential for companies and investors to seek the advice of legal and financial professionals when engaging in securities transactions.

Conclusion

Securities and securities law are crucial components of the modern financial system. Securities provide a mechanism for companies to raise capital and for investors to purchase ownership interests or debt obligations in those companies. Securities laws and regulations are designed to protect investors and ensure fair and efficient markets. The history, purpose, and scope of securities laws are complex and can be subject to change, so it is essential for market participants to stay up-to-date with the latest developments and seek professional advice when engaging in securities transactions.

Historical context and purpose of securities law

Securities law is a critical component of the legal system that regulates the issuance, sale, and transfer of securities. Securities are financial instruments that represent a form of ownership or debt, such as stocks, bonds, or derivatives. Securities law is designed to promote transparency, protect investors, and maintain fair and efficient capital markets.

To fully appreciate the scope and importance of securities law, it is essential to understand its historical context and the purpose that it serves in the modern economy.

The modern securities markets have evolved over centuries, beginning with trading in basic commodities like rice and wheat, and evolving to include more complex financial instruments such as stocks and bonds. The development of modern securities markets can be traced back to Amsterdam in the 17th century, where the Dutch East India Company issued shares to fund its operations.

In the United States, the securities markets emerged as a way for companies to raise capital during the industrialization era of the late 19th century. In the early 1900s, however, securities fraud and other forms of misconduct became rampant, leading to a series of financial scandals that severely damaged investor confidence.

In response to these issues, Congress enacted the Securities Act of 1933 and the Securities Exchange Act of 1934, which established a comprehensive regulatory framework for securities transactions in the United States. These laws required companies to register their securities offerings with the Securities and Exchange Commission (SEC) and disclose detailed information about their financial performance and business operations. The laws also created the SEC, which is charged with enforcing securities laws, regulating the securities markets, and protecting investors.

Since the passage of the Securities Act and the Securities Exchange Act, securities laws have continued to evolve in response to changing market conditions and technological innovations. For example, in the 1990s, the emergence of electronic trading platforms and the growth of online investing led to the adoption of new rules and regulations designed to address these developments.

The purpose of securities laws is to promote transparency, fairness, and efficiency in the securities markets, while protecting investors from fraudulent or deceptive practices. One of the main goals of securities laws is to ensure that investors have access to accurate and timely information about the securities they are buying, so that they can make informed investment decisions. Securities laws also aim to prevent insider trading, market manipulation, and other forms of misconduct that can undermine the integrity of the securities markets.

In addition to protecting investors, securities laws also play a critical role in promoting economic growth and development. By providing a reliable and efficient means for companies to raise capital, securities markets enable businesses to expand, create jobs, and drive innovation.

In conclusion, the historical context and purpose of securities laws provide important insights into their significance and impact on the modern economy. The securities markets are a critical component of the financial system, and securities laws are essential to promoting transparency, fairness, and efficiency in these markets.

The role of securities law in modern business

The role of securities law in modern business cannot be overstated. Securities law plays a critical role in ensuring the fair and transparent operation of the securities market, which is a cornerstone of modern finance. This section will explore the role of securities law in modern business, including its impact on the securities market, investors, and businesses.

The primary purpose of securities law is to protect investors and promote transparency in the securities market. Securities law accomplishes this goal by requiring companies to provide accurate and timely information about their financial performance and business operations to investors. This information is critical for investors to make informed decisions about whether to invest in a company's securities.

Securities law also regulates the sale and purchase of securities to prevent fraud and other abuses. This is accomplished through the registration and disclosure requirements of the Securities Act of 1933 and the Securities Exchange Act of 1934. These laws require companies to register their securities with the Securities and Exchange Commission (SEC) before they can be sold to the public. Companies must also provide ongoing disclosures about their financial performance and other material events to investors.

Securities law also provides important protections for businesses. For example, securities law helps businesses to raise capital through the sale of securities to investors. This is critical for businesses that need capital to fund their operations or to expand their businesses. Securities law also provides protections for businesses by regulating the conduct of investors and other market participants. This includes rules against insider trading, market manipulation, and other forms of securities fraud.

Overall, securities law plays a vital role in modern business by promoting transparency and fairness in the securities market, protecting investors, and providing important protections for businesses. The continued evolution of securities law will be critical to ensuring that the securities market remains a vibrant and trusted source of capital for businesses and investors alike.

EXAMPLES:

An example of the role of securities law in modern business is the requirement for public companies to file regular reports with the SEC. These reports, which include financial statements and other important information about the company, are made available to the public and provide investors with critical information about the company's financial health and operations.

Another example of the role of securities law in modern business is the regulation of insider trading. Insider trading occurs when someone with access to non-public information about a company trades on that information for personal gain. Securities law prohibits insider trading to protect investors and ensure a level playing field for all market participants.

PROBLEMS:

A company wants to raise capital by selling securities to the public. What are the registration and disclosure requirements that the company must comply with under securities law?

An investor suspects that a company's CEO is engaged in insider trading. What should the investor do to report this suspected violation of securities law?

EXERCISES:

Research a recent case involving a violation of securities law and summarize the key facts and legal issues involved.

Imagine that you are the CEO of a public company. What steps would you take to ensure that your company is in compliance with securities law?

CHAPTER 15: SECURITIES ACT OF 1933 AND 1934

The Securities Act of 1933 and the Securities Exchange Act of 1934 are two of the most important pieces of legislation in the history of the securities industry. Together, these two acts established the foundation of securities regulation in the United States and provided investors with important protections against fraud and other forms of misconduct.

In this chapter, we will provide a comprehensive introduction to the Securities Act of 1933 and the Securities Exchange Act of 1934. We will discuss the historical context that led to the passage of these laws, the key provisions of each act, and the impact they have had on the securities industry and on investors. We will also explore some of the challenges and criticisms that have been raised with respect to these laws and the ongoing efforts to update and improve them.

The Securities Act of 1933

The Securities Act of 1933 was the first major federal securities law enacted in the United States. The act was a response to the stock market crash of 1929 and the subsequent economic depression, which had eroded public confidence in the securities markets. The purpose of the Securities Act was to require issuers of securities to disclose certain information to investors before the securities were offered for sale to the public.

Under the Securities Act of 1933, companies that wished to sell securities to the public were required to register those securities with the Securities and Exchange Commission (SEC) and provide certain information to potential investors, including information about the company's financial condition, its business operations, and the risks associated with the investment. The act also established liability for false or misleading statements in the registration statement and the prospectus.

The Securities Exchange Act of 1934

The Securities Exchange Act of 1934 was passed in response to the need for ongoing regulation of the securities markets. The act created the Securities and Exchange Commission, which is charged with enforcing the securities laws and regulating the securities markets in the United States. The act also established a number of important requirements for public companies, including the requirement to file periodic reports with the SEC and to disclose certain material events.

Under the Securities Exchange Act of 1934, companies that have securities listed on a national securities exchange, such as the New York Stock Exchange or the NASDAQ, are required to file periodic reports with the SEC, including annual reports and quarterly reports. These reports must include information about the company's financial condition, its business operations, and other material events that could affect the value of the securities.

Conclusion

The Securities Act of 1933 and the Securities Exchange Act of 1934 are landmark pieces of legislation that have had a profound impact on the securities industry and on investors. These laws have helped to establish important disclosure requirements and investor protections, and have helped to promote transparency and fairness in the securities markets. However, these laws are not without their challenges and criticisms, and there is ongoing debate about the need for further reform and modernization. In the following sections, we will explore the key provisions of these laws in more detail and examine some of the ongoing debates and challenges surrounding securities regulation in the United States.

Key provisions of the Securities Act of 1933

The Securities Act of 1933, commonly referred to as the "truth in securities" law, was enacted in response to the stock market crash of 1929 and the ensuing Great Depression. The Act sought to restore investor confidence by requiring full disclosure of information about securities being sold to the public. This section will provide a detailed analysis of the key provisions of the Securities Act of 1933.

Registration Requirement

The Securities Act of 1933 requires that all securities offered for sale to the public must be registered with the Securities and Exchange Commission (SEC), unless they qualify for an exemption.

The registration process involves the submission of a registration statement to the SEC, which provides detailed information about the company, the securities being offered, and the risks involved in investing in those securities. The registration statement must be filed with the SEC before the securities can be offered for sale to the public.

The SEC will review the registration statement to ensure that all necessary information is included and that it meets the requirements of the Securities Act of 1933. Once the registration statement is approved, the company can begin offering the securities for sale to the public.

The registration requirement is intended to ensure that investors have access to all relevant information about a company and its securities before making an investment decision. It helps to promote transparency and fairness in the securities market, and to protect investors from fraud and other abuses.

There are certain exemptions to the registration requirement, which may apply in certain circumstances. These exemptions include offerings made only to accredited investors, offerings made in connection with mergers and acquisitions, and offerings of securities issued by government entities, among others.

Liability for Misstatements and Omissions

The Securities Act of 1933 imposes liability for misstatements and omissions in the registration statement, prospectus, and oral communications made in connection with the offering of securities.

Section 11 of the Act imposes strict liability on the issuer, underwriters, and certain other parties for material misstatements or omissions in the registration statement. The plaintiff need not prove that the defendant acted with any intent to deceive or defraud, but only that the registration statement contained an untrue statement of a material fact or omitted to state a material fact necessary to make the statements made not misleading. However, the defendant can avoid liability if it can demonstrate that it conducted a reasonable investigation and had reasonable grounds to believe, and did believe, that the statements were true and not misleading.

Section 12(a)(2) imposes a similar liability for material misstatements or omissions in the prospectus or oral communications made in connection with the offering of securities. Unlike Section 11, however, the plaintiff need only prove that the defendant made a material misstatement or omission, not that the defendant acted with any intent to deceive or defraud. The defendant may avoid liability by demonstrating that it did not know, and in the exercise of reasonable care could not have known, of the untruth or omission.

It's important to note that Section 11 liability applies to all purchasers of securities sold pursuant to a registration statement, while Section 12(a)(2) liability applies only to purchasers who bought the securities in the offering. Additionally, Section 11 liability is joint and several among all defendants, while Section 12(a)(2) liability is several, meaning that each defendant is only responsible for its own proportionate share of damages.

Exemptions from Registration

The Securities Act of 1933 provides for certain exemptions from the registration requirement for certain types of securities and transactions. For example, securities issued by the US government, municipal securities, and securities issued in private placements are exempt from the registration requirement. The Act also provides exemptions for transactions involving certain types of sophisticated investors, such as institutional investors and accredited investors.

The Securities Act of 1933 provides several exemptions from the requirement of registering securities with the SEC. Some of the most common exemptions are:

Private Placement Exemption: This exemption applies to the sale of securities to a limited number of accredited investors, such as wealthy individuals or institutional investors. The number of non-accredited investors is also limited, and the issuer cannot use general solicitation or advertising to offer the securities.

Regulation A Exemption: This exemption allows companies to offer and sell up to $75 million of securities in a 12-month period without having to register with the SEC. The securities must be offered through a qualified offering circular and are subject to certain disclosure requirements.

Intrastate Exemption: This exemption allows the sale of securities only within a single state, where the issuer is incorporated and has its principal place of business, and where the investors reside.

Reg S Exemption: This exemption applies to the sale of securities that occur outside of the United States, to non-U.S. investors. The issuer cannot use general solicitation or advertising to offer the securities.

Crowdfunding Exemption: This exemption allows the sale of securities through crowdfunding platforms, where small amounts of money are raised from a large number of investors. The amount that can be raised is subject to certain limits, and the issuer must comply with specific disclosure requirements.

It is important to note that while these exemptions allow the issuer to avoid registering the securities with the SEC, they do not exempt the issuer from anti-fraud provisions or from other securities laws and regulations.

Anti-Fraud Provisions

The Securities Act of 1933 includes anti-fraud provisions that prohibit the use of fraudulent, deceptive, or manipulative practices in connection with the sale of securities. These provisions are designed to protect investors from fraudulent schemes and ensure that securities are sold based on accurate and complete information. The anti-fraud provisions apply not only to the registration statement and prospectus, but also to all communications made in connection with the sale of securities.

The Securities Act of 1933 contains a number of anti-fraud provisions that are designed to protect investors from false or misleading information when purchasing securities. These provisions are found primarily in Sections 17(a) and 10(b) of the Act.

Section 17(a) of the Securities Act prohibits any person from using any device, scheme, or artifice to defraud in connection with the sale of a security. This provision applies to both registered and unregistered securities and applies to any person who sells or offers to sell a security. It prohibits any act or omission that would result in a misstatement or omission of material fact in connection with the sale of a security. This provision applies not only to affirmative misstatements, but also to situations where information is omitted that is necessary to make a statement not misleading.

Section 10(b) of the Securities Exchange Act of 1934 makes it illegal to use any manipulative or deceptive device or contrivance in connection with the purchase or sale of any security. This provision is often referred to as the "anti-fraud" provision of the Act. It prohibits any act or omission that would result in a misstatement or omission of material fact in connection with the purchase or sale of a security. Unlike Section 17(a), Section 10(b) applies only to securities traded on national securities exchanges.

Both Section 17(a) and Section 10(b) provide for civil liability for violations of the anti-fraud provisions. Any person who violates these provisions may be sued by investors who have been injured as a result of the violation. In addition, the Securities and Exchange Commission (SEC) has the authority to bring enforcement actions against individuals or companies that violate these provisions. The SEC may seek injunctions, fines, and other remedies to stop ongoing violations and punish those who have violated the anti-fraud provisions in the past.

Civil Liability for Violations

Investors who suffer losses as a result of violations of the Securities Act of 1933 may bring civil lawsuits to recover damages. The Act provides for a private right of action, which allows investors to sue issuers, underwriters, and other parties involved in the sale of securities for violations of the Act. The Act also provides for criminal penalties for willful violations of the Act.

Under the Securities Act of 1933, any person who violates any of the provisions of the Act may be held liable to any person who purchases a security that is the subject of such violation. The Act provides for civil liability for violations in several different ways:

Section 11: Any person who sells a security in violation of the registration requirements of the Act or who makes an untrue statement of material fact in the registration statement may be held liable to any person who purchases the security for the full amount paid for the security, plus interest.

Section 12(a)(1): Any person who offers or sells a security in violation of the registration requirements of the Act may be held liable to any person who purchases the security for the full amount paid for the security, plus interest.

Section 12(a)(2): Any person who offers or sells a security by means of a prospectus or oral communication that includes an untrue statement of material fact or omits to state a material fact necessary to make the statements made, in light of the circumstances under which they were made, not misleading, may be held liable to any person who purchases the security for the full amount paid for the security, plus interest.

Section 15: Any person who controls another person who violates any provision of the Act or any rule or regulation thereunder may be held liable for such violation, unless the controlling person acted in good faith and did not directly or indirectly induce the violation.

Section 17: Any person who employs any device, scheme, or artifice to defraud in connection with the offer, sale, or purchase of a security, or who engages in any transaction, practice, or course of business that operates or would operate as a fraud or deceit upon any person, may be held liable to any person who purchases or sells the security for damages caused by such conduct.

The Securities Act of 1933 provides for both public and private enforcement of its provisions. The SEC may bring an enforcement action against any person who violates the Act, and private parties who have been harmed by a violation may bring a civil action to recover damages. The Act also provides for criminal penalties for willful violations of its provisions.

Conclusion

The Securities Act of 1933 was a landmark piece of legislation that has had a profound impact on the securities industry. By requiring full disclosure of information about securities being sold to the public, the Act has helped to restore investor confidence and promote transparency in the securities markets. The Act's registration requirement, liability for misstatements and omissions, exemptions from registration, anti-fraud provisions, and civil liability for violations have all contributed to the Act's success in achieving its goals.

Registration process for securities offerings

The registration process for securities offerings is a critical aspect of the regulatory framework governing securities markets. The registration process is an essential part of the Securities Act of 1933, which was enacted in response to the market crash of 1929 and the ensuing Great Depression. The Act is designed to provide investors with reliable and accurate information about securities offerings, to ensure that they are adequately informed before investing their hard-earned money.

The Securities Act of 1933 requires that all securities offerings be registered with the Securities and Exchange Commission (SEC) before they can be offered to the public. This requirement applies to all companies that issue securities, regardless of the size of the offering or the type of security being offered. The registration process is intended to ensure that investors are provided with complete and accurate information about the securities being offered, including information about the company issuing the securities, its management team, and the risks associated with the investment.

The registration process can be a lengthy and expensive undertaking, especially for smaller companies. Companies are required to file a registration statement with the SEC that includes detailed information about the company and the securities being offered. The registration statement is reviewed by the SEC staff to ensure that it complies with all the applicable rules and regulations. The SEC may request additional information or revisions to the registration statement before it is deemed to be effective.

The registration statement includes a prospectus, which is a document that provides investors with detailed information about the securities being offered. The prospectus includes information about the company's business, its management team, the securities being offered, and the risks associated with the investment. The prospectus must be updated to reflect any material changes in the company's business or the offering before the securities can be sold.

Companies are also required to provide ongoing disclosures to investors after the registration statement has been declared effective. These ongoing disclosures include quarterly and annual financial statements, as well as other periodic reports, such as proxy statements and annual reports. The SEC monitors these disclosures to ensure that companies are providing investors with timely and accurate information about their business.

The registration process for securities offerings is a critical component of the regulatory framework governing securities markets. It is designed to ensure that investors are provided with complete and accurate information about the securities being offered, so that they can make informed investment decisions. While the registration process can be a time-consuming and expensive undertaking, it is an essential part of maintaining the integrity of the securities markets and protecting investors from fraudulent and misleading activities.

Let's take a closer look at the registration process for securities offerings.

Filing the Registration Statement

The first step in the registration process is for the company to file a registration statement with the SEC. The registration statement includes detailed information about the company, the securities being offered, and the risks associated with the investment. The registration statement is filed electronically through the SEC's Electronic Data Gathering, Analysis, and Retrieval (EDGAR) system.

The registration statement is a complex document that typically includes several sections. The first section is the prospectus, which provides investors with detailed information about the company, the securities being offered, and the risks associated with the investment. The prospectus must include a description of the business, including its history, management team, and financial performance. It must also include detailed financial statements, such as income statements, balance sheets, and cash flow statements. In addition, the prospectus must include a discussion of the risks associated with the investment, including any legal, regulatory, or competitive risks.

Another section of the registration statement is the exhibits, which provide additional information about the company and the securities being offered. This section may include copies of contracts, agreements, and other documents related to the offering.

The registration statement may also include a letter from the company's management team, which provides additional information about the business and the securities being offered. This letter typically includes a discussion of the company's strategy, financial performance, and competitive position.

Once the registration statement is filed, the SEC will review the document to ensure that it complies with all applicable rules and regulations. The SEC may request additional information from the company or ask the company to revise certain sections of the registration statement. The review process can take several months, and the company cannot begin selling securities until the registration statement is declared effective by the SEC.

From July 10, 2017, the Securities and Exchange Commission's (SEC) Division started accepting draft registration statement submissions from all issuers for nonpublic review, provided that the issuer confirms in a cover letter to the nonpublic draft submission that it will publicly file its registration statement and nonpublic draft submissions at least 15 days prior to any roadshow or anticipated effective date of the registration statement. The SEC will limit the nonpublic review to the initial submission only, but will accept draft registration statements submitted within one year of the effective date of an issuer's initial Securities Act registration statement or an issuer's Exchange Act Section 12(b) registration statement for nonpublic review. Foreign Private Issuers may elect to proceed in accordance with these procedures or those available to Emerging Growth Companies, if they qualify as one. The staff will monitor practices under these expanded processing procedures and may make modifications to limit or terminate them. Issuers may submit questions about their eligibility to use these expanded processing procedures to CFDraftPolicy@sec.gov.

Review by the SEC

After the registration statement is filed, it is reviewed by the SEC staff to ensure that it complies with all the applicable rules and regulations. The SEC staff may request additional information or revisions to the registration statement before it is deemed to be effective. The review process typically takes several months to complete.

Waiting Period

After the registration statement is filed, there is a waiting period during which the company cannot offer or sell the securities. The waiting period typically lasts 20 days for a typical securities offering. This period allows the SEC staff to review the registration statement and ensure that investors have access to all the information they need to make an informed investment decision.

Prospectus Delivery

Once the registration statement has been declared effective by the SEC, the company must deliver a prospectus to investors. The prospectus is a document that provides investors with detailed information about the securities being offered and the company issuing them. The prospectus is required to be delivered to each investor who purchases securities in the offering, regardless of whether the investor is a retail or institutional investor. The purpose of the prospectus is to ensure that investors have access to all material information about the securities being offered so that they can make informed investment decisions.

The prospectus must contain certain key information about the securities being offered, including the price at which the securities will be offered, the number of securities being offered, the use of proceeds from the offering, the risks associated with investing in the securities, and the business and financial information of the company issuing the securities. In addition to this required information, the prospectus may also contain supplementary information that the company believes will be helpful to investors, such as market data or industry trends.

The prospectus delivery requirements can be fulfilled through a variety of means, including physical delivery of a paper prospectus, electronic delivery of a prospectus, or access to a prospectus on a company's website. The SEC has specific rules and guidelines regarding how prospectuses must be delivered, including the format and content of electronic prospectuses.

It is important to note that the delivery of the prospectus is not just a legal requirement, but also a critical part of the marketing and sales process for securities offerings. The prospectus serves as a key tool for communicating with potential investors, and can help to build investor confidence and trust in the company and the securities being offered.

In addition to the initial prospectus delivery requirements, companies must also provide ongoing disclosure to investors in order to ensure that investors are informed of material changes or events that may impact their investment. This ongoing disclosure may take the form of periodic reports, such as quarterly and annual reports, as well as current reports filed with the SEC in response to specific events or circumstances, such as the resignation of a key executive or a material change in the company's business operations.

Overall, the registration process for securities offerings is a complex and heavily regulated process that requires significant attention to detail and expertise in securities law. Companies must work closely with legal and financial advisors to ensure that they comply with all relevant laws and regulations, and that they provide investors with the information they need to make informed investment decisions. By following these rules and guidelines, companies can successfully raise capital through securities offerings while also protecting the interests of investors and maintaining the integrity of the capital markets.

The Securities Act of 1934 and its impact on securities trading

The Securities Act of 1934 was enacted by the United States Congress to regulate securities trading and the activities of the securities industry. This law established the Securities and Exchange Commission (SEC) and gave it the power to enforce securities laws and regulations. In this section, we will explore the Securities Act of 1934, its history, its provisions, and its impact on securities trading.

Background

The Securities Act of 1934 was passed in response to the stock market crash of 1929, which led to the Great Depression. The stock market crash was caused by a variety of factors, including speculation, margin trading, and insider trading. The lack of regulation and oversight allowed these practices to continue unchecked, leading to a financial crisis that affected the entire country.

The Securities Act of 1934 was designed to prevent another stock market crash by regulating securities trading and the activities of the securities industry. The law was enacted on June 6, 1934, and established the SEC as the primary regulatory agency for the securities industry.

Provisions

The Securities Act of 1934 contains several provisions that are designed to regulate securities trading and the activities of the securities industry. Some of the key provisions of the law include:

Registration of Securities: The law requires issuers of securities to register with the SEC before they can be sold to the public. The registration process involves disclosing information about the company and the securities being offered, including financial statements, risk factors, and other relevant information.

Reporting Requirements: The law requires companies to file periodic reports with the SEC, including annual reports and quarterly reports. These reports provide investors with information about the company's financial performance, management, and other key aspects of the business.

Insider Trading: The law prohibits insider trading, which is the buying or selling of securities based on non-public information. Insider trading is illegal because it gives an unfair advantage to those who have access to the information.

Anti-Fraud Provisions: The law contains provisions that prohibit fraud in the sale of securities. These provisions make it illegal to make false or misleading statements or to omit material information in connection with the sale of securities.

Impact

The Securities Act of 1934 had a significant impact on securities trading and the securities industry. The law provided much-needed regulation and oversight of the securities industry, which helped to prevent another stock market crash.

One of the key impacts of the law was the establishment of the SEC. The SEC has become one of the most powerful regulatory agencies in the United States, with the power to investigate and prosecute securities fraud and other violations of securities laws.

The law also had a significant impact on the way companies raise capital. The registration process established by the law requires companies to disclose detailed information about their financial performance, management, and other aspects of their business. This information is then made available to the public, which allows investors to make informed decisions about whether to invest in the company.

The law has also had an impact on insider trading. The prohibition on insider trading has helped to level the playing field for investors by preventing those with access to non-public information from profiting unfairly.

Overall, the Securities Act of 1934 has had a positive impact on securities trading and the securities industry. The law has helped to promote transparency and accountability, which has increased investor confidence in the securities markets.

Conclusion

The Securities Act of 1934 was a landmark piece of legislation that has had a significant impact on securities trading and the securities industry. The law established the SEC as the primary regulatory agency for the securities industry and provided much-needed regulation and oversight.

The law has had a positive impact on the securities industry by promoting transparency and accountability and increasing investor confidence. The law has also had an impact on the way companies raise capital and has helped to level the playing field for investors. However, the law has also faced criticism for being too strict and costly, especially for small businesses.

One of the key provisions of the Securities Act of 1934 is the requirement for companies to file periodic reports with the SEC. This requirement ensures that companies are providing timely and accurate information to investors, which promotes transparency and helps to prevent fraudulent activity. The SEC is responsible for enforcing this requirement and has the authority to sanction companies that fail to comply.

Another important provision of the law is the requirement for insiders to disclose information about their trading activity. This provision helps to prevent insider trading and ensures that investors have access to information that may impact the value of a company's securities.

The Securities Act of 1934 has also had an impact on the way companies raise capital. The law requires companies that offer securities to register with the SEC and disclose information about their financial condition and operations. This requirement ensures that investors have access to information that may impact the value of a company's securities and helps to prevent fraudulent activity.

While the Securities Act of 1934 has had a positive impact on the securities industry, it has also faced criticism for being too strict and costly, especially for small businesses. Some critics argue that the law imposes too many requirements on companies and that these requirements can be costly and time-consuming. This can make it difficult for small businesses to raise capital and compete with larger companies.

Despite these criticisms, the Securities Act of 1934 remains an important piece of legislation that has helped to promote transparency and accountability in the securities industry. The law has also helped to protect investors from fraudulent activity and has played a role in the growth and development of the securities industry.

In conclusion, the Securities Act of 1934 has had a significant impact on securities trading and the securities industry. The law has helped to promote transparency and accountability, increase investor confidence, and prevent fraudulent activity. While the law has faced criticism for being too strict and costly, it remains an important piece of legislation that has played a vital role in the growth and development of the securities industry.

CHAPTER 16: SECURITIES EXCHANGE ACT OF 1934

The Securities Exchange Act of 1934 is a landmark piece of legislation that significantly impacted the securities industry in the United States. The law was enacted in response to the stock market crash of 1929, which had devastating effects on the American economy. The crash was the result of rampant speculation and fraud, as well as a lack of transparency and regulation in the securities industry. The Securities Exchange Act of 1934 was designed to address these issues by providing comprehensive regulation and oversight of the securities industry.

The Securities Exchange Act of 1934 established the Securities and Exchange Commission (SEC) as the primary regulatory agency for the securities industry. The SEC is responsible for enforcing federal securities laws and regulating securities markets and participants, including brokers, dealers, and exchanges. The law also requires companies that issue securities to register with the SEC and disclose certain information to the public, such as financial statements and other material information that could impact the value of their securities.

The Securities Exchange Act of 1934 has had a significant impact on the securities industry by promoting transparency, accountability, and investor protection. The law has also had a positive impact on the way companies raise capital, by providing a regulatory framework that helps to ensure fair and equitable access to capital markets.

In this chapter, we will examine the Securities Exchange Act of 1934 in detail, including its key provisions, impact on the securities industry, and ongoing significance today. We will explore the role of the SEC in regulating the securities industry, as well as the requirements for companies to register with the SEC and disclose certain information to the public. We will also analyze the impact of the law on different participants in the securities industry, including business owners and entrepreneurs, corporate executives, in-house counsel and legal departments, compliance officers, financial analysts, investment bankers, accountants, and regulators.

To understand the Securities Exchange Act of 1934 and its impact on the securities industry, it is important to examine the historical context in which it was enacted. We will explore the events leading up to the stock market crash of 1929 and the subsequent economic turmoil that followed. We will also examine the regulatory environment at the time, including the limited oversight of the securities industry and the lack of transparency and accountability in the market.

Throughout this chapter, we will provide examples, problems, and exercises to illustrate key concepts and deepen understanding of the Securities Exchange Act of 1934 and its impact on the securities industry. We will also present counterarguments and dissenting opinions in a balanced and objective way to encourage critical thinking and analysis.

Overall, the Securities Exchange Act of 1934 is a critical piece of legislation that has had a lasting impact on the securities industry in the United States. By promoting transparency, accountability, and investor protection, the law has helped to build a stronger, more resilient securities market that benefits all participants. In this chapter, we will explore the law in detail and its impact on the securities industry, providing a comprehensive understanding of its importance and relevance today.

Regulation of securities markets

The Securities Exchange Act of 1934 (the "Exchange Act") is one of the most significant pieces of securities regulation in the United States. It established the Securities and Exchange Commission (SEC) as the primary regulatory agency for the securities industry and created a framework for the regulation of securities markets. The Exchange Act was enacted in response to the stock market crash of 1929 and the Great Depression that followed. The legislation aimed to prevent future market crashes and promote transparency and fairness in the securities industry.

This section will focus on the regulation of securities markets under the Exchange Act and how it differed from the Securities Act of 1933. Specifically, we will examine the regulatory structure created by the Exchange Act, the registration and reporting requirements imposed on public companies, and the regulatory mechanisms designed to promote fairness and transparency in securities markets.

Regulation of Securities Markets under the Exchange Act

The Exchange Act established a comprehensive regulatory framework for securities markets. The Act created a centralized regulatory authority, the Securities and Exchange Commission (SEC), with broad regulatory powers over the securities industry. The SEC was given the authority to regulate securities exchanges, broker-dealers, and other market participants.

The Exchange Act required the registration of all securities exchanges, including national securities exchanges, regional exchanges, and over-the-counter (OTC) markets. The SEC was given the authority to regulate the listing standards and trading rules of these exchanges, and to monitor their compliance with the Exchange Act.

The Exchange Act also required the registration of all broker-dealers who engage in the business of buying and selling securities for customers. Broker-dealers were required to register with the SEC and comply with a series of reporting and record-keeping requirements. The Act required broker-dealers to maintain a net capital requirement, which was intended to ensure that they had sufficient capital to meet their obligations to customers.

In addition to regulating securities exchanges and broker-dealers, the Exchange Act imposed reporting and disclosure requirements on public companies. The Act required public companies to file periodic reports with the SEC, including annual reports on Form 10-K and quarterly reports on Form 10-Q. The Act also required public companies to disclose material information to the public, including information about their business operations, financial condition, and risk factors.

The Exchange Act also established rules governing insider trading and market manipulation. The Act prohibited insider trading, which is the use of material non-public information to trade securities, and provided for civil and criminal penalties for violations. The Act also prohibited market manipulation, including the use of false or misleading statements to manipulate securities prices.

Differences between the Exchange Act and the Securities Act of 1933

While the Securities Act of 1933 and the Exchange Act share similar objectives, there are several key differences between the two laws. The Securities Act of 1933 was primarily focused on regulating the issuance of securities, while the Exchange Act was focused on regulating securities markets and market participants.

The Securities Act of 1933 required the registration of all securities offerings with the SEC, and imposed liability for misstatements and omissions in registration statements. In contrast, the Exchange Act focused on regulating securities markets and market participants. The Act required the registration of securities exchanges and broker-dealers, and imposed reporting and disclosure requirements on public companies.

Another key difference between the two laws is the level of detail provided in the registration and reporting requirements. The Securities Act of 1933 required detailed disclosures about the issuer and the securities being offered, while the Exchange Act required periodic reports and disclosures about the issuer's financial condition, business operations, and risk factors.

Conclusion

In conclusion, the Securities Exchange Act of 1934 established a comprehensive regulatory framework for securities markets and market participants. The Act created a centralized regulatory authority, the Securities and Exchange Commission (SEC), which has had a significant impact on the securities industry and the broader economy.

The Act addressed several issues that were not covered by the Securities Act of 1933, such as the regulation of securities exchanges and the registration and regulation of brokers and dealers. It also established rules for disclosure, reporting, and anti-fraud measures that have improved the transparency and integrity of securities markets.

One of the most significant differences between the two laws was the focus of the Securities Act of 1933 on the initial offering of securities, while the Securities Exchange Act of 1934 focused on the ongoing trading and reporting of securities in the secondary market.

The Securities Exchange Act of 1934 also gave the SEC the authority to regulate proxy solicitation and to require public companies to file regular reports with the Commission, including annual reports, quarterly reports, and other periodic filings. This has led to greater transparency and accountability for publicly traded companies, and has helped to protect investors from fraud and other misconduct.

Overall, the Securities Exchange Act of 1934 has been instrumental in promoting transparency, accountability, and integrity in securities markets, and has helped to protect investors from fraud and other abuses. The Act has had a lasting impact on the securities industry and the broader economy, and continues to play a vital role in maintaining the integrity of securities markets today.

Reporting requirements for public companies

In the United States, public companies are required to comply with various reporting requirements to provide investors and other stakeholders with relevant and reliable information about their financial performance and operations. These reporting requirements are designed to promote transparency, accountability, and informed decision-making in the capital markets. This section will provide a detailed overview of the reporting requirements for public companies in the US, including the Securities Exchange Act of 1934 and subsequent regulations and rules.

Background:

The Securities Exchange Act of 1934 is the primary law governing securities trading in the United States. The Act established the Securities and Exchange Commission (SEC) as the primary regulatory authority for the securities industry and created a comprehensive regulatory framework for securities markets and market participants. One of the key provisions of the Act is the requirement for public companies to file periodic reports with the SEC to disclose information about their financial performance and operations.

Reporting Requirements for Public Companies:

Under the Securities Exchange Act of 1934, public companies are required to file various types of reports with the SEC, including annual reports, quarterly reports, and current reports. These reports provide investors and other stakeholders with relevant and reliable information about the company's financial performance and operations. The following sections provide a detailed overview of the different types of reports required by public companies.

Annual Reports:

Public companies are required to file an annual report with the SEC on Form 10-K. The Form 10-K is a comprehensive report that provides detailed information about the company's financial performance and operations. The report includes audited financial statements, management's discussion and analysis (MD&A) of financial condition and results of operations, and other disclosures required by SEC rules and regulations. The Form 10-K must be filed within 60 days after the end of the company's fiscal year.

Quarterly Reports:

Public companies are required to file a quarterly report with the SEC on Form 10-Q. The Form 10-Q is a shorter report than the Form 10-K and provides unaudited financial statements and a condensed version of the MD&A. The Form 10-Q must be filed within 45 days after the end of the company's fiscal quarter.

Current Reports:

Public companies are required to file current reports with the SEC on Form 8-K to disclose significant events or material information that could affect the company's financial performance or operations. Examples of events that may trigger a Form 8-K filing include a change in control of the company, a change in the company's auditor, a change in the company's executive officers or directors, or a significant acquisition or disposition of assets. The Form 8-K must be filed within four business days after the occurrence of the event.

Other Reporting Requirements:

In addition to the periodic reporting requirements described above, public companies are also subject to other reporting requirements, including proxy statements, registration statements, and beneficial ownership reports. These reports provide additional information about the company's operations, governance, and ownership structure.

Proxy Statements:

Public companies are required to file a proxy statement with the SEC in connection with their annual shareholder meetings. The proxy statement provides information about the matters to be voted on at the meeting, the company's executive compensation practices, and other information required by SEC rules and regulations.

Registration Statements:

Public companies are required to file a registration statement with the SEC in connection with the offer and sale of securities to the public. The registration statement includes detailed information about the company's operations, financial condition, and other disclosures required by SEC rules and regulations. The registration statement must be declared effective by the SEC before the company can offer and sell securities to the public.

Beneficial Ownership Reports:

Public companies are required to file beneficial ownership reports with the SEC to disclose information about the company's insiders and their ownership of the company's securities. These reports include Forms 3, 4, and 5, which provide information about the company's officers, directors, and principal shareholders, and their transactions in the company's securities.

Form 3 is filed by company insiders when they first become an officer, director, or 10% shareholder of a public company. Form 4 is filed whenever an insider buys or sells shares of the company's stock or engages in any other transaction that affects their ownership of the company's securities. Form 5 is filed at the end of each fiscal year by insiders who have not reported any transactions during the year.

The purpose of these reports is to provide transparency and disclosure to investors about insider transactions in the company's securities. This information is important for investors to evaluate the company's financial health and the management's alignment with the company's shareholders.

In addition to the beneficial ownership reports, public companies are required to file annual reports on Form 10-K, quarterly reports on Form 10-Q, and current reports on Form 8-K with the SEC.

Form 10-K is a comprehensive annual report that provides a detailed overview of the company's financial performance, management discussions and analysis of the company's operations and financial condition, and disclosures about the risks faced by the company. Form 10-Q is a quarterly report that provides updates on the company's financial performance and operations. Form 8-K is filed whenever the company experiences significant events, such as a change in control, a merger or acquisition, or a change in the company's management.

These reporting requirements ensure that public companies provide accurate and up-to-date information to investors, which is critical for making informed investment decisions. In addition to the regulatory reporting requirements, public companies may also voluntarily provide additional information to investors, such as sustainability reports or annual letters to shareholders.

Overall, the reporting requirements for public companies under the Securities Exchange Act of 1934 play a crucial role in promoting transparency and accountability in the securities markets. These requirements ensure that investors have access to accurate and timely information about public companies, which helps to promote investor confidence and market integrity.

The role of the Securities and Exchange Commission (SEC) in enforcing the Act

The Securities and Exchange Commission (SEC) is the primary regulatory agency responsible for enforcing the Securities Exchange Act of 1934. The SEC is an independent agency of the federal government, with a mandate to protect investors and maintain fair, orderly, and efficient markets.

The SEC's enforcement division is responsible for investigating and prosecuting violations of the securities laws, including the Securities Exchange Act of 1934. The enforcement division has broad authority to conduct investigations and bring enforcement actions against violators of the securities laws.

One of the key roles of the SEC's enforcement division is to conduct investigations into potential violations of the securities laws. The SEC has the authority to issue subpoenas for documents and testimony, and can bring charges against individuals and companies for violating the securities laws. The SEC's investigations can cover a wide range of activities, including insider trading, accounting fraud, market manipulation, and violations of disclosure requirements.

Once the SEC's enforcement division has completed its investigation, it may choose to bring an enforcement action against the individual or company that it believes has violated the securities laws. Enforcement actions can take a number of different forms, including civil actions, administrative proceedings, and criminal prosecutions.

Civil actions are the most common type of enforcement action brought by the SEC. In a civil action, the SEC seeks to obtain a court order requiring the defendant to stop engaging in the illegal activity and to pay fines and disgorgement of ill-gotten gains. The SEC can also seek injunctive relief, which requires the defendant to comply with certain conditions or restrictions.

Administrative proceedings are another type of enforcement action that the SEC can bring. In an administrative proceeding, the SEC brings charges against a defendant before an administrative law judge. The judge hears evidence and makes findings of fact, and can impose penalties and sanctions if the defendant is found to have violated the securities laws.

Criminal prosecutions are the most serious type of enforcement action that the SEC can bring. In a criminal prosecution, the government brings charges against the defendant for violating the securities laws. If the defendant is convicted, they may face fines, imprisonment, or both.

The SEC's enforcement division plays an important role in deterring securities fraud and other violations of the securities laws. By aggressively pursuing enforcement actions against violators, the SEC can help to protect investors and maintain the integrity of the securities markets. In addition, the SEC's enforcement actions can serve as a warning to other market participants, discouraging them from engaging in illegal activity.

However, the SEC's enforcement actions are not without controversy. Critics argue that the SEC is often slow to act and that its penalties are not severe enough to deter wrongdoing. Others argue that the SEC's enforcement actions are overly aggressive and that they can unfairly target innocent individuals and companies.

Despite these criticisms, the SEC remains a key player in enforcing the Securities Exchange Act of 1934 and other securities laws. Its enforcement division has a broad mandate to investigate and prosecute violations of the securities laws, and its actions can have a significant impact on the behavior of market participants. As such, the SEC will continue to play an important role in maintaining the integrity of the securities markets and protecting investors.

CHAPTER 17: REGULATION A AND D EXEMPTIONS

The Securities Act of 1933 requires all public offerings of securities to be registered with the Securities and Exchange Commission (SEC). However, the Act provides certain exemptions from registration, such as the Regulation A and D exemptions, which allow companies to raise capital from investors without having to register their offerings with the SEC.

Regulation A provides an exemption for offerings of up to $50 million in securities within a 12-month period, subject to certain conditions. The exemption is intended to make it easier and less costly for smaller companies to raise capital from a wider range of investors. Regulation D, on the other hand, provides several exemptions for offerings that are typically targeted at sophisticated or accredited investors, such as hedge funds and venture capitalists.

In this chapter, we will explore the details of the Regulation A and D exemptions, their requirements, and the benefits they provide for both issuers and investors. We will also examine the differences between the two exemptions and their respective limitations. Additionally, we will delve into some of the common issues that arise with these exemptions and how they can be avoided.

By the end of this chapter, you will have a solid understanding of the Regulation A and D exemptions, their purpose, and how they can be utilized to raise capital for your business while remaining compliant with SEC regulations.

Overview of Regulation A and D exemptions

Regulation A and D exemptions are two exemptions provided under the Securities Act of 1933 that allow companies to raise capital without having to register their securities with the Securities and Exchange Commission (SEC). These exemptions are popular among small businesses, startups, and entrepreneurs who are seeking to raise capital from a wide range of investors without incurring the high costs and extensive regulatory requirements associated with registering securities with the SEC.

Regulation A provides an exemption for smaller offerings of up to $50 million in a 12-month period. The exemption is divided into two tiers: Tier 1 and Tier 2. Tier 1 offerings are limited to $20 million in a 12-month period and do not require audited financial statements or ongoing reporting requirements. Tier 2 offerings are limited to $50 million in a 12-month period and require audited financial statements, ongoing reporting requirements, and additional disclosure requirements. Both tiers require the filing of an offering statement with the SEC.

Regulation D provides several exemptions for private placements of securities. Private placements are offerings of securities that are not offered to the public, but instead are offered to a select group of investors, such as accredited investors or sophisticated investors. Regulation D exempts certain private placements from the registration requirements of the Securities Act of 1933.

There are three main exemptions provided under Regulation D: Rule 504, Rule 505, and Rule 506. Rule 504 provides an exemption for offerings of up to $5 million in a 12-month period. Rule 505 provides an exemption for offerings of up to $5 million in a 12-month period, with certain restrictions on the type of investors who can participate in the offering. Rule 506 provides two exemptions, 506(b) and 506(c). Rule 506(b) allows for offerings of an unlimited amount of securities to an unlimited number of accredited investors, and up to 35 non-accredited investors, who meet certain sophistication and investment requirements. Rule 506(c) allows for offerings of an unlimited amount of securities to accredited investors only, but requires the use of general solicitation and advertising to attract investors.

Regulation A and D exemptions offer important benefits for small businesses and entrepreneurs, but they also have some limitations and risks. One of the main limitations is that these exemptions may limit the ability of companies to raise capital from a broad range of investors, as they are typically limited to a select group of investors who meet certain financial or sophistication requirements. Additionally, these exemptions may not be available in certain situations, such as in the case of fraud or misrepresentation.

Moreover, Regulation A and D exemptions require compliance with certain disclosure requirements and ongoing reporting obligations, which can be costly and time-consuming. Companies should carefully consider the costs and benefits of using these exemptions before deciding whether to use them to raise capital.

In summary, Regulation A and D exemptions provide important opportunities for small businesses, startups, and entrepreneurs to raise capital without having to register their securities with the SEC. These exemptions, however, come with limitations and risks that companies should carefully consider before using them.

Regulation D is a set of exemptions created by the Securities and Exchange Commission (SEC) under the Securities Act of 1933. These exemptions allow companies to raise capital without registering their securities with the SEC, which can be a costly and time-consuming process. The three rules under Regulation D are Rule 504, Rule 505, and Rule 506.

Rule 504 is the least restrictive of the three rules and allows companies to raise up to $5 million in a 12-month period. This rule is often used by small businesses and startups that are looking to raise capital locally. Companies can offer and sell their securities to an unlimited number of accredited and non-accredited investors, although they may have to comply with state securities laws.

Rule 505 allows companies to raise up to $5 million in a 12-month period, but imposes more restrictions than Rule 504. Companies can offer and sell their securities to an unlimited number of accredited investors, but are limited to 35 non-accredited investors. In addition, non-accredited investors must have a pre-existing relationship with the company or its executives, and must be given extensive disclosure documents. This rule is often used by small businesses and startups that want to raise capital from a smaller group of investors.

Rule 506 is the most commonly used rule under Regulation D and allows companies to raise an unlimited amount of capital from accredited investors. Under Rule 506(b), companies can offer and sell their securities to an unlimited number of accredited investors and up to 35 non-accredited investors who meet certain sophistication requirements. Companies are required to provide extensive disclosure documents to all investors and must not engage in any general solicitation or advertising to attract investors. Under Rule 506(c), companies can offer and sell their securities to an unlimited number of accredited investors, but must verify their accredited status through certain methods.

In conclusion, Regulation D provides important exemptions for companies seeking to raise capital without registering their securities with the SEC. Companies should carefully consider which rule under Regulation D is most appropriate for their needs, and should seek legal advice to ensure compliance with federal and state securities laws.

Qualifying for exemptions

Regulation A and D exemptions provide certain companies with a way to raise capital without having to register their securities with the Securities and Exchange Commission (SEC). These exemptions are designed to make it easier for companies to access capital markets, while also providing certain protections to investors. However, companies must meet certain requirements to qualify for these exemptions. This section will provide a detailed analysis of the requirements for qualifying for exemptions under Regulation A and D.

Regulation A Exemption

Regulation A is an exemption from registration requirements for small public offerings. Under Regulation A, companies can offer and sell up to $75 million of securities in a 12-month period without having to register with the SEC. This exemption is available to both public and private companies.

To qualify for Regulation A exemption, companies must meet the following requirements:

The company must be organized in and have its principal place of business in the United States or Canada.

The company cannot be a blank check company, or a shell company.

The company cannot have a class of securities registered under the Securities Exchange Act of 1934.

The company must file an offering statement with the SEC, which includes audited financial statements, a description of the company's business and properties, and information about its officers and directors.

The offering must be made through a qualified intermediary, such as a registered broker-dealer or a registered funding portal.

The offering must be made to investors who are either accredited investors, or who are not accredited investors but whose investment in the offering does not exceed 10% of their net worth or annual income.

Regulation D Exemption

Regulation D provides several exemptions from registration requirements for private offerings. These exemptions are often used by small companies and startups to raise capital from private investors. There are three main rules under Regulation D: Rule 504, Rule 505, and Rule 506.

Rule 504

Rule 504 allows companies to offer and sell up to $5 million of securities in a 12-month period without having to register with the SEC. This exemption is available to both public and private companies.

To qualify for Rule 504 exemption, companies must meet the following requirements:

✧　The company cannot be an SEC reporting company.

✧　The offering must be made in one state, or in multiple states that have a coordinated review process.

✧　The offering must be made to an unlimited number of investors, including both accredited and non-accredited investors.

✧　The company must file a notice with the SEC, which includes basic information about the offering and the company.

Rule 505

Rule 505 allows companies to offer and sell up to $5 million of securities in a 12-month period without having to register with the SEC. This exemption is available to private companies only.

To qualify for Rule 505 exemption, companies must meet the following requirements:

✧　The company cannot be an SEC reporting company.

✧　The offering must be made to no more than 35 non-accredited investors, and an unlimited number of accredited investors.

✧　The company must provide certain disclosures to investors, including financial statements and information about the risks associated with the investment.

✧　The company must file a notice with the SEC, which includes basic information about the offering and the company.

Rule 506

Rule 506 provides two different exemptions, Rule 506(b) and Rule 506(c), which allow companies to offer and sell an unlimited amount of securities to accredited investors without having to register with the SEC. This exemption is available to private companies only.

To qualify for Rule 506 exemption, companies must meet the following requirements:

❖ The company cannot use general solicitation or advertising to market the securities.

❖ The offering must be made to accredited investors only, or to a maximum of 35 non-accredited investors who have sufficient knowledge and experience in finance and business to evaluate the investment.

❖ The issuer must provide all non-accredited investors with financial and other information, similar to that which would be provided in a registered offering.

❖ The issuer must also file a Form D with the SEC within 15 days of the first sale of securities.

Under Rule 506(b), companies are allowed to offer securities to accredited investors as well as up to 35 non-accredited investors who have sufficient knowledge and experience to evaluate the investment, provided that no general solicitation or advertising is used to market the securities. In addition, the issuer must provide all non-accredited investors with financial and other information that is similar to that which would be provided in a registered offering.

Under Rule 506(c), companies are allowed to use general solicitation or advertising to offer securities, but are required to verify that all investors are accredited investors. Verification may be accomplished through various means, including income and net worth tests or obtaining written confirmation from a registered broker-dealer, attorney, or certified public accountant.

While Rule 506(c) allows for general solicitation and advertising, it requires stricter verification of investors' accreditation status. Companies that rely on this exemption must take reasonable steps to ensure that all investors are accredited investors, and failure to do so could result in the loss of the exemption and potential liability for securities law violations.

In summary, Rule 506 provides private companies with a valuable exemption from SEC registration requirements, allowing them to offer and sell securities to accredited investors without the costs and complexities of a registered offering. However, companies must meet certain requirements and restrictions in order to qualify for the exemption, and must ensure that they comply with all applicable securities laws and regulations.

Benefits and drawbacks of exemptions

Exemptions from registration requirements under the Securities Act of 1933 can provide significant benefits to companies seeking to raise capital. However, they also come with some drawbacks that companies should consider before deciding to rely on them.

Benefits:

✧   Cost savings: One of the most significant benefits of exemptions is the cost savings associated with not having to register the securities with the SEC. The registration process can be lengthy, complex, and expensive. By relying on exemptions, companies can save on legal, accounting, and other expenses.

✧   Flexibility: Exemptions offer companies more flexibility in terms of the types of investors they can raise capital from and the terms of the offering. For example, companies can raise capital from accredited investors or sophisticated institutional investors without having to comply with certain disclosure requirements or restrictions on the types of securities they can issue.

✧   Speed: Another benefit of exemptions is the speed at which companies can raise capital. Without the need to go through the registration process, companies can offer and sell securities more quickly and efficiently.

✧   Reduced regulatory burden: By relying on exemptions, companies are subject to fewer regulatory requirements and restrictions than those that must register their securities with the SEC. This can result in less regulatory scrutiny, fewer ongoing reporting requirements, and more operational flexibility.

Drawbacks:

✧   Limited pool of investors: One of the primary drawbacks of relying on exemptions is that companies are limited to raising capital from a smaller pool of investors. For example, under Rule 506 of Regulation D, companies can only sell securities to accredited investors or up to 35 non-accredited investors who meet certain financial requirements. This can limit the amount of capital companies can raise.

✧ No public advertising: Another significant drawback of relying on exemptions is that companies are prohibited from using general solicitation or advertising to market their securities. This means that companies cannot publicly advertise the offering, such as through social media or online ads.

✧ No secondary market: Exempt securities are generally not freely tradable on secondary markets, which can make it more difficult for investors to sell their securities if they need to raise capital quickly.

✧ Increased liability: Companies that rely on exemptions may face increased liability if they fail to comply with the requirements of the exemption they are relying on. For example, if a company fails to verify that its investors are accredited under Rule 506, it may lose the exemption and face legal consequences.

Conclusion:

Exemptions from registration requirements can provide significant benefits to companies seeking to raise capital, including cost savings, flexibility, speed, and reduced regulatory burden. However, companies should carefully consider the drawbacks, including the limited pool of investors, restrictions on advertising, lack of a secondary market, and increased liability. Ultimately, the decision to rely on an exemption should be made based on the company's specific needs and circumstances. Companies should work closely with legal and financial professionals to ensure that they comply with all applicable laws and regulations.

CHAPTER 18: REGULATION A+:

Regulation A+ is a relatively new exemption that was introduced in 2015 under the Jumpstart Our Business Startups (JOBS) Act. The JOBS Act was enacted to encourage small businesses to raise capital and grow their businesses by easing the regulatory burden on them. Regulation A+ is designed to help small and emerging companies raise capital by offering and selling securities to the public without having to register with the SEC. It is intended to provide an alternative to traditional IPOs and other more complex methods of raising capital, such as Regulation D exemptions.

This section will provide an in-depth analysis of Regulation A+, including its history, requirements, benefits, drawbacks, and challenges. We will explore how Regulation A+ works, who can benefit from it, and what are the potential pitfalls that companies may encounter when seeking to comply with this law. We will also examine some of the regulatory changes and challenges that have emerged since the introduction of Regulation A+, and what companies can do to ensure compliance and avoid potential legal and financial risks.

History of Regulation A+

This section will provide a brief history of Regulation A+, including its origins, purpose, and evolution. We will explore how the JOBS Act came to be, and how Regulation A+ fits into the broader regulatory landscape of securities laws and regulations. We will also examine some of the key features of Regulation A+, such as its eligibility requirements, disclosure requirements, and offering limits.

Requirements of Regulation A+

This section will delve into the specific requirements of Regulation A+, including its eligibility requirements, offering limits, and disclosure requirements. We will explore what types of companies are eligible to use Regulation A+, how much they can raise, and what types of securities they can offer. We will also examine the different stages of the Regulation A+ process, including the pre-qualification and qualification stages, and what companies need to do to comply with the SEC's disclosure requirements.

Benefits of Regulation A+

This section will explore the benefits of Regulation A+ for companies, investors, and the broader economy. We will examine how Regulation A+ can help small and emerging companies raise capital, create jobs, and spur economic growth. We will also explore how Regulation A+ can provide investors with more access to investment opportunities, and how it can help diversify their portfolios. Finally, we will examine how Regulation A+ can benefit the economy as a whole, by providing a more efficient and effective way for small and emerging companies to raise capital.

Drawbacks of Regulation A+

This section will examine some of the potential drawbacks of Regulation A+, both for companies and investors. We will explore some of the limitations of Regulation A+, such as its disclosure requirements, offering limits, and ongoing reporting obligations. We will also examine some of the potential risks that investors may face when investing in Regulation A+ offerings, such as the lack of liquidity, information asymmetry, and fraud.

Challenges of Compliance with Regulation A+

This section will examine some of the specific challenges that companies may face when seeking to comply with Regulation A+. We will explore how companies can navigate the complex regulatory landscape of securities laws and regulations, and what they can do to ensure compliance with the SEC's disclosure requirements. We will also examine some of the potential legal and financial risks that companies may face if they fail to comply with Regulation A+, and what they can do to mitigate these risks.

Conclusion

This section will conclude with a summary of the key takeaways from our analysis of Regulation A+. We will summarize the history, requirements, benefits, drawbacks, and challenges of Regulation A+, and provide some final thoughts on its potential impact on the securities markets and the broader economy. We will also provide some practical tips and advice for companies considering using Regulation A+ to raise capital.

Key Takeaways:

Regulation A+ offers companies an alternative to traditional public offerings and private placements for raising capital from a broader base of investors, including both accredited and non-accredited investors.

The exemption provides two tiers of offerings, Tier 1 and Tier 2, with different disclosure and reporting requirements.

Tier 1 offerings allow companies to raise up to $20 million in a 12-month period, while Tier 2 offerings allow up to $75 million.

While the requirements for Regulation A+ offerings are less burdensome than those for traditional public offerings, they are more onerous than those for private placements, making Regulation A+ a middle ground between the two.

Benefits of Regulation A+ include the ability to raise capital from a wider pool of investors, the potential for increased liquidity of securities, and the ability to test the waters with potential investors before launching an offering.

Drawbacks of Regulation A+ include the higher costs of compliance compared to private placements, the risk of failure to meet ongoing reporting requirements, and the possibility of market volatility due to the lower trading volumes associated with smaller issuers.

The challenges of complying with Regulation A+ include navigating the complex disclosure requirements, engaging with investors and potential investors, and understanding the potential risks associated with raising capital through this exemption.

Final Thoughts:

Overall, Regulation A+ has the potential to be a valuable tool for companies seeking to raise capital outside of traditional public offerings and private placements. However, it is important for companies to carefully consider the costs, benefits, and risks associated with this exemption, and to consult with legal and financial advisors before proceeding. While the SEC has made efforts to streamline the regulation and reduce the costs of compliance, there are still many challenges associated with using Regulation A+, particularly for smaller issuers with limited resources.

Practical Tips:

Companies should carefully evaluate the costs and benefits of both Tier 1 and Tier 2 offerings before deciding which one to pursue.

Companies should work closely with legal and financial advisors to ensure that they understand the disclosure requirements and are in compliance with all SEC regulations.

Companies should take advantage of the ability to test the waters with potential investors before launching an offering.

Companies should have a clear plan in place for engaging with investors and potential investors, and should be prepared to provide ongoing updates and information to shareholders.

Companies should be aware of the risks associated with market volatility and should have contingency plans in place to address any unforeseen events that may impact their ability to raise capital or meet ongoing reporting requirements.

Overview of Regulation A+, including its purpose, requirements, and limitations.

Regulation A+ is a provision of the Jumpstart Our Business Startups (JOBS) Act that was enacted in 2015. The goal of Regulation A+ is to provide small and emerging companies with an easier and more cost-effective way to raise capital through public offerings. This section will provide an overview of Regulation A+, including its purpose, requirements, and limitations.

Purpose of Regulation A+

Regulation A+ was designed to address the difficulties that small and emerging companies face when trying to raise capital through public offerings. These companies often struggle to meet the stringent requirements of traditional public offerings, which can be expensive and time-consuming. Regulation A+ aims to provide an alternative to these traditional offerings, by allowing companies to raise capital through a streamlined process that is less burdensome and less costly.

Requirements of Regulation A+

To take advantage of Regulation A+, companies must meet certain requirements. The most important of these are:

✧ Limits on Offering Size: Regulation A+ offers two tiers of offerings, Tier 1 and Tier 2. Tier 1 offerings can raise up to $20 million in a 12-month period, while Tier 2 offerings can raise up to $75 million in a 12-month period.

✧ Disclosure Requirements: Companies must provide detailed information about their business, including financial statements and other disclosures, to potential investors. This information must be filed with the Securities and Exchange Commission (SEC) and made available to the public.

✧   State Blue Sky Laws: Companies must comply with state securities laws, known as "Blue Sky Laws," which can vary widely from state to state.

✧   Investor Eligibility: Regulation A+ offerings are open to both accredited and non-accredited investors. However, the amount that non-accredited investors can invest is limited to a certain percentage of their income or net worth.

Limitations of Regulation A+

While Regulation A+ offers many benefits to small and emerging companies, it is not without its limitations. Some of the key limitations of Regulation A+ include:

✧   Costs: While Regulation A+ is designed to be less costly than traditional public offerings, it still requires significant legal and accounting fees to comply with the SEC's disclosure requirements.

✧   Eligibility Requirements: Companies must meet certain eligibility requirements to take advantage of Regulation A+, including limits on the size of the offering and the amount that can be raised.

✧   State Blue Sky Laws: Compliance with state securities laws can be complex and expensive, and can vary widely from state to state.

✧   Investor Protections: While Regulation A+ does offer some protections to investors, such as disclosure requirements and limits on the amount that non-accredited investors can invest, these protections may not be sufficient to fully protect investors from fraud or other risks.

Conclusion

Regulation A+ offers small and emerging companies an alternative to traditional public offerings, by providing a streamlined and less burdensome process for raising capital. However, companies must meet certain eligibility requirements and comply with state securities laws, and investors must be aware of the risks and limitations of Regulation A+ offerings. Overall, Regulation A+ has the potential to benefit both companies and investors, but it is important to carefully consider the costs and risks before deciding whether to take advantage of this provision.

Eligibility criteria for issuers seeking to use Regulation A+, such as annual revenue and investment limits.

Regulation A+ offers companies a flexible way to raise capital from both accredited and non-accredited investors. The regulation was created under Title IV of the JOBS Act and was designed to make it easier for small and medium-sized enterprises to access capital markets. To achieve these goals, the SEC has established certain eligibility requirements for companies seeking to use Regulation A+. In this section, we will discuss the eligibility criteria for issuers seeking to use Regulation A+.

Annual Revenue Limitations

One of the key requirements for companies seeking to use Regulation A+ is the annual revenue limitation. Under the SEC rules, a company cannot use Regulation A+ if it has annual revenues of over $50 million for its most recently completed fiscal year. This limitation is designed to ensure that Regulation A+ is primarily used by smaller companies seeking to raise capital.

Investment Limits

Another important requirement under Regulation A+ is the investment limits. In general, the SEC has established two tiers of offerings under Regulation A+, Tier 1 and Tier 2, each with different investment limits. Under Tier 1, investors are limited to purchasing no more than 10% of their net worth or annual income, whichever is greater. For Tier 2 offerings, investors are limited to purchasing no more than 10% of their net worth or annual income, whichever is greater, or 10% of the greater of their net worth or annual income for non-accredited investors.

Bad Actor Disqualification

Under Regulation A+, companies must also comply with bad actor disqualification rules. These rules disqualify an issuer from using Regulation A+ if it, its directors, officers, or other related parties have been convicted of certain crimes or have violated securities laws or regulations.

Other Eligibility Requirements

In addition to the above requirements, issuers seeking to use Regulation A+ must also meet other eligibility criteria. For example, the issuer must be organized in the United States or Canada, and it must not be a blank check company or an investment company. Furthermore, issuers must file a disclosure document, known as an offering circular, with the SEC and provide it to investors.

Conclusion

In conclusion, Regulation A+ offers an attractive option for smaller companies seeking to raise capital. However, issuers must meet certain eligibility requirements to use Regulation A+. These requirements include annual revenue limitations, investment limits, bad actor disqualification rules, and other eligibility criteria. By understanding these requirements, issuers can determine whether Regulation A+ is an appropriate option for their capital-raising needs.

Key differences between Regulation A+ Tier 1 and Tier 2 offerings, including different disclosure and reporting requirements and offering limits.

Regulation A+ offers companies the opportunity to raise capital by selling securities to the public without having to comply with the full registration requirements of the Securities Act of 1933. Regulation A+ is divided into two tiers, with different requirements and limitations for each. In this section, we will explore the key differences between Regulation A+ Tier 1 and Tier 2 offerings, including their different disclosure and reporting requirements and offering limits.

Regulation A+ Tier 1:

Regulation A+ Tier 1 allows companies to raise up to $20 million in a 12-month period, with no more than $6 million of that amount coming from selling security holders. This tier is subject to both federal and state securities laws and is typically used by smaller companies that are seeking to raise capital from a wider pool of investors.

One key difference between Tier 1 and Tier 2 offerings is the level of disclosure required. Tier 1 offerings require a simplified Form 1-A offering circular, which is reviewed by the SEC and must include basic information about the issuer, its business, and the securities being offered. Tier 1 offerings do not require ongoing reporting requirements after the offering is completed, but issuers must file annual and semi-annual reports with the SEC and provide them to investors upon request.

Another difference between Tier 1 and Tier 2 offerings is the ability to "test the waters" before the offering. Issuers in Tier 1 can only test the waters with potential investors to determine interest in the offering. They cannot accept any binding commitments or money from investors until the SEC qualifies the offering statement.

Regulation A+ Tier 2:

Regulation A+ Tier 2 is designed for larger companies that want to raise up to $75 million in a 12-month period, with no more than $22.5 million of that amount coming from selling security holders. Unlike Tier 1 offerings, Tier 2 offerings are exempt from state securities laws and subject only to federal securities laws.

One of the biggest advantages of Tier 2 offerings is the ability to raise a larger amount of capital. However, issuers must meet more stringent disclosure requirements, including filing a Form 1-A offering statement with the SEC that includes audited financial statements and ongoing reporting requirements after the offering is completed.

Issuers in Tier 2 can also "test the waters" with potential investors and accept binding commitments from investors before the SEC qualifies the offering statement. This allows issuers to gauge investor interest before committing to a full offering.

Conclusion:

In conclusion, Regulation A+ offers two tiers of offerings, each with its own benefits and limitations. Tier 1 offerings are subject to both federal and state securities laws, have lower offering limits, and require less extensive disclosure requirements than Tier 2 offerings. Tier 2 offerings are exempt from state securities laws, have higher offering limits, and require more extensive disclosure requirements, including ongoing reporting requirements after the offering is completed. Understanding the differences between these two tiers is essential for companies looking to raise capital through Regulation A+.

Disclosure requirements for issuers, including financial statements, business plans, and potential risks and uncertainties.

Disclosure requirements are an essential aspect of Regulation A+ offerings. As discussed in the previous sections, issuers seeking to use Regulation A+ must file offering statements with the SEC, which include detailed information about the company and the offering. In this section, we will explore the disclosure requirements for issuers, including financial statements, business plans, and potential risks and uncertainties.

Financial Statements

One of the key disclosure requirements under Regulation A+ is the submission of audited financial statements. Tier 1 and Tier 2 offerings have different requirements for financial statements, as discussed in the previous section. For Tier 1 offerings, issuers must provide financial statements that have been reviewed by an independent public accountant. The financial statements must cover the two most recent fiscal years or the period since the issuer's inception if shorter than two years.

For Tier 2 offerings, issuers must provide audited financial statements for the two most recent fiscal years. If the issuer has existed for less than two years, they must provide financial statements since inception. The financial statements must be audited by an independent public accountant registered with the Public Company Accounting Oversight Board (PCAOB).

In addition to the financial statements, issuers must also provide a balance sheet, income statement, and cash flow statement for each of the two most recent fiscal years. If the issuer has existed for less than two years, they must provide financial statements since inception.

Business Plan

Regulation A+ requires issuers to provide a detailed business plan in the offering statement. The business plan should include information on the company's products or services, management team, target market, competition, marketing strategy, and financial projections. The plan should also include a discussion of the risks and uncertainties associated with the business.

Potential Risks and Uncertainties

Regulation A+ requires issuers to provide a detailed discussion of the risks and uncertainties associated with the business. This section of the offering statement should include a description of the risks and uncertainties associated with the company's operations, competition, regulatory environment, and other factors that could affect the company's performance. The risks and uncertainties should be discussed in a manner that is understandable to investors.

Other Disclosures

In addition to the financial statements, business plan, and risks and uncertainties, Regulation A+ also requires issuers to disclose information about the use of proceeds, the offering price, dilution, and any material transactions involving the company's officers or directors. The offering statement must also include a description of the issuer's ownership and capital structure, as well as any other material information that would be necessary for investors to make an informed investment decision.

Conclusion

In summary, the disclosure requirements under Regulation A+ are designed to provide investors with the information they need to make an informed investment decision. Issuers must provide detailed financial statements, a business plan, and a discussion of potential risks and uncertainties. The disclosure requirements apply to both Tier 1 and Tier 2 offerings, although Tier 2 offerings have more extensive requirements for financial statements. The disclosure requirements are intended to balance the need for investor protection with the desire to promote capital formation, which is the primary goal of Regulation A+.

The role of intermediaries, such as broker-dealers, in facilitating Regulation A+ offerings and ensuring compliance with regulatory requirements.

Regulation A+ is a Securities and Exchange Commission (SEC) rule that allows small and medium-sized businesses to raise capital from the general public through public offerings of securities. As a result, these businesses can access financing that was previously unavailable to them, thereby spurring innovation and economic growth. However, for these offerings to be successful, they require the involvement of intermediaries, such as broker-dealers, who help issuers comply with the regulatory requirements and facilitate the sale of securities to investors. In this section, we will discuss the role of intermediaries in Regulation A+ offerings and the ways in which they ensure compliance with regulatory requirements.

Role of Intermediaries in Regulation A+ Offerings

Broker-Dealers

Broker-dealers are financial professionals who facilitate the purchase and sale of securities for their clients. In the context of Regulation A+ offerings, broker-dealers are essential intermediaries because they play a critical role in ensuring that the offering complies with the SEC's rules and regulations. Specifically, broker-dealers are responsible for reviewing the issuer's offering documents, including the offering circular, and ensuring that they comply with SEC disclosure requirements.

Additionally, broker-dealers are responsible for conducting due diligence on the issuer and the securities being offered. This includes verifying the issuer's financial statements, reviewing the issuer's business plan and operations, and identifying potential risks associated with the investment. This due diligence helps to ensure that investors receive accurate and complete information about the issuer and the securities being offered.

Furthermore, broker-dealers help to market and sell the securities being offered. This involves creating and distributing marketing materials to potential investors and facilitating the purchase and sale of securities. Broker-dealers also provide support to the issuer throughout the offering process, including providing guidance on structuring the offering and pricing the securities.

Compliance Officers

In addition to broker-dealers, issuers may also engage compliance officers to help ensure that their Regulation A+ offerings comply with SEC rules and regulations. Compliance officers are responsible for developing and implementing compliance policies and procedures that help the issuer to identify and manage regulatory risks. These policies and procedures are designed to ensure that the issuer's offering documents are accurate and complete, that the issuer has conducted appropriate due diligence on the securities being offered, and that the offering complies with SEC disclosure requirements.

Compliance officers also help to ensure that the issuer's internal controls are effective in managing regulatory risks. This involves conducting internal audits and reviews of the issuer's operations to identify potential compliance issues and developing remediation plans to address those issues.

Limitations of Intermediaries

While intermediaries play an important role in facilitating Regulation A+ offerings, there are limitations to their effectiveness. For example, intermediaries may be subject to conflicts of interest that could compromise their ability to provide objective advice to issuers and investors. Broker-dealers, in particular, may have a financial incentive to sell securities to investors, even if those securities are not suitable for the investor's financial situation or investment objectives.

Moreover, intermediaries may not always be effective in identifying and managing regulatory risks associated with Regulation A+ offerings. For example, compliance officers may lack the expertise to identify all potential compliance issues or may not have sufficient resources to adequately address those issues.

Conclusion

In conclusion, intermediaries play a critical role in facilitating Regulation A+ offerings and ensuring compliance with SEC rules and regulations. Broker-dealers help to review offering documents, conduct due diligence on the issuer and securities being offered, market and sell the securities, and provide support to the issuer throughout the offering process. Compliance officers, on the other hand, help to develop and implement compliance policies and procedures that help the issuer to identify and manage regulatory risks.

However, while intermediaries are important, they also have limitations, including potential conflicts of interest and limitations in identifying and managing regulatory risks.

Potential benefits of Regulation A+ for small and medium-sized businesses and individual investors, such as increased access to capital and a more streamlined offering process.

Regulation A+ is a securities law that was introduced in 2015 as part of the Jumpstart Our Business Startups (JOBS) Act. This regulation provides small and medium-sized businesses with an alternative path to raise capital through public offerings. Regulation A+ allows companies to offer and sell securities to the public without having to go through the traditional, time-consuming, and expensive initial public offering (IPO) process. This section will explore the potential benefits of Regulation A+ for small and medium-sized businesses and individual investors.

Increased Access to Capital

One of the most significant potential benefits of Regulation A+ is increased access to capital for small and medium-sized businesses. Before the introduction of Regulation A+, these businesses had limited options for raising capital. They could either rely on bank loans or raise money from private investors. Bank loans can be difficult to obtain, especially for new businesses or those with limited assets. Private investors, on the other hand, are typically only interested in investing in businesses that have high growth potential. This leaves many small and medium-sized businesses with limited options for raising capital.

Regulation A+ provides these businesses with a new avenue for raising capital by allowing them to offer and sell securities to the public. This means that businesses can raise money from a large pool of investors, which can include both accredited and non-accredited investors. This can be particularly beneficial for businesses that have a strong local presence or a loyal customer base, as they can now offer securities to these individuals and raise capital without having to rely on outside investors.

More Streamlined Offering Process

Another potential benefit of Regulation A+ is a more streamlined offering process. The traditional IPO process is often long and expensive, involving a significant amount of regulatory compliance and paperwork. This can be particularly challenging for small and medium-sized businesses that may not have the resources or expertise to navigate the process.

Regulation A+ offers a more streamlined offering process by providing companies with a set of rules and procedures for conducting public offerings. These rules and procedures are designed to be more flexible and less burdensome than the traditional IPO process, making it easier for small and medium-sized businesses to raise capital through public offerings.

Furthermore, Regulation A+ allows companies to use electronic filing and communication methods, which can further streamline the offering process. For example, companies can use online platforms to distribute offering materials, receive subscription agreements, and provide ongoing disclosures to investors.

Lower Costs

Another potential benefit of Regulation A+ is lower costs. The traditional IPO process can be expensive, involving significant legal and accounting fees, underwriting fees, and other costs. These costs can be particularly challenging for small and medium-sized businesses that may not have the financial resources to pay for them.

Regulation A+ offers a more cost-effective alternative to the traditional IPO process. This is because the disclosure and reporting requirements are less extensive than those required for traditional public offerings. Additionally, the regulatory compliance costs associated with Regulation A+ are typically lower than those associated with traditional public offerings.

Individual Investor Participation

Another potential benefit of Regulation A+ is increased participation by individual investors. Before the introduction of Regulation A+, many individual investors were unable to participate in public offerings due to the requirements for accredited investor status. Accredited investors are individuals or entities that meet certain wealth or income requirements, and they are the only investors that are typically allowed to participate in private offerings or public offerings conducted under Regulation D.

Regulation A+ allows non-accredited investors to participate in public offerings, which can be particularly beneficial for individual investors looking to diversify their portfolios. By participating in a Regulation A+ offering, individual investors can gain exposure to a wider range of investment opportunities, including small and medium-sized businesses that may not have been accessible to them before.

Conclusion

Regulation A+ offers small and medium-sized businesses an alternative path to raising capital through public offerings, with the potential to reach a broader pool of investors. This is particularly beneficial for companies that may have been overlooked or unable to access traditional capital markets.

Additionally, Regulation A+ provides individual investors with the opportunity to invest in companies that they believe in and support, regardless of their accredited status. This promotes democratization of investment and can help drive innovation and growth in the economy.

While Regulation A+ has its advantages, it is important to note that it still carries certain risks and limitations, particularly with regards to disclosure requirements and regulatory compliance. Issuers must provide detailed and accurate information about their business operations, financials, and potential risks and uncertainties, and failure to do so can lead to legal consequences. Moreover, there is no guarantee of success or return on investment, and investors must carefully evaluate the risks and rewards before participating in a Regulation A+ offering.

Overall, Regulation A+ has the potential to provide significant benefits to small and medium-sized businesses and individual investors, while also promoting innovation and growth in the economy. However, it is important for both issuers and investors to understand the regulatory requirements and risks involved in these offerings, and to approach them with caution and due diligence.

Potential drawbacks of Regulation A+, such as the potential costs and complexities of preparing and filing offering documents and the limited liquidity of Regulation A+ securities.

While Regulation A+ offers many potential benefits for small and medium-sized businesses and individual investors, it is important to also consider the potential drawbacks of this regulatory framework. In this section, we will explore some of the key challenges that businesses and investors may face when using Regulation A+.

Costs and complexities of preparing and filing offering documents

One of the potential drawbacks of Regulation A+ is the costs and complexities involved in preparing and filing the required offering documents. Issuers must prepare an offering circular, which is similar to a prospectus but contains less detailed information, and file it with the SEC. This process can be time-consuming and expensive, particularly for small businesses with limited resources.

In addition to the offering circular, issuers must also prepare audited financial statements and other disclosure documents, such as management's discussion and analysis (MD&A) of financial condition and results of operations. These documents must be prepared in accordance with SEC rules and regulations, which can be complex and require the assistance of legal and accounting professionals.

Furthermore, Regulation A+ requires ongoing reporting and compliance obligations for issuers, which can add to the administrative and financial burden. For Tier 2 offerings, issuers must file annual and semi-annual reports with the SEC, and may also be subject to state-level reporting requirements.

Limited liquidity of Regulation A+ securities

Another potential drawback of Regulation A+ is the limited liquidity of securities offered under this regulatory framework. Unlike securities traded on public exchanges, Regulation A+ securities are generally not listed on national exchanges and may be subject to trading restrictions. This can make it difficult for investors to sell their shares or obtain liquidity for their investments.

Additionally, the secondary market for Regulation A+ securities is still developing, and it may be challenging for investors to find buyers for their shares. As a result, investors should carefully consider their investment horizon and liquidity needs before investing in Regulation A+ offerings.

Counterargument: Despite these potential drawbacks, Regulation A+ still offers benefits for small and medium-sized businesses and individual investors that may outweigh the costs and complexities involved in using this regulatory framework. For example, businesses may be able to access capital that they would not have otherwise been able to raise, and investors may gain exposure to a wider range of investment opportunities. Furthermore, the costs and complexities of preparing and filing offering documents may be offset by the potential benefits of accessing capital markets and increasing visibility for the business.

Conclusion

Regulation A+ offers a unique regulatory framework that can provide small and medium-sized businesses with greater access to capital and individual investors with more diverse investment opportunities. However, it is important to consider the potential drawbacks of this regulatory framework, such as the costs and complexities of preparing and filing offering documents and the limited liquidity of Regulation A+ securities. Ultimately, businesses and investors should carefully weigh the potential benefits and risks of using Regulation A+ and consult with legal and financial professionals to make informed decisions.

Recent regulatory changes or proposals related to Regulation A+ and their potential impact on issuers and investors.

Since the SEC finalized Regulation A+ in 2015, the regulation has been a popular option for small and medium-sized businesses looking to raise capital through public offerings. However, the regulatory landscape is constantly evolving, and recent changes and proposals related to Regulation A+ could have significant implications for both issuers and investors. This section will explore some of the most significant regulatory changes and proposals related to Regulation A+ and their potential impact on the market.

Recent Regulatory Changes and Proposals

Amendments to Regulation A+
In November 2020, the SEC adopted amendments to Regulation A+ that aimed to simplify and improve the regulation. Some of the most significant changes include an increase in the maximum offering amount from $50 million to $75 million, the ability to submit draft offering statements confidentially, and the addition of secondary market trading eligibility for Tier 2 securities. These amendments are expected to make Regulation A+ offerings more attractive to issuers and more accessible to investors.

Proposed Changes to Accredited Investor Definition
In December 2019, the SEC proposed changes to the accredited investor definition that would expand the pool of investors eligible to participate in Regulation A+ offerings. The proposed changes would allow individuals with certain professional certifications or designations, such as a Series 7 license or a Chartered

Financial Analyst (CFA) designation, to qualify as accredited investors. Additionally, the proposal would allow individuals to qualify as accredited investors based on their status as "knowledgeable employees" of private funds. If adopted, these changes could significantly increase the number of potential investors in Regulation A+ offerings.

Proposed Amendments to Rule 701

In November 2020, the SEC proposed amendments to Rule 701, which provides exemptions from registration for securities offered by private companies to their employees, consultants, and advisors. The proposed amendments would increase the threshold for offerings under Rule 701 from $5 million to $10 million and adjust the threshold for annual sales by the issuer to be based on the company's most recent fiscal year-end instead of the prior year. These changes could impact the use of Regulation A+ by private companies that are also using Rule 701 to offer securities to their employees.

Proposed Expansion of Testing the Waters

In September 2019, the SEC proposed a rule that would expand the use of "testing the waters" for all companies, including those considering a Regulation A+ offering. Testing the waters allows issuers to gauge investor interest in a potential offering before filing an offering statement with the SEC. The proposed rule would allow issuers to use generic solicitation materials to test the waters with all potential investors, not just accredited investors. This change could significantly increase the efficiency and cost-effectiveness of Regulation A+ offerings.

Potential Impact on Issuers and Investors

The recent regulatory changes and proposals related to Regulation A+ have the potential to significantly impact both issuers and investors. The amendments to Regulation A+ are expected to make the regulation more attractive to issuers by simplifying the offering process and increasing the maximum offering amount. Additionally, the proposed changes to the accredited investor definition and Rule 701 could expand the pool of potential investors in Regulation A+ offerings, making it easier for issuers to raise capital.

However, these changes also come with potential drawbacks. For example, the increased use of testing the waters could lead to a flood of potential offerings, making it more difficult for issuers to stand out in a crowded market. Additionally, the limited liquidity of Regulation A+ securities could make it more difficult for investors to exit their positions.

Conclusion

In conclusion, recent regulatory changes and proposals related to Regulation A+ have the potential to significantly impact the market. The amendments to Regulation A+ are expected to make the regulation more attractive to issuers, while the proposed changes to the accredited investor definition could expand the pool of potential investors for Regulation A+ offerings. However, the proposed amendments to Regulation A+ could also impose additional requirements and costs on issuers, and the potential impact of these changes remains to be seen.

Overall, Regulation A+ has the potential to offer small and medium-sized businesses an alternative path to capital raising and provide individual investors with increased access to investment opportunities. However, as with any investment opportunity, there are potential drawbacks and risks associated with Regulation A+ offerings, including the potential costs and complexities of preparing and filing offering documents and the limited liquidity of Regulation A+ securities.

As the market for Regulation A+ continues to evolve, it will be important for issuers, investors, and regulatory agencies to monitor the effectiveness and impact of the regulation, as well as to remain vigilant in addressing potential risks and concerns. By working together to promote transparency, efficiency, and investor protection, the market for Regulation A+ can continue to grow and provide a valuable source of capital for small and medium-sized businesses, while also offering individual investors access to a wider range of investment opportunities.

PART 5: SECURITIES REGULATION AGENCIES AND ENFORCEMENT

The securities industry is highly regulated in order to protect investors from fraudulent or unethical practices. The regulatory framework in the United States is primarily governed by federal laws and regulations, which are enforced by various securities regulation agencies. These agencies work to ensure that the securities industry operates in a fair, transparent, and efficient manner, and that investors have access to accurate and timely information about the companies and securities in which they invest.

Part 5 of this paper will focus on the role of securities regulation agencies and enforcement in the United States. Specifically, we will discuss the primary agencies responsible for regulating the securities industry, including the Securities and Exchange Commission (SEC), the Financial Industry Regulatory Authority (FINRA), and state securities regulators. We will also examine the various tools and strategies used by these agencies to enforce securities regulations and combat fraud, including investigation, enforcement actions, and penalties.

In addition to discussing the role of securities regulation agencies and enforcement, we will also explore some of the key challenges and criticisms that have been raised in relation to the current regulatory framework. For example, some critics argue that the current system is overly complex and burdensome, making it difficult for smaller businesses to access capital and comply with regulatory requirements. Others argue that the regulatory framework is not effective enough in preventing fraudulent or unethical practices, and that more needs to be done to protect investors from these risks.

Throughout this section, we will use examples from a variety of fields, including business owners, entrepreneurs, corporate executives, in-house counsel/legal departments, compliance officers, financial analysts, investment bankers, accountants, and regulators. We will also present counterarguments and dissenting opinions in a balanced and objective way, in order to provide a comprehensive analysis of the subject matter. Overall, our goal is to provide a thorough and in-depth exploration of the role of securities regulation agencies and enforcement in the United States, and to critically examine the strengths and weaknesses of the current regulatory framework.

CHAPTER 18: SECURITIES AND EXCHANGE COMMISSION (SEC)

The Securities and Exchange Commission (SEC) is a regulatory agency of the United States federal government responsible for overseeing and regulating the securities industry. It was created by the Securities Exchange Act of 1934 in response to the stock market crash of 1929 and the ensuing Great Depression.

The mission of the SEC is to protect investors, maintain fair, orderly, and efficient markets, and facilitate capital formation. To achieve its mission, the SEC is responsible for enforcing federal securities laws, regulating the securities industry, and overseeing the operations of securities exchanges, brokers, and investment advisers.

The SEC is composed of five commissioners appointed by the President of the United States, with the advice and consent of the Senate. One of the commissioners is appointed as the Chairperson, who serves as the agency's chief executive officer.

SEC's Responsibilities

The SEC has a wide range of responsibilities, including:

Enforcement of Securities Laws: The SEC has the authority to investigate and prosecute violations of federal securities laws. It can bring civil actions against individuals and companies that violate these laws, and it can also refer cases to the Department of Justice for criminal prosecution.

Regulation of the Securities Industry: The SEC is responsible for regulating the securities industry, including securities exchanges, brokers, dealers, investment advisers, and mutual funds. It has the authority to establish rules and regulations for the industry, and it regularly conducts inspections and examinations of firms to ensure compliance with these rules.

Oversight of Securities Markets: The SEC oversees the operation of securities markets, including stock and options exchanges. It is responsible for ensuring that these markets are fair, orderly, and efficient, and it has the authority to intervene in the markets when necessary to maintain their integrity.

Investor Protection: The SEC is charged with protecting investors from fraud and other abuses in the securities markets. It does this by requiring companies to disclose important information to investors, such as financial statements and other material information, and by taking action against individuals and companies that engage in fraudulent activities.

Facilitating Capital Formation: The SEC is responsible for regulating the issuance and trading of securities, which plays a critical role in facilitating capital formation. It has the authority to establish rules and regulations for securities offerings, and it reviews and approves offerings of securities to ensure that they comply with these rules.

SEC's Impact on Businesses

The SEC's regulations and enforcement actions have a significant impact on businesses, particularly those that are publicly traded. For example, companies that are listed on a national securities exchange, such as the New York Stock Exchange or the Nasdaq Stock Market, must comply with the SEC's reporting requirements and disclosure rules.

These rules require companies to provide detailed information about their financial performance, operations, and business risks to investors on a regular basis. Failure to comply with these rules can result in significant penalties and legal liability.

In addition to its regulatory role, the SEC also plays a key role in facilitating capital formation for businesses. It does this by reviewing and approving securities offerings, such as initial public offerings (IPOs), to ensure that they comply with its rules and regulations.

The SEC's review process can be time-consuming and expensive, which can be a significant barrier to entry for small businesses seeking to raise capital through securities offerings. However, the SEC has taken steps to make this process more accessible to small businesses, such as by creating Regulation A+, which allows small companies to raise up to $50 million through a streamlined securities offering process.

Conclusion

In conclusion, the SEC plays a critical role in regulating and overseeing the securities industry in the United States. Its enforcement actions and regulations have a significant impact on businesses, particularly those that are publicly traded. At the same time, the SEC also plays an important role in facilitating capital formation for businesses, which is critical to the growth and development of the economy.

The SEC is constantly evolving to keep pace with the changing securities market and to adapt to new technologies and practices. In recent years, the SEC has focused on initiatives to enhance transparency, promote investor protection, and encourage capital formation. It has also been proactive in addressing emerging issues, such as cyber threats and the rise of digital assets.

Despite its many successes, the SEC has also faced criticism and challenges over the years. Critics have accused the agency of being too slow to act, of being overly bureaucratic, and of failing to adequately protect investors. In response, the SEC has taken steps to address these concerns and to improve its operations, such as implementing more efficient processes and investing in new technologies.

Overall, the SEC is a vital player in the securities industry and its work has a profound impact on the U.S. economy. As such, it is important for businesses, investors, and other stakeholders to stay informed about the agency's actions and to engage with it as necessary. By doing so, they can help to shape the future of securities regulation in the United States and to ensure that the markets remain fair, transparent, and efficient for years to come.

The role of the SEC in regulating securities markets

The Securities and Exchange Commission (SEC) is a federal agency responsible for regulating the securities industry in the United States. The SEC was created by the Securities Exchange Act of 1934, which was enacted in response to the stock market crash of 1929 and subsequent Great Depression. Since its creation, the SEC has played a critical role in regulating and overseeing the securities industry in the United States.

The SEC's primary mission is to protect investors, maintain fair, orderly, and efficient markets, and facilitate capital formation. To achieve these goals, the SEC has a wide range of regulatory powers and responsibilities, including:

✧   Registration of securities: The SEC requires companies to register their securities offerings before they can be sold to the public. The registration process ensures that investors receive important information about the company, its financial condition, and the risks associated with investing in its securities.

✦ Enforcement of securities laws: The SEC has the power to bring civil actions against individuals and companies that violate securities laws, such as insider trading, fraud, and market manipulation. The SEC can also bring criminal cases against individuals who engage in securities fraud.

✦ Oversight of securities markets: The SEC oversees the securities markets to ensure that they are fair, orderly, and efficient. This includes monitoring trading activity, ensuring compliance with trading rules, and investigating potential violations of securities laws.

✦ Rule-making: The SEC has the authority to create rules and regulations that govern the securities industry. These rules cover a wide range of topics, including disclosure requirements, insider trading, and accounting practices.

✦ Investor education: The SEC provides investors with information about investing, including tips on how to avoid fraud and how to evaluate investment opportunities.

The SEC's role in regulating securities markets is critical to ensuring that investors are protected and that markets function properly. One of the key ways in which the SEC protects investors is through its oversight of financial reporting by public companies. The SEC requires public companies to file regular reports with the agency that disclose financial information, such as earnings, assets, and liabilities. These reports provide investors with important information about the financial health of the company and help to ensure that markets are efficient and transparent.

In addition to its oversight of financial reporting, the SEC also plays an important role in enforcing laws that govern the behavior of market participants. The SEC has the power to investigate and bring enforcement actions against individuals and companies that violate securities laws. This includes cases of insider trading, market manipulation, and other forms of securities fraud. By holding market participants accountable for their actions, the SEC helps to maintain fair and orderly markets that are attractive to investors.

The SEC's rule-making authority is also critical to the agency's ability to regulate securities markets effectively. The SEC has the power to create rules and regulations that govern the behavior of market participants, and these rules can have a significant impact on the securities industry. For example, the SEC has created rules that require companies to disclose information about executive compensation, which helps investors evaluate the management of a company. The SEC has also created rules that require companies to disclose information about environmental, social, and governance (ESG) issues, which can help investors evaluate the long-term sustainability of a company.

Finally, the SEC's role in investor education is critical to ensuring that investors are able to make informed decisions about their investments. The SEC provides investors with a wide range of information about investing, including tips on how to avoid fraud and how to evaluate investment opportunities. By providing investors with this information, the SEC helps to level the playing field between individual investors and market professionals.

In conclusion, the SEC plays a critical role in regulating securities markets in the United States. Its powers and responsibilities include registration of securities, enforcement of securities laws, oversight of securities markets, rule-making, and investor education. By fulfilling these responsibilities, the SEC helps to protect investors and ensure the integrity of the securities markets.

One of the key roles of the SEC is to ensure that companies that issue securities to the public do so in compliance with applicable laws and regulations. This includes the registration of securities with the SEC before they can be sold to the public. The SEC reviews registration statements and other disclosures to ensure that they contain accurate and complete information about the securities being offered and the company issuing them.

The SEC also plays an important role in enforcing securities laws. It has the power to bring civil actions against companies and individuals that violate securities laws, and it can also refer cases to the Department of Justice for criminal prosecution. The SEC can also impose sanctions on companies and individuals that violate securities laws, including fines, disgorgement of profits, and injunctions.

In addition to enforcing securities laws, the SEC also has broad authority to regulate securities markets. It has the power to regulate securities exchanges and alternative trading systems, as well as broker-dealers and investment advisers. The SEC also has the authority to adopt rules and regulations to promote fair and orderly securities markets, including rules related to trading practices, disclosures, and reporting requirements.

Another important role of the SEC is to educate investors about the risks and benefits of investing in securities. The SEC provides a wide range of educational materials and resources to help investors make informed decisions, including information about investment products, investment scams, and the risks and benefits of investing in different types of securities.

Overall, the SEC plays a critical role in regulating securities markets in the United States. Its responsibilities include registration of securities, enforcement of securities laws, oversight of securities markets, rule-making, and investor education. By fulfilling these responsibilities, the SEC helps to protect investors and ensure the integrity of the securities markets.

Structure and organization of the SEC

The Securities and Exchange Commission (SEC) is a federal agency responsible for regulating the securities markets in the United States. It was created in 1934 under the Securities Exchange Act, which was passed in response to the stock market crash of 1929. Since then, the SEC has played a critical role in ensuring the integrity of the securities markets and protecting investors.

The SEC is structured as an independent agency, meaning that it is not part of any other branch of government. It is led by five commissioners who are appointed by the President and confirmed by the Senate. One of the commissioners is designated as the Chair, who serves as the agency's chief executive officer. Commissioners serve staggered five-year terms and cannot be removed by the President except for cause.

The SEC has four main divisions: Corporation Finance, Trading and Markets, Investment Management, and Enforcement. Each division is responsible for a different aspect of the SEC's mission, and they work together to ensure that the securities markets are fair, transparent, and efficient.

The Corporation Finance Division is responsible for reviewing and approving the registration statements of companies that want to issue securities to the public. This division also oversees the ongoing reporting requirements of public companies, including the filing of annual reports and proxy statements.

The Trading and Markets Division is responsible for overseeing the securities markets, including stock exchanges, broker-dealers, and alternative trading systems. This division also regulates the activities of market participants, including high-frequency traders, and monitors for potential market abuses such as insider trading and market manipulation.

The Investment Management Division is responsible for regulating investment companies such as mutual funds and exchange-traded funds (ETFs), as well as investment advisers. This division reviews and approves the registration statements of investment companies and ensures that investment advisers are providing adequate disclosure to their clients.

The Enforcement Division is responsible for investigating potential violations of securities laws and taking enforcement actions against violators. This division has the power to impose civil penalties, initiate legal proceedings, and seek criminal sanctions against individuals and companies that violate securities laws.

In addition to these four main divisions, the SEC also has a number of other offices and divisions that support its mission. These include the Office of the Chief Accountant, which provides guidance on accounting and auditing matters; the Office of Compliance Inspections and Examinations, which conducts inspections of securities firms and market participants; and the Office of Investor Education and Advocacy, which provides educational resources and assistance to investors.

The SEC also has regional offices located throughout the country, which are responsible for conducting investigations and enforcing securities laws in their respective regions. These offices work closely with other law enforcement agencies, including the Department of Justice, the Federal Bureau of Investigation, and state securities regulators, to identify and prosecute securities fraud.

Overall, the SEC's structure and organization are designed to ensure that it can effectively carry out its mission of regulating the securities markets and protecting investors. By dividing its responsibilities among different divisions and offices, the SEC is able to take a comprehensive approach to regulating the securities industry, while also responding to new developments and emerging risks in the market.

Key SEC rules and regulations

The Securities and Exchange Commission (SEC) has the power to regulate the securities industry in the United States, and it does so through a wide range of rules and regulations. These regulations help to ensure that companies that issue securities, as well as those who trade them, do so in a manner that is transparent and fair to investors. In this section, we will explore some of the key rules and regulations that the SEC has put in place to achieve these objectives.

Rule 10b-5

One of the most important SEC regulations is Rule 10b-5, which prohibits fraud and misrepresentation in the sale of securities. This rule makes it illegal for any person to make false statements or to engage in any fraudulent activity in connection with the sale or purchase of securities. This rule has been used by the SEC to prosecute a wide range of fraud cases, including insider trading, Ponzi schemes, and other forms of securities fraud.

Proxy Rules

Another important set of rules that the SEC has put in place relates to the proxy process. When shareholders of a publicly traded company wish to vote on important matters, such as the election of directors or the approval of major corporate transactions, they do so through the use of proxies. The SEC's proxy rules ensure that these voting rights are protected and that shareholders have access to the information they need to make informed decisions.

Insider Trading

The SEC also has rules in place that govern insider trading. These rules prohibit insiders, such as company executives or directors, from trading on material non-public information. This is because insiders have access to information that is not available to the general public, and trading on this information can give them an unfair advantage. The SEC's insider trading rules help to level the playing field and ensure that all investors have access to the same information.

Regulation D

Regulation D is a set of SEC rules that provides exemptions from registration for certain private securities offerings. These exemptions are intended to reduce the regulatory burden on small businesses and startups that are looking to raise capital. Regulation D allows companies to sell securities to a limited number of accredited investors without having to register the securities with the SEC. However, these offerings are subject to certain restrictions and disclosure requirements.

Sarbanes-Oxley Act

The Sarbanes-Oxley Act is a federal law that was passed in response to a series of high-profile corporate scandals, including Enron and WorldCom. The law imposes new requirements on public companies, including requirements related to financial reporting and internal controls. The SEC has been responsible for enforcing many of these requirements, and has also put in place a number of related rules and regulations.

Conclusion

The SEC has put in place a wide range of rules and regulations to help ensure that the securities industry operates in a fair and transparent manner. These rules cover everything from fraud and misrepresentation to insider trading and private securities offerings. By enforcing these rules, the SEC helps to protect investors and promote the growth and development of the securities markets. It is important for companies, investors, and other market participants to be aware of these rules and to comply with them to ensure that the securities industry remains a fair and efficient market.

CHAPTER 19: FINANCIAL INDUSTRY REGULATORY AUTHORITY (FINRA)

The role of FINRA in regulating securities firms and brokers

The financial industry is one of the most important sectors of any economy. It is responsible for the allocation of capital, which is critical for the growth and development of businesses, governments, and individuals. The financial industry includes banks, insurance companies, investment firms, and other financial institutions that provide a wide range of financial products and services to their customers.

However, with the vast amount of money involved in the financial industry, it is essential to have proper regulation to ensure that the industry operates in a fair, transparent, and efficient manner. One of the key regulators of the financial industry in the United States is the Financial Industry Regulatory Authority (FINRA).

FINRA is a non-governmental organization that oversees and regulates the securities industry in the United States. It is responsible for the regulation of all securities firms that operate in the country, including those that trade stocks, bonds, and other securities. FINRA is also responsible for the licensing of individuals and firms that work in the securities industry.

This chapter will provide an in-depth analysis of FINRA, its role in regulating the securities industry, and the rules and regulations that it enforces. It will cover the history of FINRA, its structure and organization, its powers and responsibilities, and the various rules and regulations that it enforces.

This chapter will be beneficial for students seeking a bachelor's degree in finance or business, as well as professionals working in the financial industry. It will provide a comprehensive overview of FINRA and its role in regulating the securities industry, which is essential knowledge for anyone working in this industry. It will also include examples from a variety of fields, such as business owners/entrepreneurs, corporate executives, in-house counsel/legal departments, compliance officers, financial analysts, investment bankers, accountants, and regulators.

FINRA's enforcement authority

The Financial Industry Regulatory Authority (FINRA) is a self-regulatory organization (SRO) that is responsible for regulating the activities of brokerage firms and their registered representatives in the United States. FINRA has a wide range of enforcement powers that it uses to ensure compliance with the rules and regulations that govern the securities industry. This section will provide an in-depth analysis of FINRA's enforcement authority and the various mechanisms it uses to enforce its rules and regulations.

FINRA's Enforcement Authority

FINRA has broad enforcement powers that allow it to take action against firms and individuals that violate securities laws and regulations. These enforcement powers include the ability to initiate investigations, conduct on-site examinations, bring disciplinary actions against firms and individuals, and impose sanctions and fines.

Investigations

FINRA has the authority to initiate investigations into potential violations of securities laws and regulations. FINRA's Office of Fraud Detection and Market Intelligence (OFDMI) is responsible for conducting investigations into potential violations of FINRA's rules and federal securities laws. OFDMI has the authority to compel firms and individuals to produce documents and provide testimony under oath. FINRA also has the authority to share information and coordinate investigations with other regulatory agencies, such as the Securities and Exchange Commission (SEC) and state securities regulators.

On-Site Examinations

FINRA has the authority to conduct on-site examinations of brokerage firms to ensure compliance with its rules and regulations. FINRA's Department of Member Regulation (DMR) is responsible for conducting these examinations. During these examinations, FINRA examiners review a firm's books and records, interview employees, and assess the firm's compliance with FINRA's rules and regulations. FINRA also has the authority to impose sanctions and fines if it finds that a firm has violated FINRA's rules and regulations.

Disciplinary Actions

FINRA has the authority to bring disciplinary actions against firms and individuals that violate its rules and regulations. FINRA's Department of Enforcement is responsible for bringing these actions. Disciplinary actions can include fines, suspensions, and even expulsion from the securities industry. In addition, FINRA has the authority to bar individuals from associating with FINRA-regulated firms.

Sanctions and Fines

FINRA has the authority to impose sanctions and fines on firms and individuals that violate its rules and regulations. Sanctions can include censures, fines, suspensions, and even expulsion from the securities industry. FINRA has the authority to impose fines of up to $1 million per violation. In addition, FINRA has the authority to require firms and individuals to pay restitution to customers who have been harmed by their actions.

Mechanisms for Enforcing FINRA's Rules and Regulations

FINRA uses a variety of mechanisms to enforce its rules and regulations. These mechanisms include the use of automated surveillance systems, regulatory alerts, and the filing of disciplinary actions.

Automated Surveillance Systems

FINRA uses automated surveillance systems to detect potential violations of its rules and regulations. These systems use algorithms to identify suspicious patterns of activity and alert FINRA staff to potential violations. FINRA's Market Regulation Department uses these systems to monitor trading activity in the securities markets and detect potential insider trading, market manipulation, and other forms of misconduct.

Regulatory Alerts

FINRA issues regulatory alerts to its members to inform them of potential regulatory issues and violations. These alerts provide guidance on how to comply with FINRA's rules and regulations and may include information about recent disciplinary actions and sanctions.

Disciplinary Actions

FINRA files disciplinary actions against firms and individuals that violate its rules and regulations. These actions are public and can be found on FINRA's website. Disciplinary actions can include fines, suspensions, and even expulsion from the securities industry.

Conclusion

FINRA's enforcement authority is an essential part of its role as a self-regulatory organization. As the organization responsible for overseeing the securities industry, FINRA is charged with ensuring that its members comply with federal securities laws and regulations. The enforcement actions it takes against members who violate these rules help to maintain the integrity of the securities markets and protect investors.

FINRA's enforcement program is comprehensive and covers a wide range of activities, including investigations, disciplinary proceedings, and sanctions. The organization has a highly trained and experienced staff of attorneys and investigators who work tirelessly to ensure that members are held accountable for their actions. This dedication to enforcement is critical to maintaining the confidence of investors in the securities markets and ensuring that bad actors are punished appropriately.

In recent years, FINRA has demonstrated its commitment to enforcement through several high-profile cases. These cases have resulted in significant fines and sanctions against firms and individuals who violated securities laws and regulations. They have also helped to raise awareness among FINRA's members of the importance of compliance with these rules.

Despite its successes, FINRA faces several challenges in its enforcement efforts. One of the most significant is the sheer size and complexity of the securities industry. With thousands of firms and millions of registered representatives, it can be challenging to identify and investigate wrongdoing. Additionally, the financial resources available to FINRA are limited compared to those available to the firms it regulates. This can make it difficult to pursue cases that require significant resources or take a long time to resolve.

To address these challenges, FINRA has implemented several initiatives designed to improve its enforcement capabilities. These include the use of advanced technology and data analytics to identify potential violations and the hiring of additional staff with specialized expertise in areas such as cybercrime and insider trading.

In conclusion, FINRA's enforcement authority is a critical component of its role as a self-regulatory organization. Its ability to investigate and sanction firms and individuals who violate securities laws and regulations helps to maintain the integrity of the securities markets and protect investors. While challenges remain, FINRA's commitment to improving its enforcement capabilities through the use of advanced technology and specialized staff positions it well to meet these challenges head-on.

Key FINRA rules and regulations

One of the primary goals of FINRA is to ensure that the securities markets operate fairly and honestly. To achieve this goal, FINRA has promulgated a number of rules and regulations that govern the behavior of its member firms and associated persons. These rules and regulations cover a wide range of topics, including sales practices, supervision, reporting requirements, and trading activities.

One of the most important FINRA rules is the suitability rule (Rule 2111), which requires that member firms and associated persons make recommendations that are suitable for their customers based on their financial situation, investment objectives, and other factors. This rule is designed to protect investors from being sold securities that are not appropriate for their needs.

Another important FINRA rule is the know-your-customer (KYC) rule (Rule 2090), which requires that member firms have a reasonable basis to believe that they understand the essential facts concerning their customers, including their financial situation, investment experience, and objectives. This rule is designed to ensure that member firms can make appropriate investment recommendations and provide suitable investment advice to their customers.

In addition to these rules, FINRA has also promulgated a number of rules related to the supervision of its member firms and associated persons. These rules require member firms to establish and maintain a system of supervision designed to ensure compliance with securities laws and regulations, as well as with FINRA rules.

Another key area of FINRA regulation is its rules related to reporting requirements. For example, member firms are required to report certain events, such as customer complaints, regulatory actions, and criminal charges, to FINRA. These reporting requirements help FINRA to identify potential problem areas and take appropriate action to address them.

FINRA also has rules related to trading activities, such as its rules related to the dissemination of quotations (Rule 6432), which requires that member firms provide accurate quotations for securities, and its rules related to the order audit trail system (Rule 7440), which requires member firms to maintain a record of all orders that are received and executed.

Finally, it is important to note that FINRA has the authority to take disciplinary action against member firms and associated persons that violate its rules and regulations. These disciplinary actions can include fines, suspensions, and even expulsion from FINRA membership. As such, it is important for member firms and associated persons to be familiar with FINRA's rules and regulations and to take steps to ensure compliance with them.

In conclusion, FINRA's rules and regulations play a critical role in promoting fair and honest securities markets. By requiring member firms and associated persons to act in the best interests of their customers and to comply with securities laws and regulations, FINRA helps to protect investors and promote a level playing field in the securities industry. As such, it is important for all market participants to be familiar with FINRA's rules and regulations and to take steps to ensure compliance with them.

CHAPTER 20: STATE SECURITIES REGULATORS

State securities regulators play an important role in the regulation of securities markets in the United States. Each state has its own securities laws and regulatory agencies, which work in conjunction with the federal securities laws and regulators to oversee the securities industry.

The primary goal of state securities regulators is to protect investors from fraud and other unethical practices in the securities markets. To achieve this goal, they are responsible for administering state securities laws and regulations, investigating and prosecuting violations of those laws, and providing education and outreach to investors and industry professionals.

State securities laws and regulations cover a wide range of activities related to securities, including the registration and licensing of securities firms and professionals, the sale and distribution of securities, and the reporting and disclosure of information to investors. These laws and regulations also govern the conduct of securities professionals, including brokers, investment advisers, and dealers.

In addition to enforcing state securities laws, state securities regulators also work closely with federal securities regulators, such as the Securities and Exchange Commission (SEC) and the Financial Industry Regulatory Authority (FINRA), to ensure that securities markets are fair, transparent, and efficient. They participate in the development of new regulations and rules and share information and resources with their federal counterparts.

While state securities regulators have historically played a vital role in protecting investors and maintaining the integrity of securities markets, they face a number of challenges in the modern era. The rapid pace of technological innovation, the growth of online trading platforms, and the increasing complexity of securities markets have all contributed to the challenges faced by state securities regulators.

Nonetheless, state securities regulators remain an important part of the regulatory landscape in the United States, and their work is critical to the functioning of the securities industry. As such, it is important for investors, industry professionals, and policymakers to understand the role of state securities regulators and the laws and regulations they administer.

Overview of state securities regulators

State securities regulators play a critical role in protecting investors and ensuring fair and orderly securities markets. These regulators are responsible for enforcing state securities laws and regulations, licensing securities professionals, and registering securities offerings within their respective states. This chapter provides an overview of state securities regulators, including their history, organizational structure, and key functions.

History of State Securities Regulation

The regulation of securities has a long history in the United States, dating back to the early 20th century. In 1911, Kansas became the first state to enact a securities law, followed by New York in 1919. These early state laws were designed to address fraudulent securities offerings and other abuses in the securities markets.

The Securities Act of 1933 and the Securities Exchange Act of 1934, both federal laws, were enacted in response to the stock market crash of 1929 and the subsequent Great Depression. These laws established the Securities and Exchange Commission (SEC) as the federal regulator of the securities markets and created a comprehensive regulatory framework for securities offerings and trading.

While the federal government has primary responsibility for regulating securities markets, state securities regulators play an important role in protecting investors and enforcing securities laws. State securities regulation is based on the principle of "blue sky" laws, which require securities offerings to be registered with state regulators before they can be sold to investors. The term "blue sky" refers to the idea that state securities laws protect investors from fraudulent securities offerings that are not grounded in reality, much like the idea of selling someone a piece of the "blue sky."

Organizational Structure of State Securities Regulators

State securities regulators are organized at the state level and operate within the legal framework of state securities laws and regulations. Each state has a securities division or agency responsible for administering state securities laws and regulations.

The North American Securities Administrators Association (NASAA) is a membership organization of state and provincial securities regulators in the United States, Canada, and Mexico. NASAA provides a forum for state securities regulators to share information, coordinate enforcement actions, and develop best practices for regulating securities markets.

Key Functions of State Securities Regulators

State securities regulators perform a wide range of functions related to the regulation of securities markets. Some of their key functions include:

✧ Registration of Securities Offerings: State securities regulators require securities offerings to be registered with their agencies before they can be sold to investors. The registration process requires companies to disclose information about the securities being offered, including financial information, business plans, and risk factors.

✧ Licensing of Securities Professionals: State securities regulators license securities professionals, including brokers, investment advisers, and investment adviser representatives. These professionals must meet certain qualifications and pass exams before they can obtain their licenses.

✧ Enforcement of State Securities Laws: State securities regulators have the authority to investigate and prosecute violations of state securities laws. They can take legal action against companies and individuals that engage in fraudulent or illegal securities activities, such as Ponzi schemes or insider trading.

✧ Investor Education: State securities regulators provide educational resources to investors to help them understand the risks and rewards of investing. This includes information about different types of securities, how to research investment opportunities, and how to protect themselves from investment fraud.

Conclusion

In conclusion, state securities regulators play a vital role in protecting investors and ensuring fair and orderly securities markets. Their regulatory oversight complements that of the federal government and provides an additional layer of protection for investors. By requiring securities offerings to be registered, licensing securities professionals, and enforcing state securities laws, state securities regulators help maintain the integrity of the securities markets and promote investor confidence.

Differences between state and federal securities regulation

Securities regulations are a crucial aspect of the financial industry. Securities regulation is an important function of the federal and state governments. The Securities and Exchange Commission (SEC) is the primary federal agency responsible for enforcing securities laws, while each state has its own regulatory agency responsible for regulating securities within that state. This section will explore the differences between state and federal securities regulation.

Differences between state and federal securities regulation

Registration requirements

One of the main differences between state and federal securities regulation is the registration requirements for securities offerings. Federal securities laws require registration of securities offerings with the SEC, while state securities laws require registration with state securities regulators. The registration process at the federal level is generally more rigorous and complex than at the state level.

Scope of regulation

Another difference between state and federal securities regulation is the scope of regulation. Federal securities laws apply to all securities transactions that take place across state lines or internationally. State securities laws, on the other hand, generally apply only to securities transactions that occur within the state. Therefore, securities that are only offered within one state are usually only subject to state securities laws.

Enforcement authority

Another key difference between state and federal securities regulation is the enforcement authority. The SEC has broad authority to enforce federal securities laws, including the ability to investigate potential violations, bring enforcement actions, and impose penalties and sanctions. State securities regulators, while they have similar authority to investigate and enforce securities laws, generally have more limited resources and are limited to enforcing state securities laws within their respective states.

Exemptions and exclusions

Federal and state securities laws both provide for certain exemptions and exclusions from registration requirements. However, the exemptions and exclusions available under federal and state securities laws can differ significantly. This means that securities that may be exempt from registration under federal securities laws may not be exempt under state securities laws, and vice versa.

Disclosure requirements

Finally, another important difference between state and federal securities regulation is the disclosure requirements. Federal securities laws impose more comprehensive disclosure requirements on issuers of securities than state securities laws. State securities laws, however, may impose additional disclosure requirements beyond those required by federal securities laws.

Conclusion

In conclusion, state and federal securities regulation differ in several key ways, including registration requirements, scope of regulation, enforcement authority, exemptions and exclusions, and disclosure requirements. Understanding these differences is important for anyone involved in the securities industry, including investors, issuers, brokers, and securities professionals. While federal securities laws are generally more comprehensive and complex, state securities laws play an important role in regulating securities offerings that occur within individual states.

Key state securities laws and regulations

State securities laws and regulations are a crucial part of the United States' securities regulatory framework. These laws and regulations are enacted by individual states and are designed to protect investors and maintain fair and orderly markets. State securities regulators have the authority to investigate and prosecute violations of state securities laws, as well as to regulate broker-dealers, investment advisers, and other securities professionals operating within their jurisdiction. In this section, we will provide a detailed overview of key state securities laws and regulations, including their purpose, scope, and enforcement mechanisms.

Overview of State Securities Laws and Regulations

State securities laws are commonly known as "blue sky laws" because they were originally designed to prevent fraudulent securities offerings that were described as being backed by nothing but the "blue sky." The first blue sky law was enacted in Kansas in 1911, and since then, every state has enacted its own set of securities laws and regulations. These laws and regulations are designed to supplement federal securities laws, such as the Securities Act of 1933 and the Securities Exchange Act of 1934, by providing additional investor protections at the state level.

The primary purpose of state securities laws and regulations is to protect investors from fraudulent or deceptive practices in the sale of securities. These laws typically require securities issuers and sellers to register their offerings with state securities regulators and to provide detailed disclosures about the securities being offered. State securities laws also regulate the conduct of securities professionals, such as broker-dealers and investment advisers, by imposing licensing requirements and establishing standards of conduct. In addition, state securities regulators have the authority to investigate and prosecute violations of state securities laws and to take enforcement actions against violators.

Differences between State and Federal Securities Regulation

While state and federal securities regulations share a common goal of protecting investors, there are significant differences between the two regulatory frameworks. One of the primary differences is that federal securities laws are designed to regulate interstate securities transactions, while state securities laws are designed to regulate intrastate securities transactions. This means that state securities laws apply to securities offerings that are sold only within a particular state, while federal securities laws apply to securities offerings that are sold across state lines.

Another key difference is that state securities laws are typically more prescriptive than federal securities laws. State securities laws often provide more detailed requirements for securities issuers and sellers, such as specific disclosure requirements and registration procedures. In addition, state securities regulators have broader authority to investigate and prosecute violations of state securities laws than federal regulators do. This means that state securities regulators can take enforcement actions against securities professionals who violate state securities laws, even if the conduct at issue does not violate federal securities laws.

Key State Securities Laws and Regulations

There are several key state securities laws and regulations that are important for investors and securities professionals to understand. These laws and regulations vary by state, but some of the most common include:

Uniform Securities Act (USA)

The Uniform Securities Act (USA) is a model law developed by the North American Securities Administrators Association (NASAA) that has been adopted, with modifications, by many states. The USA provides a framework for state securities regulation by establishing uniform registration, licensing, and anti-fraud provisions. The USA also provides for the investigation and enforcement of securities laws by state securities regulators.

Blue Sky Laws

As mentioned earlier, blue sky laws are state securities laws that require securities issuers and sellers to register their offerings with state securities regulators and to provide detailed disclosures about the securities being offered. Blue sky laws also regulate the conduct of securities professionals by imposing licensing requirements and establishing standards of conduct.

Investment Adviser Laws

Many states have their own investment adviser laws that require investment advisers to register with state securities regulators and to meet certain standards of conduct. These laws typically require investment advisers to provide detailed disclosures about their services and to act in the best interests of their clients.

One such example of state investment adviser law is the California Investment Advisers Act (CIAA). The CIAA requires any person who acts as an investment adviser in California to register with the state, unless an exemption applies. The act defines an investment adviser as any person who, for compensation, engages in the business of advising others, either directly or through publications or writings, as to the value of securities or as to the advisability of investing in, purchasing, or selling securities.

The CIAA requires registered investment advisers to meet certain requirements, such as providing a disclosure brochure to clients that includes information about the adviser's business, fees, and conflicts of interest. The CIAA also requires advisers to maintain certain books and records, including a record of all investment recommendations made to clients.

Another example of state investment adviser law is the Texas Securities Act. This act requires investment advisers who have more than five clients in Texas to register with the state. The act defines an investment adviser as any person who, for compensation, engages in the business of advising others, either directly or through publications or writings, as to the value of securities or as to the advisability of investing in, purchasing, or selling securities.

The Texas Securities Act requires registered investment advisers to meet certain requirements, such as providing a disclosure brochure to clients that includes information about the adviser's business, fees, and conflicts of interest. The act also requires advisers to maintain certain books and records, including a record of all investment recommendations made to clients.

In addition to these state investment adviser laws, there are also state laws that regulate investment adviser representatives. These individuals are employees or independent contractors of investment advisers who provide investment advice on behalf of the adviser.

For example, the Colorado Securities Act requires investment adviser representatives to register with the state and meet certain standards of conduct. The act defines an investment adviser representative as any person who makes any recommendations or otherwise gives investment advice regarding securities, manages accounts or portfolios of clients, determines which recommendation or advice regarding securities should be given, provides investment advice or holds themselves out as providing investment advice, receives compensation to solicit, offer, or negotiate for the sale of or for selling investment advisory services or participates in the supervisory activities of an investment adviser.

The Colorado Securities Act requires investment adviser representatives to pass certain qualification exams, such as the Series 65 exam or the Series 7 and 66 exams, and to maintain certain continuing education requirements. The act also requires investment adviser representatives to meet certain standards of conduct, including a duty to act in the best interests of their clients.

In summary, state securities laws and regulations play a crucial role in protecting investors and regulating the securities industry. Investment advisers and investment adviser representatives must comply with these laws and regulations to ensure they are acting in the best interests of their clients and meeting their obligations under the law. It is important for anyone working in the securities industry to have a thorough understanding of these laws and regulations to ensure they are operating within the boundaries of the law and avoiding any potential legal or regulatory issues.

PART 6: DISCLOSURE REQUIREMENTS AND LIABILITY

Investing can be a complex and risky endeavor. Investors must rely on information provided by others to make informed investment decisions. Securities laws impose disclosure requirements on issuers and other market participants to ensure that investors have access to the information they need to make informed investment decisions. These laws also provide for liability in cases where investors suffer losses due to a failure to disclose material information.

In Part 6 of this report, we will examine the disclosure requirements and liability provisions of federal and state securities laws. We will begin by discussing the purpose and importance of disclosure requirements in the securities industry. We will then examine the different types of disclosure required by federal and state securities laws, including registration statements, prospectuses, and periodic reports.

Next, we will explore liability under securities laws. We will discuss the different types of liability that can arise, including liability for false or misleading statements, insider trading, and market manipulation. We will also examine the defenses that are available to defendants in securities litigation.

Finally, we will examine the practical implications of disclosure requirements and liability provisions for market participants, including issuers, underwriters, investment advisers, and other market participants. We will examine the role of lawyers, accountants, and other professionals in ensuring compliance with disclosure requirements and minimizing the risk of liability. We will also discuss the impact of disclosure requirements and liability provisions on the cost of capital and the efficiency of capital markets.

By the end of this section, readers will have a thorough understanding of the importance of disclosure requirements and liability provisions in the securities industry, as well as the practical implications of these provisions for market participants. We will provide examples, problems, and exercises to illustrate the topics covered in this section.

CHAPTER 21: DISCLOSURE REQUIREMENTS FOR PUBLIC COMPANIES

Public companies have a legal obligation to disclose information to their shareholders and potential investors. Disclosure requirements are designed to ensure that investors have access to material information about a company that could affect its stock price or their investment decisions. Disclosure requirements for public companies are regulated by the Securities and Exchange Commission (SEC) under the Securities Act of 1933, the Securities Exchange Act of 1934, and the Sarbanes-Oxley Act of 2002.

The primary objective of these disclosure requirements is to ensure that investors have access to accurate and timely information about a public company's financial health and business operations. Public companies are required to disclose a wide range of information, including financial statements, executive compensation, related party transactions, risk factors, legal proceedings, and other material information.

This chapter provides an overview of the disclosure requirements for public companies under federal securities laws. It discusses the different types of disclosures that public companies are required to make, the regulatory framework governing these disclosures, and the liability that companies face for failing to comply with these requirements. The chapter also highlights some of the recent developments in disclosure requirements for public companies, including the impact of technology on disclosure practices and the growing focus on environmental, social, and governance (ESG) disclosures.

Types of Disclosures Required of Public Companies

Public companies are required to disclose a wide range of information to their shareholders and potential investors. These disclosures can be broadly classified into two categories: periodic and event-based disclosures.

Periodic Disclosures

Periodic disclosures are disclosures that companies are required to make on a regular basis. These disclosures include financial statements, management's discussion and analysis (MD&A), and other required disclosures such as executive compensation, related party transactions, and risk factors. The primary objective of these periodic disclosures is to provide investors with accurate and timely information about a company's financial health and business operations.

Financial Statements

Public companies are required to file periodic financial statements with the SEC, which include the balance sheet, income statement, statement of cash flows, and statement of changes in shareholders' equity. These financial statements must be audited by an independent public accounting firm, which provides an opinion on whether the financial statements are presented fairly in all material respects in conformity with generally accepted accounting principles (GAAP).

Management's Discussion and Analysis (MD&A)

In addition to financial statements, public companies are required to provide a narrative discussion of their financial condition and results of operations in the MD&A section of their annual report. The MD&A is intended to provide investors with management's perspective on the company's financial performance and future prospects.

Other Required Disclosures

Public companies are also required to disclose other information in their periodic filings, such as executive compensation, related party transactions, and risk factors. These disclosures are intended to provide investors with information about the company's governance practices and potential risks that could affect its financial performance.

Event-Based Disclosures

Event-based disclosures are disclosures that companies are required to make in response to specific events or changes in the company's business operations. These disclosures include, among other things, material contracts, changes in control, and legal proceedings.

Material Contracts

Public companies are required to file copies of material contracts with the SEC. Material contracts are contracts that are not entered into in the ordinary course of business and that are material to the company's business operations. Examples of material contracts include major purchase agreements, lease agreements, and joint venture agreements.

Changes in Control

Public companies are required to disclose changes in control of the company, including changes in ownership and changes in the composition of the board of directors.

Legal Proceedings

Public companies are required to disclose information about any material legal proceedings that they are involved in. This includes pending litigation, regulatory investigations, and other legal proceedings that could have a material impact on the company and its financial performance. The purpose of this disclosure requirement is to ensure that investors have access to all material information about the company, including any potential risks or liabilities that may affect its operations and financial health.

Legal proceedings can have a significant impact on a company's operations and financial performance, and failure to disclose such information can result in legal action against the company and its executives. For example, if a public company is involved in a lawsuit that could potentially result in significant damages, failure to disclose this information could be seen as a violation of securities laws.

In addition to disclosing information about legal proceedings that the company is involved in, public companies are also required to disclose any material legal proceedings that their directors, officers, or affiliates are involved in. This includes any criminal convictions or civil judgments against these individuals that are material to their ability to serve in their roles at the company.

Disclosure of legal proceedings is an important aspect of corporate governance and helps to ensure that investors have access to all material information about the company. This information can help investors make informed decisions about whether to invest in the company and can also help to reduce the risk of legal action against the company and its executives.

In addition to disclosing information about legal proceedings, public companies are also required to disclose information about any material agreements that they enter into. This includes contracts with suppliers, customers, and other third parties that are material to the company's operations and financial performance.

Material agreements can include contracts related to the sale of goods or services, joint ventures, licensing agreements, and other business arrangements. Disclosure of these agreements helps to ensure that investors have access to all material information about the company, including any potential risks or liabilities that may arise from these agreements.

Overall, disclosure requirements for public companies are designed to ensure that investors have access to all material information about the company, including financial performance, corporate governance, legal proceedings, and material agreements. Failure to comply with these disclosure requirements can result in legal action against the company and its executives and can also lead to reputational damage and loss of investor confidence.

Overview of disclosure requirements for public companies

Public companies are required to disclose a wide range of information to investors and the public. These disclosures are designed to provide transparency and to help investors make informed investment decisions. The information that public companies are required to disclose is regulated by the Securities and Exchange Commission (SEC) and is contained in a number of different forms and reports that must be filed with the SEC.

The main purpose of disclosure requirements for public companies is to provide investors with timely and accurate information about the financial condition and operating results of the company. This information allows investors to make informed investment decisions, which helps to promote efficient markets and to protect investors from fraud and other forms of misconduct.

The disclosure requirements for public companies are extensive and cover a wide range of topics. Some of the key areas that public companies are required to disclose information about include financial statements, executive compensation, related party transactions, changes in control, legal proceedings, and risk factors.

Financial Statements

One of the most important disclosure requirements for public companies is the filing of financial statements. Public companies are required to prepare and file annual and quarterly reports that contain financial statements that have been audited by an independent auditor. These financial statements must provide a comprehensive and accurate picture of the company's financial condition and operating results.

The financial statements that public companies are required to file include the balance sheet, income statement, statement of cash flows, and statement of stockholders' equity. These statements provide information about the company's assets, liabilities, revenues, expenses, cash flows, and changes in stockholders' equity.

Executive Compensation

Public companies are also required to disclose information about the compensation of their top executives. This information is contained in the company's proxy statement, which is filed with the SEC each year. The proxy statement must include detailed information about the compensation of the company's top executives, including their salaries, bonuses, stock options, and other forms of compensation.

Related Party Transactions

Public companies are also required to disclose information about any related party transactions that they are involved in. Related party transactions are transactions between the company and its officers, directors, and other related parties. These transactions must be disclosed in the company's financial statements and must be reviewed by the company's audit committee to ensure that they are fair and reasonable.

Changes in Control

Public companies are required to disclose changes in control of the company, including changes in ownership and changes in the composition of the board of directors. These disclosures are designed to ensure that investors are aware of any changes that could impact the direction and strategy of the company.

Legal Proceedings

Public companies are required to disclose information about any material legal proceedings that they are involved in. This includes pending litigation, regulatory investigations, and other legal proceedings that could have a material impact on the company. These disclosures are designed to provide investors with information about the risks that the company faces and to help investors make informed investment decisions.

Risk Factors

Public companies are also required to disclose information about the risks that they face. This information is contained in the company's annual and quarterly reports and must include a discussion of the most significant risks that the company faces. This information helps investors to understand the potential risks associated with investing in the company and to make informed investment decisions.

Conclusion

Disclosure requirements for public companies play a critical role in promoting transparency and protecting investors. The information that public companies are required to disclose provides investors with a comprehensive and accurate picture of the company's financial condition and operating results. This information helps to promote efficient markets and to protect investors from fraud and other forms of misconduct. Public companies must ensure that they comply with all of the disclosure requirements that are applicable to them, and that they provide investors with timely and accurate information.

Annual and quarterly reports

Public companies are required to file annual and quarterly reports with the Securities and Exchange Commission (SEC) to provide information to investors about their financial performance, business operations, and future prospects. These reports are an essential part of the disclosure requirements for public companies and serve as a valuable source of information for investors, analysts, and other stakeholders.

In this section, we will provide an overview of the annual and quarterly reporting requirements for public companies, including the content and timing of these reports, as well as the key disclosures that companies must include in them. We will also discuss the importance of these reports for investors and other stakeholders, as well as the potential risks associated with non-compliance with these requirements.

Annual Reports:

Public companies are required to file an annual report, known as a Form 10-K, with the SEC within 60 days of the end of their fiscal year. This report provides a comprehensive overview of the company's financial performance, including its balance sheet, income statement, cash flow statement, and other financial metrics. It also includes a detailed discussion of the company's business operations, risks, and future prospects, as well as information about its management and corporate governance.

The Form 10-K must include a number of key disclosures, including information about the company's audited financial statements, related-party transactions, legal proceedings, and risk factors. Companies must also disclose any material changes to their business or financial condition that occurred during the fiscal year, as well as any subsequent events that could have a material impact on their business.

Quarterly Reports:

In addition to annual reports, public companies are required to file quarterly reports, known as Form 10-Qs, with the SEC within 45 days of the end of each fiscal quarter. These reports provide investors with an update on the company's financial performance and business operations since the last annual report.

Form 10-Qs typically include unaudited financial statements, as well as a discussion of the company's business operations and risks. Companies must also disclose any material changes to their business or financial condition that occurred during the quarter, as well as any subsequent events that could have a material impact on their business.

Key Disclosures:

Public companies must include a number of key disclosures in their annual and quarterly reports to comply with SEC regulations. These disclosures include:

✧ Financial Statements: Companies must include audited financial statements in their annual reports, and unaudited financial statements in their quarterly reports. These statements provide information about the company's financial performance, including its revenues, expenses, assets, and liabilities.

✧ Management Discussion and Analysis (MD&A): Companies must provide a detailed discussion of their financial performance and business operations in their annual and quarterly reports. This discussion, known as the MD&A, must include an analysis of the company's financial results, as well as a discussion of any risks or uncertainties that could impact the company's future performance.

✧ Risk Factors: Companies must disclose any risks or uncertainties that could have a material impact on their business in their annual and quarterly reports. These risks could include factors such as industry competition, regulatory changes, or economic conditions.

✧ Legal Proceedings: Companies must disclose any material legal proceedings that they are involved in, including pending litigation, regulatory investigations, or other legal proceedings that could have a material impact on the company.

✧ Related-Party Transactions: Companies must disclose any transactions between the company and its officers, directors, or other related parties that could have a material impact on the company.

Importance of Annual and Quarterly Reports:

Annual and quarterly reports are an essential source of information for investors, analysts, and other stakeholders who are interested in the financial performance and business operations of public companies. These reports provide valuable insights into a company's financial health, growth prospects, and overall business strategy.

For investors, annual and quarterly reports can be used to make informed investment decisions. By analyzing the financial data provided in these reports, investors can assess a company's profitability, liquidity, and financial stability. They can also compare a company's performance to that of its competitors and industry benchmarks, which can help them determine whether to buy, hold, or sell a company's stock.

In addition to financial data, annual and quarterly reports also provide important information about a company's operations and business strategy. This includes information about new product launches, marketing initiatives, research and development efforts, and other strategic initiatives that can impact a company's future growth prospects. For example, a company's annual report may include a detailed discussion of its plans to enter new markets, acquire new businesses, or invest in new technologies.

Annual and quarterly reports also play a crucial role in promoting transparency and accountability in the corporate world. By requiring public companies to disclose certain information about their financial performance and business operations, these reports help to ensure that investors and other stakeholders have access to accurate and timely information that can inform their investment decisions. This can help to prevent fraud, insider trading, and other unethical or illegal practices that can harm investors and undermine the integrity of financial markets.

Furthermore, annual and quarterly reports can help to build trust and confidence among investors and other stakeholders. When companies provide detailed and transparent information about their financial performance and business operations, they demonstrate a commitment to accountability and responsible corporate governance. This can help to enhance their reputation and strengthen relationships with investors, customers, and other stakeholders.

In conclusion, annual and quarterly reports are a critical component of the financial reporting framework for public companies. These reports provide valuable information about a company's financial performance, growth prospects, and business strategy, which can inform investment decisions and promote transparency and accountability in the corporate world. As such, it is important for companies to prepare these reports in accordance with regulatory requirements and best practices, and to ensure that they provide accurate and meaningful information to investors and other stakeholders.

Insider trading disclosures

Insider trading is a practice in which individuals with access to non-public information about a company use that information to trade securities for their own financial gain. This practice is illegal under securities laws in most jurisdictions, including the United States. Insider trading can occur in a variety of forms, including through the buying or selling of stocks, options, or other securities. To combat this practice and ensure that the market remains fair for all investors, public companies are required to disclose information about insider trading activities.

Insider Trading Disclosures

Public companies are required to disclose certain information related to insider trading. This includes disclosures related to the ownership of securities by officers, directors, and certain other individuals affiliated with the company. These disclosures are designed to provide transparency to the market and allow investors to make informed decisions about the company's securities.

Section 16 Disclosures

Section 16 of the Securities Exchange Act of 1934 requires that officers, directors, and beneficial owners of more than 10% of a class of the company's equity securities file reports with the Securities and Exchange Commission (SEC) disclosing their ownership of the company's securities and any changes in that ownership. These reports are known as Form 4 filings and must be filed within two business days of the transaction.

In addition to Form 4 filings, officers and directors are also required to file an initial report on Form 3 when they first become affiliated with the company, and an annual report on Form 5 to report any transactions that were not reported on Form 4 during the year.

Section 16 disclosures provide investors with valuable information about the ownership of the company's securities by officers and directors. This information can be used to identify potential conflicts of interest and to monitor insider trading activity.

Rule 10b5-1 Trading Plans

In addition to Section 16 disclosures, public companies are also required to disclose information about Rule 10b5-1 trading plans. Rule 10b5-1 provides a safe harbor for executives to trade securities in their company without running afoul of insider trading laws, as long as the trades are made pursuant to a pre-established plan that meets certain criteria.

To qualify for the safe harbor provided by Rule 10b5-1, a trading plan must be entered into in good faith and not as part of a plan to evade insider trading laws. The plan must also be established at a time when the executive does not have material non-public information about the company. Once the plan is established, the executive must not have any subsequent control over the trades made pursuant to the plan.

Public companies are required to disclose the existence of Rule 10b5-1 trading plans, as well as any transactions made pursuant to those plans. These disclosures provide investors with information about the timing and amount of insider trading activity by executives.

Recent Developments

In recent years, there has been increased scrutiny of insider trading activity and the disclosures required of public companies. In particular, there has been a focus on the use of Rule 10b5-1 trading plans as a means of circumventing insider trading laws.

In response to these concerns, the SEC has proposed changes to the rules governing insider trading disclosures. These changes would require greater transparency around the use of Rule 10b5-1 trading plans, including requiring executives to disclose the specific terms of the plans when they are established.

Conclusion

Insider trading disclosures are an essential component of the securities laws and are designed to provide transparency and fairness to the market. Public companies are required to disclose information about insider trading activities by officers, directors, and other affiliated individuals. This includes Section 16 disclosures related to the ownership of securities, as well as disclosures related to Rule 10b5-1 trading plans. As the market evolves and new concerns arise, it is likely that the rules and regulations related to insider trading disclosures will continue to be refined and updated.

One recent development in this area is the increased focus on the use of technology to monitor and prevent insider trading. For example, some companies are using sophisticated data analytics tools to detect patterns and anomalies in trading activity that may indicate insider trading. Other companies are using blockchain technology to create a tamper-proof record of all trading activity, making it easier to track and investigate potential violations.

Another area of concern related to insider trading is the use of social media and other forms of electronic communication to share nonpublic information. The Securities and Exchange Commission (SEC) has issued guidance to clarify that the same rules that apply to traditional forms of communication, such as phone calls and emails, also apply to social media and other forms of electronic communication. This means that companies must be vigilant in monitoring all forms of communication to ensure that nonpublic information is not being shared inappropriately.

In addition to the legal requirements related to insider trading disclosures, many companies also have internal policies and procedures in place to prevent insider trading. These policies typically include restrictions on trading by insiders during certain periods, such as the period immediately before earnings announcements or other significant events. They may also require insiders to pre-clear any trades with the company's legal department or compliance office.

Overall, insider trading disclosures are a critical component of the securities laws, helping to ensure that the market operates fairly and transparently. Companies must be diligent in their efforts to monitor and prevent insider trading, and investors and other stakeholders should pay close attention to these disclosures when making investment decisions.

CHAPTER 22: LIABILITY FOR SECURITIES FRAUD

Securities fraud is a complex area of law that has evolved over time to address the various ways in which companies and individuals can deceive investors in the securities markets. Securities fraud can take many forms, including misrepresentations, omissions, and manipulations, and can involve a wide range of securities, including stocks, bonds, and derivatives.

The law governing securities fraud has its origins in the Securities Act of 1933 and the Securities Exchange Act of 1934, which were enacted in response to the stock market crash of 1929 and the ensuing Great Depression. These laws were designed to regulate the sale and trading of securities and to prevent fraud in the securities markets.

Since the enactment of these laws, the regulation of securities has evolved significantly, and new laws and regulations have been added to address emerging issues in the securities markets. One of the most significant developments in this area of law has been the expansion of liability for securities fraud.

Today, securities fraud liability extends to a wide range of actors, including companies, officers and directors, accountants, underwriters, analysts, and others involved in the sale and trading of securities. The law imposes strict liability for certain types of misrepresentations and omissions, and also provides for private causes of action for investors who have been harmed by securities fraud.

This chapter will provide an overview of the law governing securities fraud liability, including the types of conduct that can give rise to liability, the actors who can be held liable, and the remedies available to investors who have been harmed by securities fraud. We will also discuss some of the challenges associated with enforcing securities fraud liability, including the difficulty of proving intent and the complexity of the securities markets.

Types of Conduct Giving Rise to Liability

Securities fraud can take many forms, and liability for securities fraud can be based on a variety of different types of conduct. Some of the most common types of conduct that give rise to liability for securities fraud include:

Misrepresentations and Omissions

Misrepresentations and omissions are perhaps the most common types of conduct that give rise to liability for securities fraud. Misrepresentations occur when an individual or company makes a statement that is false or misleading in a material way. Omissions occur when an individual or company fails to disclose material information that would have been important to investors.

The law imposes strict liability for misrepresentations and omissions in certain circumstances. For example, under Section 11 of the Securities Act of 1933, a company that issues a registration statement that contains a material misstatement or omission can be held strictly liable for any damages suffered by investors who purchased securities in reliance on the registration statement.

Similarly, under Rule 10b-5 of the Securities Exchange Act of 1934, it is unlawful to make any material misrepresentation or omission in connection with the purchase or sale of securities. This rule applies to both public and private companies and imposes liability on both the individuals who make the misrepresentations or omissions and the companies that are responsible for those individuals.

Insider Trading

Insider trading occurs when an individual trades securities on the basis of material, nonpublic information. Insider trading is a type of securities fraud because it involves the use of information that is not available to the public, and it can harm other investors who are not privy to the same information.

The law imposes strict liability for insider trading in certain circumstances. For example, under Section 10(b) of the Securities Exchange Act of 1934 and Rule 10b-5, it is unlawful for any person to trade securities while in possession of material, nonpublic information.

In addition to strict liability, individuals who engage in insider trading can also face criminal prosecution and civil enforcement actions by the Securities and Exchange Commission (SEC) and private plaintiffs. Criminal penalties for insider trading can include fines, imprisonment, and disgorgement of profits. Civil penalties can include fines, injunctions, and orders to disgorge profits.

Liability for insider trading can extend beyond the individual who engaged in the trading to other parties who assisted or facilitated the trading. For example, tippees who receive material, nonpublic information from insiders can be liable for insider trading if they trade on that information or pass it along to others who then trade on it. In addition, anyone who provides information or advice to an insider who then engages in insider trading can be liable under the misappropriation theory of insider trading.

The misappropriation theory of insider trading applies to situations where an individual trades securities based on material, nonpublic information that they obtained through a breach of fiduciary duty or other relationship of trust and confidence. For example, an attorney who learns of material, nonpublic information through their representation of a client may be liable for insider trading if they trade on that information or pass it along to others who then trade on it.

Liability for insider trading can also extend to companies that fail to implement effective policies and procedures to prevent insider trading. The SEC has brought enforcement actions against companies for failing to adequately monitor trading by their employees and for failing to prevent insider trading.

In conclusion, insider trading is a serious securities fraud that can result in strict liability, criminal penalties, civil penalties, and liability for others who assist or facilitate the trading. It is important for individuals and companies to understand the legal framework surrounding insider trading and to implement effective policies and procedures to prevent it.

Definition of securities fraud

Securities fraud is a type of financial fraud that involves the intentional deception or misrepresentation of material information related to securities transactions. Securities fraud can take many forms, but the most common types involve misrepresentations or omissions of material information in connection with the sale or purchase of securities. Securities fraud can be committed by individuals or companies, and it can result in significant financial harm to investors.

The term "securities fraud" encompasses a wide range of illegal activities, including insider trading, market manipulation, Ponzi schemes, and accounting fraud. While the specific elements of securities fraud may vary depending on the type of fraud involved, there are certain common elements that are present in most cases.

One common element of securities fraud is the use of false or misleading statements or omissions of material information. These false statements may be made in connection with the sale or purchase of securities, or they may be made in connection with other financial transactions.

Another common element of securities fraud is the use of insider information. Insider trading occurs when individuals trade securities based on nonpublic information that they have obtained through their positions as officers, directors, or employees of a company. This type of trading is illegal because it gives insiders an unfair advantage over other investors who do not have access to the same information.

Market manipulation is another form of securities fraud. This occurs when individuals or companies engage in activities that artificially inflate or deflate the price of securities. For example, a company may engage in "pump and dump" schemes, where it artificially inflates the price of its own stock and then sells it at a profit.

Ponzi schemes are yet another form of securities fraud. In a Ponzi scheme, an individual or company promises investors high returns on their investments, but instead of investing the money, the individual or company uses new investors' money to pay off earlier investors. Ponzi schemes inevitably collapse when new investors stop investing and earlier investors try to withdraw their money.

Accounting fraud is another common type of securities fraud. This occurs when companies manipulate their financial statements to misrepresent their financial performance. Accounting fraud can take many forms, such as overstating revenues, understating expenses, or hiding liabilities.

Overall, securities fraud is a serious offense that can result in significant financial harm to investors. The securities laws are designed to prevent securities fraud and to provide investors with a legal remedy if they are harmed by fraudulent activities.

Types of securities fraud

Securities fraud is a serious offense that can lead to significant financial losses for investors and damage to the integrity of the financial markets. It can take many forms, and there are several types of securities fraud that investors should be aware of. In this section, we will discuss the most common types of securities fraud, including insider trading, Ponzi schemes, accounting fraud, and market manipulation.

Insider Trading

Insider trading occurs when an individual buys or sells securities based on material, non-public information. This type of trading is illegal because it gives insiders an unfair advantage over other investors who do not have access to the same information. Insiders can include company executives, directors, employees, and even family members who may be privy to confidential information. Insider trading is one of the most common types of securities fraud, and it is strictly prohibited by the Securities and Exchange Commission (SEC).

Ponzi Schemes

A Ponzi scheme is a fraudulent investment scheme in which returns are paid to earlier investors using the capital contributed by new investors. The scheme relies on a constant flow of new investors to sustain itself and often promises high returns with little or no risk. Ponzi schemes are unsustainable and will eventually collapse when there are not enough new investors to keep the scheme going. Famous examples of Ponzi schemes include the Madoff scandal and the TelexFree scam.

Accounting Fraud

Accounting fraud occurs when a company misrepresents its financial statements in order to deceive investors. This can involve inflating revenues, hiding expenses, and manipulating financial ratios to make the company appear more profitable than it actually is. Accounting fraud can be difficult to detect, but it can have serious consequences for investors and the overall financial system. Some well-known cases of accounting fraud include Enron, WorldCom, and Tyco.

Market Manipulation

Market manipulation involves intentionally influencing the price of a security in order to benefit a particular individual or group. This can involve spreading false rumors, engaging in insider trading, or artificially inflating the price of a security through buying and selling activities. Market manipulation can distort the true value of a security and harm investors who rely on accurate information to make investment decisions.

In conclusion, there are several types of securities fraud that investors should be aware of. Insider trading, Ponzi schemes, accounting fraud, and market manipulation are among the most common forms of securities fraud. Investors should be cautious and conduct thorough research before investing in any security, and should report any suspicious activity to the SEC or other regulatory bodies.

Civil and criminal liability for securities fraud

Civil and criminal liability are two types of legal consequences that can result from securities fraud. Securities fraud involves the misrepresentation or omission of material information in connection with the sale or purchase of securities. Civil liability refers to the legal responsibility for financial damages or losses incurred by another party, while criminal liability involves potential imprisonment or fines as a result of the fraudulent behavior.

Civil Liability for Securities Fraud:

Civil liability for securities fraud is enforced through private lawsuits brought by individuals or entities that have been harmed by the fraudulent behavior. The penalties for civil liability can be significant and include damages, fines, and court costs. In general, the burden of proof for civil liability is less than that of criminal liability, and a successful plaintiff can receive compensation for damages without proving that the defendant intended to defraud them.

In some cases, civil liability can also result in injunctive relief, which is a court order that requires the defendant to cease engaging in fraudulent behavior. Injunctive relief can be an effective deterrent against future misconduct and can be used to protect investors from further harm.

Criminal Liability for Securities Fraud:

Criminal liability for securities fraud can result in significant fines and imprisonment for those convicted of the offense. The penalties for criminal liability vary depending on the severity of the offense, the amount of money involved, and the defendant's criminal history. Criminal liability requires the prosecution to prove that the defendant intentionally engaged in fraudulent behavior, which can be difficult to prove beyond a reasonable doubt.

The penalties for criminal liability can include fines of up to $5 million for individuals and $25 million for corporations, as well as imprisonment for up to 20 years. Additionally, those convicted of securities fraud may be required to pay restitution to the victims of their fraud.

In some cases, criminal liability can also result in the suspension or revocation of a professional license, such as a broker's license or investment advisor's license. This can have significant consequences for individuals working in the securities industry, as it can prevent them from continuing to work in the field.

In conclusion, both civil and criminal liability can result from securities fraud, and the penalties for each can be significant. Civil liability may result in monetary damages and injunctive relief, while criminal liability can result in fines and imprisonment. It is important for individuals working in the securities industry to understand the potential legal consequences of engaging in fraudulent behavior and to take steps to prevent it from occurring.

CHAPTER 23: INSIDER TRADING AND OTHER SECURITIES VIOLATIONS

The world of finance is governed by a complex set of laws and regulations designed to prevent fraudulent activities and promote fair and efficient markets. Securities fraud is one such activity that can undermine investor confidence, destabilize financial markets, and harm the overall economy. Securities fraud refers to any unlawful activity that involves the buying or selling of securities based on false or misleading information, or the manipulation of the securities markets for personal gain.

Insider trading is a form of securities fraud that has received much attention in recent years. It occurs when an individual trades securities based on material, nonpublic information. Insider trading can involve corporate insiders, such as officers, directors, and employees, who have access to confidential information about their companies. It can also involve outsiders, such as analysts, lawyers, and consultants, who may obtain inside information through various means.

In addition to insider trading, securities fraud can take many other forms. These include market manipulation, accounting fraud, Ponzi schemes, and other types of investment fraud. Market manipulation involves the artificial inflation or deflation of securities prices through various techniques, such as "pump and dump" schemes, which involve promoting a stock to unsuspecting investors and then selling it at an inflated price.

Accounting fraud involves the manipulation of financial statements to misrepresent a company's financial health. This can include understating liabilities, overstating assets, and using improper accounting methods. Ponzi schemes involve the use of investor funds to pay off earlier investors, rather than generating legitimate returns through actual investments. Other types of investment fraud can include high-pressure sales tactics, false promises of high returns, and other forms of deception.

Securities fraud can have severe legal and financial consequences for those who engage in it. Civil penalties can include fines, disgorgement of profits, and injunctions, while criminal penalties can include imprisonment, fines, and forfeiture of assets. In addition to legal consequences, securities fraud can also lead to reputational damage, loss of employment, and other professional sanctions.

In this chapter, we will explore the various forms of securities fraud, the legal and regulatory framework that governs these activities, and the civil and criminal liability that can result from engaging in securities fraud. We will also discuss some of the challenges associated with detecting and preventing securities fraud, and the role that various actors, such as regulators, auditors, and investors, play in promoting market integrity and investor protection.

Definition of insider trading

Insider trading is a term that has received a lot of attention in recent years. It refers to the buying or selling of securities by individuals who have access to confidential information about a company. This information is not available to the public, and therefore, such trades can result in illegal gains for the insider trader. In this section, we will provide a detailed definition of insider trading, its various forms, and the legal framework surrounding it.

Definition:

Insider trading is the act of buying or selling securities based on confidential, non-public information about a company. Such information may include financial data, business strategies, mergers and acquisitions, or any other material that could affect the value of a security. The term "insider" refers to a person who has access to this information, such as an executive, director, or employee of the company.

Insider trading is generally considered illegal because it creates an unfair advantage for those who possess the confidential information. By trading on this information, insiders can earn significant profits, while the rest of the market remains unaware of the information. This can lead to a distortion of the market and a loss of confidence in the integrity of the financial system.

Forms of Insider Trading:

There are two main forms of insider trading: illegal insider trading and legal insider trading.

Illegal insider trading occurs when an insider uses confidential information to trade securities, or when an outsider trades on information obtained from an insider. This is a violation of securities laws and is subject to civil and criminal penalties.

Legal insider trading, on the other hand, is when an insider buys or sells securities in their own company, based on information that is available to the public. For example, if a CEO purchases shares in their company after a positive earnings report is released, this is considered legal insider trading.

Legal Framework:

Insider trading is regulated by various securities laws and regulations, including the Securities Exchange Act of 1934, the Insider Trading and Securities Fraud Enforcement Act of 1988, and the Sarbanes-Oxley Act of 2002.

Under these laws, insiders are required to disclose any transactions involving their company's securities to the Securities and Exchange Commission (SEC). They are also prohibited from trading on material, non-public information.

Penalties for Insider Trading:

Insider trading can result in both civil and criminal penalties. Civil penalties may include fines, disgorgement of profits, and injunctions to prevent future violations. Criminal penalties may include imprisonment and fines.

The severity of the penalties depends on various factors, including the nature and extent of the violation, the individual's level of involvement, and whether the violation was intentional or unintentional. In general, penalties for insider trading can range from fines of several hundred thousand dollars to imprisonment for several years.

Conclusion:

In conclusion, insider trading is a complex area of securities law that has far-reaching implications for the financial system. It is illegal to trade securities based on confidential, non-public information, and violators can face significant civil and criminal penalties. As such, it is essential for individuals and companies to be aware of the legal framework surrounding insider trading and to ensure that they comply with all relevant laws and regulations.

Other securities violations, such as market manipulation

The financial markets rely on trust, fairness, and transparency to function efficiently. Securities regulations aim to prevent market manipulation, insider trading, and other forms of securities fraud to maintain the integrity of the markets. The Securities and Exchange Commission (SEC) is the regulatory agency responsible for enforcing securities laws in the United States. This section will explore the definition of market manipulation and other forms of securities violations, as well as the consequences for individuals and companies found guilty of engaging in such activities.

Definition of Market Manipulation:

Market manipulation refers to a variety of activities that seek to artificially influence the price or volume of securities in the financial markets. The SEC defines market manipulation as "intentional conduct designed to deceive investors by controlling or artificially affecting the market for a security." Market manipulation can take many forms, including:

Pump and dump schemes: This is a form of market manipulation in which an individual or group of individuals purchase large quantities of a security and promote it through various means to artificially increase demand and the price of the security. Once the price has increased, the individuals sell their shares, leaving other investors holding overpriced and ultimately worthless securities.

Churning: This refers to the excessive trading of a security by a broker to generate commissions. The broker may trade the security multiple times in a short period, resulting in higher commissions for the broker but unnecessary expenses for the investor.

Spoofing: This is a tactic in which a trader places a large order for a security with no intention of executing the order. The order is intended to create the illusion of demand or supply, influencing the price of the security. Once the price has moved in the desired direction, the trader cancels the order.

Front-running: This refers to the practice of a broker executing orders for their own benefit before executing orders for their clients. The broker may use knowledge of their clients' orders to trade on the same security before the client's order is executed, resulting in a profit for the broker at the expense of the client.

Consequences for Market Manipulation:

Market manipulation is a serious offense that can have severe consequences for individuals and companies found guilty of engaging in such activities. The consequences for market manipulation can include civil and criminal penalties, including fines and imprisonment.

Civil Penalties:

Civil penalties for market manipulation can include fines, disgorgement of profits, and injunctive relief. Fines can range from thousands to millions of dollars, depending on the severity of the offense. Disgorgement of profits requires the individual or company to forfeit any profits gained from the market manipulation. Injunctive relief can require the individual or company to cease engaging in the prohibited activity.

Criminal Penalties:

Criminal penalties for market manipulation can include fines and imprisonment. The severity of the criminal penalties will depend on the nature and severity of the offense. For example, individuals found guilty of insider trading can face up to 20 years in prison and fines of up to $5 million.

Other Securities Violations:

In addition to market manipulation, there are other forms of securities violations that can result in civil and criminal penalties. These include:

Offering fraud: This refers to the misrepresentation of material facts related to the sale of securities. Offering fraud can include false statements about the company's financial performance, business prospects, or the securities being sold. Individuals found guilty of offering fraud can face fines, disgorgement of profits, and injunctive relief.

Ponzi schemes: This is a form of investment fraud in which individuals are promised high returns on their investment. However, the returns are paid using the investments of new investors, rather than through legitimate business activities. Ponzi schemes inevitably collapse when new investors are unable to be found, and existing investors are left with worthless securities. Individuals found guilty of running Ponzi schemes can face criminal charges, as well as civil lawsuits seeking to recover the losses of investors. Some high-profile Ponzi schemes include the Bernard Madoff investment scandal, which resulted in Madoff being sentenced to 150 years in prison, and the Enron scandal, which resulted in multiple individuals facing criminal charges and prison sentences.

Insider trading: As discussed earlier, insider trading involves trading securities based on non-public information that gives an individual an unfair advantage over other investors. Insider trading can result in both civil and criminal penalties, including fines, disgorgement of profits, and prison sentences.

Front running: This refers to the practice of a broker trading securities based on information obtained from a client's order before executing the client's order. This can give the broker an unfair advantage over other market participants and can be considered a violation of securities laws. Front running can result in both civil and criminal penalties.

Churning: This refers to the excessive buying and selling of securities by a broker for the purpose of generating commissions. Churning can be considered a violation of securities laws and can result in civil penalties, such as fines and disgorgement of profits, as well as disciplinary action by regulatory bodies.

In conclusion, securities violations come in various forms, and each carries its own set of civil and criminal penalties. It is important for individuals working in the securities industry to understand these laws and regulations to avoid violating them, which can result in serious legal consequences. Regulators and law enforcement agencies are vigilant in investigating and prosecuting securities violations to protect investors and maintain the integrity of the financial markets.

Consequences of securities violations

Securities violations can have severe consequences for both individuals and corporations. In addition to potential criminal charges, violations can result in civil penalties, including fines and disgorgement of profits. Civil penalties can also include injunctive relief, such as cease and desist orders or requirements for companies to implement compliance programs.

Individuals found guilty of securities violations can face significant fines and even imprisonment. For example, insider trading can result in fines of up to three times the profits gained or losses avoided as a result of the illegal activity. In addition, individuals can face up to 20 years in prison for each criminal count. Individuals found guilty of offering fraud can face fines of up to $5 million and imprisonment of up to 20 years.

Corporations can also face significant penalties for securities violations. In addition to fines and disgorgement of profits, corporations can be required to implement compliance programs and internal controls to prevent future violations. In some cases, corporations can also face debarment from certain government contracts.

In addition to legal consequences, securities violations can have significant reputational damage for individuals and corporations. Reputational harm can result in a loss of investor confidence, which can have a significant impact on a company's stock price and ability to raise capital. Reputational harm can also impact an individual's ability to work in the financial industry and can result in a loss of professional licenses.

It is important to note that the consequences of securities violations can extend beyond the immediate legal and reputational consequences. For example, offering fraud can result in significant financial harm to investors, some of whom may be retirees or others who are financially vulnerable. In addition, market manipulation can impact the overall stability of financial markets, which can have far-reaching consequences for the economy as a whole.

In order to avoid securities violations, individuals and corporations must be diligent in their compliance efforts. This can include implementing effective internal controls and compliance programs, conducting regular training on securities laws and regulations, and working with legal counsel to ensure compliance with applicable laws and regulations.

In summary, securities violations can have severe legal, reputational, and economic consequences for individuals and corporations. It is crucial for individuals and corporations to take proactive steps to prevent violations and to work closely with legal counsel to ensure compliance with securities laws and regulations.

CHAPTER 24: CONCLUSION

In this part, we have examined the legal requirements for securities disclosure and the potential liabilities that can arise when those requirements are not met. We have explored the various types of securities violations, including insider trading, market manipulation, offering fraud, and Ponzi schemes. In addition, we have discussed the consequences of violating these laws, which can range from civil penalties to criminal charges.

One of the key takeaways from this part is the importance of transparency and disclosure in the securities markets. Investors rely on accurate and timely information to make informed decisions about their investments. When companies fail to disclose material information or engage in fraudulent practices, investors are at risk of significant financial losses.

Another important takeaway is the role of regulatory bodies, such as the Securities and Exchange Commission (SEC), in enforcing securities laws and protecting investors. These agencies play a critical role in monitoring market activity, investigating potential violations, and enforcing the rules when necessary.

It is also important to note that the securities laws are constantly evolving, as new technologies and market practices emerge. This means that companies and individuals must stay up-to-date on the latest legal requirements and best practices in order to avoid potential liabilities.

Overall, the legal requirements and liabilities related to securities disclosure are complex and multifaceted. It is essential for investors, business owners, and legal professionals to have a thorough understanding of these laws in order to navigate the securities markets with confidence and integrity.

The importance of securities regulation in maintaining fair and efficient markets

Securities regulation is essential for the proper functioning of financial markets. It aims to ensure that investors are protected from fraudulent and unfair practices and that the markets operate in a fair and efficient manner. Without securities regulation, investors would be vulnerable to various forms of misconduct, such as insider trading, market manipulation, and fraudulent investment schemes. Moreover, a lack of confidence in the markets can lead to reduced investment, lower economic growth, and reduced public welfare. Therefore, in this section, we will discuss the importance of securities regulation in maintaining fair and efficient markets.

Ensuring Fair Markets:

One of the primary objectives of securities regulation is to ensure that markets are fair. This means that all investors have access to the same information, and all market participants follow the same rules. Securities regulation achieves this goal in several ways.

First, securities regulation requires companies to provide timely and accurate information to the public. This information includes financial statements, disclosures about significant events, and other information that may be relevant to investors. By ensuring that companies provide this information, securities regulation helps to ensure that all investors have access to the same information and can make informed decisions about their investments.

Second, securities regulation prohibits insider trading. Insider trading occurs when individuals with access to non-public information about a company use that information to make trades that benefit themselves. Insider trading is illegal because it gives an unfair advantage to those with access to non-public information. Securities regulation helps to prevent insider trading by requiring companies to have strict policies and procedures to prevent the misuse of non-public information.

Third, securities regulation requires all market participants to follow the same rules. This includes rules regarding the registration of securities offerings, the reporting of trades, and the disclosure of ownership. By ensuring that all market participants follow the same rules, securities regulation helps to create a level playing field for all investors.

Ensuring Efficient Markets:

Another important objective of securities regulation is to ensure that markets are efficient. Efficient markets are those in which prices reflect all available information. This means that prices accurately reflect the true value of securities, and there is no opportunity for investors to profit from mispricings.

Securities regulation helps to ensure efficient markets in several ways. First, securities regulation requires companies to provide timely and accurate information to the public. This information allows investors to make informed decisions about their investments, which helps to ensure that prices accurately reflect the true value of securities.

Second, securities regulation requires all market participants to follow the same rules. This includes rules regarding the reporting of trades and the disclosure of ownership. By ensuring that all market participants follow the same rules, securities regulation helps to prevent market manipulation and other practices that could distort prices.

Finally, securities regulation encourages competition among market participants. This competition helps to ensure that prices accurately reflect the true value of securities. Competition also helps to ensure that investors have access to a range of investment options, which allows them to diversify their portfolios and manage risk.

Conclusion:

In conclusion, securities regulation is essential for maintaining fair and efficient markets. Securities regulation helps to ensure that markets are fair by requiring companies to provide timely and accurate information, prohibiting insider trading, and ensuring that all market participants follow the same rules. Securities regulation also helps to ensure that markets are efficient by requiring companies to provide timely and accurate information, requiring all market participants to follow the same rules, and encouraging competition among market participants. By maintaining fair and efficient markets, securities regulation helps to protect investors and promote economic growth.

Future developments in securities law and regulation

Securities law and regulation play a critical role in maintaining fair and efficient financial markets. As technology continues to evolve, it is important to consider how these changes may impact the future of securities regulation. One area that has received a significant amount of attention in recent years is the emergence of cryptocurrencies and their impact on securities law.

This section will provide a detailed analysis of the future developments in securities law and regulation, specifically how cryptocurrencies are changing the Securities and Exchange Commission (SEC). We will examine the current state of regulation of cryptocurrencies and how the SEC has responded to the emergence of these digital assets. We will also explore potential future developments in securities law and regulation, including the potential impact of blockchain technology on the securities industry.

Current State of Regulation of Cryptocurrencies

The emergence of cryptocurrencies has presented a unique challenge to securities regulation. The SEC has struggled to determine how to classify cryptocurrencies, as they do not fit neatly into any existing regulatory framework. In 2017, the SEC issued a report stating that some cryptocurrencies may be considered securities under federal securities laws, while others may not. The SEC also warned that some initial coin offerings (ICOs) may be subject to securities regulation.

The SEC has taken several enforcement actions against ICOs that it believes violated federal securities laws. In 2018, the SEC launched its first enforcement action against an ICO issuer for violating securities laws. The SEC alleged that the ICO issuer had failed to register its tokens as securities and had made false and misleading statements to investors. The SEC has since launched several other enforcement actions against ICO issuers for similar violations.

In addition to enforcement actions, the SEC has also provided guidance to market participants regarding the regulatory treatment of cryptocurrencies. In 2019, the SEC issued guidance on the application of federal securities laws to digital assets. The guidance provided a framework for analyzing whether a digital asset is a security and outlined the factors that the SEC considers when making this determination.

The SEC's response to cryptocurrencies has been somewhat controversial. Critics argue that the SEC's approach is overly broad and that it may stifle innovation in the crypto industry. Proponents, however, argue that the SEC's approach is necessary to protect investors from fraudulent ICOs and other scams.

Future Developments in Securities Law and Regulation

The emergence of cryptocurrencies and blockchain technology has the potential to significantly impact securities law and regulation in the future. One potential development is the use of blockchain technology to facilitate securities transactions. Blockchain technology could potentially streamline the securities trading process, reduce costs, and increase efficiency.

However, the use of blockchain technology also presents new challenges for securities regulation. Blockchain transactions are typically anonymous and decentralized, making it difficult for regulators to monitor and enforce compliance. Regulators will need to develop new tools and techniques to monitor blockchain transactions and ensure compliance with securities laws.

Another potential development is the use of cryptocurrencies as a form of payment for securities transactions. This could potentially reduce the role of traditional financial intermediaries, such as banks and broker-dealers, in the securities trading process. However, it could also present new challenges for securities regulation, particularly with respect to money laundering and other illicit activities.

The SEC will need to continue to adapt its regulatory framework to address these new developments. This may include updating existing regulations or developing new regulations specifically tailored to cryptocurrencies and blockchain technology. The SEC will also need to continue to work closely with other regulators, both in the US and abroad, to ensure consistent and effective regulation of these emerging technologies.

Conclusion

The emergence of cryptocurrencies and blockchain technology has presented new challenges for securities regulation. The SEC has taken a cautious approach to these new developments, issuing guidance and taking enforcement actions against ICO issuers who violate federal securities laws. However, the SEC will need to continue to adapt its regulatory framework to address the unique characteristics of cryptocurrencies and the blockchain.

One of the biggest challenges for regulators is determining whether a particular cryptocurrency or token constitutes a security. The SEC has used the Howey test to determine whether a token is a security, but this test was developed in the context of traditional securities and may not be well-suited for cryptocurrencies. Some have argued that a new test specifically tailored to cryptocurrencies is needed.

Another challenge is enforcing securities laws against bad actors who operate outside the traditional financial system. Cryptocurrencies and the blockchain allow for anonymous transactions, which can make it difficult for regulators to identify and prosecute securities violators. Additionally, the global nature of the cryptocurrency market means that regulatory actions taken in one jurisdiction may not have an impact on bad actors operating in another jurisdiction.

Despite these challenges, there are opportunities for securities regulators to leverage blockchain technology to improve the efficiency and transparency of the securities markets. For example, the use of blockchain-based smart contracts could streamline the process of issuing and trading securities, reducing transaction costs and increasing transparency. Regulators could also use blockchain technology to more effectively monitor and enforce securities laws.

As the use of cryptocurrencies and blockchain technology continues to grow, it is likely that the SEC and other securities regulators will need to continue to adapt their regulatory frameworks to address new developments and challenges. This will require a careful balance between promoting innovation and protecting investors from fraud and other securities violations.

In conclusion, securities regulation plays a crucial role in maintaining fair and efficient markets. The securities laws and regulations that have been developed over the past century have helped to promote transparency, deter fraud, and protect investors. However, securities regulation is an ever-evolving field, and regulators must continue to adapt to new developments and challenges in order to effectively fulfill their mandate. The emergence of cryptocurrencies and blockchain technology represents one such challenge, and it will be interesting to see how securities regulators around the world address this new frontier in the years to come.

Recommendations for businesses and investors

Businesses and investors should be aware of the regulatory framework surrounding securities and understand the potential liabilities for noncompliance. Below are some recommendations for businesses and investors to ensure compliance with securities regulations:

Conduct Due Diligence: Businesses and investors should conduct thorough due diligence before investing in any security to ensure compliance with securities regulations. This includes reviewing disclosure documents, financial statements, and any relevant legal documents.

Consult with Legal Counsel: Businesses and investors should consult with legal counsel to understand the legal requirements for securities offerings and transactions. Legal counsel can provide guidance on compliance with securities regulations and help structure transactions to minimize legal risks.

Implement Compliance Programs: Businesses should implement compliance programs to ensure compliance with securities regulations. This includes training employees on securities regulations, maintaining accurate records, and conducting internal audits.

Stay Informed: Businesses and investors should stay informed about changes to securities regulations and industry trends. This includes regularly reviewing SEC filings, attending industry conferences, and consulting with legal counsel.

Utilize Technology: Businesses can utilize technology to streamline compliance with securities regulations. For example, blockchain technology can be used to automate compliance with securities regulations by ensuring accurate record-keeping and facilitating transactions.

Monitor for Fraud: Businesses and investors should be vigilant for signs of fraud in securities offerings and transactions. This includes reviewing financial statements, conducting background checks on issuers, and reporting suspicious activity to regulatory authorities.

Consider Alternative Investments: Investors should consider alternative investments, such as private equity or venture capital, as a way to diversify their portfolios and potentially reduce exposure to securities regulation.

Conclusion

Securities regulation plays a critical role in maintaining fair and efficient markets. The regulatory framework surrounding securities has evolved over time to address new developments and market trends. The SEC has taken a proactive approach to enforcement of securities regulations, imposing significant penalties on violators. The emergence of cryptocurrencies and blockchain technology has presented new challenges for securities regulation, and the SEC will need to continue to adapt its regulatory approach to address these developments.

Businesses and investors must be aware of the regulatory framework surrounding securities and understand the potential liabilities for noncompliance. Conducting due diligence, consulting with legal counsel, implementing compliance programs, staying informed, utilizing technology, monitoring for fraud, and considering alternative investments can help ensure compliance with securities regulations and minimize legal risks.

As securities regulation continues to evolve, it is essential for businesses and investors to stay informed and adapt to these changes to maintain compliance and ensure the integrity of the markets.

PART 7: PUBLIC OFFERING PROCESS AND REQUIREMENTS

A public offering is a process through which a company raises capital by selling shares of its stock to the public. This process can be a critical moment in the life of a company, providing it with the capital necessary to fund its growth and expansion. However, the public offering process can also be complex and challenging, requiring a thorough understanding of securities laws and regulations.

In this section, we will explore the public offering process and the legal requirements that companies must comply with when conducting a public offering. We will discuss the role of the Securities and Exchange Commission (SEC) in regulating public offerings and review the disclosure requirements that companies must meet to ensure that investors have access to the information necessary to make informed investment decisions.

We will also examine the various steps involved in the public offering process, from the initial decision to go public to the pricing of the offering and the allocation of shares to investors. We will look at the role of underwriters and other intermediaries in the process and consider the legal and ethical issues that arise in connection with their activities.

Finally, we will provide recommendations for businesses and investors that are considering a public offering, highlighting key considerations and best practices that can help to ensure a successful and compliant offering. Through a comprehensive analysis of the public offering process and requirements, this section aims to provide students with a solid foundation for understanding this critical aspect of securities regulation.

CHAPTER 25: PRE-IPO PLANNING AND PREPARATION

Chapter 25 of this securities law and regulation textbook discusses the process of pre-IPO planning and preparation for companies seeking to go public. Initial Public Offering (IPO) is a complex process that involves several steps, from choosing underwriters to drafting offering documents and obtaining regulatory approval. The pre-IPO planning and preparation stage is crucial for companies seeking to go public, as it sets the stage for a successful IPO.

This chapter will provide a comprehensive overview of the pre-IPO planning and preparation process, including the steps involved in the process, the role of underwriters, and the legal and regulatory requirements that companies must comply with. The chapter will also discuss the benefits and risks associated with going public, and provide guidance to companies on how to prepare for a successful IPO.

The chapter is organized as follows: Section 1 provides an overview of the IPO process and the role of pre-IPO planning and preparation. Section 2 discusses the benefits and risks of going public. Section 3 provides guidance on how to prepare for a successful IPO, including the role of underwriters and the legal and regulatory requirements that must be met. Finally, Section 4 concludes the chapter by summarizing the key points and providing recommendations for companies considering an IPO.

Overall, this chapter provides a valuable resource for companies seeking to go public, as well as for professionals in the securities industry, including investment bankers, attorneys, and accountants. The information presented in this chapter is based on current securities law and regulation, and reflects the best practices and industry standards for pre-IPO planning and preparation.

Overview of the pre-IPO process

The decision to go public is a major milestone for any company. The pre-IPO process is a crucial period during which a company prepares for its initial public offering (IPO). This period can be challenging, as it involves numerous legal, financial, and regulatory considerations. In this section, we will provide an overview of the pre-IPO process, highlighting the key steps that a company should take to prepare for its IPO.

Overview of the Pre-IPO Process:

Evaluating readiness for an IPO
The first step in the pre-IPO process is to evaluate whether a company is ready to go public. This involves a thorough assessment of the company's financial and operational performance, market position, and growth prospects. Companies need to have a compelling business case for going public and should have a solid track record of financial performance, growth, and profitability.

Assembling a team of advisors
Once a company decides to go public, it needs to assemble a team of advisors to guide it through the pre-IPO process. This team typically includes investment bankers, lawyers, accountants, and other financial and legal professionals. These advisors help the company navigate the complex regulatory and legal requirements of the pre-IPO process and ensure that it complies with all applicable securities laws and regulations.

Conducting a comprehensive due diligence review
Before a company can go public, it needs to conduct a comprehensive due diligence review. This involves a detailed examination of the company's financial statements, operations, corporate structure, legal and regulatory compliance, and other key aspects of its business. The due diligence process is typically conducted by the company's advisors, who work closely with the company's management team to identify any potential issues that could impact the IPO.

Developing a strong business plan and financial model
As part of the pre-IPO process, companies need to develop a strong business plan and financial model that outlines their growth strategy, competitive advantages, and key metrics. This plan should also include detailed financial projections that provide investors with a clear picture of the company's financial performance and growth prospects.

Preparing the registration statement and prospectus
The registration statement and prospectus are key documents that companies need to prepare as part of the pre-IPO process. The registration statement is filed with the Securities and Exchange Commission (SEC) and contains detailed information about the company's business, financial performance, and risks. The

prospectus is a summary of the registration statement that is provided to investors and contains key information about the offering, including the price and size of the offering, the use of proceeds, and the risks associated with investing in the company.

Conducting a roadshow and pricing the IPO

The final step in the pre-IPO process is to conduct a roadshow and price the IPO. The roadshow is a series of presentations that the company's management team makes to potential investors, highlighting the company's business, growth prospects, and financial performance. The pricing of the IPO is based on the demand for the company's shares, which is determined during the roadshow. The company's investment bankers work with the company's management team to determine the optimal price for the offering, based on the demand for the shares and the company's financial performance.

Conclusion:

The pre-IPO process is a complex and challenging period for companies that are preparing to go public. It involves numerous legal, financial, and regulatory considerations, and requires careful planning and preparation. Companies that successfully navigate the pre-IPO process are well positioned to achieve their growth objectives and create long-term value for their shareholders.

Conducting financial and legal due diligence

In preparation for an initial public offering (IPO), companies need to conduct thorough financial and legal due diligence. The process of due diligence is an essential step to ensure that the company is adequately prepared for the public offering process. Financial and legal due diligence involves the examination of the company's financial and legal records, contracts, policies, and procedures to ensure that there are no material issues or liabilities that could negatively impact the IPO. This section will provide an in-depth analysis of the financial and legal due diligence process, including its purpose, the key elements involved, and how it can be conducted effectively.

Purpose of Financial and Legal Due Diligence:

The primary purpose of conducting financial and legal due diligence is to provide a comprehensive understanding of the company's financial and legal status. Due diligence is designed to identify any potential issues or liabilities that could impact the IPO. The process aims to ensure that the information presented to investors is accurate, complete, and free from material errors or omissions. Additionally, due diligence helps to identify potential areas of risk, such as regulatory compliance issues, litigation, or outstanding liabilities. By identifying and addressing these issues early, companies can mitigate their impact on the IPO and increase their chances of a successful offering.

Key Elements of Financial and Legal Due Diligence:

Financial and legal due diligence involves a comprehensive examination of a company's financial and legal records, contracts, policies, and procedures. The following are the key elements of financial and legal due diligence:

✧ Financial Records: The examination of financial records includes a review of the company's financial statements, tax records, revenue and expense records, and any other financial records that may be relevant to the IPO.

✧ Legal Records: Legal records include a review of the company's contracts, licenses, permits, intellectual property rights, litigation records, regulatory filings, and any other legal records that may be relevant to the IPO.

✧ Policies and Procedures: The examination of policies and procedures involves a review of the company's internal controls, compliance policies, risk management policies, and any other policies or procedures that may be relevant to the IPO.

✧ Management and Organizational Structure: The examination of management and organizational structure includes a review of the company's management team, board of directors, and organizational structure to ensure that they are appropriate for the IPO.

✧ Industry and Market Research: Industry and market research involve an examination of the company's industry and market to identify potential risks, opportunities, and trends that may impact the IPO.

Effective Conduct of Financial and Legal Due Diligence:

Effective financial and legal due diligence is critical for a successful IPO. The following are some key steps that companies can take to ensure that they conduct effective due diligence:

✧ Identify Key Risks and Liabilities: Companies should identify and prioritize potential risks and liabilities that could impact the IPO. This will help to focus the due diligence process and ensure that it is comprehensive.

✧   Assemble a Team of Experts: Due diligence involves complex financial and legal analysis. Companies should assemble a team of experts, including financial analysts, lawyers, and accountants, to ensure that the due diligence process is thorough and effective.

✧   Develop a Comprehensive Due Diligence Checklist: Companies should develop a comprehensive due diligence checklist that covers all key elements of financial and legal due diligence. The checklist should be reviewed and updated regularly to ensure that it is up-to-date and comprehensive.

✧   Conduct Thorough Investigations: Companies should conduct thorough investigations of all relevant financial and legal records, contracts, policies, and procedures. They should also conduct interviews with key personnel to gather additional information.

✧   Address Identified Issues: If any issues or liabilities are identified during the due diligence process, companies should take appropriate steps to address them. This may involve implementing new policies or procedures, renegotiating contracts, or resolving outstanding disputes.

✧   Evaluate Potential Business Risks: Companies should evaluate potential business risks that could impact the IPO. This includes evaluating the competitive landscape, identifying potential regulatory issues, and analyzing market trends.

✧   Perform Financial Analysis: Companies should perform a thorough financial analysis to ensure that they are prepared for the IPO. This includes analyzing financial statements, projections, and cash flow analysis. Companies should also evaluate their valuation to ensure that it is reasonable and defensible.

✧   Assess Legal and Regulatory Compliance: Companies should assess their legal and regulatory compliance to ensure that they are in compliance with all relevant laws and regulations. This includes evaluating potential litigation and regulatory risks, as well as reviewing compliance with securities laws, tax laws, and environmental regulations.

✧   Review Corporate Governance: Companies should review their corporate governance practices to ensure that they are in line with best practices. This includes evaluating board composition, management structure, and internal controls.

✧ Engage Outside Advisors: Companies should consider engaging outside advisors, such as investment bankers, auditors, and legal counsel, to assist with the due diligence process. These advisors can provide valuable insights and expertise that can help to ensure that the IPO is successful.

In conclusion, conducting effective financial and legal due diligence is critical for a successful IPO. Companies must identify key risks and liabilities, assemble a team of experts, develop a comprehensive due diligence checklist, conduct thorough investigations, address identified issues, evaluate potential business risks, perform financial analysis, assess legal and regulatory compliance, review corporate governance, and engage outside advisors. By following these steps, companies can help to ensure that they are well-prepared for the IPO and that they are able to maximize value for their shareholders.

Developing an IPO timeline

Going public is a significant undertaking that requires careful planning and execution. One of the most critical components of the IPO process is developing a timeline that outlines the key milestones and deadlines that must be met for a successful offering.

The following is a detailed guide to developing an IPO timeline, including key considerations and best practices to ensure a smooth and successful IPO.

Determine the Optimal Timing
The timing of an IPO is critical, and companies should carefully consider market conditions, industry trends, and other external factors before moving forward. For example, companies may want to consider going public during a market upswing or when investor sentiment is positive.

Additionally, companies should consider the length of the IPO process, which can take several months or even years to complete. Therefore, they should set realistic timelines that take into account the time required for each stage of the process.

Identify Key Milestones
Once the timing has been determined, the next step is to identify the key milestones and deadlines that must be met for a successful IPO. These milestones will vary depending on the company's specific circumstances, but some common milestones include:

✧ Developing a prospectus
✧ Completing due diligence

✧ Drafting the registration statement
✧ Filing the registration statement with the SEC
✧ Conducting a roadshow
✧ Pricing the shares
✧ Closing the offering
✧ Establish a Detailed Timeline

Once the key milestones have been identified, the next step is to establish a detailed timeline that outlines the specific tasks and deadlines associated with each milestone. A detailed timeline will help the company stay on track and ensure that all required tasks are completed on time.

It is essential to assign responsibility for each task to a specific person or team and establish clear communication channels to ensure that everyone is aware of their responsibilities and deadlines.

Build in Flexibility
While it is essential to have a detailed timeline, it is also important to build in flexibility to accommodate unforeseen events or delays. The IPO process can be unpredictable, and companies should be prepared to adjust their timelines as necessary to ensure a successful offering.

Monitor Progress
Finally, companies should monitor progress regularly to ensure that they are meeting their deadlines and milestones. Monitoring progress can help identify potential issues early on and allow for timely corrective action.

Example - Proposed IPO timeline for a new company:

18-24 Months Before the Planned IPO:

Conduct a feasibility study to assess the potential success of an IPO for the company.
Evaluate and strengthen financial reporting systems and controls.
Review and update corporate governance policies and procedures.
Identify and address any potential legal or regulatory issues that could impact the IPO.
Assemble a team of experts, including investment bankers, lawyers, and accountants, to advise on the IPO process.
Develop a comprehensive business plan and financial projections for the company.

12-18 Months Before the Planned IPO:

Hire an IPO underwriter and legal counsel to advise on the IPO process.
Conduct financial and legal due diligence to identify any potential risks or liabilities.
Begin the process of selecting an exchange for the IPO.
Prepare the company's financial statements in accordance with SEC regulations.
Develop an investor relations plan to engage with potential investors and build awareness of the company.

6-12 Months Before the Planned IPO:

File a draft registration statement with the SEC and begin the review process.
Develop a roadshow presentation and schedule meetings with potential investors.
Finalize the selection of the exchange for the IPO.
Conduct any necessary pre-IPO financing rounds to ensure the company has sufficient capital to support its growth.

3-6 Months Before the Planned IPO:

File an amended registration statement with the SEC and respond to any comments or feedback.
Finalize the pricing and size of the IPO.
Conduct the roadshow to market the IPO to potential investors.
Obtain final approvals from the exchange and SEC.
1 Month Before the Planned IPO:

Finalize all legal and financial documentation.
Ensure that all regulatory requirements have been met.
Prepare for the IPO listing day.
Listing Day:

The company's shares are listed on the chosen exchange and trading begins.
The company and its underwriters monitor trading activity and respond to any issues or concerns.
The above timeline is only a general guide, and the actual IPO timeline may vary depending on the specific circumstances of the company and the market conditions at the time of the IPO.

Conclusion

Developing an IPO timeline is a critical component of a successful offering. By carefully considering market conditions, identifying key milestones, establishing a detailed timeline, building in flexibility, and monitoring progress, companies can ensure a smooth and successful IPO process.

Establishing a corporate governance structure

Establishing a strong corporate governance structure is a critical aspect of any company's success. Corporate governance refers to the system of rules, practices, and processes by which a company is directed and controlled. It provides the framework for attaining a company's objectives, while taking into account the interests of its shareholders, employees, customers, and other stakeholders. This section will provide an overview of key elements of a strong corporate governance structure.

Board of Directors:
One of the most important components of a company's corporate governance structure is its board of directors. The board is responsible for overseeing the management team, setting strategic objectives, and ensuring the company operates in compliance with applicable laws and regulations. It also ensures that the company is run in a way that is accountable and transparent to shareholders. The board should be composed of experienced and knowledgeable individuals with diverse backgrounds and expertise, who can bring different perspectives and insights to the table. Board members should be independent of management and have a fiduciary duty to act in the best interests of the company and its shareholders.

The following are proposed by-laws that would be in place for a company's Board of Directors:

✦   Size of the Board: The Board of Directors shall consist of not less than five (5) nor more than fifteen (15) members, as determined from time to time by the Board of Directors.

✦   Election of Directors: The directors of the corporation shall be elected by the shareholders at the annual meeting of shareholders.

✦   Term of Office: Each director shall hold office until the next annual meeting of shareholders or until his or her successor is duly elected and qualified, or until his or her earlier death, resignation or removal.

✧   Removal of Directors: Any director or the entire Board of Directors may be removed at any time, with or without cause, by a vote of the shareholders.

✧   Vacancies: Any vacancy occurring in the Board of Directors may be filled by a majority vote of the remaining directors, though less than a quorum, or by a sole remaining director. A director elected to fill a vacancy shall be elected for the unexpired term of his or her predecessor in office.

✧   Meetings: The Board of Directors shall hold regular meetings at least once every quarter, at such times and places as the Board of Directors may determine. Special meetings of the Board of Directors may be called by the Chairman of the Board, the Chief Executive Officer or any two directors.

✧   Quorum: A majority of the members of the Board of Directors shall constitute a quorum for the transaction of business at any meeting of the Board of Directors.

✧   Committees: The Board of Directors may, by resolution passed by a majority of the whole Board, designate one or more committees, each consisting of two or more directors, to serve at the pleasure of the Board. Such committee or committees shall have and may exercise such powers and authority of the Board of Directors in the management of the business and affairs of the corporation as may be delegated to them by the Board of Directors.

✧   Compensation: Directors shall receive such compensation for their services as may be fixed from time to time by resolution of the Board of Directors, provided that nothing herein contained shall be construed to preclude any director from serving the corporation in any other capacity and receiving compensation therefor.

✧   Standard of Care: Directors shall perform their duties as directors, including their duties as members of any committee of the Board upon which they may serve, in good faith, with the care that an ordinarily prudent person in a like position would exercise under similar circumstances, and in a manner the director reasonably believes to be in the best interests of the corporation.

✧   Conflicts of Interest: No director shall participate in the consideration or determination of any matter in which he or she has a direct or indirect personal interest that may reasonably be expected to impair his or her objectivity or independence of judgment.

✧ Indemnification: The corporation shall indemnify each person who is or was a director, officer, employee or agent of the corporation and who was or is a party or is threatened to be made a party to any threatened, pending or completed action, suit or proceeding, whether civil, criminal, administrative or investigative, by reason of the fact that he or she is or was a director, officer, employee or agent of the corporation or is or was serving at the request of the corporation as a director, officer, employee or agent of another corporation, partnership, joint venture, trust or other enterprise, against expenses, including attorneys' fees, judgments, fines and amounts paid in settlement actually and reasonably incurred by him or her in connection with such action, suit or proceeding, if he or she acted in good faith and in a manner he or she reasonably believed to be in or not opposed to the best interests of the corporation, and, with respect to any criminal action or proceeding, had no reasonable cause to believe his or her conduct was unlawful.

The indemnification provided for in this section shall not be deemed exclusive of any other rights to which those seeking indemnification may be entitled under any bylaw, agreement, vote of shareholders or disinterested directors, or otherwise. The corporation may purchase and maintain insurance on behalf of any person who is or was a director, officer, employee or agent of the corporation or is or was serving at the request of the corporation as a director, officer, employee or agent of another corporation, partnership, joint venture, trust or other enterprise against any liability asserted against him or her and incurred by him or her in any such capacity, or arising out of his or her status as such, whether or not the corporation would have the power to indemnify him or her against such liability under the provisions of this section.

The corporation shall also have the power to indemnify any person who is or was a party, or is threatened to be made a party, to any action or proceeding, other than one by or in the right of the corporation to procure a judgment in its favor, by reason of the fact that he or she is or was an employee, agent or fiduciary of the corporation or any subsidiary thereof, against expenses, including attorneys' fees, actually and reasonably incurred by him or her in connection with the defense or settlement of such action or proceeding, if he or she acted in good faith and in a manner he or she reasonably believed to be in or not opposed to the best interests of the corporation, except that no indemnification shall be made in respect of any claim, issue or matter as to which such person shall have been adjudged to be liable for negligence or misconduct in the performance of his or her duty to the corporation or its shareholders, unless and only to the extent that the court in which such action or suit was brought shall determine upon application that, despite the adjudication of liability but in view of all the circumstances of the case, such person is fairly and reasonably entitled to indemnity for such expenses as the court shall deem proper.

The indemnification provided for in this section shall be deemed to be a contract between the corporation and each director, officer, employee or agent of the corporation or other person entitled to indemnification hereunder. The rights conferred on any person by this section shall not be exclusive of any other rights that such person may have or hereafter acquire under any statute, provision of the corporation's Articles of Incorporation, Bylaws, agreement, vote of shareholders or disinterested directors, or otherwise. Any repeal or modification of this section shall not adversely affect any right or protection of a director, officer, employee, agent or other person existing at the time of such repeal or modification.

Overall, these proposed by-laws provide a robust indemnification policy for the corporation's directors, officers, employees and agents, as well as other parties who may become involved in legal proceedings related to their service for the corporation. This policy helps to ensure that those individuals who act in good faith and in the best interests of the corporation are protected from liability and potential financial loss. It also demonstrates the corporation's commitment to attracting and retaining top talent by offering such protection.

Executive Management:

Another key component of a company's corporate governance structure is its executive management team. The executive team is responsible for managing the day-to-day operations of the company and executing the strategic objectives set by the board of directors. The executive team should be composed of experienced and qualified individuals who possess the skills and expertise necessary to run the company effectively. It is important that the executive team operates in a transparent and accountable manner and is aligned with the interests of the company's shareholders.

Proposed by-laws for executive management of a company.

- ✧ Article I: Duties and Responsibilities

- ✧ Section 1. General Duties and Responsibilities

The executive management of the corporation shall consist of the Chief Executive Officer, Chief Financial Officer, Chief Operating Officer, and other officers as may be appointed by the Board of Directors. The executive management shall be responsible for the day-to-day management and operation of the corporation in accordance with the policies and directives established by the Board of Directors.

- ✧ Section 2. Specific Duties and Responsibilities

(a) Chief Executive Officer. The Chief Executive Officer shall be the principal executive officer of the corporation and shall be responsible for the overall management and direction of the corporation, including the formulation and implementation of corporate policies and strategies, and the hiring, termination, and supervision of all employees.

(b) Chief Financial Officer. The Chief Financial Officer shall be responsible for the financial management of the corporation, including the preparation of financial statements, the management of the corporation's financial resources, and the development and implementation of financial policies and procedures.

(c) Chief Operating Officer. The Chief Operating Officer shall be responsible for the day-to-day operations of the corporation, including the management of all production, distribution, and marketing activities.

✧　Article II: Compensation and Benefits

✧　Section 1. Compensation

The compensation of the executive management shall be determined by the Board of Directors and shall be reasonable and commensurate with the services performed by each officer. The compensation of the executive management shall be reviewed annually by the Board of Directors.

✧　Section 2. Benefits

The executive management shall be entitled to participate in all employee benefit plans and programs maintained by the corporation, including, but not limited to, health insurance, life insurance, and retirement plans.

✧　Article III: Removal

✧　Section 1. Removal of Executive Management

Any member of the executive management may be removed by the Board of Directors with or without cause.

✧　Section 2. Vacancies

In the event of a vacancy in any executive management position, the Board of Directors may appoint a replacement officer to serve until the next regular meeting of the shareholders, at which time a successor shall be elected.

✧　Article IV: Indemnification

✧   Section 1. Indemnification

The corporation shall indemnify each person who is or was a director, officer, employee or agent of the corporation and who was or is a party or is threatened to be made a party to any threatened, pending or completed action, suit or proceeding, whether civil, criminal, administrative or investigative, by reason of the fact that he or she is or was a director, officer, employee or agent of the corporation or is or was serving at the request of the corporation as a director, officer, employee or agent of another corporation, partnership, joint venture, trust or other enterprise, against expenses, including attorneys' fees, judgments, fines and amounts paid in settlement actually and reasonably incurred by him or her in connection with such action, suit or proceeding, if he or she acted in good faith and in a manner he or she reasonably believed to be in or not opposed to the best interests of the corporation, and, with respect to any criminal action or proceeding, had no reasonable cause to believe his or her conduct was unlawful.

✧   Section 2. Advances for Expenses

Expenses incurred in defending any action, suit or proceeding may be advanced by the corporation prior to the final disposition of such action, suit or proceeding upon receipt of an undertaking by or on behalf of the director, officer, employee or agent to repay such amount if it shall ultimately be determined that he or she is not entitled to be indemnified by the corporation as authorized in Section 1 of this Article IV.

Audit Committee:

An effective corporate governance structure also includes an audit committee. The audit committee is responsible for overseeing the company's financial reporting process and ensuring that financial statements are accurate and comply with applicable accounting standards. The audit committee also oversees the company's internal controls, risk management, and compliance functions. The committee should be composed of independent directors who possess financial expertise and are familiar with the company's industry and operations.

Proposed by-laws for an Audit Committee in a company:

Article X: Audit Committee

✧   Section 1: Purpose and Duties

The Audit Committee (the "Committee") is responsible for overseeing the financial reporting process and the audit of the financial statements of the Company. The Committee shall have the following duties and responsibilities:

(a) Select and appoint the independent auditor of the Company, subject to shareholder ratification, if required by law.

(b) Oversee the work of the independent auditor, including the resolution of disagreements between management and the auditor regarding financial reporting.

(c) Review and approve the scope of the annual audit and the audit plan.

(d) Review and discuss with management and the independent auditor the Company's financial statements, including any significant changes to accounting policies and practices.

(e) Review and discuss with management and the independent auditor the results of the annual audit, including any significant findings or recommendations.

(f) Review and discuss with management and the independent auditor the Company's internal controls over financial reporting.

(g) Establish procedures for the receipt, retention, and treatment of complaints received by the Company regarding accounting, internal accounting controls, or auditing matters, and the confidential and anonymous submission by employees of concerns regarding questionable accounting or auditing matters.

(h) Review and approve any related-party transactions required to be disclosed under applicable securities laws and regulations.

(i) Review and discuss with management and the independent auditor the Company's earnings releases, as well as financial information and earnings guidance provided to analysts and rating agencies.

(j) Prepare the report required by the rules of the Securities and Exchange Commission to be included in the Company's annual proxy statement.

✧   Section 2: Membership

(a) The Committee shall consist of at least three members, each of whom shall be appointed by the Board of Directors.

(b) All members of the Committee shall be independent directors, as defined under applicable securities laws and regulations.

(c) The Chairman of the Committee shall be designated by the Board of Directors.

✧   Section 3: Meetings

(a) The Committee shall meet at least four times per year, or more frequently as circumstances dictate.

(b) The Committee may meet with the independent auditor, internal auditors, and management, as necessary, to carry out its responsibilities.

(c) A majority of the members of the Committee shall constitute a quorum.

(d) The Committee may act by unanimous written consent.

✧ Section 4: Reporting

(a) The Committee shall report to the Board of Directors on a regular basis, summarizing the actions taken by the Committee and any significant issues arising in the course of the Committee's activities.

(b) The Committee shall prepare the report required by the rules of the Securities and Exchange Commission to be included in the Company's annual proxy statement.

✧ Section 5: Resources and Authority

(a) The Committee shall have the resources and authority necessary to discharge its duties and responsibilities, including the authority to engage independent legal, accounting, or other advisors as it determines necessary to carry out its duties.

(b) The Company shall provide appropriate funding, as determined by the Committee, for payment of:

(i) Compensation to any independent auditor engaged for the purpose of preparing or issuing an audit report or performing other audit, review, or attest services for the Company;

(ii) Compensation to any advisors employed by the Committee; and

(iii) Ordinary administrative expenses of the Committee that are necessary or appropriate in carrying out its duties.

Compensation Committee:

Another important component of a company's corporate governance structure is its compensation committee. The compensation committee is responsible for setting the compensation and benefits packages for the company's executive management team. It ensures that executive compensation is aligned with the company's strategic objectives and is fair and equitable to all stakeholders. The committee should be composed of independent directors who possess expertise in executive compensation and are familiar with the company's industry and operations.

The Compensation Committee is responsible for setting and approving the compensation packages of the company's executives and employees. The following proposed by-laws would be in place for the Compensation Committee:

✧ Composition: The Compensation Committee shall consist of at least three members who shall be appointed by the Board of Directors. The members of the Compensation Committee shall be independent directors who are free from any conflict of interest that could interfere with their ability to exercise independent judgment.

✧   Authority: The Compensation Committee shall have the authority to approve and set the compensation, benefits, and incentives for the company's executives and employees. The committee shall review and approve the compensation packages of the CEO and other top executives, including bonuses, stock options, and other benefits. The committee shall also establish performance goals and objectives for executive officers and determine whether they have been met.

✧   Meetings: The Compensation Committee shall meet at least annually to review and approve the compensation packages of the company's executives and employees. Additional meetings may be held as needed. A quorum shall consist of a majority of the members of the committee.

✧   Independent Advisors: The Compensation Committee shall have the authority to retain and terminate independent advisors, including compensation consultants, to assist in the evaluation of executive compensation. The committee shall also have the authority to approve the fees and other terms of engagement of such advisors.

✧   Reporting: The Compensation Committee shall report to the Board of Directors on a regular basis, including providing updates on executive compensation decisions and any changes made to the company's compensation policies and practices.

✧   Records: The Compensation Committee shall maintain records of its meetings and decisions, including minutes and any reports provided to the Board of Directors.

By having a Compensation Committee in place, the company can ensure that executive compensation is set fairly and objectively. This can help attract and retain top talent, while also aligning executive interests with those of the company and its shareholders.

Risk Management and Compliance:
An effective corporate governance structure includes a strong risk management and compliance framework. The company should have policies and procedures in place to identify, assess, and manage risks associated with its operations. It should also have a compliance program in place to ensure that it operates in compliance with applicable laws and regulations. The company should regularly assess its risk

management and compliance programs and make changes as necessary to ensure that they remain effective.

Proposed by-laws that would be in place for Risk Management and Compliance in a company:

Purpose

The purpose of the Risk Management and Compliance Committee (the "Committee") of the Board of Directors (the "Board") is to assist the Board in fulfilling its oversight responsibilities by reviewing and monitoring the Company's risk management and compliance activities, and making recommendations to the Board for improving these activities as necessary.

Composition

The Committee shall consist of at least three directors, each of whom shall meet the independence and experience requirements of the Securities and Exchange Commission and any other applicable regulatory agency. The Board shall appoint one of the members to serve as Chair of the Committee.

Responsibilities

The responsibilities of the Committee shall include, but not be limited to, the following:

1. Reviewing and approving the Company's risk management policies and procedures, and monitoring the effectiveness of these policies and procedures on an ongoing basis.

2. Reviewing and approving the Company's compliance policies and procedures, and monitoring the effectiveness of these policies and procedures on an ongoing basis.

3. Reviewing and assessing the Company's major risk exposures, and ensuring that management has implemented appropriate systems to identify, assess, monitor, and manage these risks.

4. Reviewing and assessing the Company's compliance with applicable laws, regulations, and ethical standards, and ensuring that management has implemented appropriate systems to monitor and manage compliance risks.

5. Reviewing and assessing the effectiveness of the Company's internal controls and risk management systems, including those related to financial

reporting and information technology, and ensuring that management has implemented appropriate systems to address any deficiencies.

6. Reviewing and assessing the Company's cybersecurity risks and controls, and ensuring that management has implemented appropriate systems to address these risks.

7. Reviewing and assessing the Company's insurance coverage, and ensuring that management has implemented appropriate systems to mitigate risk through insurance.

8. Reviewing and assessing the Company's disaster recovery and business continuity plans, and ensuring that management has implemented appropriate systems to address these risks.

9. Reviewing and assessing any material risks associated with the Company's relationships with third-party vendors, service providers, and business partners, and ensuring that management has implemented appropriate systems to manage these risks.

Reviewing and assessing the Company's risk appetite, and ensuring that management has implemented appropriate systems to align risk management and compliance activities with the Company's overall strategic objectives.

Meeting regularly with senior management to discuss risk management and compliance issues, and providing regular reports to the Board on the Committee's activities and recommendations.

Authority

The Committee shall have the authority to:

✧ Retain and terminate outside advisors, including legal counsel, risk management consultants, and other experts, as it deems necessary to carry out its duties and responsibilities.

✧ Request any information or assistance it requires from the Company's employees, including access to any books, records, or documents.

✧ Communicate directly with the Company's independent auditors and other advisors as necessary to carry out its duties and responsibilities.

✧   Recommend to the Board any actions or changes to the Company's risk management and compliance activities as it deems necessary.
Meetings

The Committee shall meet at least four times per year, or more frequently as necessary to carry out its duties and responsibilities. The Committee may meet in person or by telephone or video-conference, and shall keep written minutes of its meetings. The Chair of the Committee shall report regularly to the Board on the Committee's activities and recommendations.

By implementing a Risk Management and Compliance Committee, a multi-media company can establish a framework for identifying and managing risks, promoting compliance with laws and regulations, and aligning risk management and compliance activities with the company's strategic objectives. This can help to protect the company's reputation, enhance its financial performance, and increase stakeholder confidence.

Furthermore, by establishing clear and comprehensive bylaws for the Board of Directors, Executive Management, Audit Committee, Compensation Committee, and Risk Management and Compliance Committee, a multi-media company can ensure that it operates in a transparent and ethical manner, complies with all relevant laws and regulations, and upholds the rights and interests of all stakeholders.

It is important to note that the proposed bylaws presented here are just a starting point, and may need to be customized to meet the specific needs and circumstances of a given multi-media company. It is therefore recommended that any company seeking to establish a governance structure consult with legal and financial experts to ensure that its bylaws are comprehensive, legally sound, and appropriate for its business.

In conclusion, establishing a robust and effective corporate governance structure is critical for the long-term success and sustainability of any multi-media company. By implementing best practices in corporate governance and compliance, companies can protect their stakeholders, enhance their reputation, and achieve their strategic goals.

Conclusion

Establishing a strong corporate governance structure is critical for a company's success. It provides the framework for attaining the company's objectives, while taking into account the interests of its shareholders, employees, customers, and other stakeholders. A strong corporate governance structure includes a board of directors composed of experienced and knowledgeable individuals, an executive management team that operates transparently and is aligned with the interests of shareholders, an audit committee that oversees financial reporting, a compensation committee that ensures executive compensation is aligned with strategic objectives, and a strong risk management and compliance framework. By implementing a strong corporate governance structure, a company can increase its competitiveness, enhance its reputation, and generate long-term value for all stakeholders.

CHAPTER 26: FINANCIAL AND LEGAL DUE DILIGENCE

When a company is preparing to acquire or merge with another company, or when it is preparing for an initial public offering (IPO), it is essential to conduct a thorough due diligence process. Due diligence is the process of examining and evaluating the financial and legal aspects of a company to determine whether it is a suitable investment or acquisition target.

In this chapter, we will discuss the importance of financial and legal due diligence in the context of corporate transactions, and provide a comprehensive guide to the due diligence process. We will cover the key steps involved in conducting due diligence, including identifying the areas of risk, gathering information, and evaluating the information collected.

We will also examine the legal and financial documents that are typically reviewed during the due diligence process, including financial statements, contracts, licenses, permits, and regulatory filings. We will discuss the key issues that are typically examined during the financial and legal due diligence process, including financial performance, regulatory compliance, litigation and disputes, and intellectual property.

In addition, we will explore the role of professionals in the due diligence process, including lawyers, accountants, and financial analysts. We will discuss the qualifications and experience that are required to perform due diligence effectively, as well as the ethical considerations that are involved.

Finally, we will examine some of the common challenges that arise during the due diligence process, including issues related to data privacy and security, cultural differences, and communication barriers. We will also provide some practical tips for managing these challenges and ensuring a successful due diligence process.

Overall, this chapter will provide a comprehensive guide to financial and legal due diligence, which is essential for any company that is preparing to undertake a corporate transaction or an IPO. By following the guidelines and best practices outlined in this chapter, companies can ensure that they are well-prepared for the due diligence process, and that they are able to identify and manage the key risks associated with these transactions.

Key areas of due diligence, including financial statements, contracts, and intellectual property

When conducting a business transaction, due diligence plays a crucial role in identifying potential risks and opportunities. Due diligence is an investigation and analysis process that aims to provide a comprehensive understanding of the target company's financial, legal, and operational standing. This process is essential in mergers and acquisitions, joint ventures, and other business transactions. This chapter will explore the key areas of due diligence, including financial statements, contracts, and intellectual property.

Financial Statements:

One of the most crucial aspects of financial due diligence is the analysis of financial statements. Financial statements provide an overview of the company's financial performance and position, including its revenue, expenses, assets, and liabilities. Financial due diligence involves a detailed examination of these statements, including the balance sheet, income statement, and cash flow statement.

When reviewing financial statements, the due diligence team should examine the quality of the financial reporting and the accuracy of the numbers presented. The team should also assess the company's financial position by analyzing its liquidity, solvency, and profitability. Additionally, the team should consider the impact of any potential liabilities, such as pending lawsuits, regulatory fines, or tax obligations, on the company's financial position.

Contracts:

Another critical area of due diligence is the review of contracts. Contracts are legally binding agreements that dictate the terms of a business relationship, such as employment agreements, vendor contracts, and customer agreements. Due diligence involves a thorough analysis of these contracts to identify any potential legal or financial risks.

When reviewing contracts, the due diligence team should examine the contract terms, including the scope of work, payment terms, and termination clauses. The team should also review any warranties, guarantees, or indemnification clauses included in the contract. Additionally, the team should identify any potential breach of contract risks, such as non-payment, failure to deliver goods or services, or failure to meet contractual obligations.

Intellectual Property:

Intellectual property (IP) is a crucial asset for many companies, particularly in technology, media, and entertainment industries. IP includes patents, trademarks, copyrights, trade secrets, and other intangible assets that provide the company with a competitive advantage. Due diligence involves a thorough analysis of the company's IP portfolio to identify any potential legal or financial risks.

When reviewing the company's IP portfolio, the due diligence team should examine the ownership and validity of the IP assets. The team should also assess the company's IP protection strategies, including any pending patent applications or trademark registrations. Additionally, the team should identify any potential infringement risks, such as the use of unlicensed third-party IP.

Conclusion:

In conclusion, due diligence is a critical process in any business transaction. Financial statements, contracts, and intellectual property are key areas of due diligence that require a thorough examination to identify potential risks and opportunities. Financial due diligence involves a detailed analysis of the company's financial statements to assess its financial position and potential liabilities. Contract review involves the examination of the terms of the agreements to identify any potential legal or financial risks. Intellectual property due diligence involves a thorough analysis of the company's IP portfolio to identify any potential infringement risks or opportunities. By conducting comprehensive due diligence, companies can minimize risks and maximize the value of their investments.

Working with auditors and legal counsel

Working with auditors and legal counsel is an essential part of conducting financial and legal due diligence in any business transaction. In this section, we will explore the role of auditors and legal counsel in due diligence, as well as best practices for working with these professionals to ensure a successful due diligence process.

Role of Auditors in Due Diligence

Auditors play a crucial role in financial due diligence by reviewing a company's financial statements and records to ensure that they are accurate, complete, and in compliance with accounting standards. Auditors also examine a company's internal controls and financial reporting processes to identify any weaknesses or areas of potential risk.

During the due diligence process, auditors will typically request access to a company's financial records, including balance sheets, income statements, cash flow statements, and other relevant financial data. They will also conduct interviews with key personnel, review internal controls and risk management processes, and perform tests to ensure the accuracy and completeness of financial information.

Working with auditors during due diligence requires a high level of collaboration and communication between the auditors and the company's management team. The company's management team should be prepared to provide the auditors with access to all relevant financial records and information, and to answer any questions that the auditors may have about the company's financial position and reporting processes.

It is important for the company's management team to be transparent and forthcoming with the auditors, as any attempts to withhold or misrepresent information can result in serious legal and financial consequences.

Role of Legal Counsel in Due Diligence

Legal counsel plays a critical role in legal due diligence by reviewing a company's contracts, agreements, and other legal documents to ensure that they are accurate, complete, and in compliance with relevant laws and regulations. Legal counsel also identifies potential legal risks and liabilities associated with the company's operations and advises on strategies to mitigate those risks.

During the due diligence process, legal counsel will typically request access to a company's legal documents and agreements, including employment contracts, supplier agreements, customer contracts, and other relevant legal documents. They will also conduct interviews with key personnel and review internal policies and procedures to identify potential legal risks and liabilities.

Working with legal counsel during due diligence requires a high level of collaboration and communication between the legal team and the company's management team. The company's management team should be prepared to provide legal counsel with access to all relevant legal documents and information, and to answer any questions that the legal team may have about the company's legal position and compliance with relevant laws and regulations.

It is important for the company's management team to be transparent and forthcoming with legal counsel, as any attempts to withhold or misrepresent information can result in serious legal and financial consequences.

Best Practices for Working with Auditors and Legal Counsel

To ensure a successful due diligence process, it is important to establish clear lines of communication and collaboration between the company's management team, auditors, and legal counsel. Some best practices for working with these professionals include:

✦ Establish clear objectives and timelines for the due diligence process.

✦ Provide auditors and legal counsel with access to all relevant documents and information in a timely manner.

✦ Be transparent and forthcoming with auditors and legal counsel, and address any questions or concerns they may have promptly and honestly.

✦ Maintain open lines of communication throughout the due diligence process, and provide regular updates on progress and any significant findings.

✦ Collaborate with auditors and legal counsel to identify potential risks and liabilities, and develop strategies to mitigate those risks.

✦ Ensure that all parties involved in the due diligence process understand their roles and responsibilities, and work together to achieve the common goal of conducting a thorough and successful due diligence.

Conclusion

Working with auditors and legal counsel is a critical component of conducting financial and legal due diligence in any business transaction.

Identifying and addressing potential issues

In any business transaction, whether it's a merger, acquisition, or investment, it's important to identify and address potential issues before finalizing the deal. Failure to do so can result in costly legal and financial consequences. Due diligence is the process of examining and analyzing a company's financial and legal records to identify any potential issues that could affect the transaction. One important aspect of due diligence is working with auditors and legal counsel. This section will provide a detailed analysis of working with auditors and legal counsel in the due diligence process, including their roles and responsibilities, key considerations, and potential issues that may arise.

Roles and Responsibilities:

Auditors and legal counsel play important roles in the due diligence process. The primary role of auditors is to examine a company's financial records and provide an independent assessment of its financial health. Auditors can help identify potential issues related to financial reporting, tax compliance, and internal controls. Legal counsel, on the other hand, is responsible for examining a company's legal records and identifying any potential legal issues. This includes reviewing contracts, litigation, intellectual property, and regulatory compliance. Legal counsel can also provide advice on the structure of the transaction, potential liabilities, and risk management strategies.

Key Considerations:

When working with auditors and legal counsel in the due diligence process, there are several key considerations to keep in mind. First, it's important to establish clear communication channels and timelines. This includes identifying key points of contact and establishing deadlines for the completion of the audit and legal review. Second, it's important to establish clear roles and responsibilities for each party involved. This includes defining the scope of the audit and legal review, as well as determining who will be responsible for addressing any potential issues that are identified. Finally, it's important to establish clear expectations for the outcome of the due diligence process. This includes defining the criteria for success and determining how any potential issues will be addressed.

Potential Issues:

Despite the importance of due diligence, potential issues can arise during the process. One common issue is a lack of transparency from the target company. This can make it difficult to obtain accurate financial and legal information and may raise concerns about the target company's overall integrity. Another potential issue is conflicts of interest. For example, if the same audit firm is working for both the buyer and seller, there may be concerns about the impartiality of the audit. Similarly, if legal counsel has a pre-existing relationship with one of the parties, there may be concerns about their ability to provide impartial advice. Finally, potential issues can arise if there are language barriers, cultural differences, or other challenges that impact the ability of auditors and legal counsel to effectively communicate and collaborate.

Conclusion:

In conclusion, working with auditors and legal counsel is a critical component of the due diligence process. Auditors and legal counsel play important roles in identifying and addressing potential issues that could impact a business transaction. To ensure a successful due diligence process, it's important to establish clear communication channels and timelines, define roles and responsibilities, and establish clear expectations for the outcome of the process. It's also important to be aware of potential issues, such as lack of transparency, conflicts of interest, and communication challenges, and to take steps to address them proactively. By taking these steps, businesses can help ensure a successful and profitable transaction.

CHAPTER 27: IPO ALTERNATIVES AND CONSIDERATIONS

Initial Public Offerings (IPOs) have long been a sought-after method of raising capital for companies. However, the process can be complicated, time-consuming, and expensive, which can deter some businesses from pursuing an IPO. Additionally, the IPO process requires a level of transparency and scrutiny that some companies may not be comfortable with.

Fortunately, there are alternative methods for companies to access capital that may be more suitable for their needs. In this chapter, we will explore several of these alternatives and the considerations that companies should take into account when deciding whether to pursue an IPO or one of these alternative methods.

The first section will provide an overview of the traditional IPO process, including the benefits and drawbacks. We will examine the various steps involved in an IPO, from preparing financial statements to conducting roadshows, and discuss the regulatory requirements that companies must comply with.

The second section will explore the various alternatives to IPOs, including direct listings, special purpose acquisition companies (SPACs), and crowdfunding. We will examine the benefits and drawbacks of each of these alternatives and discuss the regulatory requirements that companies must comply with when pursuing them.

The third section will provide guidance for companies in deciding which method is right for them. We will examine the factors that companies should consider when choosing between an IPO and an alternative method, including the company's financial goals, its risk tolerance, and its desired level of control.

Finally, the fourth section will explore the legal and financial considerations that companies must take into account when pursuing an IPO or an alternative method. We will examine the role of legal counsel and financial advisors in the process, as well as the costs and risks associated with each method.

Overall, this chapter will provide students with a comprehensive understanding of the various methods that companies can use to access capital, and the benefits and drawbacks of each. By the end of this chapter, students will be equipped to advise clients on the best method for raising capital based on their individual needs and circumstances.

Alternative methods of raising capital

Capital is an essential element of any business, and raising capital is a critical aspect of a company's growth and development. Traditionally, companies have relied on public offerings and private placements to raise capital. However, alternative methods of raising capital have gained popularity in recent years, providing companies with more options to raise capital. This section explores alternative methods of raising capital, including venture capital, crowdfunding, initial coin offerings, and private equity.

Venture Capital

Venture capital is a type of financing that is provided to startup companies that have high growth potential but lack access to traditional financing sources. Venture capitalists invest in exchange for equity in the company, and they typically have significant influence over the company's operations and decision-making.

Venture capital funding is usually provided in stages, with each stage representing a significant milestone in the company's growth. The first stage is seed funding, which is provided to a company to help it develop a product or service. The second stage is series A funding, which is provided to help the company scale its operations. The third stage is series B funding, which is provided to help the company expand into new markets. Finally, the fourth stage is series C funding, which is provided to help the company prepare for an initial public offering (IPO) or acquisition.

Venture capital funding is highly competitive, and companies must have a strong business plan, a clear growth strategy, and a talented management team to attract investment. Venture capital firms typically invest in a specific industry or sector, and they look for companies that have the potential to disrupt the market.

Crowdfunding

Crowdfunding is a method of raising capital that involves raising small amounts of money from a large number of individuals. Crowdfunding platforms, such as Kickstarter and Indiegogo, allow companies to showcase their products or services and solicit funding from the public.

Crowdfunding can be divided into four main types: rewards-based crowdfunding, donation-based crowdfunding, equity-based crowdfunding, and debt-based crowdfunding. Rewards-based crowdfunding involves providing backers with a reward or incentive for their investment, such as a discount on the product or service being developed. Donation-based crowdfunding involves soliciting donations from the public for a particular cause or project. Equity-based crowdfunding involves selling shares in the company to the public in exchange for funding, and debt-based crowdfunding involves raising capital through loans from the public.

Crowdfunding can be an effective way for companies to raise capital, particularly for those that are unable to secure traditional financing sources. However, companies must be careful to comply with securities laws and regulations, particularly when selling equity or debt securities to the public.

Initial Coin Offerings

Initial coin offerings (ICOs) are a method of raising capital that involves the sale of digital tokens or coins to investors in exchange for cryptocurrencies, such as Bitcoin or Ethereum. ICOs have gained popularity in recent years as a way for blockchain-based startups to raise capital.

ICOs are typically used to fund the development of a new cryptocurrency or blockchain platform. Investors purchase tokens or coins with the expectation that they will increase in value as the platform or currency gains popularity.

ICOs are largely unregulated, and there is a significant risk of fraud or scams associated with these offerings. Investors must be careful to conduct thorough due diligence and research before investing in an ICO.

Private Equity

Private equity is a type of investment that involves the purchase of a company or a significant stake in a company by a private equity firm. Private equity firms typically invest in mature companies that have a stable cash flow and a proven business model.

Private equity firms use a combination of debt and equity financing to acquire companies, and they often work closely with the management team to improve the company's operations and profitability. Private equity firms typically exit their investments within three to seven years, either through an IPO or sale to another company.

Private equity can be a viable alternative for companies looking to raise capital, especially if they are not interested in going public. However, it is not without risks. Private equity firms may be highly selective in the companies they choose to invest in, and they may require a high degree of control over the company's operations in exchange for their investment.

Another alternative method of raising capital is through venture capital. Venture capital is a type of financing that is typically provided to startups or early-stage companies with high growth potential. Venture capital firms invest in these companies in exchange for equity, and they often provide valuable resources and support to help the company grow and succeed.

Venture capital firms may provide strategic advice, introduce the company to potential customers and partners, and provide additional financing as needed. Venture capital investments are typically high-risk, high-reward, as many startups fail to achieve the growth and success they anticipate. However, for those that do succeed, the potential returns can be significant.

Another alternative method of raising capital is through crowdfunding. Crowdfunding is a method of raising capital through the collective efforts of a large number of individuals, typically via the internet. Crowdfunding can be used to raise funds for a wide range of projects, from creative endeavors like films and music albums to social causes and charitable initiatives.

Crowdfunding can be a cost-effective way to raise capital, as it often requires little upfront investment. However, it can be challenging to attract enough interest and support to reach the funding goal, and the company may need to offer incentives or rewards to attract backers.

Finally, companies can also raise capital through debt financing. Debt financing involves borrowing money from lenders, such as banks or other financial institutions, in exchange for a promise to repay the principal and interest over time. Debt financing can be an attractive option for companies that have a stable cash flow and a proven track record of profitability.

However, debt financing also carries risks. Companies that take on too much debt may struggle to meet their repayment obligations, which can lead to financial distress and even bankruptcy. Additionally, lenders may require collateral, such as real estate or other assets, to secure the loan, which can limit the company's flexibility and future growth potential.

In conclusion, there are many alternative methods of raising capital beyond an initial public offering. Each method has its advantages and disadvantages, and companies should carefully consider their options before making a decision. Private equity, venture capital, crowdfunding, and debt financing can all be viable options, depending on the company's specific needs and goals.

Pros and cons of going public

Going public through an initial public offering (IPO) is a major decision for any company. It can provide a company with access to significant amounts of capital, as well as increased visibility and prestige. However, going public also has potential drawbacks and risks that must be carefully considered.

Pros of Going Public:

✧ Access to Capital: One of the most significant benefits of going public is access to capital. Going public allows a company to raise a significant amount of capital by issuing shares of stock to the public. This capital can be used to fund expansion, research and development, or other initiatives that can help the company grow and succeed.

✧ Increased Visibility: Going public can increase a company's visibility and raise its profile in the market. This increased visibility can attract new customers, partners, and investors, as well as enhance the company's reputation.

✧ Prestige: Being a public company can add prestige to a company's brand and help establish it as a leader in its industry.

✧ Liquidity: Going public provides shareholders with liquidity, meaning they can sell their shares of stock on the public market at any time.

✧ Access to Better Talent: Public companies are often viewed as more stable and secure than private companies. This perception can help attract top talent who may be hesitant to join a private company.

Cons of Going Public:

✧ Cost: Going public can be expensive. Companies must pay significant fees to underwriters, lawyers, and accountants to prepare for and execute an IPO. Public companies must also comply with a host of regulations and reporting requirements, which can be costly and time-consuming.

✧ Loss of Control: Going public requires a company to relinquish some control to its shareholders. Shareholders can exert influence over the company through their voting rights and may pressure management to make decisions that are not in the best interests of the company.

✧ Increased Regulation: Public companies are subject to increased regulation and scrutiny from regulators and the public. This can be a significant burden for companies that are used to operating in a more flexible and agile environment.

✧ Short-Term Focus: Public companies are often under pressure to deliver short-term results to satisfy shareholders and investors. This pressure can lead to a focus on quarterly earnings and other short-term metrics, rather than long-term strategic goals.

✧ Disclosure: Public companies must disclose a significant amount of information about their business, including financial performance, strategic plans, and other sensitive information. This information is available to competitors, customers, and other stakeholders, which can create risks for the company.

Conclusion:

Going public is a major decision for any company. While it can provide significant benefits, it also comes with potential drawbacks and risks that must be carefully considered. Companies should weigh the pros and cons of going public and carefully assess their readiness before pursuing an IPO. In addition, companies should work closely with their legal and financial advisors to ensure that they are prepared to meet the rigorous demands of being a public company.

Considerations for companies contemplating an IPO

When a company considers going public, it must carefully weigh the advantages and disadvantages of this decision. Going public is a significant step that involves a complex process, and it is not suitable for every company. In this section, we will discuss some of the key considerations that companies should take into account when contemplating an IPO.

✧ Costs: Going public can be expensive. The costs associated with an IPO include underwriting fees, legal and accounting fees, and other expenses such as printing and mailing costs. Additionally, the company will need to comply with ongoing reporting and disclosure requirements, which can also be costly.

Some of the estimated costs associated with an IPO can include:

➢ Underwriting fees: Underwriters are investment banks that help companies to sell their shares to the public. Underwriters typically charge a fee for their services, which can range from 2-7% of the total offering amount.

➢ Legal and accounting fees: Companies that are considering an IPO will need to work with legal and accounting professionals to prepare the necessary documents and comply with regulatory requirements. These professionals typically charge a fee for their services.

➢ Printing and mailing costs: Companies will need to prepare and distribute prospectuses and other documents to potential investors. These costs can add up quickly, especially if the company is planning to issue a large offering.

➢ Compliance costs: Once a company goes public, it will be subject to ongoing reporting and disclosure requirements imposed by regulatory bodies such as the SEC. Compliance with these requirements can be expensive, as it may require the company to hire additional staff or invest in new technology.

✧ Disclosure and transparency: As a public company, a firm will need to provide more detailed financial and operational information to its shareholders, potential investors, and the Securities and Exchange Commission (SEC). This can reduce the company's competitive advantage by disclosing proprietary information and increase the risk of litigation if the company fails to meet expectations.

Examples of disclosure and transparency:

➢ Quarterly and annual financial reports: Publicly traded companies are required to file regular financial reports with the Securities and Exchange Commission (SEC). These reports must include detailed information about the company's financial performance, including revenue, expenses, profits, and losses. This information must be accurate and transparent so that investors can make informed decisions about whether to buy or sell shares of the company's stock.

➢ Insider trading: Company insiders, such as executives and board members, are required to disclose any trades of the company's stock that they make. This helps to ensure that they are not using their inside knowledge of the company's performance to make unfair profits.

➢ Corporate social responsibility (CSR) reports: Many companies now issue CSR reports that detail their efforts to operate in a socially responsible manner. These reports may include information about the company's environmental impact, labor practices, and charitable giving. By being transparent about their CSR efforts, companies can build trust with their stakeholders.

➢ Board of Directors meetings: The minutes from board meetings are often made public so that investors and other stakeholders can see how the company's leaders are making decisions. This can help to ensure that the board is acting in the best interests of the company and its shareholders.

➢ Executive compensation: Companies are required to disclose how much their top executives are paid, including salaries, bonuses, and stock options. This helps investors to understand how the company's leadership is compensated and whether their incentives are aligned with those of shareholders.

✧ Management distractions: An IPO can be a time-consuming and distracting process for management, as they must spend significant time preparing for the IPO and managing investor relations after going public. This may lead to a loss of focus on the company's day-to-day operations, which could ultimately affect the company's performance.

Examples of management distractions that can occur during the IPO process:

➢ Time and Resource Allocation: The process of going public can be time-consuming and requires significant resources from the company's management team. This can take away from the focus on day-to-day operations and other important business initiatives.

➢ Increased Scrutiny: Once a company goes public, it is subject to increased regulatory scrutiny and reporting requirements. This can be a significant distraction for management, as they may need to spend more time on compliance-related activities.

➢ Investor Relations: Going public requires a greater focus on investor relations and communication with shareholders. This can be a distraction for management, as they need to balance their time between running the business and communicating with investors.

➤ Stock Price Volatility: Once a company goes public, its stock price may become volatile, which can be a distraction for management. They may need to spend more time addressing investor concerns and managing market expectations.

➤ Increased Pressure: Going public can put additional pressure on the management team to meet financial targets and deliver results. This can be a distraction, as they may need to focus more on short-term performance rather than long-term strategic planning.

✧ Market volatility: Public companies are subject to market forces that can cause their stock prices to fluctuate significantly. This can be challenging for companies that are not accustomed to managing public expectations and may have difficulty reacting to sudden changes in the market.

Market volatility refers to the tendency of financial markets to fluctuate rapidly and unpredictably, often in response to economic, political, or other external events. Here are some examples of market volatility:

➤ Economic downturns: During a recession, stock prices may fall sharply due to decreased investor confidence and lower earnings expectations for companies.

➤ Political events: Elections, changes in government policies, and geopolitical tensions can cause fluctuations in financial markets.

➤ Natural disasters: Natural disasters such as earthquakes, hurricanes, and pandemics can disrupt supply chains, lead to lower consumer demand, and create market volatility.

➤ Company-specific events: Poor earnings reports, management scandals, and product recalls can all cause a company's stock price to fluctuate.

➤ Interest rate changes: When central banks change interest rates, it can affect borrowing costs, inflation expectations, and ultimately impact financial markets.

➤ Currency fluctuations: Changes in currency exchange rates can impact a company's revenue and expenses, which in turn can affect the company's stock price.

It is important to note that market volatility can create both opportunities and risks for investors. Some investors may use market volatility to make strategic investments or to trade for short-term profits, while others may avoid investing during periods of high volatility in favor of more stable investments.

✦ Corporate governance: Going public requires companies to adopt a more formal and structured approach to corporate governance. This includes establishing a board of directors with independent directors and implementing formal policies and procedures to ensure compliance with regulatory requirements.

Examples of corporate governance:

➤ Board of Directors: The board of directors is responsible for overseeing the management of the company and making strategic decisions. They are elected by the shareholders and are accountable to them.

➤ Executive compensation: Executive compensation refers to the pay and benefits received by top executives of a company. It is typically approved by the board of directors and is often linked to the company's performance.

➤ Shareholder rights: Shareholders have certain rights, such as the right to vote on major decisions and the right to inspect the company's books and records.

➤ Audit committees: Audit committees are responsible for overseeing the company's financial reporting and ensuring that the company is complying with relevant laws and regulations.

➤ Ethics and compliance: Companies are expected to have a strong ethical culture and to comply with laws and regulations. This may include having a code of conduct, whistleblower policies, and training programs for employees.

> ➤ Risk management: Companies need to identify and manage risks that could impact their business. This may include risks related to cybersecurity, supply chain disruptions, and natural disasters.

Overall, corporate governance is important because it helps to ensure that companies are run in a responsible and transparent manner, which can ultimately benefit shareholders and other stakeholders.

✧ Capital raising: An IPO can provide a company with access to significant amounts of capital that can be used to fund growth and expansion. However, going public also means that the company will be subject to the whims of the market and may not be able to raise additional capital at favorable terms.

✧ Exit strategy: Going public is often seen as a key milestone in a company's life cycle. However, it is important to consider the potential consequences of going public, such as the impact on the company's ownership structure and the ability of the founders and early investors to exit their investments.

✧ Investor expectations: Going public means that a company will have to meet the expectations of a wide range of investors. This can be challenging, as investors may have different goals and expectations for the company's growth and profitability.

✧ Regulatory compliance: As a public company, a firm will be subject to a variety of regulatory requirements, including reporting and disclosure requirements under the Securities Exchange Act of 1934, the Sarbanes-Oxley Act of 2002, and the Dodd-Frank Wall Street Reform and Consumer Protection Act of 2010.

✧ Reputation risk: Public companies are subject to greater scrutiny from the media, analysts, and investors, which can impact the company's reputation. This can be particularly damaging if the company experiences a financial or operational setback.

In conclusion, an IPO is a significant decision for any company and should not be taken lightly. Companies considering an IPO should carefully weigh the potential benefits and drawbacks of going public and seek advice from legal, financial, and other advisors before making a final decision. While an IPO can provide a company with access to significant amounts of capital, it also involves significant costs and regulatory compliance requirements that may not be suitable for all companies.

CHAPTER 28: PREPARING FOR PUBLIC SCRUTINY

When a company decides to go public, it is not only raising capital but also inviting a significant amount of scrutiny from the public, investors, and regulators. Going public requires a great deal of preparation and planning, and the company must be prepared to meet the heightened expectations and obligations that come with being a public company.

In this chapter, we will discuss the steps a company should take to prepare for public scrutiny, including:

- ✧ Understanding the regulatory requirements for public companies
- ✧ Developing a comprehensive compliance program
- ✧ Establishing a strong corporate governance structure
- ✧ Enhancing financial reporting and internal controls
- ✧ Building investor relations and public relations programs
- ✧ Addressing potential risks and contingencies
- ✧ Understanding the Regulatory Requirements for Public Companies

When a company goes public, it must comply with a complex web of securities laws and regulations, including the Securities Act of 1933, the Securities Exchange Act of 1934, and the Sarbanes-Oxley Act of 2002. These laws and regulations impose strict disclosure, reporting, and governance requirements on public companies and their management teams.

One of the key requirements for public companies is the filing of periodic reports with the Securities and Exchange Commission (SEC), including quarterly and annual reports, proxy statements, and other disclosures. These reports must be accurate, complete, and timely, and failure to comply can result in significant legal and financial penalties.

In addition to SEC filings, public companies must also comply with a range of other regulatory requirements, including those related to accounting and auditing, insider trading, executive compensation, and corporate social responsibility. Failure to comply with these requirements can result in reputational damage, legal liability, and negative impacts on the company's financial performance.

Developing a Comprehensive Compliance Program

To meet the regulatory requirements for public companies, it is essential for companies to develop a comprehensive compliance program that includes policies, procedures, and controls for all areas of the business. This program should be designed to prevent, detect, and remediate violations of securities laws and regulations.

A robust compliance program should include:

✧   Policies and procedures for identifying, evaluating, and mitigating risks
✧   Employee training and awareness programs
✧   Systems and controls for monitoring and testing compliance
✧   Reporting mechanisms for employees and other stakeholders to report potential violations
✧   Regular assessments and reviews of the effectiveness of the compliance program
✧   Establishing a Strong Corporate Governance Structure

Corporate governance refers to the system of rules, practices, and processes by which a company is directed and controlled. A strong corporate governance structure is essential for public companies to maintain transparency, accountability, and integrity in their operations.

To establish a strong corporate governance structure, companies should:

✧   Develop a clear and comprehensive code of conduct and ethics
✧   Establish an independent board of directors with appropriate skills, experience, and diversity
✧   Create committees of the board, such as audit, compensation, and nominating committees, to oversee key areas of the business
✧   Implement policies and procedures for related-party transactions, conflict of interest, and whistleblower protection
✧   Conduct regular assessments of the effectiveness of the corporate governance structure
✧   Enhancing Financial Reporting and Internal Controls

Public companies must comply with strict financial reporting requirements, including generally accepted accounting principles (GAAP) and SEC rules and regulations. They must also establish and maintain effective internal controls over financial reporting (ICFR) to ensure the accuracy and reliability of financial statements.

To enhance financial reporting and internal controls, companies should:

✧ Implement robust accounting policies and procedures that comply with GAAP and SEC rules and regulations

✧ Conduct regular internal and external audits to ensure compliance and identify potential weaknesses

✧ Establish and maintain effective ICFR, including policies and procedures for financial reporting, risk management, and fraud prevention

✧ Maintain a strong tone at the top by setting the right ethical culture and promoting transparency and accountability throughout the organization

✧ Provide adequate training and resources to employees to ensure they understand their roles and responsibilities in financial reporting and internal controls

✧ Engage competent and independent auditors to conduct audits of financial statements and internal controls

✧ Stay up to date with changes in GAAP, SEC rules and regulations, and other accounting and reporting standards to ensure ongoing compliance

✧ Ensure effective communication with stakeholders, including investors, analysts, regulators, and other interested parties, to promote transparency and build trust

Effective financial reporting and internal controls are essential for maintaining public trust and confidence in a company. They help to prevent financial misstatements and irregularities, reduce the risk of fraud and misconduct, and provide stakeholders with reliable information for making informed decisions.

In addition to enhancing financial reporting and internal controls, companies preparing for public scrutiny should also consider other key areas, such as:

Corporate governance: Public companies are subject to greater scrutiny and oversight from regulators, investors, and other stakeholders. They must have strong corporate governance practices in place to ensure effective oversight, accountability, and transparency. This includes establishing a board of directors with appropriate skills and experience, adopting sound governance policies and procedures, and maintaining effective communication with shareholders and other stakeholders.

Risk management: Going public exposes companies to a range of risks, including market, operational, reputational, and legal risks. Companies must have robust risk management strategies in place to identify, assess, and mitigate these risks. This includes establishing a risk management framework, conducting regular risk assessments, and implementing risk mitigation measures.

Compliance: Public companies must comply with a range of laws and regulations, including securities laws, accounting standards, and corporate governance requirements. Failure to comply can result in significant penalties and reputational damage. Companies must establish a robust compliance program to ensure ongoing compliance with all applicable laws and regulations.

Investor relations: Public companies must engage with a wide range of stakeholders, including investors, analysts, and the media. They must have effective investor relations strategies in place to communicate their performance, strategy, and outlook, and to respond to stakeholder inquiries and concerns.

Preparing for public scrutiny requires careful planning, investment, and commitment from senior management and the board of directors. By enhancing financial reporting and internal controls, adopting sound corporate governance practices, managing risk effectively, and ensuring ongoing compliance and effective investor relations, companies can build trust and confidence with stakeholders and maximize their chances of success in the public markets.

Understanding the regulatory environment

When a company goes public, it becomes subject to a complex web of regulations and regulatory bodies. It is essential for companies to understand the regulatory environment in which they operate and to comply with all relevant laws and regulations.

The Securities and Exchange Commission (SEC) is the primary regulatory body for public companies in the United States. The SEC is responsible for enforcing federal securities laws and regulating the securities industry. The SEC oversees the registration of securities offerings, disclosure requirements, insider trading, and other aspects of the securities market.

Companies must file various reports with the SEC, including registration statements, annual reports (Form 10-K), quarterly reports (Form 10-Q), and periodic reports of insider trading (Form 4). These reports provide investors with information about the company's financial condition and operations.

In addition to the SEC, companies must comply with a range of other regulatory bodies, including state securities regulators, stock exchanges, and industry-specific regulators. For example, companies in the financial industry are subject to regulation by the Federal Reserve, the Office of the Comptroller of the Currency, and the Federal Deposit Insurance Corporation.

Non-compliance with regulatory requirements can result in significant fines, legal liability, and damage to the company's reputation. To avoid these risks, companies must establish and maintain effective compliance programs that ensure compliance with all relevant laws and regulations.

Compliance programs typically include the following components:

Policies and procedures: Companies should establish written policies and procedures that outline the steps employees must take to comply with relevant laws and regulations. These policies and procedures should be reviewed regularly and updated as necessary.

Training and education: Companies should provide training and education to employees on relevant laws and regulations, as well as the company's policies and procedures. This training should be provided to all employees, including senior executives and members of the board of directors.

Monitoring and auditing: Companies should establish processes for monitoring and auditing compliance with relevant laws and regulations. This includes conducting regular internal audits and reviews of the company's compliance program.

Reporting and investigation: Companies should establish processes for reporting and investigating potential violations of relevant laws and regulations. This includes providing employees with a mechanism for reporting potential violations and conducting prompt and thorough investigations of reported violations.

By establishing and maintaining effective compliance programs, companies can reduce their exposure to regulatory risks and ensure that they operate in a manner that complies with all relevant laws and regulations.

Developing a communications strategy

Going public requires careful planning and execution of a communications strategy to ensure that key stakeholders are informed and engaged throughout the process. A communications strategy should be developed and implemented well before the IPO and should be designed to achieve several key objectives, including:

Building awareness and interest in the company: A communications strategy should help build awareness and interest in the company among potential investors, customers, and other stakeholders. This can be achieved through various communication channels, including traditional media, social media, and direct outreach.

Managing expectations: Going public is a complex process, and it is important to manage the expectations of stakeholders. A communications strategy should be designed to set realistic expectations about the IPO process, the company's performance, and the potential risks and rewards associated with investing in the company.

Addressing potential concerns: Going public can raise concerns among stakeholders, including customers, employees, and investors. A communications strategy should be designed to address these concerns and provide reassurance that the company is well-prepared for the transition to public ownership.

Enhancing transparency: Public companies are subject to a range of reporting and disclosure requirements, and it is important to establish a culture of transparency from the outset. A communications strategy should be designed to promote transparency and openness, which can help build trust among stakeholders.

Developing a communications strategy requires a thorough understanding of the company's target audience and the communication channels that will be most effective in reaching them. The strategy should be tailored to the specific needs of the company and should be reviewed and updated on an ongoing basis to ensure that it remains relevant and effective.

Key components of a communications strategy include:

Messaging: A clear and consistent messaging platform is essential for a successful communications strategy. Messaging should be developed around the company's core values, mission, and objectives, and should be tailored to the specific needs and interests of each stakeholder group.

Communication channels: A communications strategy should include a range of communication channels, including traditional media, social media, email, and direct outreach. Each channel should be evaluated for its effectiveness in reaching the target audience and should be used strategically to deliver key messages.

Key performance indicators: A communications strategy should be designed to achieve specific objectives, and key performance indicators (KPIs) should be established to measure progress. KPIs may include metrics such as media coverage, social media engagement, and website traffic.

Crisis management: A communications strategy should include a plan for managing potential crises or issues that may arise during the IPO process. This should include a clear escalation process, designated spokespersons, and prepared statements.

In summary, developing a communications strategy is a critical component of a successful IPO. A well-designed strategy can help build awareness and interest in the company, manage expectations, address concerns, and promote transparency. Key components of a communications strategy include messaging, communication channels, KPIs, and crisis management.

Preparing for investor questions and concerns

Preparing for investor questions and concerns is an important part of communicating effectively with stakeholders. Investors are typically interested in a company's financial performance, strategic direction, and potential risks and opportunities. As such, companies must be prepared to answer questions and provide information that satisfies investors' concerns.

Here are some key steps companies can take to prepare for investor questions and concerns:

Develop a comprehensive investor relations strategy
The first step in preparing for investor questions and concerns is to develop a comprehensive investor relations strategy. This should include identifying the key stakeholders and their interests, as well as developing a clear and consistent messaging framework that can be used across all communication channels.

Anticipate potential questions and concerns
Companies should also anticipate potential questions and concerns that investors may have, based on their previous interactions with the company, their industry knowledge, and their expectations for the future. Common questions and concerns may include the company's financial performance, competition, regulatory environment, growth prospects, and risk management practices.

Prepare for different types of investors
Investors can have different levels of expertise and understanding of the company and its industry, so it's important to be prepared to address questions and concerns from different types of investors. This may include institutional investors, retail investors, and analysts, each with different expectations and levels of knowledge.

Provide timely and accurate information

To satisfy investors' concerns, companies must provide timely and accurate information that is relevant and meaningful. This may include financial statements, annual reports, earnings releases, and other disclosures, as well as information on corporate governance practices and risk management strategies.

Be transparent and open

Companies must also be transparent and open in their communications with investors, providing honest and candid answers to questions and concerns. This helps build trust and credibility with investors, and can ultimately enhance the company's reputation and financial performance.

Prepare for potential challenges

Despite the best efforts of companies to prepare for investor questions and concerns, challenges can still arise. For example, unexpected financial results or a change in market conditions can lead to difficult questions from investors. Companies should be prepared to address these challenges with transparency and professionalism, and to work proactively with stakeholders to address any issues or concerns that arise.

In summary, preparing for investor questions and concerns requires a comprehensive and proactive approach to investor relations. By developing a clear strategy, anticipating potential questions and concerns, providing timely and accurate information, being transparent and open, and preparing for potential challenges, companies can enhance their relationships with stakeholders and build a strong foundation for long-term success.

Preparing for the backlash from social media and mitigating the fallout

In today's digital age, social media has become a powerful tool for businesses to connect with their customers and stakeholders. However, it also creates a risk of backlash and negative publicity if companies fail to address customer concerns and complaints in a timely and appropriate manner. In this section, we will explore strategies for preparing for the backlash from social media and mitigating the fallout.

Develop a crisis communication plan

A crisis communication plan is a comprehensive document that outlines the company's communication strategy in the event of a crisis or negative publicity. It should include a list of potential crisis scenarios, key stakeholders and their contact information, designated spokespeople, messaging templates, and a timeline for response. The plan should be regularly reviewed and updated to ensure its relevance and effectiveness.

Monitor social media channels

Monitoring social media channels is essential for early detection of potential issues and negative publicity. This includes regularly reviewing comments, mentions, and reviews on social media platforms such as Twitter, Facebook, Instagram, and LinkedIn. Social media monitoring tools such as Hootsuite and Sprout Social can be used to automate and streamline the monitoring process.

Respond quickly and appropriately

A prompt and appropriate response is critical to mitigating the fallout from negative publicity on social media. Responding quickly shows that the company takes customer concerns seriously and is committed to resolving issues. Responses should be empathetic, transparent, and provide a clear course of action for addressing the issue.

Apologize if necessary

Apologizing is an essential part of crisis communication. If the company is at fault or has caused harm to its customers, an apology can go a long way towards repairing damaged relationships. A sincere and authentic apology should acknowledge the issue, take responsibility, express remorse, and outline steps being taken to address the problem.

Provide updates and follow-up

Providing regular updates and follow-up is essential for maintaining transparency and trust with stakeholders. Updates should be provided on the progress of resolving the issue and any additional steps being taken to prevent similar incidents in the future. Follow-up should include a survey or other mechanism for gathering feedback from stakeholders to ensure that their concerns have been adequately addressed.

Train employees

Employee training is critical to ensuring that they are prepared to respond appropriately to negative publicity on social media. Employees should be trained to identify potential issues, escalate them to the appropriate personnel, and respond appropriately to customer concerns. Training should also include a review of the company's crisis communication plan and messaging templates.

Work with social media influencers

Social media influencers can be valuable allies in mitigating the fallout from negative publicity. Companies should identify relevant social media influencers in their industry and engage them in their crisis communication plan. Influencers can help amplify the company's messaging and provide a trusted voice to reassure their followers that the company is taking the issue seriously.

Conduct a post-mortem

Once the crisis has been resolved, it is essential to conduct a post-mortem to review the company's response and identify areas for improvement. This includes reviewing the effectiveness of the crisis communication plan, analyzing the company's response on social media, and identifying opportunities for preventing similar incidents in the future.

In conclusion, social media can be a powerful tool for businesses, but it also creates risks of negative publicity and backlash. By developing a crisis communication plan, monitoring social media channels, responding quickly and appropriately, apologizing if necessary, providing updates and follow-up, training employees, working with social media influencers, and conducting a post-mortem, companies can effectively prepare for and mitigate the fallout from negative publicity on social media.

In today's digital age, businesses and their leaders are under constant public scrutiny. With the widespread use of social media and instant communication, negative news or controversies can spread rapidly and can quickly damage a company's reputation and bottom line. Therefore, it is crucial for businesses to have a solid plan in place to deal with public scrutiny and backlash.

One of the main themes of this chapter is the importance of having a well-thought-out communications strategy. Businesses need to be proactive in communicating with the public and stakeholders, ensuring that their message is consistent and clear. This involves understanding the needs and concerns of different stakeholders and tailoring the message accordingly. A good communications strategy should also be able to address negative feedback and criticisms in a professional and respectful manner.

Another key takeaway from this chapter is the need to be prepared for investor questions and concerns. Investors play a crucial role in any business, and it is important to be able to address their questions and concerns in a confident and knowledgeable manner. This requires a deep understanding of the business and its operations, as well as the ability to anticipate and prepare for potential issues.

Finally, this chapter highlights the importance of mitigating the fallout from social media. Social media platforms can be a double-edged sword for businesses. While they can be an effective tool for communication and promotion, they can also quickly amplify negative news or controversies. Businesses need to be able to monitor social media activity and respond quickly and effectively to any negative feedback or criticisms.

In conclusion, preparing for public scrutiny is an essential part of any business strategy. By developing a communications strategy, preparing for investor questions and concerns, and mitigating the fallout from social media, businesses can effectively manage public scrutiny and minimize any negative impact. It is crucial for businesses to be proactive in their approach to public scrutiny and to be prepared for any potential issues that may arise. With careful planning and preparation, businesses can navigate the complex world of public scrutiny and emerge stronger and more resilient.

PART 8: IPO PROCESS AND REQUIREMENTS

The Initial Public Offering (IPO) process is a critical milestone in the life of a company. Going public can provide access to a significant amount of capital, which can be used to fund growth and expansion, as well as increase the visibility of a company. The process of going public is highly regulated, with specific requirements and guidelines that must be followed to ensure compliance with securities laws and regulations. In this section, we will explore the IPO process and the various requirements that companies must meet to successfully complete an IPO.

Chapter 29: The IPO Process

The IPO process typically involves several stages, including preparation, planning, and execution. In the preparation stage, a company must assess its readiness to go public, including its financial performance, management team, and corporate governance structure. The company must also engage various advisors, such as underwriters, legal counsel, and accounting firms, to assist with the IPO process.

During the planning stage, the company will develop a prospectus, which is a legal document that provides details about the company's business, financial performance, risks, and other important information. The prospectus must be filed with the Securities and Exchange Commission (SEC) and must comply with various securities laws and regulations.

In the execution stage, the company will conduct a roadshow, which involves presenting the company to potential investors to generate interest in the IPO. Once the roadshow is complete, the company will set a price for the shares and sell them to investors through the underwriters.

Chapter 30: IPO Requirements

To successfully complete an IPO, a company must meet various requirements related to financial performance, corporate governance, and regulatory compliance. These requirements are designed to ensure that investors have access to accurate and reliable information about the company, as well as to protect investors from fraudulent or misleading practices.

Financial Performance Requirements

Companies must meet certain financial performance requirements to be eligible for an IPO. For example, companies must have a minimum level of revenue and profitability, as well as a certain level of financial reporting and accounting controls. Companies must also comply with generally accepted accounting principles (GAAP) and disclose their financial performance in their prospectus.

Corporate Governance Requirements

Companies must meet various corporate governance requirements to be eligible for an IPO. These requirements are designed to ensure that companies have effective management and oversight structures in place, as well as to protect the interests of shareholders. For example, companies must have a board of directors with a majority of independent directors, as well as an audit committee and a compensation committee.

Regulatory Compliance Requirements

Companies must also comply with various regulatory requirements related to the IPO process. For example, companies must file a registration statement with the SEC, which must include a prospectus and other important information about the company. Companies must also comply with various securities laws and regulations, including the Securities Act of 1933 and the Securities Exchange Act of 1934.

Conclusion

The IPO process can be a complex and challenging process, requiring significant resources and expertise. Companies must meet various requirements related to financial performance, corporate governance, and regulatory compliance to successfully complete an IPO. However, the benefits of going public can be significant, including access to capital and increased visibility. In the next chapters, we will explore the various legal and regulatory requirements that companies must comply with to successfully complete an IPO.

CHAPTER 29: CHOOSING AN UNDERWRITER AND PRICING THE OFFERING

The process of going public can be a critical juncture for companies seeking to grow and expand their business. The decision to go public involves a complex set of considerations, including the selection of an underwriter and the pricing of the offering. Choosing an underwriter is a crucial step in the initial public offering (IPO) process, as the underwriter plays a significant role in shaping the offering and its success.

In this chapter, we will discuss the role of underwriters in the IPO process and examine the key factors that companies should consider when choosing an underwriter. We will also delve into the pricing of the offering, which is a critical factor in determining the success of the IPO. Proper pricing ensures that the offering is attractive to investors and that the company receives adequate proceeds from the offering.

We will explore the factors that underwriters consider when pricing an offering, including market conditions, industry trends, and the financial health of the company. We will also discuss various pricing strategies and the potential risks associated with each strategy.

The process of going public can be complex and daunting for companies. It is critical to have a thorough understanding of the underwriting process and pricing strategies to ensure the success of the IPO. This chapter will provide a comprehensive overview of the underwriting process and pricing strategies, providing readers with the knowledge they need to make informed decisions when choosing an underwriter and pricing the offering.

The Role of Underwriters in the IPO Process

Underwriters are financial intermediaries that play a crucial role in the IPO process. They are responsible for purchasing the shares from the company and then reselling them to investors. This process, known as underwriting, helps to ensure that the IPO is successful and that the company receives adequate proceeds from the offering.

Underwriters also provide a range of other services, including due diligence, legal advice, and market research. Due diligence involves a thorough review of the company's financials, operations, and management team to ensure that it is a sound investment. Legal advice includes drafting the prospectus, the document that provides investors with key information about the offering, and ensuring that the offering is in compliance with securities laws. Market research involves analyzing market conditions and investor demand to determine the appropriate pricing for the offering.

Choosing an Underwriter

Choosing the right underwriter is critical to the success of the IPO. The underwriter plays a critical role in shaping the offering and its success. There are several factors that companies should consider when choosing an underwriter, including:

Reputation: The reputation of the underwriter is a critical factor in the success of the IPO. Companies should look for underwriters with a strong reputation for successfully managing IPOs in their industry.

Experience: The experience of the underwriter is also critical. Companies should look for underwriters with extensive experience in managing IPOs in their industry.

Services: The services provided by the underwriter are also important. Companies should look for underwriters that provide a range of services, including due diligence, legal advice, and market research.

Fees: The fees charged by the underwriter are also an important consideration. Companies should look for underwriters that charge reasonable fees and provide value for their services.

Pricing the Offering

Pricing the offering is a critical factor in determining the success of the IPO. Proper pricing ensures that the offering is attractive to investors and that the company receives adequate proceeds from the offering. There are several factors that underwriters consider when pricing an offering, including:

Market conditions: Underwriters consider market conditions when pricing an offering. They analyze factors such as interest rates, inflation, and economic growth to determine the appropriate pricing for the offering.

Industry trends: Underwriters also consider industry trends when pricing an offering. They analyze factors such as the company's position within the industry, its competitive landscape, and the growth potential of the industry as a whole.

Financial health of the company: The financial health of the company is also a critical factor in pricing the offering. Underwriters examine the company's financial statements, including revenue, profit margins, and cash flow, to determine the appropriate pricing for the offering.

Pricing strategies: There are several pricing strategies that underwriters may use when pricing an offering. These include fixed pricing, Dutch auction, and book building.

Fixed pricing is the most common pricing strategy, in which the underwriter and the company agree on a fixed price for the shares. This price is then offered to investors, and the underwriter assumes the risk of any unsold shares.

In a Dutch auction, investors submit bids for the shares, and the underwriter determines the price based on the highest bid that will sell all of the shares.

In book building, the underwriter gathers indications of interest from potential investors and then sets the price based on the demand for the shares.

Each pricing strategy has its own advantages and disadvantages, and the appropriate pricing strategy will depend on the specific circumstances of the offering.

Potential Risks Associated with Underwriting and Pricing

There are several potential risks associated with underwriting and pricing an IPO. These risks include:

Overvaluation: If the offering is priced too high, it may not be attractive to investors, and the company may not receive the expected proceeds from the offering. Overvaluation can also lead to a subsequent drop in the stock price after the IPO.

Undervaluation: If the offering is priced too low, the company may not receive the full value of the shares and may miss out on potential proceeds. Undervaluation can also lead to a perception that the company is not as strong as it actually is, which can impact its ability to raise capital in the future.

Market conditions: Market conditions can have a significant impact on the success of the IPO. If market conditions are unfavorable, it may be difficult to attract investors, regardless of the pricing strategy.

Regulatory compliance: The IPO process is highly regulated, and companies must ensure that they are in compliance with all relevant securities laws. Failure to comply with these laws can result in legal and financial consequences.

Conclusion

Choosing an underwriter and pricing the offering are critical steps in the IPO process. Companies must carefully consider the reputation and experience of potential underwriters, as well as the services provided and the fees charged. Underwriters must consider market conditions, industry trends, and the financial health of the company when pricing the offering, and select the appropriate pricing strategy based on the specific circumstances of the offering.

The process of going public is complex and involves a range of risks. However, with careful consideration and planning, companies can successfully navigate the IPO process and achieve their growth and expansion goals.

Role of underwriters in the IPO process

The initial public offering (IPO) is a significant event for a company, as it provides an opportunity for the company to raise capital by issuing new shares of stock to the public. Underwriters play a crucial role in the IPO process by providing a range of services that facilitate the offering. In this section, we will explore the role of underwriters in the IPO process.

Definition of Underwriters:

Underwriters are financial institutions, usually investment banks, that help companies prepare and execute an IPO. They purchase shares from the issuer and then sell them to the public, providing liquidity for the shares. Underwriters also provide a range of services, including due diligence, underwriting, and marketing the offering.

Due Diligence:

Before an IPO, underwriters perform due diligence to ensure that the company is a suitable candidate for an IPO. Due diligence involves a thorough review of the company's financial statements, business operations, and legal and regulatory compliance. Underwriters must identify any potential risks that could negatively impact the offering or the company's reputation.

Underwriting:

Once the underwriters have completed their due diligence, they determine the IPO's price and the number of shares to be sold. Underwriters must balance the demand for the shares with the price at which the shares will be offered. They must also ensure that the company's valuation is fair and reasonable.

Marketing the Offering:

Underwriters play a critical role in marketing the offering to potential investors. They typically work with the company to create a prospectus, which is a legal document that provides detailed information about the offering. Underwriters use the prospectus to generate interest in the offering and identify potential investors. They also help the company meet with institutional investors and other potential buyers to generate demand for the offering.

Stabilization:

After the IPO, underwriters may engage in stabilization activities to support the stock's price. Stabilization involves purchasing additional shares of the company's stock to help support the price. Underwriters may also provide support by providing market-making services, such as buying and selling shares to maintain liquidity.

Legal Liability:

Underwriters are subject to legal liability if they fail to perform their duties adequately. They must ensure that the company's financial statements are accurate, and they must disclose any potential risks associated with the offering. If the underwriters fail to disclose any material information, they can be held liable for any losses suffered by investors.

Conclusion:

Underwriters play a crucial role in the IPO process. They provide a range of services, including due diligence, underwriting, and marketing the offering.

Underwriters must ensure that the company is a suitable candidate for an IPO, determine the IPO's price and the number of shares to be sold, and generate interest in the offering. Underwriters are subject to legal liability if they fail to perform their duties adequately.

Factors to consider when choosing an underwriter

An initial public offering (IPO) is a significant event for a company. It is an opportunity for the company to raise capital, increase its visibility, and create liquidity for its shareholders. One of the most crucial decisions a company has to make during the IPO process is choosing an underwriter. The underwriter plays a critical role in the IPO process, and their performance can significantly impact the success of the IPO. In this section, we will explore the factors that a company should consider when choosing an underwriter for their IPO.

✧   Reputation and Experience

The reputation and experience of an underwriter are essential considerations when selecting an underwriter. The company should choose an underwriter with a strong reputation in the market and a proven track record of successful IPOs. The underwriter should have extensive experience in the industry and a deep understanding of the company's business model, financials, and growth potential. The underwriter should also have a solid understanding of the regulatory requirements for IPOs and be able to guide the company through the process.

✧   Underwriter Fees

Underwriting fees are an essential consideration when choosing an underwriter. The underwriter fee is typically a percentage of the total proceeds from the IPO. The fee can vary depending on the size and complexity of the offering. Companies should compare underwriting fees offered by different underwriters and negotiate the best possible deal. However, the company should not choose an underwriter solely based on their fee structure. The company should consider the value that the underwriter brings to the table and the potential impact on the success of the IPO.

✧   Market Expertise

The underwriter's market expertise is another critical factor to consider. The underwriter should have a deep understanding of the market in which the company operates. They should have a strong network of investors, analysts, and brokers who can help promote the IPO and generate interest in the company. The underwriter should also have experience in marketing IPOs to institutional investors and retail investors.

✧   Support Services

The underwriter should provide the company with a range of support services during the IPO process. This includes financial advisory services, due diligence, and legal and accounting support. The underwriter should also provide support in the preparation of the registration statement, the roadshow, and the pricing of the offering. The underwriter should be available to answer any questions that the company may have during the process and be responsive to their needs.

✧   Reputation of the Underwriter's Syndicate

The reputation of the underwriter's syndicate is another essential consideration. The syndicate is made up of the underwriter and a group of other investment banks that help distribute the securities to investors. The underwriter should have a strong and reputable syndicate that can help generate demand for the company's shares. The syndicate should have a solid understanding of the market and be able to provide valuable insights to the company.

✧   Research Capabilities

The underwriter should have strong research capabilities and provide the company with comprehensive research reports on the industry and the company's financials. The research reports should provide investors with valuable insights into the company's business model, growth potential, and competitive landscape. The research reports should be well-written, easy to understand, and provide investors with the information they need to make informed investment decisions.

In conclusion, choosing an underwriter is a critical decision for a company considering an IPO. The company should consider the reputation and experience of the underwriter, the underwriting fees, the underwriter's market expertise, the support services provided by the underwriter, the reputation of the underwriter's syndicate, and the underwriter's research capabilities. By considering these factors, the company can choose an underwriter that will provide the support and guidance they need to have a successful IPO.

Pricing the offering and setting the initial public offering price (IPO price)

The initial public offering (IPO) price is the price at which shares of a company are sold to the public for the first time. Determining the IPO price is a critical step in the IPO process, as it sets the initial valuation of the company and determines the amount of capital that can be raised. The IPO price is typically set by the underwriters, who are responsible for pricing the offering and ensuring that the shares are sold to investors. This section will discuss the factors that underwriters consider when pricing an IPO and setting the IPO price.

Factors to Consider when Pricing an IPO

Company Valuation

The first factor that underwriters consider when pricing an IPO is the valuation of the company. Company valuation is the process of determining the worth of a company based on its financial performance, assets, and other factors. The valuation is used to determine the number of shares that will be offered in the IPO and the price per share. The underwriters work with the company's management to determine the valuation based on factors such as the company's revenue, earnings, growth prospects, market position, and industry trends.

Market Conditions

The second factor that underwriters consider when pricing an IPO is the market conditions. The underwriters will look at the current state of the stock market, the performance of similar companies in the industry, and the demand for the company's shares. If the market conditions are favorable, the underwriters may price the IPO higher to take advantage of investor demand. If the market conditions are unfavorable, the underwriters may price the IPO lower to ensure that the shares are sold.

Investor Demand

The third factor that underwriters consider when pricing an IPO is investor demand. Underwriters will gauge investor interest in the company by conducting a roadshow, where they meet with potential investors to pitch the company and gauge their interest. Based on the level of investor demand, the underwriters may adjust the IPO price. If there is high demand, the underwriters may price the IPO higher, and if there is low demand, they may price the IPO lower.

Company Performance

The fourth factor that underwriters consider when pricing an IPO is the company's performance. Underwriters will look at the company's financial performance over the past few years and the growth prospects for the future. If the company has a strong track record of growth and profitability, the underwriters may price the IPO higher to reflect the company's value. If the company has a weaker track record or is in a highly competitive industry, the underwriters may price the IPO lower to reflect the risks associated with the investment.

Comparable Company Analysis

The fifth factor that underwriters consider when pricing an IPO is the comparable company analysis. Underwriters will look at the performance and valuation of similar companies in the industry to determine the appropriate valuation for the company. This analysis helps underwriters determine if the company is overvalued or undervalued compared to its peers.

Setting the IPO Price

Once the underwriters have considered the factors mentioned above, they will set the IPO price. The IPO price is typically set at a slight discount to the expected market price to ensure that the shares are sold. If the IPO is oversubscribed, meaning that there is more demand for shares than there are shares available, the underwriters may raise the IPO price to reflect the high demand. If the IPO is undersubscribed, meaning that there is less demand for shares than there are shares available, the underwriters may lower the IPO price to ensure that the shares are sold.

Conclusion

Pricing an IPO and setting the IPO price is a complex process that involves consideration of multiple factors. Underwriters play a critical role in this process, as they are responsible for pricing the offering and ensuring that the shares are sold to investors. By considering factors such as market conditions, the company's financial performance and growth prospects, and the pricing strategies of comparable companies, underwriters can help determine the most appropriate IPO price for a particular company.

It is important to note that pricing an IPO is not an exact science, and there is always a certain level of uncertainty involved. This is because the IPO market is influenced by a wide range of external factors, such as economic conditions, political events, and investor sentiment. Additionally, the company's performance and growth prospects may change in the period between the pricing of the IPO and its actual listing on the stock exchange, which can affect investor demand and the company's valuation.

One way in which underwriters can help mitigate these uncertainties is by using a pricing mechanism that allows for price stabilization in the days following the IPO. This mechanism, known as the greenshoe option or overallotment option, allows underwriters to sell additional shares at the IPO price in the event of high demand, which can help support the price and prevent it from falling below the IPO price.

In conclusion, pricing an IPO is a complex process that involves careful consideration of multiple factors, including market conditions, the company's financial performance and growth prospects, and the pricing strategies of comparable companies. Underwriters play a critical role in this process, as they are responsible for pricing the offering and ensuring that the shares are sold to investors. While there is always a certain level of uncertainty involved in pricing an IPO, underwriters can help mitigate these uncertainties through the use of pricing mechanisms such as the greenshoe option. Ultimately, the success of an IPO depends on a variety of factors, including the company's performance and growth prospects, market conditions, and investor sentiment, and careful planning and execution are essential for a successful outcome.

PART 10: FINANCIAL STATEMENT ANALYSIS AND REPORTING

Financial statement analysis is an important tool that is used by investors, lenders, analysts, and other stakeholders to assess the financial health and performance of a business. Financial statements provide information on the financial position, performance, and cash flows of a business, which is essential for making informed decisions about investments, loans, and other financial transactions.

This part of the book provides a comprehensive overview of financial statement analysis and reporting. We will begin by discussing the purpose and importance of financial statement analysis, and how it can be used to assess the financial health of a business. We will then delve into the various components of financial statements, including the income statement, balance sheet, and cash flow statement, and explain how to analyze each component.

We will also cover financial ratios, which are an important tool for analyzing financial statements. Financial ratios are used to compare different aspects of a business, such as profitability, liquidity, and solvency. We will discuss the different types of financial ratios and how to calculate them, as well as how to interpret the results.

In addition, we will cover the different types of financial statements, including annual reports, quarterly reports, and other regulatory filings. We will explain the purpose and requirements of each type of financial statement, as well as the different regulations and reporting standards that apply.

Throughout this part of the book, we will use real-world examples and case studies to illustrate key concepts and principles. We will also provide exercises and problems to help readers develop their skills in financial statement analysis and reporting.

Overall, this part of the book is designed to provide readers with a comprehensive understanding of financial statement analysis and reporting, and how it can be used to assess the financial health and performance of a business. Whether you are an investor, lender, analyst, or business owner, this knowledge is essential for making informed decisions about financial transactions.

CHAPTER 30: INTRODUCTION TO FINANCIAL STATEMENT ANALYSIS

Financial statement analysis is a fundamental tool for evaluating the financial health and performance of a company. In this chapter, we will provide an overview of the key concepts and techniques used in financial statement analysis, and explore their application in a variety of contexts.

At its core, financial statement analysis involves the review and interpretation of a company's financial statements, which provide a snapshot of its financial position and performance over a specific period of time. These statements include the balance sheet, income statement, and cash flow statement, each of which provides unique insights into different aspects of a company's financial performance.

By analyzing these statements, financial analysts can gain a deep understanding of a company's profitability, liquidity, solvency, and overall financial health. This analysis can be used to identify trends, highlight areas of strength and weakness, and inform investment and strategic decision-making.

In this chapter, we will explore the different methods and tools used in financial statement analysis, including ratio analysis, trend analysis, and benchmarking. We will also discuss the importance of understanding the broader economic and industry context in which a company operates, and how this context can inform our interpretation of financial statements.

Finally, we will examine some of the key challenges and limitations of financial statement analysis, including the potential for misleading or incomplete information, and the need for judgment and interpretation in the analysis process.

By the end of this chapter, you should have a solid understanding of the principles and techniques used in financial statement analysis, as well as a practical understanding of how to apply these concepts in a real-world setting. Whether you are an aspiring financial analyst, business owner, or simply interested in gaining a deeper understanding of the financial performance of companies, this chapter will provide a valuable foundation for further study and analysis.

Overview of financial statements

Financial statements are a critical tool that companies use to communicate their financial performance to stakeholders. Financial statements provide valuable information about a company's financial health, including its assets, liabilities, equity, revenues, and expenses. This information is used by investors, creditors, analysts, and regulators to assess the company's financial condition and make informed decisions.

There are three primary financial statements that companies prepare and publish: the income statement, the balance sheet, and the cash flow statement. Each statement provides a unique perspective on a company's financial performance.

Income Statement

The income statement, also known as the profit and loss statement, summarizes a company's revenues and expenses over a specific period. The income statement starts with revenue, which is the income earned by the company from its operations. From revenue, the cost of goods sold is deducted to arrive at gross profit. Then, operating expenses are subtracted from gross profit to arrive at operating income, also known as earnings before interest and taxes (EBIT). Interest expenses and taxes are then deducted to arrive at net income.

The income statement provides valuable information about a company's profitability and ability to generate earnings from its operations. It also provides information on the company's expenses and how they relate to revenue.

Balance Sheet

The balance sheet provides a snapshot of a company's financial position at a specific point in time. The balance sheet reports the company's assets, liabilities, and equity. Assets are what the company owns or has a right to, and they are listed in order of liquidity. Liabilities are what the company owes, and they are also listed in order of liquidity. Equity represents the residual value of the assets after deducting liabilities.

The balance sheet provides valuable information about a company's liquidity, solvency, and financial structure. It provides information on the company's assets, liabilities, and equity and how they have changed over time.

Cash Flow Statement

The cash flow statement provides information on a company's cash inflows and outflows over a specific period. It reports the company's operating, investing, and financing activities. Operating activities refer to cash flows from the company's core business operations. Investing activities refer to cash flows from investments in long-term assets, such as property, plant, and equipment. Financing activities refer to cash flows from debt and equity financing activities, such as borrowing money or issuing stock.

The cash flow statement provides valuable information about a company's liquidity and ability to generate cash from its operations. It also provides information on the company's investing and financing activities and how they relate to the company's overall financial health.

Conclusion

Financial statements are an essential tool for assessing a company's financial performance. The income statement, balance sheet, and cash flow statement provide unique perspectives on a company's profitability, financial position, and cash flow. Understanding financial statements is critical for investors, creditors, analysts, and regulators to make informed decisions about a company's financial health.

Importance of financial statement analysis

Financial statement analysis is an essential tool for businesses, investors, and creditors. Financial statements provide a summary of a company's financial transactions, which are important for decision-making. The analysis of financial statements can be used to assess a company's profitability, liquidity, and solvency, among other things. The importance of financial statement analysis cannot be overstated, as it helps businesses make informed decisions about their future plans.

Importance of Financial Statement Analysis:

Helps in Decision Making:
Financial statement analysis provides important information to decision-makers, including investors, creditors, and managers. Financial statements can be used to assess the profitability, liquidity, and solvency of a company, which helps decision-makers make informed decisions about their investment, lending, and management activities.

Assessment of Financial Performance:

Financial statement analysis allows investors to evaluate a company's financial performance over time. By analyzing financial statements, investors can identify trends and patterns in a company's financial performance, which can help them make more informed investment decisions.

Evaluation of Creditworthiness:
Creditors use financial statement analysis to evaluate a company's creditworthiness. By analyzing a company's financial statements, creditors can assess its ability to repay its debt obligations. This helps creditors make informed decisions about lending money to a company.

Regulatory Compliance:
Businesses are required to comply with various regulatory requirements, including financial reporting standards. Financial statement analysis helps businesses ensure that they are in compliance with these requirements. By analyzing financial statements, businesses can identify any discrepancies or errors in their financial reporting, which helps them correct these issues before they become a problem.

Improvement of Financial Performance:
Financial statement analysis can also be used to identify areas of a business that need improvement. By analyzing financial statements, businesses can identify areas of inefficiency, which can then be addressed to improve the company's financial performance.

Examples:

A common example of the importance of financial statement analysis can be seen in the case of a business seeking a loan. Before a lender approves a loan application, they will typically require the business to provide financial statements. The lender will then use financial statement analysis to assess the creditworthiness of the business and determine whether they are likely to repay the loan.

Another example is the use of financial statement analysis by investors. Investors may use financial statements to assess the profitability of a company and its potential for future growth. Based on this analysis, investors can make informed decisions about buying or selling stock in the company.

Exercises:

Why is financial statement analysis important for businesses, investors, and creditors?

How can financial statement analysis be used to evaluate a company's creditworthiness?

What are some common examples of the importance of financial statement analysis in business?

Conclusion:

In conclusion, financial statement analysis is an essential tool for businesses, investors, and creditors. Financial statements provide important information about a company's financial transactions, which can be used to make informed decisions about investment, lending, and management activities. The analysis of financial statements helps businesses identify areas for improvement, investors assess a company's potential for future growth, and creditors assess a company's creditworthiness. The importance of financial statement analysis cannot be overstated, and it is a critical skill for anyone working in the financial sector.

Understanding the different financial statements

Financial statements provide critical information to investors, creditors, regulators, and other stakeholders about the financial position, performance, and cash flows of a company. Understanding financial statements is essential for decision-making in a business context. Financial statements typically include the income statement, balance sheet, statement of cash flows, and statement of changes in equity. Each of these statements provides unique insights into a company's financial health, and an understanding of how they interrelate is key to interpreting financial statements.

This section aims to provide a comprehensive overview of each financial statement and how to analyze them effectively.

Income Statement:

The income statement, also known as the profit and loss statement, provides a summary of a company's revenues, expenses, and net income over a specific period. The statement begins with the revenue section, where the company's sales or services are recorded. The cost of goods sold is then subtracted from the revenue to calculate the gross profit. The next section is the operating expenses, which include selling, general, and administrative expenses. The sum of the gross profit and the operating expenses yields operating income.

Non-operating income and expenses, such as interest income and expenses, gains or losses on investments, and other items not related to the company's core operations, are also included in the income statement. The bottom line of the income statement shows the net income, which is the profit or loss after accounting for all expenses and income.

The income statement provides critical information about a company's profitability and performance, which is essential for investors and creditors. By analyzing trends in revenue, cost of goods sold, and operating expenses over time, stakeholders can identify areas where the company is excelling or struggling.

Balance Sheet:

The balance sheet, also known as the statement of financial position, provides a snapshot of a company's assets, liabilities, and equity at a specific point in time. The balance sheet is divided into two sections: assets and liabilities and equity. The assets section lists all of the company's resources, including cash, accounts receivable, inventory, and property, plant, and equipment. The liabilities and equity section lists all of the company's obligations, including accounts payable, short-term and long-term debt, and equity.

The balance sheet equation is assets = liabilities + equity. This equation demonstrates that a company's assets are financed through either debt or equity. A high debt-to-equity ratio can indicate that a company is taking on too much debt and is risky for creditors and investors.

The balance sheet provides a snapshot of a company's financial position at a specific point in time. By analyzing trends in the balance sheet over time, stakeholders can identify areas where the company is growing or facing challenges.

Statement of Cash Flows:

The statement of cash flows provides information about the cash inflows and outflows of a company over a specific period. The statement is divided into three sections: operating activities, investing activities, and financing activities. The operating activities section shows the cash flow from the company's core operations, such as cash received from customers and cash paid to suppliers. The investing activities section shows the cash flow from investments, such as the purchase or sale of property, plant, and equipment. The financing activities section shows the cash flow from financing activities, such as the issuance or repayment of debt and equity.

The bottom line of the statement of cash flows shows the net change in cash and cash equivalents during the period. Positive cash flows indicate that a company has generated more cash than it has spent, while negative cash flows indicate that a company has spent more cash than it has generated.

The statement of cash flows is essential for understanding a company's liquidity and its ability to meet its financial obligations. By analyzing trends in cash flows over time, stakeholders can identify areas where the company is generating cash or experiencing challenges.

Statement of Changes in Equity:

The Statement of Changes in Equity (also known as the Statement of Stockholders' Equity) is another important financial statement that provides insights into a company's financial position. This statement details the changes in equity for a particular period, including the beginning balance of equity, net income, additional capital contributions, dividends paid, and any other adjustments made to equity accounts.

The statement of changes in equity is particularly important for investors, as it helps them understand how the company's retained earnings and shareholder equity have changed over time. It also provides information about how a company is funding its operations, whether through issuing new shares or retaining earnings.

For example, if a company has issued new shares, it would reflect as an increase in equity on the statement of changes in equity. Alternatively, if a company paid out dividends, this would result in a decrease in equity. By analyzing the statement of changes in equity, stakeholders can also identify any unusual changes in equity that may warrant further investigation.

It is important to note that the statement of changes in equity may not be required for all types of companies or in all jurisdictions. However, it is a common financial statement for publicly traded companies and those subject to generally accepted accounting principles (GAAP).

In summary, the statement of changes in equity provides crucial information about a company's financial position, particularly for investors and analysts. By analyzing changes in equity over time, stakeholders can gain insights into how the company is financing its operations and any significant changes to its equity accounts.

CHAPTER 31: ANALYSIS OF INCOME STATEMENTS, BALANCE SHEETS, AND CASH FLOW STATEMENTS

Overview of income statements, balance sheets, and cash flow statements
Financial statements are a fundamental aspect of any business, providing insights into its financial health and performance. Analysis of income statements, balance sheets, and cash flow statements is crucial for making informed decisions about investment, credit, and other financial matters.

Chapter 31 will cover the basics of analyzing these financial statements and understanding the various metrics and ratios used to evaluate a company's performance. This chapter will provide an overview of income statements, balance sheets, and cash flow statements, including their purpose, content, and importance.

The chapter will begin with an explanation of the different types of financial statements and how they are used in financial analysis. It will then delve into the details of each statement, discussing their components, structure, and significance.

The first section of the chapter will focus on the income statement, which shows a company's revenues and expenses over a specific period. It will explain the various components of an income statement, such as revenues, cost of goods sold, gross profit, operating expenses, and net income.

The second section will discuss the balance sheet, which provides a snapshot of a company's financial position at a specific point in time. It will explain the different components of a balance sheet, such as assets, liabilities, and equity.

The third section will focus on the statement of cash flows, which shows a company's cash inflows and outflows over a specific period. It will explain the different categories of cash flows, including operating, investing, and financing activities.

The final section of the chapter will discuss the importance of analyzing financial statements in making investment and credit decisions. It will provide an overview of the various metrics and ratios used in financial analysis, such as liquidity ratios, profitability ratios, and debt ratios.

Overall, this chapter will provide a comprehensive introduction to the analysis of income statements, balance sheets, and cash flow statements, covering their purpose, content, and importance. It will serve as a valuable resource for students seeking to understand the basics of financial statement analysis and its applications in the real world.

Analysis of revenue, expenses, assets, and liabilities

Understanding financial statements is crucial for evaluating the performance of a company. Financial statements provide a snapshot of a company's financial health by presenting information about its revenue, expenses, assets, and liabilities. Analyzing these statements allows investors, creditors, and other stakeholders to make informed decisions about investing in or doing business with the company.

One of the key components of financial statement analysis is the analysis of revenue, expenses, assets, and liabilities. This chapter will provide an in-depth analysis of these components and discuss the different techniques used to analyze them.

Analysis of Revenue:

Revenue is the amount of money a company earns from selling its products or services. Analyzing revenue is essential for understanding a company's growth and profitability. There are several methods for analyzing revenue, including trend analysis, comparative analysis, and ratio analysis.

Trend analysis involves comparing revenue over time to identify patterns and trends. This analysis can help identify whether revenue is growing, declining, or remaining stable. Comparative analysis involves comparing a company's revenue with that of its competitors to identify strengths and weaknesses. Ratio analysis involves calculating ratios such as gross margin, net profit margin, and return on investment to evaluate a company's revenue performance.

Analysis of Expenses:

Expenses refer to the costs incurred by a company in the process of generating revenue. Analyzing expenses is crucial for understanding a company's profitability and cost efficiency. There are several methods for analyzing expenses, including trend analysis, comparative analysis, and ratio analysis.

Trend analysis involves comparing expenses over time to identify patterns and trends. This analysis can help identify whether expenses are increasing, decreasing, or remaining stable. Comparative analysis involves comparing a company's expenses with that of its competitors to identify strengths and weaknesses. Ratio analysis involves calculating ratios such as expense ratio, operating expense ratio, and return on investment to evaluate a company's expense performance.

Analysis of Assets:

Assets are resources owned by a company that can be used to generate revenue. Analyzing assets is crucial for understanding a company's ability to generate cash flow and create value for its shareholders. There are several methods for analyzing assets, including trend analysis, comparative analysis, and ratio analysis.

Trend analysis involves comparing assets over time to identify patterns and trends. This analysis can help identify whether assets are growing, declining, or remaining stable. Comparative analysis involves comparing a company's assets with that of its competitors to identify strengths and weaknesses. Ratio analysis involves calculating ratios such as return on assets, asset turnover, and debt-to-equity ratio to evaluate a company's asset performance.

Analysis of Liabilities:

Liabilities refer to the obligations a company owes to others, including creditors and suppliers. Analyzing liabilities is crucial for understanding a company's financial obligations and ability to meet them. There are several methods for analyzing liabilities, including trend analysis, comparative analysis, and ratio analysis.

Trend analysis involves comparing liabilities over time to identify patterns and trends. This analysis can help identify whether liabilities are growing, declining, or remaining stable. Comparative analysis involves comparing a company's liabilities with that of its competitors to identify strengths and weaknesses. Ratio analysis involves calculating ratios such as debt-to-equity ratio, debt-to-asset ratio, and interest coverage ratio to evaluate a company's liability performance.

Conclusion:

Analyzing revenue, expenses, assets, and liabilities is crucial for understanding a company's financial performance. The different methods for analyzing these components provide a comprehensive view of a company's financial health and allow stakeholders to make informed decisions about investing in or doing business with the company. By utilizing the tools and techniques discussed in this chapter, stakeholders can gain a deeper understanding of financial statements and make more informed decisions.

Understanding the relationship between the different financial statements

Financial statements are essential documents that provide stakeholders with valuable information about a company's financial performance. They help in making informed decisions about investing, lending, or doing business with a company. The three primary financial statements are the income statement, balance sheet, and statement of cash flows. Each statement provides unique information that is crucial for analyzing a company's financial health.

Understanding the relationship between the different financial statements is crucial for effective financial analysis. In this section, we will discuss how the income statement, balance sheet, and statement of cash flows are interrelated and how they can be used together to provide a comprehensive picture of a company's financial performance.

Relationship between the Income Statement and the Balance Sheet:

The income statement and the balance sheet are closely related as they provide information about a company's financial performance over a specific period. The income statement shows a company's revenues, expenses, gains, and losses, resulting in its net income or net loss for a period. On the other hand, the balance sheet provides a snapshot of a company's financial position at a specific point in time, including its assets, liabilities, and equity.

The income statement and the balance sheet are connected through retained earnings, which is the cumulative net income earned by a company since its inception, less any dividends paid to shareholders. Retained earnings are reported on the balance sheet as part of the equity section. The net income reported on the income statement increases retained earnings, while net losses decrease retained earnings.

Moreover, the income statement and the balance sheet are also connected through the concept of accrual accounting. Accrual accounting requires companies to record revenues and expenses when they are earned or incurred, regardless of when cash is received or paid. As a result, the income statement reports revenues and expenses that have been earned or incurred during the period, even if cash has not yet been received or paid. The balance sheet reports the cash and accounts receivable balances related to the revenues reported on the income statement and the accounts payable balances related to the expenses reported on the income statement.

Relationship between the Income Statement and the Statement of Cash Flows:

The income statement and the statement of cash flows are also closely related as they both provide information about a company's financial performance. However, they focus on different aspects of a company's financial performance. The income statement reports a company's revenues, expenses, gains, and losses, resulting in its net income or net loss for a period. The statement of cash flows, on the other hand, provides information about a company's cash inflows and outflows during a period.

The income statement and the statement of cash flows are connected through the concept of cash-based accounting. Cash-based accounting requires companies to record revenues and expenses when cash is received or paid. As a result, the statement of cash flows reports the cash inflows and outflows related to the revenues and expenses reported on the income statement.

Moreover, the statement of cash flows provides additional information that is not reported on the income statement, such as cash flows related to investing and financing activities. By analyzing the statement of cash flows, stakeholders can gain insights into a company's ability to generate cash, invest in its operations, and meet its financial obligations.

Relationship between the Balance Sheet and the Statement of Cash Flows:

The balance sheet and the statement of cash flows are also closely related as they both provide information about a company's financial position. However, they focus on different aspects of a company's financial position. The balance sheet provides a snapshot of a company's financial position at a specific point in time, including its assets, liabilities, and equity. The statement of cash flows, on the other hand, provides information about a company's cash inflows and outflows during a period.

The balance sheet and the statement of cash flows are connected through the concept of cash-based accounting. Cash-based accounting is a method of accounting that recognizes revenue and expenses when cash is received or paid out, respectively. This means that changes in a company's cash balance will be reflected in both the balance sheet and the statement of cash flows.

For example, if a company generates revenue from the sale of goods or services and receives payment in cash, this will be reflected as an increase in both the cash balance on the balance sheet and the cash inflow section of the statement of cash flows. Conversely, if a company pays its suppliers in cash for materials or services, this will be reflected as a decrease in both the cash balance on the balance sheet and the cash outflow section of the statement of cash flows.

The balance sheet and the statement of cash flows can also be used together to assess a company's liquidity and solvency. Liquidity refers to a company's ability to meet its short-term financial obligations, while solvency refers to a company's ability to meet its long-term financial obligations. By analyzing the changes in a company's cash balance and its cash inflows and outflows over time, stakeholders can gain insight into the company's ability to generate cash and meet its financial obligations.

For example, if a company has a large cash balance on its balance sheet but is experiencing negative cash flows from operations, this may indicate that the company is relying on external financing to fund its operations. On the other hand, if a company has a relatively low cash balance on its balance sheet but is generating positive cash flows from operations, this may indicate that the company is effectively managing its cash and generating sufficient cash to meet its financial obligations.

In summary, the relationship between the balance sheet and the statement of cash flows is essential for understanding a company's financial position and performance. Both financial statements provide important information about a company's assets, liabilities, and equity, as well as its cash inflows and outflows. By analyzing these financial statements together, stakeholders can gain insight into a company's liquidity, solvency, and cash management practices.

CHAPTER 32: RATIO ANALYSIS AND INTERPRETATION

Ratio analysis is a powerful tool used by businesses, investors, and financial analysts to interpret and understand financial statements. Ratios are calculated by dividing one financial statement item by another, providing a means to compare and evaluate different aspects of a company's financial performance. This chapter will provide an overview of the most commonly used financial ratios and how they can be interpreted to gain insight into a company's financial health.

Understanding Financial Ratios

There are many different financial ratios that can be calculated from a company's financial statements. These ratios can be grouped into several categories, including liquidity ratios, solvency ratios, profitability ratios, and market value ratios.

Liquidity ratios measure a company's ability to meet its short-term financial obligations. The most commonly used liquidity ratios are the current ratio and the quick ratio. The current ratio measures a company's ability to pay its current liabilities with its current assets, while the quick ratio measures the company's ability to pay its current liabilities with its most liquid assets (excluding inventory).

Solvency ratios, also known as leverage ratios, measure a company's ability to meet its long-term financial obligations. The debt-to-equity ratio, the interest coverage ratio, and the debt-to-assets ratio are all examples of solvency ratios.

Profitability ratios measure a company's ability to generate profits. The return on equity (ROE) ratio, the return on assets (ROA) ratio, and the gross profit margin are all examples of profitability ratios.

Market value ratios measure a company's market value relative to its financial performance. The price-to-earnings ratio (P/E ratio) and the market-to-book ratio are examples of market value ratios.

Interpreting Financial Ratios

Interpreting financial ratios is an important part of analyzing a company's financial performance. It is not enough to simply calculate the ratios; they must be compared to industry averages and historical trends to gain insight into a company's financial health.

A ratio that is higher than the industry average may indicate that a company is performing well in that area. Conversely, a ratio that is lower than the industry average may indicate that a company is underperforming in that area.

However, it is important to note that financial ratios should not be viewed in isolation. They should be considered alongside other factors, such as a company's operating environment, management decisions, and economic conditions.

Example of Financial Ratio Analysis

To illustrate the use of financial ratio analysis, consider the following example. ABC Company has provided the following financial statements:

Income Statement:

Revenue: $1,000,000
Cost of Goods Sold: $600,000
Gross Profit: $400,000
Operating Expenses: $250,000
Net Income: $150,000

Balance Sheet:

Assets:
Cash: $100,000
Accounts Receivable: $150,000
Inventory: $200,000
Total Current Assets: $450,000
Property, Plant, and Equipment: $400,000
Total Assets: $850,000

Liabilities:
Accounts Payable: $100,000
Short-term Debt: $50,000
Total Current Liabilities: $150,000
Long-term Debt: $300,000
Total Liabilities: $450,000

Equity:
Common Stock: $200,000
Retained Earnings: $200,000
Total Equity: $400,000

Using these financial statements, we can calculate several financial ratios:

Current Ratio = Current Assets / Current Liabilities = $450,000 / $150,000 = 3.0
Quick Ratio = (Current Assets - Inventory) / Current Liabilities = ($450,000 - $200,000) / $150,000 = 1.0
Debt Equity Ratio = Total Liabilities / Total Equity = $300,000 / $400,000 = 0.75
Debt Ratio = Total Debt / Total Assets = $150,000 / $750,000 = 0.20
Profit Margin = Net Income / Revenue = $50,000 / $500,000 = 0.10
Return on Assets (ROA) = Net Income / Total Assets = $50,000 / $750,000 = 0.0667 or 6.67%
Return on Equity (ROE) = Net Income / Total Equity = $50,000 / $400,000 = 0.125 or 12.5%

These financial ratios provide useful information to investors and analysts to assess a company's financial performance and health. The current ratio and quick ratio measure a company's ability to pay off its current liabilities with its current assets. A current ratio of 3.0 indicates that the company has $3 in current assets for every $1 in current liabilities, which is considered a good liquidity position. The quick ratio of 1.0 indicates that the company can cover its current liabilities with its quick assets, which are assets that can be easily converted to cash.

The debt-to-equity ratio measures a company's leverage or the amount of debt it uses to finance its operations relative to its equity. A debt-to-equity ratio of 0.75 indicates that the company has $0.75 in debt for every $1 in equity, which suggests that the company has a conservative capital structure.

The debt ratio measures the proportion of a company's assets that are financed with debt. A debt ratio of 0.20 indicates that 20% of the company's assets are financed with debt, while the remaining 80% is financed with equity. A lower debt ratio indicates that the company has a lower risk of default.

The profit margin measures a company's profitability by comparing its net income to its revenue. A profit margin of 0.10 indicates that the company earns $0.10 in profit for every dollar of revenue generated. A higher profit margin is generally preferred as it indicates that the company is more efficient in generating profits from its operations.

The return on assets (ROA) measures how efficiently a company uses its assets to generate profits. A higher ROA indicates that the company is more efficient in generating profits from its assets. The return on equity (ROE) measures the return that shareholders earn on their investments. A higher ROE indicates that the company is generating more profits from the capital invested by its shareholders.

In conclusion, financial ratio analysis is an important tool for investors and analysts to evaluate a company's financial performance and health. By analyzing a company's financial statements and calculating financial ratios, investors and analysts can gain insight into a company's liquidity, leverage, profitability, and efficiency. These financial ratios can help investors and analysts make informed decisions about whether to invest in a company or not.

Overview of financial ratios and their importance

Financial ratios are commonly used tools that help investors, analysts, and managers evaluate a company's financial performance and health. They are calculated using data from a company's financial statements, which provide a snapshot of the company's financial position at a specific point in time or over a period of time.

Financial ratios can be classified into several categories, including liquidity ratios, solvency ratios, profitability ratios, and market value ratios. Each of these categories measures a different aspect of a company's financial performance, and investors and analysts use them to gain insights into the company's financial health.

Liquidity Ratios

Liquidity ratios measure a company's ability to meet its short-term obligations. These ratios are important to investors and creditors because they indicate whether a company has sufficient cash and other liquid assets to cover its short-term liabilities.

The most commonly used liquidity ratios include the current ratio and the quick ratio. The current ratio measures a company's ability to pay its short-term obligations with its current assets. The quick ratio, also known as the acid-test ratio, measures a company's ability to pay its short-term obligations without relying on the sale of inventory.

Solvency Ratios

Solvency ratios measure a company's ability to meet its long-term obligations. These ratios are important to investors and creditors because they indicate whether a company has sufficient assets to cover its long-term liabilities.

The most commonly used solvency ratios include the debt-to-equity ratio and the interest coverage ratio. The debt-to-equity ratio measures a company's leverage, or the extent to which it is financed by debt versus equity. The interest coverage ratio measures a company's ability to meet its interest payments on outstanding debt.

Profitability Ratios

Profitability ratios measure a company's ability to generate profits from its operations. These ratios are important to investors and analysts because they indicate whether a company is able to generate sufficient profits to sustain its operations and provide a return to its shareholders.

The most commonly used profitability ratios include the return on equity and the return on assets. The return on equity measures a company's profitability relative to the amount of equity invested in the company by its shareholders. The return on assets measures a company's profitability relative to the total assets it has.

Market Value Ratios

Market value ratios measure a company's market value relative to its earnings, assets, or other financial metrics. These ratios are important to investors because they indicate whether a company's stock is undervalued or overvalued relative to its financial performance.

The most commonly used market value ratios include the price-to-earnings ratio and the price-to-book ratio. The price-to-earnings ratio measures a company's stock price relative to its earnings per share. The price-to-book ratio measures a company's stock price relative to its book value per share.

Importance of Financial Ratios

Financial ratios are important tools that help investors and analysts evaluate a company's financial performance and health. They provide insights into a company's liquidity, solvency, profitability, and market value, which can help investors make informed decisions about whether to invest in a company's stock or debt.

Financial ratios are also important to managers because they can help identify areas where a company is performing well and areas where improvement is needed. For example, if a company's liquidity ratios indicate that it has insufficient cash to meet its short-term obligations, the company's management may need to take steps to improve its cash flow, such as reducing expenses or increasing revenue.

Examples and Exercises

Example 1:

ABC Company has current assets of $500,000 and current liabilities of $200,000. Calculate the company's current ratio.

Solution:

Current Ratio = Current Assets / Current Liabilities = $500,000 / $200,000 = 2.
Example 2:

DEF Corporation has total assets of $1,500,000, total liabilities of $500,000, and total equity of $1,000,000. Calculate the debt-to-equity ratio.

Solution:

Debt-to-Equity Ratio = Total Liabilities / Total Equity = $500,000 / $1,000,000 = 0.5

Example 3:

GHI Inc. has net income of $100,000, total assets of $1,000,000, and total equity of $500,000. Calculate the return on equity (ROE) ratio.

Solution:

Return on Equity (ROE) Ratio = Net Income / Total Equity = $100,000 / $500,000 = 0.2 or 20%

Example 4:

JKL Corporation has a current ratio of 2.0 and inventory turnover ratio of 4. Calculate the company's quick ratio.

Solution:

Inventory Turnover Ratio = Cost of Goods Sold / Average Inventory
Assuming the Cost of Goods Sold is $1,000,000 and the average inventory is $250,000:
Inventory Turnover Ratio = $1,000,000 / $250,000 = 4
Quick Ratio = (Current Assets - Inventory) / Current Liabilities = ($500,000 - $250,000) / $250,000 = 1

Exercise 1:

MNO Inc. has current assets of $750,000 and current liabilities of $500,000. Calculate the company's current ratio.

Exercise 2:

PQR Company has total assets of $2,000,000, total liabilities of $1,000,000, and total equity of $1,000,000. Calculate the debt-to-equity ratio.

Exercise 3:

STU Corporation has net income of $150,000, total assets of $1,500,000, and total equity of $500,000. Calculate the return on equity (ROE) ratio.

Exercise 4:

VWX Inc. has a current ratio of 3.0 and inventory turnover ratio of 5. Calculate the company's quick ratio.

Calculation and interpretation of liquidity, profitability, and solvency ratios

Financial ratios are tools that help investors and analysts assess a company's financial health and performance. They provide a way to quantify a company's financial statements and compare them to other companies or industry standards. The three main categories of financial ratios are liquidity ratios, profitability ratios, and solvency ratios. This section will provide an overview of these categories, as well as examples and exercises to help students understand how to calculate and interpret these ratios.

Liquidity Ratios:

Liquidity ratios measure a company's ability to meet its short-term financial obligations. In other words, they assess whether a company has enough liquid assets to pay off its current liabilities. The two main liquidity ratios are the current ratio and the quick ratio.

The current ratio is calculated by dividing a company's current assets by its current liabilities. A current ratio of 1 or higher indicates that a company has enough current assets to cover its current liabilities. A current ratio of less than 1 may indicate that a company may have difficulty paying off its current liabilities.

The quick ratio, also known as the acid-test ratio, is a more conservative liquidity ratio that takes into account a company's inventory. It is calculated by subtracting inventory from current assets and then dividing by current liabilities. The quick ratio provides a more accurate picture of a company's ability to pay off its short-term obligations because inventory may not be easily convertible into cash.

Example:

XYZ Company has current assets of $500,000 and current liabilities of $200,000. Calculate the company's current ratio and quick ratio.

Solution:

Current Ratio = Current Assets / Current Liabilities = $500,000 / $200,000 = 2.5
Quick Ratio = (Current Assets - Inventory) / Current Liabilities = ($500,000 - $100,000) / $200,000 = 2.0

Interpretation:

Both the current ratio and quick ratio indicate that XYZ Company has enough liquid assets to cover its current liabilities. However, the quick ratio is more conservative and provides a more accurate picture of XYZ Company's ability to pay off its short-term obligations.

Profitability Ratios:

Profitability ratios measure a company's ability to generate profits relative to its sales, assets, or equity. They provide insight into a company's efficiency and effectiveness in generating profits. The two main profitability ratios are the gross profit margin and the net profit margin.

The gross profit margin is calculated by dividing a company's gross profit by its revenue. Gross profit is calculated by subtracting the cost of goods sold from revenue. The gross profit margin indicates how much profit a company is generating from its sales before accounting for other expenses such as operating expenses, interest, and taxes.

The net profit margin is calculated by dividing a company's net profit by its revenue. Net profit is calculated by subtracting all expenses, including operating expenses, interest, and taxes, from revenue. The net profit margin indicates how much profit a company is generating from its sales after accounting for all expenses.

Example:

ABC Company has revenue of $1,000,000, cost of goods sold of $600,000, operating expenses of $200,000, interest expense of $50,000, and taxes of $50,000. Calculate the company's gross profit margin and net profit margin.

Solution:

Gross Profit Margin = (Revenue - Cost of Goods Sold) / Revenue = ($1,000,000 - $600,000) / $1,000,000 = 0.4 or 40%
Net Profit Margin = Net Profit / Revenue = ($1,000,000 - $600,000 - $200,000 - $50,000 - $50,000) / $1,000,000 = 0.1 or 10%

Interpretation:

ABC Company's gross profit margin of 40% means that the company is earning 40 cents on every dollar of sales after accounting for the direct costs of producing those goods. A higher gross profit margin indicates that the company is generating more revenue than it is spending on production costs, which is a positive sign for investors and may indicate that the company has a competitive advantage.

ABC Company's net profit margin of 10% means that the company is earning 10 cents on every dollar of sales after accounting for all expenses, including operating expenses, interest, and taxes. This indicates that the company is able to generate a profit on its sales, which is essential for long-term success. A higher net profit margin is generally preferred, as it indicates that the company is more efficient at managing its costs and generating profits.

Investors and analysts often use profitability ratios to evaluate a company's ability to generate profits relative to its revenue and expenses. A company with a high gross profit margin and net profit margin is typically viewed as more profitable and efficient than a company with low profit margins. However, it is important to consider the industry and economic conditions when evaluating profitability ratios, as different industries and market conditions may affect profitability.

Comparison with industry benchmarks and historical trends

Comparison with industry benchmarks and historical trends is an important step in analyzing financial ratios. It provides insights into how a company is performing relative to its competitors and over time.

Industry benchmarks are ratios that represent the average or median ratios for companies within a particular industry. These benchmarks are useful in understanding how a company compares to its peers and identifying areas where it may be performing better or worse than the industry average. Some common sources of industry benchmarks include trade associations, industry reports, and financial databases.

Historical trends, on the other hand, compare a company's financial ratios over time. This allows for the identification of trends and changes in a company's financial performance. For example, a company's profitability may be increasing over time, while its liquidity is decreasing. This information can be useful in understanding the company's financial trajectory and identifying areas for improvement.

When comparing a company's financial ratios to industry benchmarks and historical trends, it is important to consider the specific circumstances of the company. Factors such as company size, market conditions, and strategic decisions can all affect a company's financial ratios. Additionally, benchmarks and trends can vary across industries and over time.

To illustrate the importance of comparing financial ratios to industry benchmarks and historical trends, consider the following example:

Example:

XYZ Corporation operates in the retail industry and has a current ratio of 2.5. The industry benchmark for the retail industry is 3.0. Is XYZ Corporation's current ratio above or below the industry average?

Solution:

XYZ Corporation's current ratio of 2.5 is below the industry benchmark of 3.0. This suggests that the company may be less liquid than its competitors in the retail industry.

To further analyze this situation, XYZ Corporation should compare its current ratio to its historical trends. If the company's current ratio has been decreasing over time, this could indicate a concerning trend that requires further investigation. Alternatively, if the company's current ratio has remained relatively stable over time, this may suggest that the current ratio is not a significant concern and that the company's financial performance is generally in line with historical trends.

Overall, comparing financial ratios to industry benchmarks and historical trends is an essential step in analyzing a company's financial performance. This information can provide valuable insights into a company's strengths and weaknesses, as well as identifying areas for improvement. However, it is important to consider the specific circumstances of the company and to use benchmarks and trends as one of many tools for financial analysis.

PART 11: FINANCIAL REPORTING STANDARDS AND REGULATIONS

Financial reporting is an essential aspect of corporate governance, enabling stakeholders to gain insights into the financial performance of companies. To ensure that financial reporting is reliable, relevant, and comparable across companies, various financial reporting standards and regulations have been put in place by regulatory bodies, such as the Financial Accounting Standards Board (FASB) and the International Accounting Standards Board (IASB). These standards and regulations help to ensure that companies adhere to a set of rules and principles when preparing their financial statements.

This part of the textbook will provide an overview of financial reporting standards and regulations, including the Generally Accepted Accounting Principles (GAAP), International Financial Reporting Standards (IFRS), and the regulatory bodies responsible for setting these standards. It will also explore the importance of financial reporting standards and regulations, the challenges faced by companies in complying with them, and the consequences of non-compliance.

Generally Accepted Accounting Principles (GAAP)

The Generally Accepted Accounting Principles (GAAP) is a set of accounting principles, standards, and procedures that companies in the United States must adhere to when preparing their financial statements. The Financial Accounting Standards Board (FASB) is the primary body responsible for developing and maintaining GAAP.

The chapter will provide an overview of GAAP, including its history, the primary principles and concepts it encompasses, and how it is enforced. It will also explore the benefits of GAAP, such as providing comparability and consistency across financial statements, and the challenges that companies face in complying with GAAP.

International Financial Reporting Standards (IFRS)

International Financial Reporting Standards (IFRS) are a set of accounting standards developed and maintained by the International Accounting Standards Board (IASB) for companies outside of the United States. IFRS aims to provide a common global language for business affairs to ensure that financial statements are transparent, comparable, and of high quality.

The chapter will provide an overview of IFRS, including its history, the primary principles and concepts it encompasses, and how it is enforced. It will also explore the benefits of IFRS, such as providing comparability and consistency across financial statements, and the challenges that companies face in complying with IFRS.

Regulatory Bodies and Enforcement

Regulatory bodies play a crucial role in setting financial reporting standards and enforcing compliance with these standards. The chapter will provide an overview of the regulatory bodies responsible for setting financial reporting standards, such as the FASB and IASB, and how they operate. It will also explore the enforcement mechanisms used to ensure compliance with financial reporting standards, such as audits, inspections, and penalties for non-compliance.

Importance and Challenges of Financial Reporting Standards and Regulations

The chapter will explore the importance of financial reporting standards and regulations, including the benefits they provide to stakeholders, such as transparency, comparability, and accountability. It will also discuss the challenges that companies face in complying with these standards, such as the complexity of the standards, the cost of compliance, and the need for continuous updates.

Consequences of Non-Compliance

The consequences of non-compliance with financial reporting standards and regulations can be severe, including financial penalties, damage to a company's reputation, and legal action. The chapter will explore the consequences of non-compliance and the measures companies can take to ensure compliance.

Conclusion

Financial reporting standards and regulations play a critical role in ensuring that financial statements are reliable, relevant, and comparable across companies. The introduction of this part of the textbook provides an overview of financial reporting standards and regulations, including the GAAP, IFRS, and the regulatory bodies responsible for setting these standards. It also explores the importance of financial reporting standards and regulations, the challenges faced by companies in complying with these standards, and the benefits of adhering to these standards.

One of the main benefits of complying with financial reporting standards and regulations is the increased transparency and trust it provides to stakeholders. Investors and creditors rely on accurate and reliable financial statements to make informed decisions about where to allocate their resources. Companies that follow these standards and regulations are seen as more trustworthy and reliable, which can result in increased investor confidence and higher stock prices.

In addition, complying with financial reporting standards and regulations can also help companies identify areas where they can improve their operations and increase profitability. By analyzing financial statements in accordance with established standards, companies can identify areas of weakness and take steps to address them. This can lead to more efficient operations, better decision making, and increased profitability.

However, there are also challenges associated with complying with financial reporting standards and regulations. For example, the complex nature of these standards can make them difficult to interpret and apply. Additionally, companies may face significant costs associated with complying with these standards, such as the cost of hiring additional staff or implementing new accounting systems.

Despite these challenges, it is essential for companies to comply with financial reporting standards and regulations. Failure to do so can result in legal and financial consequences, such as fines, lawsuits, and loss of investor confidence. Therefore, it is important for companies to stay up-to-date with the latest standards and regulations and to make sure that their financial statements are accurate, reliable, and comparable.

Overall, financial reporting standards and regulations play a crucial role in promoting transparency and trust in financial reporting. By adhering to these standards, companies can improve their operations, increase investor confidence, and avoid legal and financial consequences. As such, it is essential for companies to understand and comply with these standards and regulations.

CHAPTER 33: GENERALLY ACCEPTED ACCOUNTING PRINCIPLES (GAAP)

Chapter 33 of this textbook provides an overview of Generally Accepted Accounting Principles (GAAP), the set of accounting standards used in the United States. GAAP provides a framework for companies to prepare financial statements that are accurate, reliable, and comparable. This chapter will discuss the history and development of GAAP, its key principles, and how it is used in practice. We will also examine some of the challenges and criticisms of GAAP, as well as its relationship to other accounting standards such as International Financial Reporting Standards (IFRS).

History and Development of GAAP

The history of GAAP can be traced back to the early 20th century, when the Securities and Exchange Commission (SEC) was created to regulate the securities industry in the United States. The SEC recognized the need for a set of accounting standards that could be used by companies to prepare their financial statements in a consistent and comparable manner. In 1934, the SEC authorized the creation of the Committee on Accounting Procedure (CAP), which was responsible for developing accounting standards for companies in the United States.

Over time, the role of the CAP was taken over by the Accounting Principles Board (APB), which was established in 1959. The APB continued to develop and issue accounting standards until 1973, when it was replaced by the Financial Accounting Standards Board (FASB). The FASB is the primary body responsible for setting accounting standards in the United States today.

Key Principles of GAAP

GAAP is based on a set of key principles that govern how financial statements should be prepared. These principles include:

Entity concept: Financial statements should reflect the financial position and performance of a specific company, rather than the interests of its owners or other stakeholders.

Going concern concept: Financial statements should assume that the company will continue to operate for the foreseeable future, and that it will not be forced to liquidate its assets.

Historical cost concept: Assets and liabilities should be recorded at their original cost, rather than at their current market value.

Revenue recognition principle: Revenue should be recognized when it is earned, regardless of when payment is received.

Matching principle: Expenses should be recognized in the same period as the revenue they help to generate.

Full disclosure principle: Companies should provide complete and transparent information in their financial statements, including any relevant information that might affect the decisions of investors or other stakeholders.

Consistency principle: Financial statements should be prepared in a consistent manner from one period to the next, so that they can be compared over time.

Use of GAAP in Practice

GAAP is used by companies in the United States to prepare their financial statements in a consistent and comparable manner. This allows investors and other stakeholders to compare the financial performance of different companies, and to make informed decisions about where to invest their money.

In addition to companies, GAAP is also used by accounting professionals, financial analysts, and regulators. Accounting professionals use GAAP to prepare financial statements, while financial analysts use GAAP to analyze and evaluate the financial performance of companies. Regulators, such as the SEC, use GAAP to ensure that companies are following the appropriate accounting standards when preparing their financial statements.

Challenges and Criticisms of GAAP

Despite its importance in the accounting profession, GAAP has faced a number of challenges and criticisms over the years. One of the main criticisms of GAAP is that it can be overly complex and difficult to understand, particularly for non-accountants. This can make it challenging for investors and other stakeholders to interpret financial statements and make informed decisions.

Another criticism of GAAP is that it can be slow to adapt to changing business environments and new technologies. This can result in accounting standards that are outdated or no longer relevant. For example, GAAP does not currently have specific guidelines for accounting for digital assets, which can create ambiguity and confusion for companies operating in the digital space.

Critics have also argued that GAAP is too focused on compliance and meeting regulatory requirements, rather than on providing useful information to investors and other stakeholders. This can lead to a lack of transparency and a focus on meeting accounting standards rather than on presenting a clear and accurate picture of a company's financial performance.

There is also a growing concern that GAAP is not well-suited to the needs of small and medium-sized enterprises (SMEs). SMEs often face unique challenges in financial reporting, including limited resources and expertise, and may struggle to comply with complex GAAP requirements. Some argue that GAAP should be adapted to better serve the needs of SMEs, or that alternative reporting frameworks should be developed specifically for this group.

Despite these challenges and criticisms, GAAP remains the dominant accounting framework in the United States and is widely used around the world. Its importance in the accounting profession is unlikely to diminish in the near future, although ongoing efforts to improve and update the framework will be necessary to address some of the challenges it faces.

Example:

To better understand the challenges and criticisms of GAAP, consider the example of a small software startup. The company has limited resources and expertise in accounting, but is required to comply with GAAP in order to produce accurate financial statements. However, many of the GAAP requirements are complex and difficult to understand, making it challenging for the company to accurately report its financial performance. In addition, the company operates in the digital space and holds significant amounts of digital assets, which are not currently covered by GAAP. This creates ambiguity and uncertainty around how the company should report these assets, which could lead to inaccuracies or misunderstandings among stakeholders. Despite these challenges, the company must continue to comply with GAAP in order to maintain its credibility and meet regulatory requirements.

Overview of GAAP and its role in financial reporting

Generally Accepted Accounting Principles (GAAP) are a set of accounting rules and standards that provide guidance on how to prepare and present financial statements. These principles are used by companies to ensure that their financial statements are reliable, relevant, and comparable across companies. GAAP is a set of guidelines that are used in the United States to standardize financial accounting practices.

GAAP has a significant impact on financial reporting, as it provides a framework for companies to report their financial information in a way that is consistent with industry standards. It is important for companies to follow GAAP because it helps ensure that their financial statements are accurate and provide relevant information to stakeholders such as investors, lenders, and regulators.

The Financial Accounting Standards Board (FASB) is responsible for setting and updating GAAP in the United States. The FASB is an independent organization that works with the accounting profession to develop and update accounting standards. In addition to the FASB, the Securities and Exchange Commission (SEC) also has the authority to set and enforce accounting standards for publicly traded companies in the United States.

GAAP provides guidance on a variety of accounting topics, including revenue recognition, inventory valuation, and depreciation. It also provides guidance on how to prepare financial statements such as balance sheets, income statements, and statements of cash flows. Following GAAP ensures that financial statements are prepared in a consistent and understandable manner, which allows for accurate comparison of financial statements across companies and industries.

The Role of GAAP in Financial Reporting

GAAP plays a crucial role in financial reporting as it provides a standard set of guidelines for companies to follow when preparing financial statements. By following GAAP, companies can ensure that their financial statements are accurate and provide relevant information to stakeholders such as investors, lenders, and regulators. This is important because financial statements are used to make informed decisions about the company's financial health.

GAAP also provides a framework for auditing financial statements. Auditors use GAAP to ensure that a company's financial statements are presented fairly and accurately. This is important because auditors provide an independent verification of a company's financial statements, which can increase the reliability of the information presented.

Challenges and Criticisms of GAAP

Despite its importance in the accounting profession, GAAP has faced a number of challenges and criticisms over the years. One of the main criticisms of GAAP is that it can be overly complex and difficult to understand, particularly for non-accountants. This can make it challenging for investors and other stakeholders to interpret financial statements and make informed decisions.

Another criticism of GAAP is that it can be slow to adapt to changing business environments and new technologies. This can result in accounting standards that are outdated or no longer relevant. For example, GAAP does not currently provide guidance on how to account for cryptocurrency transactions, which can make it difficult for companies to accurately report these transactions.

In addition, some critics argue that GAAP can be too rules-based, which can result in companies following the letter of the law rather than the spirit of the law. This can lead to financial statements that are technically compliant with GAAP but may not provide a true representation of the company's financial position.

Overall, GAAP plays a crucial role in financial reporting by providing a standardized set of guidelines for companies to follow when preparing financial statements. While it is not perfect, it is important for companies to follow GAAP in order to ensure that their financial statements are accurate and provide relevant information to stakeholders.

Principles-based vs. rules-based accounting

Accounting standards play an important role in the preparation and presentation of financial statements. They provide guidance on how financial information should be presented, the measurement and recognition of items in the financial statements, and the disclosure requirements. There are two primary approaches to accounting standards: principles-based and rules-based accounting. This section will provide an in-depth analysis of these two approaches and compare their advantages and disadvantages.

Principles-Based Accounting

Principles-based accounting is a framework that relies on a set of overarching principles to guide the preparation and presentation of financial statements. The principles provide a framework for judgment and require that management exercise professional judgment when making accounting decisions. The principles are usually broad and provide flexibility to the preparers of financial statements to apply them in a way that is appropriate for their particular circumstances.

Advantages of Principles-Based Accounting

The primary advantage of principles-based accounting is its flexibility. The principles provide a framework for judgment and allow for the exercise of professional judgment in the application of the principles. This enables preparers to tailor their accounting policies and practices to their specific circumstances, which can result in more accurate and relevant financial statements.

Another advantage of principles-based accounting is that it encourages a focus on substance over form. The principles emphasize the economic substance of transactions rather than their legal form. This can result in more meaningful financial statements that accurately reflect the underlying economic reality of the transactions.

Disadvantages of Principles-Based Accounting

One of the criticisms of principles-based accounting is that it can be subjective. The broad principles require the exercise of professional judgment, which can result in different interpretations and applications of the principles. This can lead to inconsistency in the application of the principles and a lack of comparability between financial statements.

Another criticism of principles-based accounting is that it can result in a lack of guidance. The principles are often broad and do not provide detailed guidance on how to account for specific transactions. This can result in uncertainty and difficulty in the application of the principles.

Rules-Based Accounting

Rules-based accounting is a framework that relies on a set of specific rules to guide the preparation and presentation of financial statements. The rules are usually detailed and provide specific guidance on how to account for specific transactions.

Advantages of Rules-Based Accounting

The primary advantage of rules-based accounting is that it provides specific guidance on how to account for specific transactions. This can result in more consistent and comparable financial statements. The rules provide a clear framework for the preparation and presentation of financial statements and can reduce uncertainty and subjectivity in the application of accounting principles.

Another advantage of rules-based accounting is that it can provide a higher degree of transparency. The specific rules provide detailed guidance on how items are measured and recognized in the financial statements, which can result in more meaningful financial statements.

Disadvantages of Rules-Based Accounting

One of the criticisms of rules-based accounting is that it can be inflexible. The detailed rules can result in a lack of flexibility in the preparation and presentation of financial statements. This can lead to financial statements that do not accurately reflect the underlying economic reality of transactions.

Another criticism of rules-based accounting is that it can result in a focus on form over substance. The specific rules can result in a focus on the legal form of transactions rather than their economic substance. This can result in financial statements that do not accurately reflect the underlying economic reality of transactions.

Conclusion

In conclusion, both principles-based and rules-based accounting have advantages and disadvantages. Principles-based accounting provides flexibility and encourages a focus on substance over form, but can be subjective and lack guidance. Rules-based accounting provides specific guidance and transparency, but can be inflexible and result in a focus on form over substance. Ultimately, the choice between principles-based and rules-based accounting will depend on the specific circumstances and the goals of the preparers of financial statements.

Key GAAP standards and their application

There are a number of key GAAP standards that are widely used in financial reporting. These standards provide guidelines for how financial information should be recorded, presented, and disclosed, with the aim of ensuring that financial statements are consistent, accurate, and useful to investors and other stakeholders. In this section, we will discuss some of the most important GAAP standards and their application in financial reporting.

Revenue Recognition

One of the most fundamental GAAP standards is revenue recognition. This standard provides guidance on when and how to recognize revenue in financial statements. The principle underlying this standard is that revenue should be recognized when it is earned, and not when cash is received. This means that revenue should be recognized when a company has delivered goods or services to a customer, and the customer has accepted them.

The application of the revenue recognition standard can be complex and requires careful analysis of the specific transaction in question. For example, if a company sells a product with a warranty, the revenue recognition may need to be adjusted to account for the expected costs of the warranty. Similarly, if a company receives advance payments from customers, the revenue recognition may need to be deferred until the goods or services are delivered.

Inventory Valuation

Another key GAAP standard is inventory valuation. This standard provides guidance on how to value inventory in financial statements, with the aim of ensuring that inventory is reported at its true economic value. There are several methods that can be used to value inventory, including the first-in, first-out (FIFO) method, the last-in, first-out (LIFO) method, and the weighted average method.

The choice of inventory valuation method can have a significant impact on financial statements, particularly for companies that hold large amounts of inventory. For example, the LIFO method tends to result in lower reported profits and lower tax liabilities in periods of inflation, while the FIFO method tends to result in higher reported profits and higher tax liabilities.

Depreciation

Depreciation is another key GAAP standard that provides guidance on how to account for the wear and tear on assets over time. This standard requires that companies record the decline in value of fixed assets over their useful life, using a systematic method that reflects the pattern of consumption of the asset. This means that the cost of the asset is spread out over its useful life, rather than being recorded as a one-time expense.

The application of the depreciation standard can also be complex, as it requires careful estimation of the useful life of an asset and the appropriate method of depreciation. For example, some assets may have a predictable pattern of consumption, such as a vehicle that is driven a certain number of miles each year.

Other assets, such as a building, may have a less predictable pattern of consumption and require more judgment in determining the appropriate method of depreciation.

Goodwill Impairment

Goodwill impairment is a GAAP standard that provides guidance on how to account for the decline in value of goodwill, which is the value of a company's intangible assets, such as its reputation and brand name. Under this standard, companies are required to periodically test for impairment of their goodwill, and record any impairment charges in their financial statements.

The application of the goodwill impairment standard can be challenging, as it requires companies to make subjective judgments about the value of their intangible assets. For example, a company may need to assess whether changes in the competitive landscape or the regulatory environment have affected the value of its brand name or customer relationships.

Lease Accounting

Lease accounting is a GAAP standard that provides guidance on how to account for leases in financial statements. Under this standard, leases are classified as either operating leases or finance leases, depending on the nature of the lease agreement. Operating leases are recorded as rent expense in the income statement over the lease term, while finance leases are recorded as both an asset and a liability on the balance sheet.

The new lease accounting standard, ASC 842, was issued by the FASB in 2016 and became effective for public companies in 2019. The standard requires lessees to recognize most leases on their balance sheets as right-of-use assets and lease liabilities. It also eliminates the concept of operating leases and requires all leases to be classified as finance leases or operating leases based on specific criteria.

The impact of the new lease accounting standard on financial statements can be significant, especially for companies with significant lease agreements. For example, the recognition of lease liabilities and right-of-use assets can significantly increase a company's reported debt and total assets, respectively.

Revenue recognition is another key GAAP standard that provides guidance on how to account for revenue in financial statements. The standard provides a framework for recognizing revenue based on the transfer of control of goods or services to customers. Under this standard, revenue is recognized when it is earned and measurable, and when the company has fulfilled its obligations to the customer.

The new revenue recognition standard, ASC 606, was issued by the FASB in 2014 and became effective for public companies in 2018. The standard requires companies to recognize revenue based on a five-step model, which includes identifying the contract, identifying the performance obligations, determining the transaction price, allocating the transaction price to the performance obligations, and recognizing revenue when the performance obligations are satisfied.

The new revenue recognition standard is designed to provide a more comprehensive and consistent approach to revenue recognition across industries and types of transactions. It also requires more extensive disclosures about the nature and timing of revenue recognition, which can help investors and other stakeholders better understand a company's revenue streams and performance.

Another important GAAP standard is the accounting for business combinations. This standard provides guidance on how to account for mergers, acquisitions, and other types of business combinations in financial statements. Under this standard, the acquirer must recognize the assets acquired and liabilities assumed at fair value on the acquisition date, and must also recognize any goodwill arising from the acquisition.

The accounting for business combinations can be complex, and requires significant judgment and analysis. The new accounting standard for business combinations, ASC 805, was issued by the FASB in 2007 and became effective in 2009. The standard requires a more detailed analysis of the fair value of assets and liabilities acquired, as well as more extensive disclosures about the nature and impact of the acquisition on the acquirer's financial statements.

Finally, the accounting for income taxes is another key GAAP standard that provides guidance on how to account for income taxes in financial statements. Under this standard, companies must recognize the current and deferred tax effects of transactions and events in the financial statements.

The new accounting standard for income taxes, ASC 740, was issued by the FASB in 1992 and has undergone several updates since then. The standard requires companies to provide detailed disclosures about their income tax accounting policies, as well as the impact of changes in tax laws and rates on their financial statements.

In summary, GAAP provides a comprehensive framework for financial reporting that is designed to ensure consistency and transparency in financial statements. Key GAAP standards, such as lease accounting, revenue recognition, accounting for business combinations, and accounting for income taxes, provide guidance on how to account for specific types of transactions and events in financial statements. The application of these standards requires significant judgment and analysis, and can have a significant impact on a company's reported financial results.

CHAPTER 34: INTERNATIONAL FINANCIAL REPORTING STANDARDS (IFRS)

The globalization of business has led to the need for global accounting standards that can be adopted by companies operating in different countries. International Financial Reporting Standards (IFRS) is a set of accounting standards developed by the International Accounting Standards Board (IASB) to provide a common financial reporting language that can be understood by investors and stakeholders across borders.

This chapter will introduce the reader to the key concepts and principles of IFRS, its adoption and implementation, and its impact on financial reporting practices in different countries. We will start by discussing the history and development of IFRS, followed by an overview of its structure and content. We will then examine the advantages and disadvantages of adopting IFRS, as well as the challenges associated with its implementation. Finally, we will discuss the impact of IFRS on financial reporting practices in different countries, and the ongoing efforts to achieve convergence between IFRS and US Generally Accepted Accounting Principles (GAAP).

History and Development of IFRS

IFRS is the result of the efforts of the International Accounting Standards Board (IASB), which was established in 2001 as an independent, private-sector organization. The IASB is responsible for the development and publication of IFRS, which is now used in more than 140 countries worldwide. The IASB works closely with national accounting standard-setters, regulatory bodies, and other stakeholders to ensure that IFRS reflects the needs of the global business community.

The origins of IFRS can be traced back to the 1960s, when the International Accounting Standards Committee (IASC) was established to promote the harmonization of accounting standards across different countries. The IASC developed a set of accounting standards known as International Accounting

Standards (IAS), which were later adopted by many countries around the world. However, the IASC faced criticism for its lack of independence and for its failure to keep pace with the changing needs of the global business environment.

In response to these criticisms, the IASB was established as an independent standard-setting body, with a mandate to develop high-quality, globally accepted accounting standards. Since its establishment, the IASB has worked to develop and improve IFRS, with the aim of promoting transparency, comparability, and consistency in financial reporting.

Structure and Content of IFRS

IFRS is a principles-based set of accounting standards, which means that it is based on a set of fundamental principles and concepts that guide the preparation and presentation of financial statements. The principles-based approach of IFRS provides more flexibility in the application of accounting standards, allowing companies to tailor their financial reporting practices to their specific circumstances.

IFRS covers a wide range of financial reporting topics, including the presentation of financial statements, revenue recognition, leases, financial instruments, and business combinations. Each standard within IFRS provides guidance on how to account for a specific financial reporting topic, and includes detailed requirements and disclosures that companies must follow.

Advantages and Disadvantages of Adopting IFRS

The adoption of IFRS has a number of potential advantages and disadvantages for companies, investors, and other stakeholders. One of the key advantages of adopting IFRS is the increased comparability and consistency of financial reporting across different countries, which can facilitate cross-border investment and enhance transparency in financial reporting.

Other potential advantages of adopting IFRS include the reduced costs of preparing financial statements, as well as the increased credibility of financial reporting due to the use of globally accepted accounting standards. However, there are also potential disadvantages associated with the adoption of IFRS, such as the costs of implementing new accounting systems and the need for additional training and resources to comply with the new standards.

Challenges of Implementing IFRS

The implementation of IFRS can pose significant challenges for companies, particularly those that have operated under a different set of accounting standards for many years. One of the biggest challenges is the need to overhaul existing accounting systems and processes to align with the new standards. This can be a time-consuming and costly process, especially for large, complex organizations that have multiple business units and operations in multiple countries.

In addition to the technical challenges associated with implementing IFRS, there are also cultural and organizational challenges. Companies may need to change the way they think about financial reporting and adopt new practices and procedures to ensure compliance with the new standards. This can require significant investment in training and resources, as well as changes to internal governance and control structures.

Another challenge of implementing IFRS is the potential for differences in interpretation and application of the standards. While IFRS is intended to provide a common set of accounting standards that can be applied globally, there may be differences in how the standards are interpreted and applied in different countries and regions. This can lead to inconsistencies in financial reporting and make it difficult for investors and analysts to compare the financial performance of different companies.

Conclusion

In conclusion, the adoption of IFRS represents a significant shift in the way that companies prepare and present their financial statements. While there are many potential benefits associated with the adoption of IFRS, such as increased transparency, comparability, and credibility of financial reporting, there are also challenges that need to be addressed, such as the costs and complexities of implementing the new standards, and the potential for differences in interpretation and application of the standards.

Overall, companies that are considering the adoption of IFRS need to carefully weigh the potential benefits and challenges associated with this decision, and ensure that they have the resources and expertise necessary to successfully implement the new standards. With the right approach, however, the adoption of IFRS can be a positive step towards improving the quality and transparency of financial reporting, and enhancing the credibility of companies in the global marketplace.

Overview of IFRS and its adoption around the world

The International Financial Reporting Standards (IFRS) is a set of accounting standards developed by the International Accounting Standards Board (IASB) with the goal of establishing a common financial reporting language across the world. IFRS is designed to provide transparency, comparability, and consistency in financial reporting, and to improve the quality and reliability of financial statements.

IFRS has been adopted by more than 140 countries around the world, including the European Union, Australia, Canada, Japan, and South Africa. The United States is the notable exception, as it still relies on the Generally Accepted Accounting Principles (GAAP).

The adoption of IFRS is driven by the need to have globally recognized accounting standards, which is essential in a global economy where businesses and investors operate across borders. Adoption of IFRS also enhances the comparability of financial statements, which is critical for investors when making investment decisions.

Benefits of Adopting IFRS

One of the primary benefits of adopting IFRS is the potential for increased comparability and transparency of financial statements across different countries and industries. This allows investors and stakeholders to make informed decisions based on reliable and consistent financial information.

Another advantage of IFRS is that it allows companies to access international capital markets more easily. By adopting IFRS, companies can attract a wider pool of investors who are familiar with the global accounting standards. Additionally, IFRS can help to reduce the cost of capital for companies, as investors may be more willing to invest in companies that use globally recognized accounting standards.

IFRS can also lead to a reduction in costs associated with preparing financial statements. Since IFRS is a single set of accounting standards that can be applied globally, companies can avoid the need to prepare multiple sets of financial statements to comply with different national standards. This can result in cost savings for companies and make it easier to manage financial reporting.

Challenges of Adopting IFRS

While there are numerous benefits to adopting IFRS, there are also several challenges that companies may face when implementing the standards. One of the primary challenges is the cost of implementation, which can be significant, especially for large multinational companies with complex accounting systems. Companies may need to invest in new technology, software, and training to ensure compliance with IFRS.

Another challenge is the need to make significant changes to existing accounting policies and practices. The transition from national accounting standards to IFRS may require companies to make adjustments to their financial statements, which can be a time-consuming and resource-intensive process.

Another potential challenge of IFRS adoption is the need to ensure consistency in the application of the standards across different countries and regions. While IFRS is a globally recognized set of accounting standards, there may still be variations in the way that the standards are applied in different jurisdictions. This can create inconsistencies in financial reporting and make it more difficult for investors to compare companies across different regions.

Conclusion

IFRS is a set of accounting standards that is designed to provide transparency, comparability, and consistency in financial reporting. It has been adopted by more than 140 countries around the world, and its adoption is driven by the need to have globally recognized accounting standards. While there are numerous benefits to adopting IFRS, there are also several challenges that companies may face when implementing the standards. However, the benefits of IFRS adoption outweigh the challenges, as it can lead to increased comparability and transparency in financial reporting, which is essential for investors and stakeholders in a global economy.

Key differences between IFRS and US GAAP

International Financial Reporting Standards (IFRS) and US Generally Accepted Accounting Principles (GAAP) are the two most commonly used sets of accounting standards in the world. While there are many similarities between the two, there are also significant differences in their treatment of various accounting issues. This section will provide an overview of the key differences between IFRS and US GAAP, with a focus on financial statement presentation and measurement.

Financial Statement Presentation

One of the key differences between IFRS and US GAAP is their approach to financial statement presentation. Under IFRS, companies are required to present a statement of comprehensive income, which combines the results of the income statement and other comprehensive income items, such as unrealized gains and losses on available-for-sale financial assets. This statement is presented as a primary statement, alongside the statement of financial position and statement of cash flows.

In contrast, US GAAP requires companies to present the results of other comprehensive income items in a separate statement of comprehensive income, which is presented as a part of the financial statements but is not considered a primary statement. US GAAP also requires the presentation of a statement of stockholders' equity, which is not required under IFRS.

Measurement

Another key difference between IFRS and US GAAP is their approach to measurement. Under IFRS, the principles-based approach is used, which focuses on the substance of transactions and requires judgment in determining the appropriate accounting treatment. This approach allows for greater flexibility in financial reporting, but it also requires more judgment and can lead to greater variability in financial reporting.

In contrast, US GAAP is a more rules-based system, with specific rules and guidelines for accounting treatment. This approach provides greater certainty and consistency in financial reporting, but it can also be more rigid and inflexible.

Revenue Recognition

Another area of difference between IFRS and US GAAP is revenue recognition. Under IFRS, revenue is recognized when it is probable that economic benefits will flow to the company and the amount of revenue can be reliably measured. In addition, revenue is recognized over time when the company is transferring control of goods or services to a customer.

Under US GAAP, revenue is recognized when there is persuasive evidence of an arrangement, delivery has occurred or services have been rendered, the price is fixed or determinable, and collectability is reasonably assured. Additionally, revenue can only be recognized over time in limited circumstances, such as construction contracts.

Leases

IFRS and US GAAP also have different approaches to the accounting for leases. Under IFRS, leases are classified as either finance leases or operating leases. Finance leases are treated similarly to the accounting for purchased assets, while operating leases are treated as rental agreements.

Under US GAAP, leases are also classified as either finance leases or operating leases. However, the accounting treatment for operating leases is changing, with the adoption of a new standard that requires companies to recognize operating leases on the balance sheet as both a right-of-use asset and a lease liability.

Inventory Valuation

Inventory valuation is another area where IFRS and US GAAP have different approaches. Under IFRS, inventory is generally valued at the lower of cost or net realizable value, with cost being determined using various methods such as first-in, first-out (FIFO), last-in, first-out (LIFO), and weighted average cost.

Under US GAAP, inventory is valued at the lower of cost or market, with cost being determined using methods such as FIFO, LIFO, and average cost. However, the use of LIFO is more common in the US than in other countries, and is not allowed under IFRS.

Conclusion

In conclusion, while there are many similarities between IFRS and US GAAP, there are also significant differences in their treatment of various accounting standards. Understanding these differences is crucial for businesses and investors who operate in multiple jurisdictions and are required to comply with different accounting standards. It is important to note that IFRS is the standard used by most countries around the world, with the exception of the United States, which continues to use US GAAP.

The trend towards global convergence of accounting standards is likely to continue, and it is expected that the differences between IFRS and US GAAP will continue to decrease over time. In fact, the Financial Accounting Standards Board (FASB) and the International Accounting Standards Board (IASB) have been working together to reduce the differences between the two sets of standards.

In the meantime, businesses that operate in both IFRS and US GAAP jurisdictions will need to be prepared to deal with the differences between the two standards. This may involve implementing accounting systems that can handle both standards or training staff to understand and comply with both sets of standards.

Overall, the adoption of IFRS around the world has created a more unified and transparent accounting system, which has increased the comparability and reliability of financial statements. As globalization continues, the adoption of IFRS is likely to become even more important, as investors and regulators seek to ensure that financial information is consistent and comparable across borders.

In order to stay up-to-date with the latest developments in IFRS and US GAAP, it is important to consult with professionals who are knowledgeable in both sets of standards. This may include accountants, financial analysts, investment bankers, and lawyers who specialize in international accounting and finance. By staying informed about changes and updates to these standards, businesses and investors can ensure that they are able to comply with regulations and make informed decisions based on accurate financial information.

Impact of IFRS on financial reporting and analysis

International Financial Reporting Standards (IFRS) have become increasingly popular around the world, with many countries adopting these standards as the basis for financial reporting. The widespread adoption of IFRS has had a significant impact on financial reporting and analysis. This section will discuss the impact of IFRS on financial reporting and analysis, including the benefits and challenges associated with this change.

Benefits of IFRS on Financial Reporting and Analysis

One of the primary benefits of IFRS on financial reporting and analysis is increased comparability. Since IFRS is a globally accepted accounting standard, it provides a common language for financial reporting across different countries and regions. This allows investors and analysts to more easily compare the financial performance of companies operating in different jurisdictions, facilitating better investment decision-making.

Another benefit of IFRS is increased transparency. IFRS requires companies to disclose more information about their financial performance than many local accounting standards, providing investors with a more complete picture of a company's financial position. This transparency can help to build trust between companies and investors, ultimately leading to increased investment and economic growth.

IFRS can also improve the accuracy and reliability of financial reporting. IFRS provides detailed guidance on how to prepare financial statements, reducing the risk of errors and inconsistencies in financial reporting. This can enhance the reliability of financial statements, making them more useful to investors and analysts in making investment decisions.

Challenges of IFRS on Financial Reporting and Analysis

Despite the benefits of IFRS on financial reporting and analysis, there are also some challenges associated with the adoption of these standards. One challenge is the cost of implementation. Adopting IFRS may require significant changes to a company's accounting systems, which can be costly and time-consuming. This can be particularly challenging for smaller companies with limited resources.

Another challenge is the need for additional training and resources to comply with the new standards. IFRS is a complex and detailed accounting standard, and companies may need to invest in additional training and resources to ensure compliance. This can be a significant burden for companies, particularly those operating in multiple jurisdictions.

Impact of IFRS on Financial Analysis

The adoption of IFRS has also had a significant impact on financial analysis. One of the key impacts has been the increased comparability of financial statements across different countries and regions. This has made it easier for investors and analysts to compare the financial performance of companies operating in different jurisdictions, facilitating better investment decision-making.

IFRS has also led to changes in the way financial analysis is conducted. With the adoption of IFRS, companies are required to disclose more information about their financial performance, including more detailed information about the underlying assumptions and estimates used in preparing financial statements. This has led to an increased focus on the quality of earnings and the underlying drivers of financial performance.

In addition, IFRS has led to a shift in the way financial analysis is conducted, with greater emphasis on cash flow analysis. IFRS requires companies to prepare a statement of cash flows, which provides information about a company's sources and uses of cash. This information is critical for financial analysis, as it helps to assess a company's ability to generate cash and meet its financial obligations.

Conclusion

The adoption of IFRS has had a significant impact on financial reporting and analysis. While there are challenges associated with the adoption of these standards, including the cost of implementation and the need for additional training and resources, the benefits of increased comparability, transparency, and accuracy of financial reporting are significant. The increased comparability of financial statements across different jurisdictions has facilitated better investment decision-making, while the increased transparency has helped to build trust between companies and investors. Overall, the adoption of IFRS has led to a more efficient and effective global financial reporting and analysis system.

CHAPTER 35: SARBANES-OXLEY ACT OF 2002

The Sarbanes-Oxley Act of 2002 (SOX) is a landmark piece of legislation that was enacted by the United States Congress in response to the accounting scandals that rocked corporate America in the early 2000s, including the Enron and WorldCom scandals. SOX was intended to increase the transparency and accountability of public companies, restore investor confidence in the financial markets, and prevent corporate fraud and accounting irregularities. The act introduced a number of significant changes to the regulatory framework governing financial reporting and corporate governance, and had far-reaching consequences for businesses operating in the United States and abroad.

This chapter provides an overview of the key provisions of SOX and their implications for businesses, investors, and other stakeholders. We begin by examining the background and context of the legislation, including the events that led to its enactment and the political and economic climate in which it was introduced. We then discuss the major provisions of the act, including those related to financial reporting, internal controls, auditor independence, and corporate governance. We also examine the impact of SOX on financial reporting and analysis, as well as the criticisms and controversies that have arisen in response to the act.

The chapter concludes with a discussion of the ongoing debate over the effectiveness of SOX and the future of financial regulation in the United States and globally. We argue that while SOX has made significant progress in improving the quality and reliability of financial reporting, there is still much work to be done to ensure that businesses operate in a transparent, ethical, and sustainable manner. We also highlight the importance of continued research and education in the field of financial reporting and analysis, and the role that professionals in accounting, finance, law, and other related fields can play in shaping the future of corporate governance and regulation.

Overview of the Sarbanes-Oxley Act and its objectives

The Sarbanes-Oxley Act of 2002 (SOX) is a US federal law that was enacted in response to the accounting scandals of the early 2000s, including the Enron and WorldCom scandals. The Act is named after its two primary sponsors, Senator Paul Sarbanes and Representative Michael Oxley, and is intended to improve corporate governance, financial reporting, and auditing standards in publicly traded companies in the US.

Overview of the Sarbanes-Oxley Act

The Sarbanes-Oxley Act of 2002 (SOX) is a sweeping piece of legislation that affects all publicly traded companies in the US. The Act is divided into eleven titles, each of which addresses a specific aspect of corporate governance, financial reporting, or auditing. Some of the key provisions of the Act include:

Title I: Public Company Accounting Oversight Board (PCAOB)

Title I of the Act created the Public Company Accounting Oversight Board (PCAOB), which is responsible for overseeing the audits of public companies in the US. The PCAOB has the authority to set auditing standards, conduct inspections of audit firms, and enforce compliance with auditing regulations.

Title II: Auditor Independence

Title II of the Act establishes new rules governing the independence of auditors. The Act prohibits auditors from providing certain non-audit services to their clients, such as bookkeeping and financial system design. It also requires companies to disclose their relationships with their auditors and the fees paid to them.

Title III: Corporate Responsibility

Title III of the Act holds corporate executives responsible for the accuracy and completeness of their company's financial statements. The Act requires CEOs and CFOs to certify the accuracy of their company's financial statements and imposes penalties for false certifications.

Title IV: Enhanced Financial Disclosures

Title IV of the Act requires companies to provide more detailed and timely disclosures of their financial information. The Act requires companies to disclose off-balance sheet transactions, pro forma financial information, and other material financial information.

Title V: Analyst Conflicts of Interest

Title V of the Act addresses conflicts of interest among securities analysts. The Act requires securities analysts to disclose any conflicts of interest that may affect their recommendations, and it prohibits companies from retaliating against analysts who issue negative reports.

Title VI: Commission Resources and Authority

Title VI of the Act provides additional resources and authority to the Securities and Exchange Commission (SEC) to enforce the securities laws. The Act increases the SEC's budget and staffing levels and gives the agency new enforcement powers.

Title VII: Studies and Reports

Title VII of the Act requires several studies and reports to be prepared on various aspects of corporate governance, financial reporting, and auditing. For example, the Act requires a study on the use of off-balance sheet transactions and a report on the role of credit rating agencies in the securities markets.

Title VIII: Corporate and Criminal Fraud Accountability

Title VIII of the Act creates several new criminal offenses related to securities fraud, including mail and wire fraud, securities fraud, and insider trading. The Act also increases the penalties for existing securities fraud offenses.

Title IX: White-Collar Crime Penalty Enhancements

Title IX of the Act increases the penalties for white-collar crimes, such as securities fraud and mail and wire fraud. The Act also provides for forfeiture of any proceeds obtained through white-collar crime.

Title X: Corporate Tax Returns

Title X of the Act requires companies to file their tax returns with the SEC and prohibits companies from deducting fines and penalties imposed for violations of securities laws.

Title XI: Corporate Fraud and Accountability

Title XI of the Act establishes new protections for whistleblowers who report securities fraud. The Act prohibits companies from retaliating against whistleblowers and provides for monetary rewards for individuals who provide information leading to successful enforcement actions.

Objectives of the Sarbanes-Oxley Act

The Sarbanes-Oxley Act of 2002 was enacted in response to the widespread accounting fraud and corporate scandals that occurred in the early 2000s, including the collapse of Enron, WorldCom, and Tyco International. The Act aimed to restore investor confidence in the financial markets by improving the accuracy and reliability of corporate disclosures and strengthening corporate governance and accountability.

The Act had several key objectives:

Improving financial reporting: The Act aimed to improve the accuracy and reliability of financial reporting by requiring companies to establish and maintain effective internal controls over financial reporting. Section 404 of the Act requires companies to include a report in their annual reports that provides an assessment of the effectiveness of their internal controls over financial reporting.

Strengthening corporate governance: The Act sought to strengthen corporate governance by requiring companies to have independent directors on their boards, establishing audit committees composed of independent directors, and requiring companies to disclose their corporate governance practices.

Enhancing auditor independence: The Act aimed to enhance auditor independence by prohibiting auditors from providing certain non-audit services to their audit clients, such as bookkeeping, financial system design, and internal audit outsourcing.

Increasing penalties for corporate fraud: The Act increased the penalties for securities fraud and other corporate crimes, including fines, imprisonment, and disgorgement of ill-gotten gains.

Protecting whistleblowers: The Act established new protections for whistleblowers who report securities fraud, including prohibitions against retaliation and monetary rewards for individuals who provide information leading to successful enforcement actions.

Impact of the Sarbanes-Oxley Act

The Sarbanes-Oxley Act had a significant impact on financial reporting and corporate governance in the United States. The Act led to increased scrutiny of corporate disclosures and a renewed focus on the quality of financial reporting. It also led to changes in corporate governance practices, including the composition and role of boards of directors and the establishment of audit committees.

The Act also had significant compliance costs for companies, particularly smaller public companies, which faced substantial expenses in implementing new internal controls over financial reporting and complying with the Act's disclosure requirements. Some critics of the Act argued that it imposed excessive regulatory burdens on companies and that its compliance costs outweighed its benefits.

Despite these criticisms, the Sarbanes-Oxley Act is generally viewed as having had a positive impact on corporate accountability and investor confidence. The Act's emphasis on transparency, accountability, and good governance has helped to restore investor trust in the financial markets and has contributed to a more level playing field for investors and companies alike.

Key provisions related to financial reporting and internal controls

The Sarbanes-Oxley Act of 2002 (SOX) introduced significant changes to the financial reporting and internal controls of publicly traded companies. The objectives of these changes were to increase transparency and accountability, and to reduce the risk of fraud and financial misconduct.

Key Provisions of the Act Related to Financial Reporting
The key provisions related to financial reporting include requirements for increased financial disclosure, more rigorous internal controls, and greater oversight by boards of directors and independent auditors. These provisions were intended to improve the accuracy and reliability of financial reporting and to increase the confidence of investors in the financial markets.

Section 302: Corporate Responsibility for Financial Reports
This section of the Act requires that the CEO and CFO of a public company certify the accuracy of the company's financial statements and disclosures in each quarterly and annual report. The certification must state that the financial statements and disclosures fairly present, in all material respects, the financial condition, results of operations, and cash flows of the company. The CEO and CFO must also attest to the effectiveness of the company's internal controls over financial reporting.

Section 404: Management Assessment of Internal Controls
Section 404 requires that public companies assess and report on the effectiveness of their internal controls over financial reporting. This section requires

that the company's management assess the design and operation of its internal controls and that the company's auditor evaluate the effectiveness of those controls.

Section 409: Real-Time Disclosure

This section requires that public companies disclose material changes in their financial condition or operations in "real time." This means that companies must disclose material information in a timely manner, rather than waiting for quarterly or annual financial reports.

Section 802: Criminal Penalties for Altering Documents

This section establishes criminal penalties for knowingly altering or destroying documents or other records to impede or obstruct any official proceeding, such as an investigation by the SEC or other regulatory agency.

Section 906: Criminal Penalties for CEO and CFO Misconduct

This section imposes criminal penalties on CEOs and CFOs who knowingly certify false financial statements.

Key Provisions of the Act Related to Internal Controls

The key provisions related to internal controls include requirements for more rigorous internal controls and greater oversight by boards of directors and independent auditors. These provisions were intended to reduce the risk of fraud and financial misconduct.

Section 404: Management Assessment of Internal Controls

Section 404 requires that public companies assess and report on the effectiveness of their internal controls over financial reporting. This section requires that the company's management assess the design and operation of its internal controls and that the company's auditor evaluate the effectiveness of those controls.

Section 301: Public Company Audit Committees

This section requires that public companies have audit committees that are composed entirely of independent directors. The audit committee is responsible for overseeing the company's financial reporting and internal controls, and for appointing and overseeing the work of the company's independent auditors.

Section 302: Corporate Responsibility for Financial Reports

This section requires that the CEO and CFO of a public company certify the accuracy of the company's financial statements and disclosures in each quarterly and annual report. The certification must state that the financial statements and disclosures fairly present, in all material respects, the financial condition, results of operations, and cash flows of the company. The CEO and CFO must also attest to the effectiveness of the company's internal controls over financial reporting.

Section 404(b): Auditor Attestation

This section requires that the company's independent auditor provide an attestation report on management's assessment of the effectiveness of the company's internal controls over financial reporting.

Section 802: Criminal Penalties for Altering Documents

This section establishes criminal penalties for knowingly altering or destroying documents or other records to impede or obstruct any investigation by a federal agency, any matter within the jurisdiction of any such agency, or any bankruptcy case. The penalties for such acts can include fines and imprisonment for up to 20 years.

The Act also creates new requirements for financial reporting and internal controls to enhance the accuracy and reliability of financial statements. Companies are required to establish and maintain adequate internal controls to ensure the accuracy of their financial statements, and to regularly assess the effectiveness of these controls. The Act requires companies to provide certifications of their financial statements by the CEO and CFO, stating that the financial statements are accurate and complete and that the company's internal controls are effective.

Additionally, the Act requires that public accounting firms be independent from the companies they audit, to ensure that they provide objective and impartial assessments of the company's financial statements. This requirement aims to prevent conflicts of interest and to promote the accuracy and reliability of financial reporting.

The Act also establishes the Public Company Accounting Oversight Board (PCAOB), a non-profit corporation that oversees the auditing of public companies. The PCAOB is responsible for setting auditing standards, inspecting public accounting firms, and enforcing compliance with the Act's requirements. The PCAOB also has the authority to discipline public accounting firms that fail to comply with its standards or engage in unethical behavior.

To further promote transparency and accountability in financial reporting, the Act requires companies to disclose any material changes to their financial condition or operations on a timely basis. Companies are also required to disclose any off-balance-sheet transactions, arrangements, obligations, or other relationships that may have a material impact on their financial condition or results of operations.

The Act establishes new criminal penalties for securities fraud and other violations of securities laws. It creates a new offense of securities fraud, which includes any scheme or artifice to defraud, any false statements or omissions of material fact, or any act or omission that operates as a fraud or deceit upon any person in connection with the purchase or sale of securities. The penalties for securities fraud include fines and imprisonment for up to 25 years.

Moreover, the Act requires companies to disclose whether they have adopted a code of ethics for their senior financial officers, and if not, to explain why not. It also requires companies to disclose any waivers of their code of ethics for senior financial officers, and to explain why such waivers were granted.

In summary, the Sarbanes-Oxley Act of 2002 was enacted to address the issues of corporate fraud and financial reporting, which were brought to the forefront by high-profile corporate scandals such as Enron and WorldCom. The Act established new requirements for financial reporting and internal controls, created new criminal penalties for securities fraud and other violations of securities laws, and established the PCAOB to oversee the auditing of public companies. The Act also aims to promote transparency and accountability in financial reporting and to protect whistleblowers who report securities fraud.

Impact on financial reporting and corporate governance

The Sarbanes-Oxley Act of 2002 has had a significant impact on financial reporting and corporate governance. The Act was enacted in response to the financial scandals of Enron, WorldCom, and other high-profile companies, and its provisions are aimed at preventing such scandals from happening again. In this section, we will discuss the impact of the Act on financial reporting and corporate governance.

Financial Reporting:

One of the primary objectives of the Sarbanes-Oxley Act is to improve the accuracy and reliability of financial reporting. The Act contains several provisions that affect financial reporting, including:

Section 302: Corporate Responsibility for Financial Reports: This section requires the CEO and CFO to certify the accuracy of the financial statements and disclosures in the company's annual and quarterly reports. The certification must be based on their knowledge, and the results of any internal control evaluations or audits.

Section 404: Management Assessment of Internal Controls: This section requires companies to establish and maintain effective internal controls over financial reporting. The company's management must evaluate the effectiveness of its internal controls and provide a report on its assessment as part of the annual report.

Section 802: Criminal Penalties for Altering Documents: This section establishes criminal penalties for knowingly altering or destroying documents or other records to impede or obstruct any official proceeding or investigation.

The implementation of these provisions has had a significant impact on financial reporting. Companies must now devote more time and resources to internal controls, and the certification requirements have increased the personal liability of CEOs and CFOs for the accuracy of financial reports. These requirements have also led to increased scrutiny by auditors and regulators, which has improved the overall quality of financial reporting.

Corporate Governance:

The Sarbanes-Oxley Act also contains provisions aimed at improving corporate governance. These provisions include:

Section 301: Public Company Audit Committees: This section requires that public companies have an audit committee that is composed of independent directors. The audit committee is responsible for overseeing the company's financial reporting and accounting practices, as well as the work of the company's auditors.

Section 404: Management Assessment of Internal Controls: This section also requires companies to establish and maintain effective internal controls over financial reporting. The company's management must evaluate the effectiveness of its internal controls and provide a report on its assessment as part of the annual report.

Section 406: Code of Ethics for Senior Financial Officers: This section requires public companies to adopt a code of ethics for their senior financial officers, such as the CFO and controller. The code of ethics must be in writing, and the company must disclose any waivers of the code that have been granted.

These provisions have led to increased accountability and transparency in corporate governance. The audit committee is now a more prominent feature of corporate governance, and the requirement for a code of ethics for senior financial officers has led to more ethical behavior in the financial industry. The increased focus on internal controls has also improved the accuracy and reliability of financial reporting.

Impact on Small Businesses:

The Sarbanes-Oxley Act has had a significant impact on large public companies, but its impact on small businesses has been less clear. Small businesses are exempt from some of the Act's provisions, but they must still comply with the Act's requirements for financial reporting and internal controls. This can be a significant burden for small businesses, which may not have the same resources as larger companies.

Conclusion:

The Sarbanes-Oxley Act of 2002 has had a significant impact on financial reporting and corporate governance. The Act's provisions have led to increased accountability and transparency in corporate governance, and the increased focus on internal controls has improved the accuracy and reliability of financial reporting. While the Act has been criticized for imposing a significant burden on businesses, its provisions have also helped to restore public trust in the financial industry. Overall, the Sarbanes-Oxley Act represents an important step in the effort to prevent corporate fraud and ensure the integrity of financial reporting.

However, the impact of the Act is not limited to the companies it regulates. The Act has also had significant implications for the legal and accounting professions. Legal departments and compliance officers are now required to provide more rigorous oversight of their company's financial reporting and internal controls, while auditors and accountants must adhere to more stringent auditing and reporting standards. The increased scrutiny on financial reporting and corporate governance has also led to a greater demand for qualified professionals in these fields.

Furthermore, the Sarbanes-Oxley Act has influenced the development of similar legislation around the world. Many countries have implemented their own versions of the Act, or have adopted similar regulations, in an effort to improve corporate governance and financial reporting standards.

In conclusion, the Sarbanes-Oxley Act has had a significant impact on the financial industry, corporate governance, and the legal and accounting professions. While it has been met with some criticism for the burden it places on businesses, the Act's provisions have helped to restore public trust in the financial industry and improve the accuracy and reliability of financial reporting. The Act has also had a ripple effect around the world, influencing the development of similar legislation in other countries. As such, it represents an important milestone in the ongoing effort to prevent corporate fraud and ensure the integrity of financial reporting.

PART 12: INTERNATIONAL ACCOUNTING STANDARDS

The globalization of business and capital markets has led to increased demand for high-quality financial reporting and accounting standards. In response to this demand, the International Accounting Standards Board (IASB) was established in 2001 to develop and promote the use of international financial reporting standards (IFRS). The use of IFRS is now mandatory or permitted in over 140 jurisdictions, including the European Union, Canada, Australia, and India.

Part 12 of this textbook provides an overview of international accounting standards, including the history and development of IFRS, the structure and function of the IASB, and the key principles and requirements of IFRS. The section also discusses the benefits and challenges of adopting IFRS, as well as the ongoing efforts to promote convergence between IFRS and US Generally Accepted Accounting Principles (GAAP).

History and Development of IFRS

The history of international accounting standards dates back to the early 1970s when the International Accounting Standards Committee (IASC) was established to develop a set of global accounting standards. In 2001, the IASB was established as an independent standard-setting body to replace the IASC. The IASB is responsible for developing and promoting the use of IFRS, which are designed to provide a common language for financial reporting that can be used by companies, investors, and other stakeholders around the world.

Since the establishment of the IASB, the use of IFRS has grown rapidly, with over 140 jurisdictions now requiring or permitting the use of IFRS. The European

Union adopted IFRS in 2005, and since then, many other countries, including Canada, Australia, and India, have followed suit.

Structure and Function of the IASB

The IASB is an independent, private-sector organization that is overseen by a group of trustees. The trustees are responsible for appointing the members of the IASB and ensuring that the organization operates effectively and transparently.

The IASB has 14 members who are appointed for a renewable term of five years. The members of the IASB are selected for their technical expertise and experience in financial reporting and accounting.

The IASB is responsible for developing and maintaining IFRS, which are designed to provide a single set of high-quality global accounting standards. The IASB's work is guided by a set of due process procedures, which are designed to ensure that its standards are developed in a transparent and consultative manner.

Key Principles and Requirements of IFRS

IFRS are designed to provide a common language for financial reporting that can be used by companies, investors, and other stakeholders around the world. The key principles and requirements of IFRS are as follows:

Fair presentation: IFRS requires that financial statements present fairly the financial position, performance, and cash flows of an entity.

Substance over form: IFRS requires that transactions be accounted for based on their economic substance, rather than their legal form.

Materiality: IFRS requires that financial statements include all material items, including those that may be individually immaterial but are material when aggregated with other items.

Comparability: IFRS requires that financial statements be comparable to those of other entities, both within the same industry and across different industries.

Consistency: IFRS requires that accounting policies be applied consistently from period to period.

Going concern: IFRS requires that financial statements be prepared on a going-concern basis, unless management intends to liquidate the entity or cease trading.

Benefits and Challenges of Adopting IFRS

The adoption of IFRS can bring many benefits, including improved financial reporting, increased comparability, and reduced costs of compliance for multinational companies. Adopting IFRS can also facilitate cross-border investment and improve access to capital markets, as investors can easily compare financial statements across different countries. Additionally, IFRS can help improve the accuracy and reliability of financial statements by providing a standardized framework for financial reporting.

However, there are also challenges associated with adopting IFRS. One of the biggest challenges is the cost of implementing and complying with IFRS, especially for smaller companies. IFRS requires significant changes to accounting systems, processes, and controls, and may require companies to hire additional staff or outside consultants. Additionally, IFRS may require changes to existing contracts, such as debt agreements, and could lead to increased legal and regulatory risks.

Another challenge is the potential loss of flexibility and the need to adjust to a standardized approach. Companies that have operated under different accounting standards for many years may struggle to adjust to the standardized approach of IFRS, which may require changes to long-established practices and systems. Additionally, IFRS may not be well-suited to certain industries or business models, which could result in additional complexity and cost.

Furthermore, cultural and linguistic differences can pose a challenge in the adoption of IFRS. IFRS is written in English, and companies whose primary language is not English may face additional translation and interpretation challenges. Additionally, IFRS may not fully consider the unique cultural and regulatory contexts in certain countries, which could create additional complexities.

Despite these challenges, the trend towards global convergence of accounting standards is expected to continue, as the benefits of a standardized approach to financial reporting become increasingly apparent. The adoption of IFRS has already been widely adopted across many countries, and it is likely that more countries will follow suit in the coming years.

CHAPTER 36: GLOBAL FINANCIAL REPORTING STANDARDS

In today's globalized economy, businesses and investors operate in a complex and dynamic environment. As companies expand their operations beyond their home countries, they face a range of challenges, including varying financial reporting requirements and regulations. The lack of uniformity in financial reporting standards creates a significant barrier to the free flow of capital and makes it difficult for investors to make informed decisions. To address this issue, a movement towards global financial reporting standards has emerged in recent years.

This chapter will provide an overview of the global financial reporting standards, including the International Financial Reporting Standards (IFRS), and their impact on the financial reporting landscape. We will explore the history of global financial reporting standards and their adoption by different countries around the world. Additionally, we will examine the benefits and challenges of adopting these standards, as well as the role of regulatory bodies in promoting their use.

Overview of Global Financial Reporting Standards

Global financial reporting standards refer to a set of accounting principles and guidelines that are used to prepare financial statements in a consistent and standardized manner across different countries and jurisdictions. These standards aim to enhance transparency and comparability of financial information, providing investors and other stakeholders with the ability to make informed decisions.

One of the most widely recognized sets of global financial reporting standards is the International Financial Reporting Standards (IFRS), developed by the International Accounting Standards Board (IASB). The IFRS is a set of accounting standards used in over 120 countries worldwide, including the European Union, Australia, Canada, India, and Japan. The IFRS is designed to provide a common language for financial reporting and to ensure consistency in accounting practices across borders.

History of Global Financial Reporting Standards

The push towards global financial reporting standards can be traced back to the 1960s when the International Accounting Standards Committee (IASC) was formed. The IASC was a predecessor to the IASB, responsible for issuing a set of accounting standards called the International Accounting Standards (IAS). The IAS were intended to promote the development of a single set of high-quality, globally accepted accounting standards.

In 2001, the IASC was replaced by the IASB, which was charged with developing and promoting the adoption of the IFRS. The IFRS is designed to be principles-based, providing guidance on how to account for transactions and events in financial statements, rather than prescribing specific rules.

Adoption of Global Financial Reporting Standards

The adoption of global financial reporting standards varies widely by country and jurisdiction. Some countries, such as Australia and the European Union, have fully adopted the IFRS, while others, such as the United States, have not.

In the United States, the Generally Accepted Accounting Principles (GAAP) are used for financial reporting. However, the Securities and Exchange Commission (SEC) allows foreign private issuers to prepare their financial statements using the IFRS without reconciliation to GAAP. In addition, the Financial Accounting Standards Board (FASB) and the IASB have been working towards the convergence of GAAP and IFRS to create a single set of global financial reporting standards.

Benefits of Adopting Global Financial Reporting Standards

The adoption of global financial reporting standards can provide several benefits. One of the most significant benefits is the increased comparability of financial statements across borders. This comparability allows investors and other stakeholders to make informed decisions about investing and doing business in different countries.

In addition, the adoption of global financial reporting standards can improve transparency and reduce the cost of preparing financial statements for multinational companies. It can also help to promote economic growth by increasing access to capital, facilitating cross-border investments, and reducing barriers to trade.

Challenges of Adopting Global Financial Reporting Standards

While there are many benefits to adopting global financial reporting standards, there are also challenges. Onemajor challenge is the issue of enforcement. Although the adoption of global financial reporting standards can help to improve transparency, without effective enforcement mechanisms, there is no guarantee that companies will actually comply with these standards. The lack of enforcement can also undermine the credibility of the standards, as investors and other stakeholders may question the reliability and accuracy of the financial information provided by companies.

Another challenge is the potential loss of national sovereignty in setting accounting standards. Many countries have long-established accounting traditions and may be reluctant to adopt a global standard that does not reflect their unique economic and cultural factors. This can lead to resistance from regulators, accounting professionals, and other stakeholders who fear that adopting global financial reporting standards will result in the loss of their national identity.

Another challenge is the cost of implementation. Adopting global financial reporting standards requires significant investment in training, software, and other resources. Smaller companies may find it particularly challenging to bear the cost of implementing these standards. Furthermore, it can be difficult to achieve a uniform implementation of the standards across different countries, which can create additional costs and complexity.

Finally, cultural and linguistic differences can also pose a challenge. Accounting standards are not only technical in nature but also reflect cultural norms and values. Differences in language, culture, and legal systems can make it difficult to develop and implement global financial reporting standards that are acceptable to all stakeholders.

Despite these challenges, the adoption of global financial reporting standards is becoming increasingly important as the world economy becomes more integrated. Companies are increasingly operating across borders, and investors are demanding more transparent and comparable financial information. The benefits of global financial reporting standards are clear, and many countries are making progress towards their adoption.

Overview of global financial reporting standards

Financial reporting is the process of providing information about the financial performance and position of a company to stakeholders, including investors, creditors, and regulators. Global financial reporting standards are a set of rules and guidelines that govern the preparation and presentation of financial statements. These standards help to ensure that financial statements are consistent, comparable, and reliable across different countries and industries.

The main purpose of global financial reporting standards is to promote transparency and accountability in financial reporting. By adopting these standards, companies can provide more accurate and reliable financial information to their stakeholders. This, in turn, helps to build trust and confidence in the financial markets, which is essential for economic growth.

Global financial reporting standards are developed by various organizations, including the International Accounting Standards Board (IASB), which sets International Financial Reporting Standards (IFRS), and the Financial Accounting Standards Board (FASB), which sets Generally Accepted Accounting Principles (GAAP) in the United States. Other countries have their own accounting standards, such as Japan's Generally Accepted Accounting Principles (J-GAAP) and China's Accounting Standards for Business Enterprises (ASBE).

Benefits of Adopting Global Financial Reporting Standards

Adopting global financial reporting standards can bring many benefits to companies, investors, and regulators. Some of the key benefits are discussed below:

Improved Comparability

One of the main benefits of adopting global financial reporting standards is the improved comparability of financial statements across different companies and industries. When companies use the same set of accounting standards, it becomes easier for investors to compare financial statements and make informed investment decisions. This can lead to more efficient capital allocation and higher levels of investment, which can help to drive economic growth.

Reduced Costs

Adopting global financial reporting standards can also reduce the cost of preparing financial statements for multinational companies. Companies that operate in multiple countries may have to prepare multiple sets of financial statements to comply with different accounting standards. This can be time-consuming and expensive. By adopting a single set of global financial reporting standards, companies can streamline their financial reporting processes and reduce their compliance costs.

Improved Transparency

Another benefit of adopting global financial reporting standards is improved transparency in financial reporting. When companies use the same set of accounting standards, it becomes easier for stakeholders to understand and interpret financial statements. This can help to build trust and confidence in the financial markets, which is essential for economic growth.

Increased Access to Capital

Adopting global financial reporting standards can also increase access to capital for companies. When companies use the same set of accounting standards, it becomes easier for investors to understand and compare financial statements. This can lead to increased investment in the company and improved access to capital.

Facilitate Cross-border Investments

Adopting global financial reporting standards can also facilitate cross-border investments. When investors have access to consistent and reliable financial information, they are more likely to invest in companies in other countries. This can help to promote economic growth and reduce barriers to trade.

Challenges of Adopting Global Financial Reporting Standards

While there are many benefits to adopting global financial reporting standards, there are also challenges. Some of the key challenges are discussed below:

Costs of Implementation

Adopting global financial reporting standards can be expensive, especially for companies that operate in multiple countries. Companies may need to invest in new accounting systems and train their staff to comply with the new standards. This can be time-consuming and expensive, especially for smaller companies.

Complexity of Standards

Global financial reporting standards can be complex, especially for companies that operate in multiple jurisdictions. Companies may need to comply with different accounting standards in different countries, which can be confusing and time-consuming. This can make it difficult for companies to provide consistent and reliable financial information to their stakeholders.

Lack of Flexibility

Global financial reporting standards may not be flexible enough to accommodate the needs of all companies. Different industries have unique characteristics and may require different accounting treatments. For example, revenue recognition for a software company may differ from that of a manufacturing company. Therefore, companies may find it challenging to apply a one-size-fits-all approach to financial reporting.

Furthermore, global financial reporting standards are generally principles-based, which means they provide a framework for reporting rather than specific rules. This can make it difficult for companies to interpret and apply the standards, which may lead to inconsistent reporting.

Costs of Adoption

Adopting global financial reporting standards can be expensive for companies, especially smaller ones. Companies may need to invest in new software, training, and personnel to comply with the standards. Additionally, the process of adopting the standards may require significant time and effort from management and employees.

In some cases, companies may also need to restate their financial statements to comply with the new standards. This can be a costly and time-consuming process, and it may also affect the company's reputation if restatements are seen as a sign of poor financial management.

Conclusion

Global financial reporting standards have become increasingly important in today's globalized economy. They provide a common language for financial reporting, which can increase transparency and comparability of financial information. This, in turn, can lead to increased investor confidence and improved access to capital.

However, there are also challenges associated with adopting global financial reporting standards. These include the complexity of the standards, the lack of flexibility, and the costs of adoption. Companies need to carefully consider these factors when deciding whether to adopt global financial reporting standards and how to implement them.

Importance of international accounting standards

International accounting standards play a critical role in global financial reporting. The adoption of these standards has become increasingly important in today's globalized economy as more and more companies operate across borders. The following section will explore the importance of international accounting standards and their impact on financial reporting.

Enhancing Transparency and Consistency

One of the key benefits of international accounting standards is that they enhance transparency and consistency in financial reporting. By using a common set of accounting standards, companies can provide investors, analysts, and other stakeholders with more reliable and comparable financial information. This can help to build trust in financial reporting and improve decision-making for investors.

Facilitating Cross-Border Comparisons

Another key benefit of international accounting standards is that they facilitate cross-border comparisons. Companies operating in multiple countries often face the challenge of complying with different accounting standards in each jurisdiction. International accounting standards help to overcome this challenge by providing a common set of accounting rules that can be used across different countries. This allows for more accurate and meaningful comparisons between companies operating in different jurisdictions, which is essential for investors and analysts.

Reducing Costs

The adoption of international accounting standards can also help to reduce costs for companies. Complying with multiple sets of accounting standards can be time-consuming and expensive for multinational companies. By adopting a common set of accounting standards, companies can streamline their accounting processes and reduce compliance costs.

Promoting Economic Growth

International accounting standards also play a role in promoting economic growth. By providing a common set of accounting standards, these standards can help to increase transparency and trust in financial reporting, which can in turn increase access to capital and facilitate cross-border investments. This can help to drive economic growth by providing companies with the resources they need to invest in new projects, expand their operations, and create jobs.

Enhancing Corporate Governance

International accounting standards can also enhance corporate governance. These standards require companies to provide more detailed and transparent financial information, which can help to prevent fraud and improve accountability. This can help to build trust in companies and their management, which is essential for investors and other stakeholders.

Conclusion

In conclusion, international accounting standards are an essential component of global financial reporting. The adoption of these standards can enhance transparency and consistency, facilitate cross-border comparisons, reduce costs, promote economic growth, and enhance corporate governance. As more and more companies operate across borders, the importance of these standards will only continue to grow. It is essential for companies to stay up to date with the latest international accounting standards and ensure that they are complying with these standards in all jurisdictions where they operate.

Adoption of international accounting standards around the world

Adoption of international accounting standards around the world has been a topic of discussion for many years. The importance of having a globally accepted set of accounting standards has been recognized by both governments and businesses alike. This section will discuss the adoption of international accounting standards around the world, including its benefits, challenges, and current status.

Benefits of Adopting International Accounting Standards

The adoption of international accounting standards can provide several benefits to businesses and governments around the world. One of the key benefits is the standardization of financial reporting, which can provide comparability and consistency across different jurisdictions. This can help to promote transparency and increase investor confidence in financial statements. It can also make it easier for businesses to operate in multiple jurisdictions, as they do not need to comply with different accounting standards in different countries.

Another benefit of adopting international accounting standards is the reduction of costs associated with preparing financial statements. This is particularly true for multinational companies, which can reduce the cost of complying with different accounting standards in different countries. It can also reduce the cost of auditing financial statements, as auditors do not need to be familiar with multiple accounting standards.

Finally, the adoption of international accounting standards can help to promote economic growth by increasing access to capital, facilitating cross-border investments, and reducing barriers to trade. This is because investors and businesses can make more informed investment decisions based on comparable financial information across different jurisdictions.

Challenges of Adopting International Accounting Standards

Despite the benefits of adopting international accounting standards, there are also challenges. One of the key challenges is the complexity of the standards. International accounting standards can be complex, especially for companies that operate in multiple jurisdictions. Companies may need to comply with different accounting standards in different countries, which can be confusing and time-consuming. This can make it difficult for companies to provide consistent and reliable financial information to their stakeholders.

Another challenge is the lack of flexibility in international accounting standards. The standards may not be flexible enough to accommodate the needs of all businesses, particularly those in emerging markets. This can make it difficult for these businesses to comply with the standards, which can in turn limit their access to capital.

Current Status of Adoption

The adoption of international accounting standards has been a slow process, but progress has been made in recent years. The International Accounting Standards Board (IASB) has been working to develop and update international accounting standards since its inception in 2001. The IASB has developed a set of standards called International Financial Reporting Standards (IFRS), which have been adopted by many countries around the world.

As of 2021, over 120 countries require or permit the use of IFRS for financial reporting. The European Union has required the use of IFRS since 2005, while countries such as Canada, Australia, and Japan have also adopted the standards. In the United States, the Financial Accounting Standards Board (FASB) and the IASB have been working to converge their accounting standards, but as of now, the U.S. has not adopted IFRS.

Conclusion

In conclusion, the adoption of international accounting standards can provide several benefits to businesses and governments around the world, including standardization of financial reporting, cost reduction, and promotion of economic growth. However, there are also challenges associated with the adoption of these standards, including their complexity and lack of flexibility. Despite these challenges, progress has been made in the adoption of international accounting standards, and many countries have already adopted or are in the process of adopting these standards.

CHAPTER 37: OVERVIEW OF IFRS AND ITS ADOPTION AROUND THE WORLD

International Financial Reporting Standards (IFRS) are a set of accounting standards developed by the International Accounting Standards Board (IASB) for the purpose of providing a common global language for business affairs. The adoption of IFRS has gained momentum over the years, with more and more countries choosing to adopt these standards as their primary accounting principles.

This chapter provides an overview of IFRS and its adoption around the world. It begins by discussing the history and development of IFRS, followed by an analysis of its benefits and challenges. The chapter then examines the adoption of IFRS in various countries and regions, including the European Union, the United States, and Asia. Finally, the chapter concludes with a discussion of the future of IFRS and its role in the global accounting landscape.

History and Development of IFRS

The development of IFRS can be traced back to the 1960s, when a group of professional accounting bodies from around the world came together to establish the International Accounting Standards Committee (IASC). The IASC was created with the goal of developing a set of global accounting standards that could be used by companies operating in multiple countries.

Over the years, the IASC developed a series of International Accounting Standards (IAS), which were designed to provide a common set of accounting principles for companies operating in different countries. However, these standards were not widely adopted, and many countries continued to use their own accounting standards.

In 2001, the IASC was replaced by the International Accounting Standards Board (IASB), which was given the task of developing a single set of high-quality, globally accepted accounting standards. The IASB developed a new set of accounting standards known as IFRS, which were designed to replace the existing IAS and provide a single set of high-quality accounting standards that could be used by companies around the world.

Benefits of IFRS Adoption

The adoption of IFRS can provide several benefits to companies, investors, and other stakeholders. One of the primary benefits of IFRS adoption is increased comparability of financial statements. Since IFRS is a globally accepted set of accounting standards, companies that adopt these standards are able to provide financial statements that are comparable with those of other companies around the world. This comparability can make it easier for investors to evaluate and compare companies, and can help to increase market efficiency.

Another benefit of IFRS adoption is increased transparency in financial reporting. IFRS requires companies to disclose more information in their financial statements, which can help to reduce information asymmetry between companies and their stakeholders. This increased transparency can help to improve the quality of financial reporting and increase investor confidence.

In addition, IFRS adoption can help to reduce the cost of preparing financial statements for multinational companies. Since IFRS is a single set of accounting standards that can be used by companies around the world, it can reduce the need for companies to prepare separate financial statements for different countries. This can help to reduce the cost and complexity of financial reporting for multinational companies.

Challenges of IFRS Adoption

Despite the benefits of IFRS adoption, there are also several challenges that must be considered. One of the primary challenges is the complexity of IFRS. IFRS is a complex set of accounting standards that can be difficult for companies to understand and implement, especially for companies that operate in multiple jurisdictions. This complexity can make it difficult for companies to provide consistent and reliable financial information to their stakeholders.

Another challenge of IFRS adoption is the lack of flexibility in the standards. IFRS is a principles-based set of accounting standards, which means that there is more room for interpretation and judgment than in a rules-based system. However, this can also lead to inconsistency in the application of the standards, which can create challenges for companies and investors. Additionally, there may be cultural and legal differences that can make it difficult to implement IFRS consistently across different countries.

A lack of resources can also be a challenge for companies looking to adopt IFRS. Smaller companies may not have the resources to hire experts or invest in the necessary technology and training to comply with the standards. This can create a barrier to entry for smaller companies, potentially limiting their ability to access international capital markets.

Furthermore, IFRS adoption can also create challenges for regulators and standard-setting bodies. These organizations must ensure that the standards are consistently applied and enforced across different jurisdictions. This requires cooperation and coordination among various regulatory bodies, which can be challenging in practice.

Finally, there may be resistance to IFRS adoption from some stakeholders. This can include companies that have already invested significant resources in complying with local accounting standards, or investors who are accustomed to using local accounting information to make investment decisions. Resistance to IFRS adoption may also come from regulators who are concerned about losing control over local accounting standards and practices.

To address these challenges, companies and regulators must work together to ensure a smooth transition to IFRS. This may involve investing in technology and training to improve compliance, coordinating efforts among regulatory bodies, and engaging with stakeholders to address concerns and build support for the adoption of IFRS. Ultimately, the benefits of IFRS adoption are likely to outweigh the challenges, as companies and investors gain access to more consistent and reliable financial information across international markets.

History and development of IFRS

International Financial Reporting Standards (IFRS) have become increasingly important in the global business environment. They are accounting standards developed and maintained by the International Accounting Standards Board (IASB) that provide a common language for financial reporting across countries and industries. In this section, we will examine the history and development of IFRS, including the key players involved and the major events that have shaped their evolution.

Development of IFRS:

The development of IFRS began in the late 1960s, when the International Accounting Standards Committee (IASC) was established. The IASC was an independent organization that was responsible for developing and promoting international accounting standards. The IASC produced a series of standards called International Accounting Standards (IAS) that were intended to be used by companies in the preparation of their financial statements.

In 2001, the IASC was restructured and renamed the International Accounting Standards Board (IASB). The IASB is an independent, privately-funded organization that is responsible for the development and publication of IFRS. IFRS are now widely used by companies around the world, particularly those listed on stock exchanges in Europe, Asia, and South America.

The IASB is responsible for developing and maintaining IFRS, which are principles-based standards that provide guidance on the preparation and presentation of financial statements. The IASB operates through a due process that involves extensive consultation with stakeholders, including regulators, investors, and accounting professionals.

The IFRS are designed to be globally applicable, so they take into account the different legal, economic, and cultural contexts in which companies operate. This approach has made IFRS increasingly popular with multinational companies that operate in multiple jurisdictions.

Benefits of IFRS:

IFRS provide a number of benefits to companies and stakeholders. One of the main benefits is that they provide a common language for financial reporting that facilitates comparisons between companies and countries. This can help investors make more informed decisions and improve market efficiency.

IFRS also promote transparency and accountability by requiring companies to disclose more information about their financial performance and operations. This can help reduce the risk of fraud and improve corporate governance.

Another benefit of IFRS is that they can help reduce the costs of compliance for multinational companies. By using a common set of accounting standards, companies can avoid the need to comply with multiple sets of rules and regulations in different countries.

Finally, IFRS can help promote economic growth and development by improving the quality of financial information and facilitating cross-border investment.

Challenges of IFRS Adoption:

Despite the benefits of IFRS adoption, there are also several challenges that must be considered. One of the primary challenges is the complexity of IFRS. IFRS is a complex set of accounting standards that can be difficult for companies to understand and implement, especially for companies that operate in multiple jurisdictions. This complexity can make it difficult for companies to provide consistent and reliable financial information to their stakeholders.

Another challenge of IFRS adoption is the lack of flexibility in the standards. IFRS is a principles-based set of accounting standards, which means that there is more room for interpretation and judgment than in a rules-based system. However, this can also lead to inconsistencies in the application of the standards, which can make it difficult for investors and other stakeholders to compare financial information across companies.

Conclusion:

IFRS have become increasingly important in the global business environment, and their adoption has brought significant benefits to companies and stakeholders. However, the complexity of the standards and the lack of flexibility in their application are also significant challenges that must be addressed. As the global economy continues to evolve, it is likely that the role and importance of IFRS will continue to grow, and companies will need to adapt to these changes to remain competitive in a rapidly changing business environment.

Adoption of IFRS in different countries and regions

The adoption of International Financial Reporting Standards (IFRS) has been a topic of discussion among various countries and regions for the past few decades. IFRS was developed by the International Accounting Standards Board (IASB) as a way to create a globally accepted set of accounting standards that could be used by companies worldwide. Since the first set of IFRS was issued in 2001, many countries and regions have adopted these standards as their primary accounting framework. In this section, we will examine the adoption of IFRS in different countries and regions and the challenges that have arisen from this process.

Europe

The adoption of IFRS in Europe began in 2002 with the European Union's (EU) endorsement of IFRS as the mandatory accounting framework for all publicly traded companies in the EU. This endorsement was based on the premise that a common set of accounting standards would facilitate cross-border investment and improve financial reporting quality. The EU's adoption of IFRS was a significant milestone in the global adoption of IFRS, and it has been widely recognized as a success.

However, the adoption of IFRS in Europe has not been without challenges. One of the significant challenges has been the need to harmonize IFRS with local GAAP. Many EU member states had their own set of accounting standards, and the transition to IFRS required a significant effort to align these standards with the new framework. Another challenge has been the implementation of IFRS, which has required significant resources and training for companies and their employees.

Asia-Pacific

In Asia-Pacific, the adoption of IFRS has been less widespread than in Europe, although it is growing rapidly. Some countries, such as Australia, New Zealand, and Hong Kong, have already fully adopted IFRS. Other countries, such as Japan and South Korea, have adopted IFRS only for specific sectors, such as financial institutions. China has adopted IFRS for its publicly traded companies, but it has modified certain aspects of IFRS to align with its unique legal and economic environment.

One of the significant challenges of IFRS adoption in Asia-Pacific has been the need to harmonize IFRS with local GAAP. Many countries in the region have their own set of accounting standards, and the transition to IFRS has required a significant effort to align these standards with the new framework. Another challenge has been the lack of resources and training available for companies and their employees, particularly in countries where IFRS adoption is still in its early stages.

Africa

In Africa, the adoption of IFRS has been mixed. Some countries, such as South Africa, have fully adopted IFRS, while others have only partially adopted the standards or have yet to do so. The African Union has endorsed IFRS as the preferred accounting framework for member states, and many countries are expected to fully adopt IFRS in the coming years.

One of the significant challenges of IFRS adoption in Africa has been the lack of resources and training available for companies and their employees. Many companies in the region are small and medium-sized enterprises (SMEs), and they may not have the resources to implement IFRS effectively. Another challenge has been the need to harmonize IFRS with local GAAP, particularly in countries where the legal and economic environment is different from that in developed countries.

Americas

In the Americas, the adoption of IFRS has been less widespread than in Europe, but it is growing. Some countries, such as Canada and Mexico, have fully adopted IFRS, while others, such as the United States, have yet to do so. The United States allows foreign companies listed on U.S. exchanges to use IFRS, but it has not yet adopted the standards for domestic companies.

One of the main reasons for the slower adoption of IFRS in the Americas is the strong influence of the U.S. Generally Accepted Accounting Principles (GAAP) in the region. GAAP is a set of accounting standards used primarily in the United States, but also in some other countries in the Americas, such as Mexico. The familiarity and reliance on GAAP by companies and investors in the region has made the adoption of IFRS more challenging.

However, in recent years, there has been increased interest and discussion about the potential benefits of adopting IFRS in the Americas. For example, in Canada, the adoption of IFRS was seen as a way to improve financial reporting and increase transparency, while also aligning with international standards. Similarly, in Mexico, the adoption of IFRS was seen as a way to attract foreign investment and increase the country's competitiveness.

In South America, the adoption of IFRS has also been mixed. Some countries, such as Brazil, have fully adopted IFRS, while others, such as Argentina and Venezuela, have not. In Brazil, the adoption of IFRS was driven in part by the need to attract foreign investment and improve financial reporting. However, the transition to IFRS has not been without challenges, particularly for smaller companies with limited resources.

In Central America and the Caribbean, the adoption of IFRS has also been relatively slow. Some countries, such as Jamaica, have made progress towards adopting IFRS, but others, such as Haiti, have yet to do so. One of the challenges in the region is the lack of resources and infrastructure for implementing and enforcing the standards.

Overall, the adoption of IFRS in the Americas has been slower and more varied than in other regions. However, there are signs of increasing interest and momentum towards adopting the standards, particularly in Canada and Mexico. As the benefits of IFRS become more widely recognized and the global trend towards convergence continues, it is likely that more countries in the region will move towards adopting IFRS in the coming years.

Benefits and challenges of adopting IFRS

The adoption of International Financial Reporting Standards (IFRS) has become a global phenomenon in the past few decades. IFRS is a set of accounting standards that provide a common language for financial reporting. The aim of IFRS adoption is to improve the quality, comparability, and transparency of financial information. However, as with any significant change, there are also challenges that must be considered. In this section, we will discuss the benefits and challenges of adopting IFRS.

Benefits of Adopting IFRS

Improved Comparability

One of the primary benefits of IFRS adoption is the improved comparability of financial information. IFRS provides a standardized set of accounting principles, which ensures that companies report their financial information in a consistent and transparent manner. This standardization allows investors, analysts, and other stakeholders to compare the financial performance of different companies across different countries and industries, facilitating investment decisions and risk assessments.

Increased Transparency

IFRS adoption also increases the transparency of financial information by requiring companies to disclose more information about their financial activities. This increased transparency can enhance the credibility of financial information and reduce the likelihood of financial fraud and errors. It also enables investors and analysts to better understand a company's financial performance and make informed investment decisions.

Access to Global Capital Markets

Another benefit of IFRS adoption is the access to global capital markets. Companies that adopt IFRS can more easily raise capital in international markets, as investors and analysts are more likely to invest in companies that report financial information in accordance with IFRS. This can provide companies with greater opportunities for growth and expansion.

Improved Financial Management

IFRS adoption can also improve financial management within companies. The standardized accounting principles and increased transparency of financial information can help companies to identify areas for improvement and make better-informed financial decisions.

Challenges of Adopting IFRS

Complexity

One of the primary challenges of IFRS adoption is the complexity of the standards. IFRS is a complex set of accounting standards that can be difficult for companies to understand and implement, especially for companies that operate in multiple jurisdictions. This complexity can make it difficult for companies to provide consistent and reliable financial information to their stakeholders.

Lack of Flexibility

Another challenge of IFRS adoption is the lack of flexibility in the standards. IFRS is a principles-based set of accounting standards, which means that there is more room for interpretation and judgment than in a rules-based system. However, this can also lead to inconsistencies in the application of the standards, making it difficult for investors and analysts to compare the financial performance of different companies.

Costs of Adoption

IFRS adoption can also be costly for companies, particularly for those that have to transition from a different accounting framework. The costs of adoption can include training and education for employees, changes to accounting systems and processes, and additional compliance costs.

Cultural Differences

IFRS adoption can also face resistance due to cultural differences between countries. The cultural differences can impact the interpretation and implementation of IFRS, resulting in inconsistencies in financial reporting across different countries.

Conclusion

IFRS adoption has become increasingly important in the global economy, with the benefits of improved comparability, increased transparency, access to global capital markets, and improved financial management. However, there are also challenges that must be considered, including the complexity and lack of flexibility of the standards, the costs of adoption, and cultural differences. Companies must carefully consider these factors when deciding whether to adopt IFRS, and ensure that they have the necessary resources and expertise to implement the standards effectively.

CHAPTER 38: COMPARISON WITH US GAAP

IFRS and US GAAP are two sets of accounting standards that are widely used around the world. While both standards aim to provide a consistent and transparent framework for financial reporting, there are some significant differences between them. This chapter will provide a comprehensive overview of the main differences between IFRS and US GAAP, including differences in the recognition and measurement of assets, liabilities, equity, and revenue.

Recognition and Measurement of Assets

One of the main differences between IFRS and US GAAP is the way that assets are recognized and measured. Under IFRS, the recognition criteria for assets are more flexible than under US GAAP. IFRS allows assets to be recognized when it is probable that future economic benefits will flow to the entity and when the asset can be measured reliably. In contrast, US GAAP generally requires that an asset must have a determinable value and be owned or controlled by the entity before it can be recognized.

Measurement of assets also differs between IFRS and US GAAP. For example, under IFRS, property, plant, and equipment can be measured using either the cost model or the revaluation model. The cost model is similar to US GAAP, which requires assets to be recorded at cost less accumulated depreciation. The revaluation model, however, allows assets to be measured at fair value, which may result in higher values for the assets. Under US GAAP, property, plant, and equipment are generally recorded at cost less accumulated depreciation.

Recognition and Measurement of Liabilities

Similar to assets, the recognition and measurement criteria for liabilities also differ between IFRS and US GAAP. IFRS allows liabilities to be recognized when it is probable that future economic outflows will be required to settle the obligation and when the obligation can be measured reliably. US GAAP, on the other hand, generally requires that the obligation must be fixed and determinable before it can be recognized.

Measurement of liabilities also differs between IFRS and US GAAP. For example, under IFRS, provisions and contingencies can be recognized if it is probable that an outflow of resources will be required to settle the obligation and the amount of the obligation can be measured reliably. Under US GAAP, provisions and contingencies are generally recognized only if the amount can be reasonably estimated and the obligation is probable.

Recognition and Measurement of Equity

IFRS and US GAAP also differ in the recognition and measurement of equity. Under IFRS, equity includes all changes in equity during the reporting period, including transactions with owners. US GAAP, however, distinguishes between comprehensive income and other changes in equity.

Measurement of equity also differs between IFRS and US GAAP. For example, under IFRS, financial instruments can be measured at fair value, with changes in fair value recognized in equity. Under US GAAP, financial instruments are generally measured at cost, with changes in fair value recognized in the income statement.

Recognition and Measurement of Revenue

Finally, the recognition and measurement of revenue also differ between IFRS and US GAAP. IFRS and US GAAP both use a principles-based approach to revenue recognition, but there are some differences in the specific criteria for recognizing revenue. For example, under IFRS, revenue can be recognized when it is probable that future economic benefits will flow to the entity and when the amount of revenue can be measured reliably. Under US GAAP, revenue can be recognized when persuasive evidence of an arrangement exists, delivery has occurred or services have been rendered, the price is fixed or determinable, and collectability is reasonably assured.

Conclusion

In conclusion, while IFRS and US GAAP both aim to provide a consistent and transparent framework for financial reporting, there are some significant differences between the two standards. These differences are driven by various factors, such as cultural, historical, and legal contexts, as well as the needs and preferences of different stakeholders. As the world becomes more interconnected and globalized, the need for a common language of accounting has become more urgent. While IFRS has made significant progress in achieving this goal, there is still a long way to go before it can be fully adopted by all countries and organizations. In the meantime, companies that operate across borders or seek to attract foreign investment must be aware of the differences between IFRS and US GAAP and be prepared to navigate them.

For students and professionals in the accounting and finance field, it is essential to have a solid understanding of both IFRS and US GAAP, as well as their similarities and differences. This knowledge can help them make informed decisions and recommendations, and enable them to work effectively in a global context. It can also provide them with a competitive advantage in the job market, as employers increasingly value international expertise and cross-cultural skills.

Overall, the comparison between IFRS and US GAAP highlights the complexities and challenges of accounting standardization in a globalized world. While there is no one-size-fits-all solution, it is important to continue the dialogue and collaboration among different stakeholders to promote harmonization and convergence of accounting standards. By doing so, we can create a more efficient, transparent, and accountable financial reporting system that benefits businesses, investors, and society as a whole.

Key differences between IFRS and US GAAP

IFRS and US GAAP have several key differences in their accounting standards. Understanding these differences is essential for companies operating in multiple countries and for investors who need to interpret financial statements prepared under these different standards.

Conceptual Framework

IFRS and US GAAP have different conceptual frameworks. IFRS follows a principles-based approach, whereas US GAAP follows a rules-based approach. The principles-based approach of IFRS allows companies to exercise more judgment and flexibility in their accounting policies, whereas US GAAP provides more detailed and prescriptive guidance.

Revenue Recognition

IFRS and US GAAP have different criteria for recognizing revenue. Under IFRS, revenue is recognized when there is a transfer of risks and rewards of ownership, whereas under US GAAP, revenue is recognized when the risks and rewards of ownership have been transferred and collection is reasonably assured.

Inventory Valuation

IFRS and US GAAP have different methods for valuing inventory. Under IFRS, inventory is valued at the lower of cost and net realizable value, while under US GAAP, inventory is valued at the lower of cost or market.

Treatment of Leases

IFRS and US GAAP have different methods for accounting for leases. Under IFRS, leases are classified as either finance or operating leases, while under US GAAP, leases are classified as either capital leases or operating leases.

Financial Statement Presentation

IFRS and US GAAP have different rules for the presentation of financial statements. Under IFRS, there is no requirement to present separate statements for comprehensive income and changes in equity, while under US GAAP, separate statements for comprehensive income and changes in equity are required.

Intangible Assets

IFRS and US GAAP have different rules for accounting for intangible assets. Under IFRS, intangible assets are initially recognized at cost, while under US GAAP, intangible assets can be initially recognized at fair value.

Research and Development Costs

IFRS and US GAAP have different treatments for research and development costs. Under IFRS, research and development costs are expensed as incurred, while under US GAAP, research and development costs can be capitalized under certain circumstances.

Taxation

IFRS and US GAAP have different rules for accounting for income taxes. Under IFRS, deferred tax assets and liabilities are recognized based on the temporary differences between the carrying amount of assets and liabilities in the financial statements and their tax basis, while under US GAAP, deferred taxes are recognized based on the balance sheet approach.

Goodwill

IFRS and US GAAP have different rules for accounting for goodwill. Under IFRS, goodwill is tested annually for impairment, while under US GAAP, goodwill is tested for impairment at least annually, or more frequently if certain indicators exist.

Contingencies

IFRS and US GAAP have different rules for accounting for contingencies. Under IFRS, contingencies are recognized when it is probable that a liability has been incurred and the amount can be reliably estimated, while under US GAAP, contingencies are recognized when it is probable that a loss has been incurred and the amount can be reasonably estimated.

Conclusion

IFRS and US GAAP have some significant differences in their accounting standards. Understanding these differences is crucial for companies operating in multiple countries and for investors who need to interpret financial statements prepared under these different standards. Companies that operate in multiple jurisdictions should carefully consider the potential impact of adopting one or the other standards, and investors should be aware of the potential impact of differences in accounting policies on the comparability of financial statements across different companies and countries.

Implications for financial reporting and analysis

The adoption of IFRS and US GAAP has significant implications for financial reporting and analysis. Understanding the key differences between the two frameworks is crucial for businesses and investors alike. In this section, we will discuss the implications of adopting IFRS or US GAAP for financial reporting and analysis.

Financial Reporting

One of the main implications of adopting IFRS or US GAAP is the impact it has on financial reporting. The frameworks have different requirements for the presentation of financial statements, including the balance sheet, income statement, and statement of cash flows. For example, under IFRS, companies are required to present a statement of changes in equity, which is not required under US GAAP.

IFRS and US GAAP also differ in terms of the treatment of certain items. For instance, under IFRS, companies can use the revaluation model to measure property, plant, and equipment (PPE) at fair value, while US GAAP requires the cost model. Additionally, under IFRS, companies must report the fair value of all financial instruments on their balance sheet, while US GAAP has specific requirements for the treatment of financial instruments.

Financial Analysis

The adoption of IFRS or US GAAP can also have significant implications for financial analysis. Differences in the accounting treatment of certain items can impact key financial ratios and metrics. For example, the use of the revaluation model for PPE under IFRS can result in higher asset values and lower depreciation charges, which can inflate profitability ratios such as return on assets (ROA) and return on equity (ROE).

Additionally, differences in the presentation of financial statements can impact financial analysis. For example, the use of the statement of changes in equity under IFRS can provide additional information on changes in shareholders' equity, which can be useful for investors and analysts.

International Comparability

Another implication of the adoption of IFRS or US GAAP is the impact it has on international comparability. IFRS is used in over 120 countries, while US GAAP is only used in the United States. Adopting IFRS can enhance comparability between companies in different countries, making it easier for investors to make informed investment decisions.

However, there are also concerns that the adoption of IFRS can result in a loss of comparability between companies within the same country. This is because companies may have different interpretations of the standards or apply them differently, resulting in differences in financial reporting.

Cost of Adoption

Finally, the adoption of IFRS or US GAAP can have significant costs for companies. The adoption process can be time-consuming and costly, requiring significant resources to implement changes to accounting systems and processes. Additionally, ongoing training and maintenance costs may be required to ensure compliance with the standards.

For smaller companies, the costs of adoption may be prohibitive, resulting in a competitive disadvantage relative to larger companies. In some cases, companies may choose to adopt only certain aspects of the standards, resulting in a hybrid approach to financial reporting.

Conclusion

In conclusion, the adoption of IFRS or US GAAP has significant implications for financial reporting and analysis. Understanding the key differences between the two frameworks is crucial for businesses and investors alike. While the adoption of IFRS can enhance international comparability, it can also result in a loss of comparability within the same country. Additionally, the adoption process can be costly, particularly for smaller companies. Ultimately, the decision to adopt IFRS or US GAAP will depend on a variety of factors, including the company's size, industry, and international presence.

Efforts towards convergence of IFRS and US GAAP

Efforts towards convergence of IFRS and US GAAP have been ongoing for several years. While both frameworks share the same objective of providing a consistent and transparent framework for financial reporting, the differences between the two have resulted in challenges for multinational companies that operate in both jurisdictions.

The efforts towards convergence began in 2002 when the Financial Accounting Standards Board (FASB) and the International Accounting Standards Board (IASB) signed the Norwalk Agreement. The agreement aimed to develop high-quality, compatible accounting standards that could be used globally. The agreement was followed by a series of projects aimed at reducing the differences between IFRS and US GAAP.

One of the main goals of the convergence effort was to create a set of high-quality global accounting standards that could be adopted by countries around the world. This would simplify financial reporting for multinational companies and reduce the costs associated with maintaining multiple sets of financial statements. However, progress towards convergence has been slow, and there are still many significant differences between the two frameworks.

One of the key differences between IFRS and US GAAP is the treatment of inventory. IFRS requires the use of the first-in, first-out (FIFO) or weighted-average cost methods for inventory valuation, while US GAAP allows for the use of the last-in, first-out (LIFO) method. This can result in significant differences in the reported value of inventory and can impact financial ratios and performance measures used by investors and analysts.

Another major difference between the two frameworks is the treatment of goodwill. Under IFRS, goodwill is tested for impairment annually and is written down if its recoverable amount falls below its carrying amount. Under US GAAP, goodwill is tested for impairment annually or whenever events or changes in circumstances suggest that the carrying amount may not be recoverable.

The differences between IFRS and US GAAP can have significant implications for financial reporting and analysis. Multinational companies must navigate the different requirements of each framework and ensure that their financial statements comply with the relevant standards. This can result in additional costs and resources being allocated to financial reporting.

Furthermore, the differences in accounting standards can impact the comparability of financial statements across jurisdictions. Investors and analysts must understand the differences between IFRS and US GAAP and adjust their analysis accordingly. This can be particularly challenging for analysts who cover multinational companies that report under both frameworks.

Efforts towards convergence have been ongoing, but progress has been slow. The FASB and IASB have completed several joint projects aimed at reducing the differences between the two frameworks, but significant differences still remain. In 2019, the IASB published an exposure draft proposing amendments to IFRS 1 First-time Adoption of International Financial Reporting Standards. The amendments would provide relief to companies adopting IFRS for the first time by allowing them to apply certain exemptions to retrospective application of IFRS, including exemptions related to financial instruments and hedge accounting.

In conclusion, efforts towards convergence of IFRS and US GAAP have been ongoing for several years. While progress has been made, significant differences remain between the two frameworks. Multinational companies must navigate the different requirements of each framework and ensure that their financial statements comply with the relevant standards. Investors and analysts must also understand the differences between the two frameworks and adjust their analysis accordingly. As the world becomes increasingly globalized, the need for high-quality, compatible accounting standards has never been more important.

CHAPTER 39: CHALLENGES AND OPPORTUNITIES IN GLOBAL FINANCIAL REPORTING

In recent years, globalization has become an increasingly important aspect of business operations. Companies are expanding their operations across borders and engaging in cross-border transactions, which has led to a greater demand for consistent and transparent financial reporting standards. The International Financial Reporting Standards (IFRS) and the United States Generally Accepted Accounting Principles (US GAAP) are the two most widely used accounting standards in the world. While these standards share many similarities, there are significant differences that can create challenges for companies operating globally. In this chapter, we will explore the key differences between IFRS and US GAAP, the implications for financial reporting and analysis, and efforts towards convergence.

Key Differences between IFRS and US GAAP:

The key differences between IFRS and US GAAP can be grouped into five categories: (1) revenue recognition, (2) inventory valuation, (3) impairment of long-lived assets, (4) research and development costs, and (5) consolidation and joint ventures.

Revenue recognition:
One of the most significant differences between IFRS and US GAAP is the recognition of revenue. Under IFRS, revenue is recognized when it is probable that economic benefits will flow to the company, and the amount of revenue can be measured reliably. In contrast, US GAAP follows a more prescriptive approach, with detailed rules on when and how revenue should be recognized. For example, US GAAP requires the use of the percentage-of-completion method for long-term contracts, whereas IFRS allows for the use of either the percentage-of-completion method or the completed contract method.

Inventory valuation:
Another key difference between IFRS and US GAAP is the valuation of inventory. Under IFRS, inventory is valued at the lower of cost or net realizable value, while

under US GAAP, inventory is valued at the lower of cost or market. Additionally, IFRS allows for the use of the first-in, first-out (FIFO) or weighted average cost method, while US GAAP requires the use of the FIFO or specific identification method.

Impairment of long-lived assets:

IFRS and US GAAP also differ in the treatment of impairment of long-lived assets. Under IFRS, long-lived assets are reviewed for impairment when there is an indication of impairment, while under US GAAP, long-lived assets are reviewed for impairment annually, regardless of whether there is an indication of impairment. Additionally, IFRS requires the reversal of impairment losses if the conditions that caused the impairment no longer exist, while US GAAP generally prohibits the reversal of impairment losses.

Research and development costs:

Another area of difference between IFRS and US GAAP is the treatment of research and development costs. Under IFRS, research costs are expensed as incurred, while development costs can be capitalized if certain criteria are met. In contrast, US GAAP requires the capitalization of both research and development costs if certain criteria are met.

Consolidation and joint ventures:

Finally, IFRS and US GAAP differ in the treatment of consolidation and joint ventures. Under IFRS, a company must consolidate its subsidiaries unless certain criteria are met, while under US GAAP, the consolidation of subsidiaries depends on the level of control the company has over its subsidiaries. Additionally, IFRS allows for the use of proportionate consolidation for joint ventures, while US GAAP generally requires the use of the equity method.

Implications for financial reporting and analysis:

The key differences between IFRS and US GAAP can create challenges for companies operating globally, as they may need to comply with both sets of standards. This can be costly and time-consuming, as companies may need to maintain separate accounting systems and hire additional staff to handle the complexities of reporting under multiple standards. It can also create confusion for investors and other stakeholders, who may struggle to compare financial information across companies and jurisdictions.

One significant implication of these differences is that they can impact the valuation of assets and liabilities. For example, differences in how inventory is valued under IFRS and US GAAP can result in significant variations in reported earnings and asset values. This can make it difficult for investors to accurately compare the financial performance of companies operating in different jurisdictions.

Another important consideration is the impact of these differences on financial risk management. In particular, differences in accounting for derivatives and hedging activities under IFRS and US GAAP can create significant challenges for companies engaged in these activities. For example, companies may need to maintain multiple sets of records and reporting systems to comply with both standards, which can increase the risk of errors and misreporting.

Furthermore, differences in financial reporting standards can also impact corporate governance and transparency. For example, differences in the treatment of related party transactions under IFRS and US GAAP can affect the accuracy and completeness of financial statements, which can undermine investor confidence and trust in the financial reporting process.

To address these challenges, companies operating globally may need to adopt a variety of strategies, such as:

Standardization: Adopting a single set of accounting standards across all jurisdictions in which the company operates can help to simplify reporting and reduce compliance costs. This approach is often favored by multinational corporations, which may have the resources to adopt a common reporting standard.

Local adaptation: Adapting financial reporting practices to local requirements can help to ensure compliance with local regulations and reduce the risk of non-compliance. However, this approach can be costly and may require significant resources to maintain multiple reporting systems.

Hybrid approach: Adopting a hybrid approach that combines elements of both standardization and local adaptation can help to balance the need for global consistency with the need for local compliance. This approach can be particularly effective for companies operating in multiple jurisdictions with varying reporting requirements.

In addition to these strategies, efforts towards convergence of IFRS and US GAAP have been ongoing for many years. These efforts aim to reduce the differences between the two sets of standards and promote greater consistency and transparency in financial reporting.

One of the major steps towards convergence was the Memorandum of Understanding (MoU) signed in 2002 between the Financial Accounting Standards

Board (FASB) and the International Accounting Standards Board (IASB). The MoU established a framework for the two organizations to work together towards the development of a single set of high-quality global accounting standards.

Since then, the FASB and IASB have been working together to converge their respective standards, with the aim of achieving a single set of high-quality global accounting standards that can be used by companies operating in all jurisdictions. However, progress towards convergence has been slow and there are still significant differences between IFRS and US GAAP.

Despite these challenges, efforts towards convergence are continuing, and it is likely that the differences between IFRS and US GAAP will continue to narrow in the coming years. This will help to promote greater consistency and transparency in financial reporting, and make it easier for companies operating globally to comply with reporting requirements and for investors to compare financial information across companies and jurisdictions.

In conclusion, the key differences between IFRS and US GAAP have significant implications for financial reporting and analysis, particularly for companies operating globally. These differences can impact the valuation of assets and liabilities, financial risk management, corporate governance, and transparency. To address these challenges, companies may need to adopt a variety of strategies, such as standardization, local adaptation, or a hybrid approach. Additionally, efforts towards convergence of IFRS and US GAAP are ongoing, which will help to promote greater consistency and comparability in financial reporting across borders.

The convergence of IFRS and US GAAP has been a topic of discussion for many years, as it can significantly reduce the costs and complexities of multinational financial reporting. The main goal of convergence is to establish a single set of high-quality, globally accepted accounting standards that can be used by companies worldwide. While progress towards convergence has been slow, there have been some significant developments in recent years.

One of the key steps towards convergence was the issuance of the Joint Statement by the SEC and the European Commission in 2002, which expressed the commitment of both regulators towards the development of a single set of high-quality accounting standards. Since then, there have been several significant developments, including the formation of the International Accounting Standards Board (IASB) in 2001, which has been working towards the development of high-quality, globally accepted accounting standards.

Another important development was the adoption of the IFRS by the European Union in 2005, which required all publicly traded companies to use IFRS for their consolidated financial statements. This led to a significant increase in the adoption of IFRS globally, as many multinational companies chose to adopt IFRS in order to comply with the EU regulations.

However, despite these efforts, there are still significant differences between IFRS and US GAAP. One of the key areas of difference is the treatment of fair value accounting. While both standards allow for the use of fair value accounting, there are significant differences in the application of this principle. IFRS tends to be more principles-based and allows for greater discretion in the use of fair value accounting, while US GAAP tends to be more rules-based and prescriptive.

Another area of difference is the treatment of inventory costs. Under IFRS, companies can use either the first-in, first-out (FIFO) or weighted-average cost method to value their inventory, while US GAAP requires the use of the FIFO method. This can result in significant differences in the reported financial results of companies that operate in both jurisdictions.

In conclusion, while efforts towards convergence of IFRS and US GAAP are ongoing, there are still significant differences between the two standards. Companies that operate globally need to be aware of these differences and develop appropriate strategies to address the challenges that they create. By doing so, companies can improve their financial reporting and analysis, enhance transparency, and promote greater consistency and comparability in financial reporting across borders.

Challenges in implementing global financial reporting standards

Global financial reporting standards aim to provide consistency, transparency, and comparability in financial reporting across different countries and regions. However, the implementation of these standards can be challenging due to several factors, including legal and regulatory differences, cultural differences, and differences in accounting practices. This section will explore some of the challenges that companies and regulators face in implementing global financial reporting standards.

Legal and Regulatory Differences:

One of the primary challenges in implementing global financial reporting standards is the differences in legal and regulatory requirements across countries and regions. For example, in the United States, the Securities and Exchange Commission (SEC) is responsible for regulating financial reporting for public companies, while in the European Union, the European Securities and Markets Authority (ESMA) plays a similar role. These regulatory bodies have different requirements and enforcement mechanisms, which can create confusion and compliance challenges for multinational companies operating in multiple jurisdictions.

Cultural Differences:

Cultural differences can also pose challenges for companies in implementing global financial reporting standards. For example, in some cultures, there may be less emphasis on financial reporting and transparency, which can make it challenging to implement global standards that require a high level of disclosure. Additionally, cultural differences in the interpretation of accounting standards can lead to differences in accounting practices, even when the standards themselves are the same.

Accounting Practices:

Another challenge in implementing global financial reporting standards is differences in accounting practices. For example, the treatment of intangible assets can vary widely between countries, which can impact the valuation of assets and liabilities. Similarly, the recognition and measurement of revenue can also vary, which can impact financial statements and ratios.

Education and Training:

A lack of education and training on global financial reporting standards can also pose a challenge for companies and regulators. Many accounting professionals and financial analysts may be familiar with the standards in their own country, but not with global standards, which can make it challenging to implement these standards across borders. Additionally, the complexity of global standards can make it challenging for companies to understand and comply with the requirements, especially smaller companies that may not have the resources to hire specialized staff.

Technology:

The use of technology can help overcome some of the challenges of implementing global financial reporting standards. For example, digital tools can be used to standardize accounting practices and facilitate compliance across borders. However, technology can also create new challenges, such as cybersecurity risks and data privacy concerns, which must be addressed to ensure the integrity and security of financial information.

Conclusion:

Implementing global financial reporting standards can be a challenging task for companies and regulators due to legal and regulatory differences, cultural differences, differences in accounting practices, lack of education and training, and technological challenges. To overcome these challenges, companies may need to adopt strategies such as standardization, local adaptation, or a hybrid approach. Additionally, regulators can help by promoting greater consistency and harmonization of accounting standards across jurisdictions, while also providing guidance and education on global financial reporting standards. Overall, successful implementation of global financial reporting standards can help promote transparency, comparability, and efficiency in financial reporting, which is essential for the functioning of global capital markets.

Exercises:

Discuss the legal and regulatory differences that can create challenges for companies operating globally.

Explain how cultural differences can impact the implementation of global financial reporting standards.

Identify some of the accounting practices that can vary between countries and regions, and explain how these differences can impact financial reporting.

Discuss some of the strategies that companies can adopt to overcome the challenges of implementing global financial reporting standards.

Explain how technology can help overcome some of the challenges of implementing global financial reporting standards, and identify some of the new challenges that technology can create.

Opportunities for improved transparency and comparability

The adoption of global financial reporting standards has the potential to improve transparency and comparability in financial reporting. Transparency is crucial for investors, regulators, and other stakeholders to make informed decisions about a company's financial health and performance. Comparability, on the other hand, enables stakeholders to compare the financial performance of companies across different countries and industries. However, the implementation of these standards poses several challenges, such as differing legal and regulatory environments, cultural differences, and variations in accounting practices. This section will explore the opportunities for improved transparency and comparability that arise from the implementation of global financial reporting standards, as well as the challenges that must be overcome to realize these benefits.

Opportunities for improved transparency:

Improved transparency is one of the primary benefits of the adoption of global financial reporting standards. These standards require companies to disclose more information about their financial performance and operations, including the use of fair value measurements, the disclosure of related party transactions, and the use of consolidated financial statements. This increased disclosure can help to reduce information asymmetry between companies and their stakeholders, leading to more informed investment decisions and increased confidence in financial markets.

In addition to improving transparency for investors and regulators, global financial reporting standards can also enhance transparency for other stakeholders, such as customers, suppliers, and employees. For example, companies may use their financial statements to demonstrate their financial stability to customers, which can help to build trust and confidence in their products and services. Similarly, suppliers may use financial statements to evaluate a company's financial strength and ability to pay for goods and services on time. Finally, employees may use financial statements to evaluate a company's financial performance and stability, which can impact their decisions about their employment and career development.

Opportunities for improved comparability:

Another benefit of global financial reporting standards is improved comparability across companies and industries. Comparability enables stakeholders to evaluate the financial performance of companies across different countries and industries, which can be useful for benchmarking and performance evaluation purposes. For example, investors may use financial statements to compare the financial performance of companies in the same industry, or to evaluate the financial performance of a company relative to its peers.

Global financial reporting standards also facilitate the comparison of financial performance across different countries. This is particularly important for multinational companies that operate in multiple countries and report their financial performance in different currencies. The use of a common set of financial reporting standards can help to standardize financial reporting practices across countries and reduce the need for currency translation adjustments, which can improve the comparability of financial statements.

Challenges in implementing global financial reporting standards:

While there are many potential benefits to the adoption of global financial reporting standards, there are also several challenges that must be overcome to realize these benefits. One of the biggest challenges is the differing legal and regulatory environments across countries. Different countries have different legal and regulatory frameworks that impact financial reporting practices, such as tax laws, securities regulations, and company law. These differences can create confusion and ambiguity when implementing global financial reporting standards.

Cultural differences can also pose a challenge to the implementation of global financial reporting standards. For example, in some cultures, it may be considered inappropriate to disclose certain financial information, such as executive compensation or related party transactions. This can make it difficult to implement global financial reporting standards in a consistent and uniform manner across different cultures.

Finally, variations in accounting practices can also pose a challenge to the implementation of global financial reporting standards. Accounting practices can differ across countries due to historical, cultural, or legal factors. These differences can make it difficult to compare financial statements across countries and industries, and can create additional costs for companies that must comply with multiple sets of accounting standards.

Conclusion:

The adoption of global financial reporting standards has the potential to improve transparency and comparability in financial reporting, which can benefit a wide range of stakeholders, including investors, regulators, customers,and suppliers. The use of a common set of accounting standards can help to reduce the costs of financial reporting and analysis for multinational companies, making it easier for them to operate across different countries and regions. Additionally, it can increase the efficiency and effectiveness of the financial reporting process, making it easier for companies to identify and manage financial risks, and improve corporate governance.

Despite the challenges in implementing global financial reporting standards, there are also opportunities for improvement. One such opportunity is the use of technology to facilitate the implementation and monitoring of these standards. For example, blockchain technology can be used to create a secure, transparent, and decentralized system for financial reporting and auditing, reducing the risk of fraud and error.

Another opportunity is the increasing collaboration and cooperation among international organizations and regulatory bodies. The convergence of IFRS and US GAAP, as well as the development of other regional accounting standards, such as the Asian-Pacific Economic Cooperation (APEC) Framework, can help to promote greater consistency and comparability in financial reporting across different regions.

Furthermore, the use of integrated reporting, which combines financial and non-financial information, can provide a more comprehensive picture of a company's performance and its impact on society and the environment. This can help to promote greater transparency and accountability, and can be particularly beneficial for companies that operate in industries with high social and environmental impact, such as energy, mining, and agriculture.

In conclusion, the adoption of global financial reporting standards presents both challenges and opportunities for companies operating in a globalized economy. While there are significant differences between IFRS and US GAAP, efforts towards convergence and the development of other regional standards can help to promote greater consistency and comparability in financial reporting. Additionally, the use of technology and integrated reporting can further enhance transparency and accountability, ultimately benefiting a wide range of stakeholders. As such, companies should carefully consider the potential benefits and challenges of adopting global financial reporting standards, and develop strategies to effectively navigate the changing landscape of financial reporting and analysis.

Future trends and developments in international accounting standards.

International accounting standards have been evolving over the years to keep pace with the changing business environment and global economic conditions. The convergence of different accounting standards, such as IFRS and US GAAP, has been a major trend in recent years. However, there are still several challenges that need to be addressed to achieve greater consistency and comparability in financial reporting across different jurisdictions. In this section, we will explore some of the future trends and developments in international accounting standards and their potential impact on the global business community.

The Rise of Sustainability Reporting

One of the major trends in international accounting standards is the growing focus on sustainability reporting. Many companies are now recognizing the need to report on their social and environmental impact, in addition to their financial performance. This trend is being driven by a number of factors, including increased stakeholder demand, regulatory requirements, and investor expectations. To address these challenges, various international organizations have developed frameworks and guidelines for sustainability reporting, such as the Global Reporting Initiative (GRI) and the Sustainability Accounting Standards Board (SASB).

The Impact of Technology on Financial Reporting

Another trend in international accounting standards is the increasing use of technology in financial reporting. Advances in technology, such as cloud computing and blockchain, are transforming the way companies manage their financial information and communicate with stakeholders. These technologies offer opportunities to improve the accuracy, timeliness, and transparency of financial reporting, while also reducing costs and enhancing data security. However, they also pose significant challenges, such as the need for data standardization, cybersecurity risks, and regulatory compliance.

The Role of Integrated Reporting

Integrated reporting is a concept that seeks to integrate financial and non-financial information in a single report to provide a more holistic view of a company's performance. This approach emphasizes the importance of non-financial factors, such as environmental and social impact, in driving long-term value creation. The International Integrated Reporting Council (IIRC) has developed a framework for integrated reporting, which has been adopted by several companies and organizations around the world. However, the implementation of integrated reporting requires significant changes in organizational culture, systems, and processes.

The Need for Simplification and Clarity

Despite efforts to converge accounting standards, there is still a significant degree of complexity and diversity in financial reporting requirements across different jurisdictions. This can create challenges for companies operating globally, as they may need to comply with multiple sets of standards and regulations. To address these challenges, there is a growing need for simplification and clarity in financial reporting, particularly for small and medium-sized enterprises (SMEs). Several initiatives, such as the IFRS for SMEs standard and the AICPA's Financial Reporting Framework for SMEs, have been developed to provide simpler and more flexible reporting requirements for SMEs.

The Role of Standard-Setters and Regulators

The development and implementation of international accounting standards are largely driven by standard-setters and regulators. The International Accounting

Standards Board (IASB) is the primary organization responsible for developing and promoting IFRS, while the Financial Accounting Standards Board (FASB) is responsible for US GAAP. However, the roles and responsibilities of these organizations are subject to ongoing debate and scrutiny. Some stakeholders argue that standard-setters and regulators should focus on promoting greater consistency and comparability in financial reporting, while others argue that they should prioritize the needs of specific industries and jurisdictions.

Conclusion

International accounting standards are constantly evolving to meet the changing needs of the global business community. The rise of sustainability reporting, the impact of technology on financial reporting, the role of integrated reporting, the need for simplification and clarity, and the role of standard-setters and regulators are some of the major trends and developments that are shaping the future of international accounting standards. To succeed in this dynamic environment, companies and professionals must remain vigilant and informed about emerging trends and developments, and be prepared to adapt to changes in the regulatory environment.

One of the major trends in international accounting standards is the rise of sustainability reporting. As companies face increasing pressure from stakeholders to address environmental, social, and governance (ESG) concerns, sustainability reporting has become an important tool for communicating a company's performance in these areas. Many countries have already implemented mandatory reporting requirements for ESG information, and standard-setters are working to develop a comprehensive framework for sustainability reporting.

Technology is also having a significant impact on financial reporting. As companies increasingly rely on automated systems to gather and process financial data, there is a growing need for accounting standards to address the unique challenges posed by technology. Standard-setters are exploring ways to incorporate new technologies, such as artificial intelligence and blockchain, into financial reporting frameworks.

Integrated reporting is another area of focus for standard-setters. Integrated reporting seeks to provide a more comprehensive view of a company's performance by incorporating both financial and non-financial information into a single report. This approach aims to promote a more holistic understanding of a company's value creation and its impact on society and the environment.

In addition to these trends, there is a growing recognition of the need for simplification and clarity in accounting standards. The complexity of current accounting standards can make it difficult for companies to comply with regulations and for stakeholders to understand financial statements. Standard-setters are working to develop more concise and user-friendly accounting standards that reduce complexity and increase transparency.

Finally, the role of standard-setters and regulators is evolving as they work to promote greater consistency and convergence in accounting standards. While the ultimate goal is a single set of global accounting standards, achieving this is a complex and time-consuming process. Standard-setters are working to develop a framework for international convergence while also addressing the unique needs and concerns of individual countries and regions.

In conclusion, the future of international accounting standards is characterized by a number of major trends and developments, including sustainability reporting, technology, integrated reporting, simplification and clarity, and standard-setting and regulatory convergence. To succeed in this dynamic environment, companies and professionals must remain informed about emerging trends and be prepared to adapt to changes in the regulatory landscape.

PART 13: HOW WEB 3 AND BITCOIN ARE CHANGING THE FINANCIAL LANDSCAPE AND LAWS AS IT RELATES TO BUSINESS:

In recent years, the emergence of new technologies and the increasing use of cryptocurrencies have disrupted the traditional financial landscape, raising important questions about how businesses and regulators should adapt to these changes. Web 3, which refers to the next generation of the internet, promises to be more decentralized, secure, and private, potentially transforming the way we interact with information, assets, and each other. Bitcoin, the first and most famous cryptocurrency, has challenged the traditional notion of money, introducing a new form of digital asset that is not controlled by any central authority.

The implications of these developments for businesses are significant. On the one hand, Web 3 and Bitcoin can offer new opportunities for innovation, efficiency, and financial inclusion, potentially reducing transaction costs, increasing transparency, and empowering individuals and communities. On the other hand, they can also pose new risks and challenges, such as regulatory uncertainty, market volatility, cybersecurity threats, and social impact.

To navigate these issues, businesses need to understand the fundamentals of Web 3 and Bitcoin, as well as the legal, financial, and ethical implications of their adoption. This part of the textbook aims to provide a comprehensive overview of the current state of Web 3 and Bitcoin, their potential impact on businesses and society, and the key legal and regulatory issues that need to be addressed.

The first chapter will introduce the basic concepts of Web 3, including its underlying technologies, such as blockchain, decentralized applications (dapps), and smart contracts. It will also discuss the main features of Web 3, such as decentralization, privacy, and interoperability, and their potential benefits and challenges for businesses.

The second chapter will focus on Bitcoin, providing an overview of its history, architecture, and economics. It will also analyze the key drivers and barriers of Bitcoin adoption, such as user experience, scalability, and security, and their implications for businesses.

The third chapter will explore the legal and regulatory landscape of Web 3 and Bitcoin, discussing the current state of affairs and the major challenges and opportunities for businesses and regulators. It will cover topics such as the classification of cryptocurrencies, the taxation of digital assets, the protection of consumer rights, and the prevention of money laundering and terrorist financing.

The fourth chapter will examine the ethical and social implications of Web 3 and Bitcoin, addressing issues such as privacy, security, environmental impact, and social justice. It will also discuss the potential role of businesses in promoting ethical and responsible practices in the Web 3 and Bitcoin ecosystem.

Throughout this part of the textbook, we will use real-world examples and case studies to illustrate the concepts and challenges discussed, and provide exercises and questions to help students deepen their understanding and critical thinking skills. By the end of this part, students should have a solid foundation of knowledge and skills to navigate the evolving landscape of Web 3 and Bitcoin and its implications for business and society.

CHAPTER 40: EXPLANATION OF WHAT WEB 3 AND BITCOIN ARE

Chapter 40 of this textbook aims to provide a comprehensive understanding of two of the most significant developments in the financial landscape of the 21st century: Web 3 and Bitcoin. In recent years, these two concepts have garnered increasing attention from scholars, policymakers, and business leaders alike. The rise of Web 3 and Bitcoin has the potential to revolutionize the way we conduct business, transfer funds, and interact with one another online.

Web 3, also known as the decentralized web, is the next stage of internet evolution. It aims to address the limitations of the current internet infrastructure by creating a decentralized network that is secure, open, and transparent. Web 3 is built on the blockchain, a distributed ledger technology that enables secure and transparent transactions without the need for intermediaries such as banks, governments, or corporations. With Web 3, users can take back control of their data and digital identities, and interact with each other without the need for third-party intermediaries.

Bitcoin, on the other hand, is a digital currency that was created in 2009 by an unknown person using the pseudonym Satoshi Nakamoto. Bitcoin is built on the blockchain technology and operates as a decentralized digital currency. Unlike traditional currencies, Bitcoin is not issued or controlled by any central authority, and its value is determined by the market demand and supply. Bitcoin transactions are recorded on a public ledger called the blockchain, which is distributed across a network of computers. This makes Bitcoin transactions secure, transparent, and virtually impossible to hack or manipulate.

In this chapter, we will provide an overview of Web 3 and Bitcoin, including their history, technical underpinnings, and potential implications for businesses, governments, and individuals. We will also discuss some of the key legal and regulatory issues that arise with the use of Web 3 and Bitcoin, such as data privacy, cybersecurity, and financial regulation. Finally, we will explore some of the challenges and opportunities that Web 3 and Bitcoin present for businesses, investors, and policymakers, and provide some guidance on how to navigate this rapidly evolving landscape.

Overview of how they are changing the financial landscape

Web 3 and Bitcoin are revolutionizing the financial landscape by introducing new ways of conducting financial transactions, storing value, and creating new financial instruments. These technologies have the potential to transform traditional finance by providing greater transparency, security, and efficiency. This chapter provides an overview of Web 3 and Bitcoin and explains how they are changing the financial landscape.

Explanation of Web 3:

Web 3 refers to the third generation of the internet, which is characterized by decentralized, peer-to-peer networks. In Web 3, users have greater control over their data and can interact with each other directly without intermediaries. The technology underlying Web 3 is blockchain, which is a decentralized ledger that records transactions and stores data in a secure and transparent manner.

Blockchain enables the creation of decentralized applications (dApps), which are software programs that run on the blockchain. These applications can perform a variety of functions, such as enabling peer-to-peer transactions, creating new financial instruments, and managing digital assets.

Web 3 also includes other emerging technologies, such as artificial intelligence, machine learning, and the internet of things (IoT). These technologies enable greater automation, data analytics, and real-time monitoring of financial transactions.

Explanation of Bitcoin:

Bitcoin is a decentralized digital currency that operates on a peer-to-peer network. It was created in 2009 by an unknown person or group using the pseudonym Satoshi Nakamoto. Bitcoin is based on blockchain technology and is designed to be a decentralized and trustless system.

Bitcoin transactions are verified by network nodes through cryptography and recorded on a public ledger called the blockchain. Transactions are validated by network nodes through cryptography and recorded on a public ledger called the blockchain. Bitcoin is not backed by any government or physical commodity, and its value is determined by market demand.

✧ Bitcoin has several unique features that make it attractive to users. These include:

✧   Decentralization: Bitcoin is not controlled by any central authority, making it resistant to censorship and manipulation.

✧   Security: Bitcoin transactions are secured by cryptography, making them virtually tamper-proof.

✧   Privacy: Bitcoin transactions are pseudonymous, meaning that they do not reveal the identities of the parties involved in the transaction.

✧   Transparency: The blockchain provides a transparent and public ledger of all Bitcoin transactions.

Overview of how Web 3 and Bitcoin are changing the financial landscape:

Web 3 and Bitcoin are transforming the financial landscape in several ways. These include:

✧   Disintermediation: Web 3 and Bitcoin enable direct, peer-to-peer transactions without the need for intermediaries such as banks or payment processors. This reduces transaction costs and increases the speed and efficiency of transactions.

✧   Democratization of finance: Web 3 and Bitcoin enable anyone with an internet connection to participate in financial transactions, regardless of their location or financial status. This has the potential to increase financial inclusion and reduce inequality.

✧   Creation of new financial instruments: Web 3 and Bitcoin enable the creation of new financial instruments, such as cryptocurrencies, stablecoins, and decentralized finance (DeFi) protocols. These instruments have the potential to provide greater financial innovation and flexibility.

✧   Greater transparency and security: Web 3 and Bitcoin provide greater transparency and security through their use of blockchain technology. The blockchain provides a tamper-proof and transparent record of all transactions, making it easier to detect fraud and other forms of financial crime.

Conclusion:

Web 3 and Bitcoin are transforming the financial landscape by providing greater transparency, security, and efficiency. These technologies have the potential to democratize finance, reduce transaction costs, and increase financial inclusion. As these technologies continue to evolve, they are likely to create new opportunities and challenges for businesses, regulators, and consumers. It is important for businesses to understand these technologies and their implications for the financial landscape in order to stay competitive and compliant.

Brief overview of how laws related to business are being impacted

The rise of Web 3 and Bitcoin has significantly impacted the financial landscape and changed the way businesses operate. With the emergence of these new technologies, there has been a need to update existing laws and regulations to keep up with the changes in the industry. This section will provide a brief overview of how laws related to business are being impacted by Web 3 and Bitcoin.

✧   Cryptocurrency Regulations:
As Bitcoin and other cryptocurrencies have gained popularity, governments around the world have struggled to come up with a cohesive approach to regulating them. Some countries have banned cryptocurrencies altogether, while others have implemented strict regulations. For example, in the United States, the IRS treats Bitcoin as property for tax purposes, while the SEC has taken a stricter stance and has classified some cryptocurrencies as securities.

✧   Smart Contracts:
Smart contracts are self-executing contracts with the terms of the agreement written directly into code. They are often used in decentralized finance (DeFi) and other applications built on blockchain technology. The legal implications of smart contracts are still being explored, but they have the potential to streamline many legal processes and reduce the need for intermediaries.

✧   Privacy Concerns:
Web 3 technologies prioritize privacy and decentralization, which has raised concerns among regulators and lawmakers. Privacy-focused cryptocurrencies like Monero and Zcash are difficult to trace, making them appealing to criminals. As a result, many governments have implemented regulations aimed at preventing the use of cryptocurrency for illegal activities.

✧   Cross-Border Transactions:
Web 3 and Bitcoin have made cross-border transactions faster and cheaper than ever before. This has led to a need for new regulations to address issues related to international money transfers. Some countries have implemented strict regulations to prevent money laundering and terrorism financing, while others have embraced the potential of these new technologies to drive economic growth.

✧   Tokenization:

Tokenization refers to the process of creating digital tokens that represent real-world assets, such as real estate or stocks. Tokenization has the potential to revolutionize the way assets are traded and managed, but it also raises legal questions. For example, who owns the tokenized asset, and who is responsible for its management?

Conclusion:

Web 3 and Bitcoin are rapidly changing the financial landscape and challenging existing laws and regulations. While these new technologies offer many benefits, they also raise legal questions and concerns. Businesses and regulators must work together to develop new frameworks that balance innovation with protection for consumers and investors.

CHAPTER 41: THE EMERGENCE OF WEB 3 AND BITCOIN

Over the last few decades, technology has significantly transformed the financial industry. One of the most exciting developments in recent times is the emergence of Web 3 and Bitcoin. These two technologies have the potential to change the way we conduct business, store value, and even interact with each other.

Web 3 refers to the next generation of the internet, which is decentralized, open-source, and community-driven. It aims to create a more democratic and transparent digital world where individuals can take control of their data, identity, and online activities. In contrast, Bitcoin is a digital currency that operates on a decentralized network, allowing for secure, transparent, and fast peer-to-peer transactions without the need for intermediaries.

This chapter will provide a comprehensive overview of Web 3 and Bitcoin. We will discuss the key concepts and principles behind these technologies, their potential impact on the financial landscape, and the legal and regulatory challenges they pose.

Web 3: The Next Generation of the Internet

The current internet is dominated by a few large tech companies that control user data and content. Users are forced to accept terms and conditions that give these companies unrestricted access to their personal information, including browsing history, location, and search queries. This centralized model raises serious privacy and security concerns, as well as limitations on innovation and user autonomy.

Web 3 aims to address these issues by creating a more decentralized, transparent, and secure internet. It leverages blockchain technology, a distributed ledger system that allows for secure and immutable transactions without the need for intermediaries. This enables users to control their data and identity, eliminate the risk of censorship and manipulation, and foster more open and collaborative communities.

One of the key features of Web 3 is the use of smart contracts. These are self-executing contracts that are encoded on the blockchain and can automate complex business processes. Smart contracts enable secure and transparent transactions between parties without the need for intermediaries, reducing transaction costs and increasing efficiency.

Another essential element of Web 3 is the development of decentralized applications or dApps. These are applications that operate on a peer-to-peer network, allowing users to interact directly with each other without the need for intermediaries. dApps enable more user control, privacy, and transparency, and have the potential to disrupt traditional business models and industries.

Web 3 also aims to address issues related to digital identity. The current internet lacks a secure and reliable digital identity system, which makes it easy for hackers to steal personal information and commit fraud. Web 3 aims to create a decentralized and user-controlled digital identity system that is secure and resistant to fraud.

Bitcoin: A Decentralized Digital Currency

Bitcoin is a digital currency that operates on a decentralized network, allowing for secure and transparent peer-to-peer transactions without the need for intermediaries. Bitcoin was created in 2009 by an unknown person or group of people using the pseudonym Satoshi Nakamoto. It is based on a decentralized ledger system called blockchain, which enables secure and immutable transactions.

One of the key features of Bitcoin is its decentralized nature. It is not controlled by any government or financial institution, making it resistant to censorship and manipulation. Transactions are verified by a network of nodes, and new bitcoins are created through a process called mining, which involves solving complex mathematical problems.

Bitcoin transactions are anonymous, but all transactions are recorded on the blockchain, making them transparent and immutable. This means that Bitcoin can be used for secure and private transactions without the need for intermediaries.

Bitcoin has several advantages over traditional currencies. It is faster and cheaper to transfer, and it can be used anywhere in the world without the need for currency conversion. It is also inflation-resistant, as the supply of bitcoins is limited to 21 million.

Legal and Regulatory Challenges

The emergence of Web 3 and Bitcoin has brought significant legal and regulatory challenges to the financial landscape. As decentralized technologies continue to develop, traditional regulatory frameworks struggle to keep up with the rapidly evolving landscape. This section will provide an overview of the legal and regulatory challenges posed by the emergence of Web 3 and Bitcoin.

Legal and Regulatory Challenges

One of the key legal and regulatory challenges facing Web 3 and Bitcoin is the lack of clarity around regulatory requirements. Traditional regulatory frameworks were designed with centralized entities in mind, and often struggle to adequately address the unique features of decentralized systems. As such, regulatory bodies are faced with the challenge of applying existing regulations to emerging technologies that may not fit neatly into existing frameworks.

In the United States, the Securities and Exchange Commission (SEC) has taken a particularly active role in regulating cryptocurrency and other decentralized technologies. The SEC has indicated that many tokens issued through initial coin offerings (ICOs) are securities, and therefore subject to regulation under existing securities laws. However, the application of existing securities laws to decentralized systems is complex, and has led to significant legal uncertainty in the industry.

Another key legal challenge facing Web 3 and Bitcoin is the issue of jurisdiction. Decentralized systems are global in nature, and can be accessed from anywhere in the world. This creates a significant challenge for regulatory bodies, as they are often limited by jurisdictional boundaries. In addition, the lack of a central authority in decentralized systems makes it difficult to determine which regulatory body should be responsible for overseeing the system.

In some cases, regulators have attempted to assert their authority over decentralized systems by targeting individuals or companies involved in the development or operation of the system. For example, in 2019 the SEC filed a lawsuit against Telegram, alleging that the company's ICO for its TON blockchain platform violated securities laws. The lawsuit ultimately led to the cancellation of the ICO and the return of investor funds.

Another legal challenge facing Web 3 and Bitcoin is the issue of data protection and privacy. Decentralized systems often rely on the use of personal data to function, and as such are subject to data protection regulations such as the General Data Protection Regulation (GDPR) in the European Union. However, the decentralized nature of these systems makes it difficult to determine who is responsible for data protection, and how data protection regulations should be applied.

Finally, the use of decentralized systems such as Bitcoin can raise significant issues related to money laundering and other forms of financial crime. The pseudonymous nature of Bitcoin transactions makes it difficult to trace the flow of funds, and can create opportunities for criminal activity. As such, regulatory bodies around the world are increasingly focused on implementing measures to prevent money laundering and other financial crimes in the context of decentralized systems.

Conclusion

The emergence of Web 3 and Bitcoin has brought significant legal and regulatory challenges to the financial landscape. These challenges are complex and multifaceted, and require careful consideration from regulatory bodies, industry participants, and other stakeholders. As decentralized technologies continue to develop, it is likely that the legal and regulatory landscape will continue to evolve in response.

Brief history of the development of Bitcoin

Bitcoin was created in 2008 by an individual or group of individuals using the pseudonym "Satoshi Nakamoto." It was released as an open-source software program and presented as a peer-to-peer electronic cash system. The software is designed to allow users to send and receive payments without the need for intermediaries like banks or other financial institutions.

The first Bitcoin transaction took place in January 2009 when Nakamoto sent 10 Bitcoins to developer Hal Finney. Over time, the use of Bitcoin grew, with more individuals and businesses accepting it as a form of payment. In 2010, the first real-world transaction occurred when a user used 10,000 Bitcoins to purchase two pizzas.

As Bitcoin grew in popularity, so did its value. In 2013, the price of one Bitcoin reached $1,000 for the first time, attracting more attention from investors and businesses. However, this growth was not without controversy. Bitcoin has been associated with illegal activities, such as money laundering and drug trafficking, and has been criticized for its lack of regulation and potential use in financial crimes.

Despite these challenges, Bitcoin continues to be used and traded around the world. Its decentralized nature and the anonymity it offers to its users make it an attractive option for those seeking privacy and autonomy in their financial transactions.

As the use and acceptance of Bitcoin grew, so did the development of other cryptocurrencies. These digital currencies use blockchain technology, a decentralized ledger that records transactions, and offer alternative ways to store and transfer value.

However, the emergence of Bitcoin and other cryptocurrencies has also presented legal and regulatory challenges. Governments and regulatory bodies around the world have struggled to define and regulate these new forms of currency. The lack of a centralized authority has made it difficult to apply existing financial regulations to the use of cryptocurrencies, and some governments have even banned their use altogether.

In summary, the development of Bitcoin has been a significant milestone in the evolution of digital currencies. Its decentralized nature and potential for anonymity have made it an attractive option for some users, but its association with illegal activities and lack of regulation have also made it controversial. The emergence of Bitcoin and other cryptocurrencies has presented legal and regulatory challenges, with governments and regulatory bodies struggling to define and regulate these new forms of currency.

Explanation of how Web 3 builds on the principles of decentralization

The emergence of Web 3 represents a new phase in the evolution of the internet. Unlike its predecessors, Web 3 is designed to be decentralized, giving users more control over their data and online interactions. At the heart of Web 3 is blockchain technology, which allows for the creation of decentralized applications and services that operate without the need for intermediaries or central authorities.

Decentralization is a key principle of Web 3, and it builds on the foundations laid by Bitcoin and other blockchain-based cryptocurrencies. The original vision behind Bitcoin was to create a decentralized, trustless, and censorship-resistant digital currency that could operate without the need for banks or other intermediaries. Bitcoin achieved this through the use of a decentralized ledger known as the blockchain, which records transactions in a transparent and immutable manner.

Web 3 takes this concept of decentralization to the next level by extending the principles of the blockchain beyond cryptocurrencies and into other areas of the internet. The idea is to create a more open and democratic internet where users have greater control over their data and online interactions.

One of the key benefits of decentralization is that it eliminates the need for intermediaries and central authorities. In traditional centralized systems, intermediaries such as banks, social media platforms, and search engines have significant power and control over users' data and online interactions. This can lead to issues such as censorship, data breaches, and the misuse of personal information.

With Web 3, users have more control over their data and interactions because they can participate in decentralized networks that operate without the need for intermediaries. This is achieved through the use of decentralized applications (dApps) and protocols that operate on the blockchain. These dApps and protocols allow for the creation of decentralized marketplaces, social networks, messaging platforms, and other services that operate without the need for intermediaries or central authorities.

Another key benefit of decentralization is that it enables greater transparency and accountability. In centralized systems, intermediaries and central authorities have significant power and control over users' data and online interactions. This can lead to issues such as opaque decision-making processes, lack of transparency, and the potential for corruption.

With Web 3, users have more transparency and accountability because they can participate in decentralized networks that operate on the blockchain. These networks are transparent and immutable, meaning that users can verify the authenticity and integrity of transactions and interactions. This allows for greater trust and confidence in online interactions, as users can be sure that their data and interactions are secure and trustworthy.

In summary, Web 3 represents a new era in the evolution of the internet, one that is characterized by decentralization, transparency, and accountability. Decentralization is a key principle of Web 3, and it builds on the foundations laid by Bitcoin and other blockchain-based cryptocurrencies. By enabling greater control over data and interactions, Web 3 has the potential to create a more open and democratic internet that is better suited to the needs and interests of its users.

Overview of key features of Web 3, including smart contracts and decentralized applications

As we discussed in the previous section, Web 3 is built on the principles of decentralization, open source, and peer-to-peer networking. These principles allow for the creation of decentralized applications (dApps) that operate without the need for a central authority. This is made possible through the use of smart contracts, which are self-executing programs that run on the blockchain.

Smart contracts are computer programs that automatically execute the terms of a contract when certain conditions are met. These conditions are encoded into the smart contract, and when they are met, the contract is executed without the need for human intervention. This makes smart contracts ideal for use in a variety of industries, including finance, real estate, and supply chain management.

One of the key benefits of smart contracts is their transparency. Because smart contracts are stored on the blockchain, they are immutable and transparent. This means that anyone can view the code of a smart contract and verify that it is functioning as intended. This level of transparency helps to reduce fraud and increase trust in the system.

Another benefit of smart contracts is their ability to automate complex processes. Smart contracts can be programmed to execute a wide range of functions, from simple transactions to more complex processes like insurance claims or supply chain management. By automating these processes, smart contracts can reduce the need for intermediaries and streamline operations.

In addition to smart contracts, Web 3 also supports the development of decentralized applications (dApps). These are applications that are built on a decentralized network, such as the blockchain. Because dApps operate on a decentralized network, they are not subject to the same limitations as traditional applications. For example, dApps can operate without the need for a central server, and they can be accessed from anywhere in the world.

There are a wide range of dApps currently in development, including decentralized marketplaces, social networks, and gaming platforms. These applications are built on top of the blockchain, which provides a secure and transparent platform for their operation.

One of the challenges of developing dApps is the need to ensure that they are user-friendly. Because dApps operate on a decentralized network, they can be more complex to use than traditional applications. However, developers are working to create user-friendly interfaces that make it easier for users to interact with dApps.

Another challenge of developing dApps is the need to ensure their security. Because dApps operate on a decentralized network, they are more vulnerable to attack than traditional applications. To address this challenge, developers are working to create secure coding practices and implement robust security measures.

In summary, Web 3 is built on the principles of decentralization, open source, and peer-to-peer networking. These principles allow for the creation of decentralized applications that operate without the need for a central authority. Smart contracts and decentralized applications are two key features of Web 3 that enable this decentralization. Smart contracts are self-executing programs that automatically execute the terms of a contract when certain conditions are met. Decentralized applications are applications that operate on a decentralized network, such as the blockchain. While there are challenges associated with the development of dApps, developers are working to create user-friendly and secure applications that can revolutionize a wide range of industries.

CHAPTER 43: IMPACTS ON THE FINANCIAL LANDSCAPE

The emergence of cryptocurrencies and blockchain technology has revolutionized the financial landscape. From Bitcoin to Ethereum, these digital assets have gained significant popularity over the years, creating new investment opportunities and challenging traditional financial institutions. In this chapter, we will examine the impacts of cryptocurrencies and blockchain technology on the financial landscape, including the challenges and opportunities they present. We will also discuss the legal and regulatory frameworks governing the use of cryptocurrencies and blockchain technology, as well as the risks associated with investing in them.

Impacts of Cryptocurrencies on the Financial Landscape

Cryptocurrencies have disrupted traditional financial systems in many ways. One of the most significant impacts is their ability to provide an alternative to the traditional banking system. Cryptocurrencies enable users to make peer-to-peer transactions without the need for intermediaries, such as banks. This eliminates the need for traditional banking services and reduces transaction costs. Furthermore, cryptocurrencies are not subject to the same regulations and restrictions as traditional currencies, which can make them more accessible to people who have been excluded from traditional banking services.

Another impact of cryptocurrencies on the financial landscape is the rise of Initial Coin Offerings (ICOs). An ICO is a fundraising method that allows companies to raise capital by issuing digital tokens or coins to investors. These tokens can represent anything from equity in the company to access to a particular product or service. ICOs have become an increasingly popular way for companies to raise capital, with billions of dollars being raised through this method in recent years.

Cryptocurrencies have also created new investment opportunities for individuals and institutions. Bitcoin, for example, has become a popular investment option, with some investors seeing significant returns on their investments. Additionally, cryptocurrencies have created new jobs in the financial industry, such as cryptocurrency traders and blockchain developers.

Impacts of Blockchain Technology on the Financial Landscape

Blockchain technology, the underlying technology behind cryptocurrencies, has also had a significant impact on the financial landscape. One of the most significant impacts is its ability to provide a secure and transparent system for recording transactions. Blockchain technology uses a decentralized ledger that is distributed across multiple computers, making it difficult to hack or manipulate. This provides greater security and transparency, which can reduce the risk of fraud and corruption.

Blockchain technology has also enabled the development of smart contracts. Smart contracts are self-executing contracts that can be programmed to automatically enforce the terms of an agreement. This can eliminate the need for intermediaries, such as lawyers or banks, to oversee and enforce contracts, reducing transaction costs and increasing efficiency.

Furthermore, blockchain technology has the potential to revolutionize the way financial institutions operate. For example, it can enable the creation of decentralized autonomous organizations (DAOs), which are organizations that operate without a central authority. DAOs can be governed by smart contracts and operate on a blockchain, making them more transparent and democratic than traditional organizations.

Legal and Regulatory Frameworks

The emergence of cryptocurrencies and blockchain technology has presented unique legal and regulatory challenges. Governments and regulatory bodies around the world are struggling to develop frameworks that balance innovation with consumer protection and financial stability. One of the biggest challenges is the lack of uniformity in regulatory approaches, with some countries adopting a more permissive approach to cryptocurrencies and others taking a more restrictive stance.

In the United States, for example, the Securities and Exchange Commission (SEC) has taken a cautious approach to cryptocurrencies, issuing guidance that many ICOs are considered securities and therefore subject to securities laws. Other countries, such as Malta and Switzerland, have adopted more permissive regulatory frameworks, encouraging the development of blockchain and cryptocurrency projects in their jurisdictions.

Risks and Challenges

Despite the potential benefits of cryptocurrencies and blockchain technology, they also present significant risks and challenges. One of the biggest risks is the ...volatility of cryptocurrencies. The prices of cryptocurrencies are highly volatile, and can fluctuate widely in a short period of time. This makes it difficult to use them as a store of value or a unit of account. For example, Bitcoin's price has experienced several booms and busts over the years. In 2017, its price increased from around $1,000 to nearly $20,000, before dropping to around $3,000 in 2018. This volatility can make it difficult for businesses and consumers to use cryptocurrencies for everyday transactions, and can discourage investment.

Another challenge is the lack of regulation and oversight. Cryptocurrencies operate outside of traditional financial systems, and are not subject to the same regulations and oversight as traditional financial instruments. This makes them vulnerable to fraud, money laundering, and other illegal activities. For example, in 2014, the Mt. Gox exchange, which was then the largest Bitcoin exchange, filed for bankruptcy after losing 850,000 Bitcoins, worth around $500 million at the time. The exchange was accused of mismanagement and fraud, and its collapse caused widespread losses for investors.

In addition, the anonymous and decentralized nature of cryptocurrencies can make them attractive to criminals and terrorists. Cryptocurrencies can be used to finance illegal activities, such as drug trafficking, human trafficking, and terrorism. The lack of transparency in cryptocurrency transactions can make it difficult for law enforcement agencies to track and prosecute criminals who use them.

Another risk is the potential for hacks and cyber attacks. Cryptocurrency exchanges and wallets can be vulnerable to hacks and cyber attacks, which can result in the theft of cryptocurrencies. In 2018, for example, the Japanese cryptocurrency exchange Coincheck was hacked, resulting in the theft of around $530 million worth of cryptocurrency. Such incidents can erode investor confidence in cryptocurrencies, and lead to a decline in their value.

Finally, there is the risk of technological obsolescence. Cryptocurrencies and blockchain technology are still in their infancy, and it is possible that newer and better technologies could emerge that make them obsolete. This could result in a decline in the value of cryptocurrencies and blockchain-based businesses, and could lead to significant losses for investors.

In summary, while cryptocurrencies and blockchain technology have the potential to revolutionize the financial landscape, they also present significant risks and challenges. These include volatility, lack of regulation and oversight, susceptibility to illegal activities, vulnerability to hacks and cyber attacks, and the risk of technological obsolescence. As such, investors and businesses should approach these technologies with caution, and should take steps to mitigate these risks.

Explanation of how Bitcoin and other cryptocurrencies are changing the nature of money

Bitcoin, the first and most well-known cryptocurrency, was created in 2009 by an anonymous individual or group using the pseudonym Satoshi Nakamoto. Since then, thousands of other cryptocurrencies have been created, each with its unique features and purposes. At their core, cryptocurrencies are digital assets designed to function as a medium of exchange, store of value, or unit of account.

One of the most significant ways in which Bitcoin and other cryptocurrencies are changing the nature of money is by introducing a decentralized model of currency. Unlike traditional fiat currencies that are controlled by governments and central banks, cryptocurrencies are created and maintained by decentralized networks of users.

Decentralization means that no single entity has control over the currency, and all users have equal rights and responsibilities in maintaining the network. Transactions are verified and recorded on a public ledger called a blockchain, which is distributed across the network and accessible to anyone with an internet connection.

This model of currency has several advantages over traditional fiat currencies. First, it eliminates the need for intermediaries, such as banks and payment processors, which can reduce transaction fees and increase the speed and efficiency of transactions. Second, it provides greater privacy and security since users do not need to disclose personal information to make transactions. Finally, it allows for greater financial inclusion, as anyone with an internet connection can access and use cryptocurrencies, regardless of their geographic location or socioeconomic status.

However, cryptocurrencies also present several challenges and risks. One of the most significant challenges is the issue of scalability. Currently, most cryptocurrencies have limited transaction processing capabilities, which can lead to long processing times and high transaction fees during periods of high network activity.

Another challenge is the issue of regulation. Since cryptocurrencies are decentralized and operate outside of traditional financial systems, they present challenges for regulators and law enforcement agencies. There are concerns about their potential use for illegal activities, such as money laundering and terrorist financing, as well as their impact on financial stability and consumer protection.

Despite these challenges, cryptocurrencies have already had a significant impact on the financial landscape. They have attracted significant investment from individuals and institutions alike, with some seeing them as a potential hedge against inflation and a store of value similar to gold. They have also spawned a new industry of cryptocurrency exchanges, wallets, and other service providers, creating new opportunities for entrepreneurs and investors.

Overall, the rise of Bitcoin and other cryptocurrencies has sparked a renewed interest in the nature of money and its role in modern society. It has opened up new possibilities for financial innovation and inclusion, while also presenting significant challenges and risks. As the cryptocurrency market continues to evolve, it is essential for policymakers, regulators, and industry participants to work together to ensure that these technologies are used responsibly and ethically, to the benefit of all.

Example Problem:

What is the main advantage of the decentralized model of currency introduced by Bitcoin and other cryptocurrencies?
a. Eliminates the need for intermediaries
b. Provides greater privacy and security
c. Allows for greater financial inclusion
d. All of the above

You are a financial analyst at a large investment bank, and your boss asks you to evaluate the potential risks and benefits of investing in cryptocurrencies. Using what you have learned in this section, create a report outlining your analysis, including at least three potential benefits and three potential risks of investing in cryptocurrencies.

Imagine you are the CEO of a small business, and you are considering accepting Bitcoin as a form of payment. What are some potential advantages and disadvantages of doing so? How might the decentralized nature of Bitcoin affect your decision?

You are a compliance officer at a large corporation, and you are tasked with developing a set of guidelines for employees who wish to invest in cryptocurrencies. What are some key considerations to keep in mind when developing these guidelines, and how might they differ from traditional investment guidelines?

You are a regulator at a government agency, and you are tasked with drafting regulations for the use of cryptocurrencies in your country. What are some potential benefits and risks of cryptocurrencies, and how might you balance these considerations when drafting regulations?

You are a business lawyer advising a client who is interested in launching a new cryptocurrency. What legal considerations should your client be aware of, and what steps can they take to mitigate potential risks? How might the decentralized nature of cryptocurrencies affect their legal strategy?

Example Problem:

The decentralized model of currency introduced by Bitcoin and other cryptocurrencies has several advantages. Which of the following is not one of these advantages?
a. Eliminates the need for intermediaries
b. Provides greater privacy and security
c. Allows for greater financial inclusion
d. Is controlled by a central authority

Discussion of the potential benefits and risks of these changes

Blockchain technology and cryptocurrencies have the potential to significantly disrupt traditional financial institutions and processes. While they offer several benefits, they also present significant risks and challenges. In this section, we will explore the potential benefits and risks of these changes.

Potential Benefits

✧   Efficiency and Cost Savings

One of the most significant potential benefits of blockchain technology is increased efficiency and cost savings. Blockchain technology enables the secure and efficient transfer of assets and information without the need for intermediaries, such as banks or financial institutions. By eliminating intermediaries, blockchain technology reduces transaction fees, which can result in significant cost savings for individuals and businesses.

Blockchain technology can also increase efficiency by automating processes, reducing the need for manual intervention, and increasing transaction speed. Smart contracts, for example, can be programmed to execute automatically once specific conditions are met, reducing the need for intermediaries and increasing efficiency.

✧   Greater Financial Inclusion

Blockchain technology and cryptocurrencies have the potential to increase financial inclusion by providing access to financial services to individuals who are currently unbanked or underbanked. According to the World Bank, approximately 1.7 billion adults worldwide do not have access to a bank account. Blockchain technology can provide these individuals with access to financial services by enabling peer-to-peer transactions without the need for a financial institution.

Cryptocurrencies can also enable cross-border transactions without the need for a traditional bank, making it easier and more affordable for individuals and businesses to conduct international transactions.

✧   Transparency and Security

Blockchain technology can increase transparency and security in financial transactions. Blockchain is a distributed ledger that records transactions in a transparent and immutable way. Once a transaction is recorded on the blockchain, it cannot be altered, providing a high level of security.

In addition, blockchain technology can increase transparency by enabling participants to view and verify transactions. This increased transparency can help to reduce fraud and corruption, as all participants have access to the same information.

Potential Risks

➢   Regulatory Risks

One of the most significant risks associated with cryptocurrencies and blockchain technology is regulatory risk. Governments around the world are still grappling with how to regulate cryptocurrencies and blockchain technology, and regulatory frameworks are still evolving.

The lack of regulatory clarity can create uncertainty for businesses and individuals that want to participate in the cryptocurrency market, and it can also make it easier for bad actors to engage in illicit activities, such as money laundering and terrorism financing.

> ➢ Volatility and Lack of Stability

Cryptocurrencies are highly volatile and lack stability, making them a risky investment. Cryptocurrency prices can fluctuate widely in a short period, making it difficult for investors to make informed decisions.

In addition, cryptocurrencies are not backed by any government or financial institution, which can make them more susceptible to market manipulation and fraud. This lack of stability and regulation can make cryptocurrencies a risky investment for individuals and businesses.

> ➢ Security Risks

While blockchain technology can increase security in financial transactions, it is not immune to security risks. Hackers have targeted cryptocurrency exchanges and wallets, resulting in the loss of millions of dollars in cryptocurrency.

In addition, because blockchain technology is decentralized, there is no central authority to regulate transactions or resolve disputes. This lack of oversight can make it difficult to recover lost or stolen cryptocurrency, leaving investors with little recourse in the event of a security breach.

> ➢ Adoption and Technical Challenges

Finally, the adoption of blockchain technology and cryptocurrencies faces significant technical challenges. The technology is still relatively new, and there are many technical hurdles that need to be overcome to make blockchain and cryptocurrency more accessible to individuals and businesses.

For example, the scalability of blockchain technology remains a significant challenge. As more people use blockchain networks, the size of the ledger grows, making it more difficult and time-consuming to validate transactions. Bitcoin, for example, has a transaction processing capacity of just seven transactions per second, which is far lower than the average transaction processing capacity of traditional payment systems such as Visa and Mastercard. This can result in delays and high transaction fees, making it less attractive for everyday transactions.

Another technical challenge is the issue of interoperability. Currently, there are many different blockchain networks and cryptocurrencies, each with its own set of rules and protocols. This can make it difficult to transfer value between different networks, which can limit the potential for widespread adoption. While there are efforts underway to develop interoperability protocols, it remains a significant challenge.

Furthermore, blockchain technology and cryptocurrencies are highly technical in nature, which can make them difficult for the average person to use and understand. This can be a significant barrier to adoption, as many people may not be willing or able to invest the time and effort required to understand how the technology works and how to use it effectively.

In addition to these technical challenges, there are also significant regulatory and legal challenges that must be addressed before blockchain and cryptocurrency can be widely adopted. Many countries have yet to develop clear regulatory frameworks for cryptocurrencies, which can create uncertainty for businesses and individuals operating in the space.

Furthermore, the anonymity and decentralized nature of cryptocurrencies can make them attractive to criminal elements, which has led to concerns around money laundering, terrorist financing, and other illicit activities. This has prompted many governments and regulatory bodies to introduce stricter regulations and compliance requirements for cryptocurrency exchanges and other businesses operating in the space.

Overall, while blockchain technology and cryptocurrencies have the potential to revolutionize the way we think about and use money, they also present significant risks and challenges. Addressing these challenges will be essential for realizing the full potential of this technology and ensuring that it is used in a safe, secure, and responsible manner.

Impacts of Cryptocurrencies on the Financial Landscape

The rise of cryptocurrencies has had significant impacts on the financial landscape. While traditional financial systems have existed for centuries, cryptocurrencies have challenged the traditional ways of conducting financial transactions. This section will discuss the impacts of cryptocurrencies on the financial landscape.

Increased Accessibility

One of the most significant impacts of cryptocurrencies on the financial landscape is increased accessibility. Cryptocurrencies provide individuals with the ability to conduct financial transactions without the need for intermediaries such as banks. This has made it easier for individuals who do not have access to traditional banking services to participate in the financial system.

For example, in many developing countries, individuals may not have access to traditional banking services due to a lack of infrastructure. Cryptocurrencies provide these individuals with a way to participate in the global economy. Additionally, cryptocurrencies can be used to send money across borders quickly and at a lower cost than traditional methods.

Increased Financial Inclusion

Cryptocurrencies have also increased financial inclusion by providing individuals with access to financial services that were previously unavailable to them. This is particularly important for individuals who live in areas where traditional banking services are scarce or non-existent.

Cryptocurrencies can also be used to provide microfinance loans to individuals who do not have access to traditional banking services. This has the potential to lift individuals out of poverty by providing them with the capital they need to start their own businesses or invest in education.

Increased Security

Cryptocurrencies have also increased security in the financial landscape. Traditional financial systems are vulnerable to cyber attacks, which can result in the loss of sensitive financial information. Cryptocurrencies, on the other hand, use advanced cryptographic algorithms to secure transactions and protect user data.

Additionally, cryptocurrencies provide users with the ability to remain anonymous. This has made it more difficult for criminals to conduct illegal activities such as money laundering and terrorism financing.

Disrupting Traditional Financial Institutions

The rise of cryptocurrencies has disrupted traditional financial institutions. Cryptocurrencies have challenged the traditional ways of conducting financial transactions and have provided individuals with an alternative to traditional banking services.

This disruption has led to the creation of new financial products and services, such as decentralized finance (DeFi) platforms. These platforms allow individuals to lend, borrow, and trade cryptocurrencies without the need for intermediaries such as banks.

However, this disruption has also created challenges for traditional financial institutions, which must adapt to the changing financial landscape or risk becoming obsolete.

Increased Regulation

The rise of cryptocurrencies has also led to increased regulation in the financial landscape. Regulators around the world have recognized the potential risks associated with cryptocurrencies, such as money laundering, terrorism financing, and fraud.

As a result, many countries have implemented regulations to govern the use of cryptocurrencies. These regulations vary by country but generally include measures to prevent illegal activities and protect consumers.

However, increased regulation has also created challenges for the adoption of cryptocurrencies. Some individuals may be hesitant to use cryptocurrencies due to concerns about government regulation and oversight.

Conclusion

In conclusion, the impacts of cryptocurrencies on the financial landscape have been significant. Cryptocurrencies have increased accessibility, financial inclusion, security, and have disrupted traditional financial institutions. However, these changes have also created challenges, including technical hurdles, increased regulation, and concerns about the stability of cryptocurrencies.

It is clear that cryptocurrencies will continue to play a significant role in the financial landscape in the coming years. As such, it is important for individuals and businesses to stay informed about the changing financial landscape and the impacts of cryptocurrencies on traditional financial systems.

Legal and Regulatory Frameworks

Blockchain technology and cryptocurrencies have disrupted the traditional financial system and have created new opportunities for businesses and individuals. However, with these new opportunities come new challenges for regulators and lawmakers. In this section, we will discuss the legal and regulatory frameworks that apply to blockchain technology and cryptocurrencies.

Legal Frameworks:

The legal frameworks that apply to blockchain technology and cryptocurrencies vary from country to country. In some countries, such as the United States, blockchain technology and cryptocurrencies are subject to a patchwork of state and federal regulations. In other countries, such as Japan, there are comprehensive regulatory frameworks in place that apply to blockchain technology and cryptocurrencies.

In the United States, the regulatory framework for blockchain technology and cryptocurrencies is still in its infancy. The primary federal regulator for cryptocurrencies is the Securities and Exchange Commission (SEC), which has issued guidance on the classification of cryptocurrencies as securities. The Commodity Futures Trading Commission (CFTC) also has jurisdiction over cryptocurrencies, and has issued guidance on the classification of cryptocurrencies as commodities.

State-level regulation of cryptocurrencies has been more active, with some states enacting legislation that provides legal recognition for blockchain technology and cryptocurrencies. For example, Wyoming has enacted a series of laws that provide a comprehensive legal framework for blockchain technology and cryptocurrencies.

In other countries, such as Japan, there are comprehensive legal frameworks in place for blockchain technology and cryptocurrencies. The Japanese government has recognized cryptocurrencies as a legitimate form of payment, and has enacted laws and regulations that provide a framework for the operation of cryptocurrency exchanges.

Regulatory Frameworks:

The regulatory frameworks that apply to blockchain technology and cryptocurrencies also vary from country to country. In some countries, such as China, the government has taken a hardline stance on cryptocurrencies, banning cryptocurrency exchanges and initial coin offerings (ICOs). In other countries, such as Switzerland, the government has taken a more hands-off approach, allowing for the development of a vibrant cryptocurrency industry.

In the United States, the regulatory framework for blockchain technology and cryptocurrencies is still evolving. The SEC has taken enforcement action against a number of ICOs that it deems to be securities offerings, and has indicated that it may take further action to regulate the cryptocurrency industry.

The CFTC has taken a more hands-off approach, allowing for the development of cryptocurrency derivatives markets. However, the agency has also indicated that it may take further action to regulate the cryptocurrency industry if necessary.

In other countries, such as Japan, the regulatory framework for blockchain technology and cryptocurrencies is more developed. The Japanese government has enacted laws and regulations that provide a framework for the operation of cryptocurrency exchanges and has established a regulatory agency to oversee the industry.

Conclusion:

The legal and regulatory frameworks that apply to blockchain technology and cryptocurrencies are still in the process of being developed. In some countries, comprehensive legal and regulatory frameworks have been established, while in others, the frameworks are still evolving. As blockchain technology and cryptocurrencies continue to disrupt the traditional financial system, regulators and lawmakers will need to develop legal and regulatory frameworks that strike a balance between protecting consumers and promoting innovation.

CHAPTER 44: IMPACTS ON BUSINESS AND COMMERCE

The emergence of blockchain technology and cryptocurrencies has created significant impacts on the business and commerce landscape. In this chapter, we will explore the effects of these new technologies on various aspects of business, including transactions, financing, and regulation. We will discuss the potential benefits and drawbacks of blockchain and cryptocurrency adoption and examine the legal and regulatory frameworks that businesses and entrepreneurs must navigate.

Transactions

The adoption of blockchain technology has the potential to revolutionize the way businesses conduct transactions. Blockchain technology enables the creation of decentralized and immutable ledgers that can be used to record transactions between parties without the need for intermediaries such as banks or financial institutions.

The use of blockchain technology in transactions can increase efficiency and reduce costs. Transactions can be completed faster and with fewer errors, resulting in lower transaction fees and processing times. Moreover, blockchain technology provides a high level of security and transparency, making it easier for businesses to detect and prevent fraud.

Cryptocurrencies, on the other hand, offer a new way to transfer value without relying on traditional banking systems. Cryptocurrencies such as Bitcoin and Ethereum can be used to send and receive payments without intermediaries. This can be particularly beneficial for businesses operating in areas where traditional banking systems are not available or are unreliable.

However, the adoption of cryptocurrencies in transactions also presents challenges. Cryptocurrencies are highly volatile, with prices fluctuating significantly over short periods. This can make it difficult for businesses to price goods and services and manage cash flows. Moreover, the lack of regulation and oversight in the cryptocurrency market can create risks for businesses that use cryptocurrencies, such as money laundering and fraud.

Financing

The emergence of blockchain technology and cryptocurrencies has also created new opportunities for businesses to raise capital. Initial coin offerings (ICOs) have become a popular way for blockchain-based startups to raise funds. ICOs enable investors to purchase digital tokens that represent ownership in a project or company. These tokens can be traded on cryptocurrency exchanges, providing liquidity for investors.

ICOs have become a significant source of funding for blockchain-based startups. In 2017, ICOs raised over $6 billion in capital, surpassing traditional venture capital funding for the first time. However, ICOs also present risks for investors. Many ICOs are unregulated and lack transparency, making it difficult for investors to evaluate the potential risks and rewards of investing.

In addition to ICOs, blockchain technology has also enabled the creation of decentralized finance (DeFi) platforms. DeFi platforms provide a way for individuals to lend and borrow funds without relying on traditional financial institutions. DeFi platforms can be accessed by anyone with an internet connection, making them accessible to individuals who may not have access to traditional banking services.

DeFi platforms have the potential to disrupt traditional banking and finance systems, but they also present risks. DeFi platforms are highly speculative and can be subject to significant price volatility. Moreover, the lack of regulation and oversight in the DeFi market can create risks for investors and borrowers.

Regulation

The emergence of blockchain technology and cryptocurrencies has created significant challenges for regulators and policymakers. Blockchain technology and cryptocurrencies operate outside of traditional financial systems, making it difficult for regulators to monitor and oversee their use.

The lack of regulation and oversight in the blockchain and cryptocurrency markets has created opportunities for fraud and abuse. For example, the lack of regulation has allowed for the creation of fraudulent ICOs, which have defrauded investors out of millions of dollars. Moreover, the lack of oversight has created opportunities for money laundering and other illegal activities.

To address these risks, regulators and policymakers around the world have begun to develop legal and regulatory frameworks for blockchain and cryptocurrencies. These frameworks aim to protect investors and consumers while promoting innovation and growth in the blockchain and cryptocurrency markets.

However, the development of legal and regulatory frameworks for blockchain and cryptocurrencies is still in its early stages, and there are significant challenges that must be addressed. One of the main challenges is the lack of uniformity in the regulations around the world. As blockchain and cryptocurrency markets are global, it is difficult to create a uniform set of regulations that can be applied in all jurisdictions.

Another challenge is the fast-paced nature of the technology. The legal and regulatory frameworks must keep up with the rapid changes and advancements in blockchain and cryptocurrency. This requires continuous updates and adjustments to the regulatory landscape, which can be challenging for regulators and policymakers.

Furthermore, the decentralized nature of blockchain technology and cryptocurrencies creates a unique regulatory challenge. Traditional regulatory approaches may not be applicable to these technologies, and new approaches may need to be developed. Additionally, the lack of central control in the blockchain network makes it difficult to identify and regulate bad actors.

In addition, the lack of clarity around the classification of cryptocurrencies creates confusion for businesses and investors. For example, some cryptocurrencies may be classified as securities, while others may not. This creates uncertainty around the legal and regulatory requirements that businesses and investors must comply with.

The lack of clarity around taxation is also a challenge. Different jurisdictions have different approaches to the taxation of cryptocurrencies, which can create confusion for businesses and investors operating in multiple jurisdictions. The lack of clarity can also make it difficult for regulators to enforce tax compliance.

Finally, the lack of standardization in the blockchain and cryptocurrency industry presents a challenge for regulators and policymakers. There are many different blockchain and cryptocurrency platforms, each with their own unique features and capabilities. This can make it difficult for regulators and policymakers to develop regulations that can be applied uniformly across the industry.

Despite these challenges, the development of legal and regulatory frameworks for blockchain and cryptocurrencies is essential for the growth and development of the industry. The frameworks aim to protect investors and consumers while promoting innovation and growth in the blockchain and cryptocurrency markets. However, it is important for regulators and policymakers to carefully consider the challenges and complexities of the technology when developing these frameworks. They must balance the need for regulation with the need for innovation and growth, and ensure that the regulations are effective in achieving their intended goals.

Explanation of how Web 3 and Bitcoin are changing the nature of business transactions

Web 3 and Bitcoin are changing the nature of business transactions in several ways. Firstly, Web 3 technologies enable decentralized and peer-to-peer transactions, which eliminate the need for intermediaries such as banks or payment processors. This results in faster and more cost-efficient transactions, as well as increased transparency and security.

Bitcoin, as the first and most well-known cryptocurrency, has also revolutionized the way businesses conduct financial transactions. Bitcoin transactions are recorded on a decentralized ledger known as the blockchain, which is secure, transparent, and tamper-proof. This eliminates the need for a centralized authority to verify transactions, such as a bank or a government agency.

In addition to its decentralized nature, Bitcoin also offers several advantages over traditional forms of payment. Bitcoin transactions are fast, cheap, and irreversible, which makes them ideal for conducting international transactions and micropayments. They also offer a high degree of anonymity, which can be beneficial for businesses that want to protect their privacy and avoid being tracked by competitors or government agencies.

Moreover, Web 3 technologies and Bitcoin are enabling the creation of new types of business models that were previously impossible. For example, decentralized applications (dApps) are being built on top of blockchain technology, which enable peer-to-peer transactions without the need for intermediaries. This creates new opportunities for businesses to interact with customers and suppliers in novel ways, and to create new revenue streams.

However, the adoption of Web 3 technologies and Bitcoin is not without its challenges. One major challenge is the lack of regulatory clarity around these technologies, which has resulted in a patchwork of legal and regulatory frameworks that vary widely from country to country. This has created uncertainty for businesses that want to adopt these technologies, as they are unsure of how to comply with existing laws and regulations.

Another challenge is the issue of scalability. As more businesses adopt Web 3 technologies and Bitcoin, the volume of transactions on the blockchain will increase exponentially. This could lead to issues with scalability, such as slow transaction times and high fees. While several solutions have been proposed to address this issue, such as increasing the block size or implementing off-chain scaling solutions, these solutions are not without their own tradeoffs and challenges.

Despite these challenges, it is clear that Web 3 technologies and Bitcoin are changing the nature of business transactions in profound ways. As businesses continue to adopt these technologies, they will need to navigate the legal and regulatory landscape, and find innovative solutions to address scalability and other challenges. However, the potential benefits of these technologies, including increased efficiency, transparency, and security, are likely to outweigh the challenges, and will drive continued adoption and innovation in the years to come.

Overview of how blockchain technology can facilitate secure and efficient supply chain management

In recent years, the use of blockchain technology has emerged as a promising solution for secure and efficient supply chain management. The traditional supply chain management system relies on a complex network of intermediaries, manual record-keeping, and inefficient communication channels. Blockchain technology has the potential to streamline the supply chain management process by providing a secure, transparent, and decentralized system that eliminates the need for intermediaries, automates record-keeping, and enhances communication channels.

This section will provide an overview of how blockchain technology can facilitate secure and efficient supply chain management. It will begin by defining supply chain management and discussing its challenges. Then, it will explore how blockchain technology can address these challenges by providing a secure, transparent, and decentralized system for supply chain management. Finally, it will examine some of the applications of blockchain technology in supply chain management.

Supply Chain Management:

Supply chain management refers to the process of managing the flow of goods and services from the point of origin to the point of consumption. It involves the coordination of multiple stakeholders, including suppliers, manufacturers, distributors, retailers, and customers. The goal of supply chain management is to ensure that the right product is delivered to the right place at the right time and at the right cost.

The traditional supply chain management system is a complex network of intermediaries, manual record-keeping, and inefficient communication channels. This system is prone to errors, delays, fraud, and inefficiencies. For example, manual record-keeping can lead to errors and discrepancies in inventory management, resulting in overstocking or understocking of products. Inefficient communication channels can cause delays in the delivery of goods and services, leading to customer dissatisfaction and lost revenue.

Challenges in Supply Chain Management:

There are several challenges in supply chain management that blockchain technology can address. These challenges include:

Lack of Transparency: The traditional supply chain management system lacks transparency, making it difficult to trace the origin and journey of a product. This lack of transparency can lead to counterfeiting, fraud, and other illegal activities.

Inefficient Record-Keeping: Manual record-keeping is prone to errors and discrepancies, leading to inefficiencies in inventory management and other supply chain processes.

High Costs: The traditional supply chain management system involves multiple intermediaries, resulting in high costs and reduced profit margins.

Slow Communication Channels: The traditional supply chain management system relies on inefficient communication channels, leading to delays in the delivery of goods and services.

How Blockchain Technology Can Address These Challenges:

Blockchain technology can address these challenges by providing a secure, transparent, and decentralized system for supply chain management. Here are some of the ways in which blockchain technology can facilitate secure and efficient supply chain management:

Transparency: Blockchain technology provides a transparent and immutable record of transactions. This means that every transaction on the blockchain is recorded and cannot be altered, providing complete transparency and traceability.

Decentralization: Blockchain technology is a decentralized system that eliminates the need for intermediaries. This reduces costs and increases efficiency by removing unnecessary intermediaries from the supply chain.

Efficiency: Blockchain technology automates record-keeping and other supply chain processes, reducing the potential for errors and increasing efficiency.

Security: Blockchain technology uses advanced cryptography to secure transactions and data, making it difficult for hackers to tamper with the data.

Applications of Blockchain Technology in Supply Chain Management:

Blockchain technology has several applications in supply chain management. Here are some of the ways in which blockchain technology is being used in supply chain management:

Traceability: Blockchain technology can be used to trace the origin and journey of a product. This helps to prevent counterfeiting, fraud, and other illegal activities.

Smart Contracts: Blockchain technology can be used to create smart contracts that automate supply chain processes. For example, a smart contract can automatically trigger a payment when a shipment of goods arrives at a particular location or when a certain set of conditions are met.

Inventory Management: Blockchain technology can be used to track inventory in real-time, making it easier to manage and optimize supply chain operations. This can help businesses reduce waste, lower costs, and improve efficiency.

Transparency: Blockchain technology can provide transparency in supply chain management by making information about products and their journey available to all parties involved. This can help to build trust between suppliers, manufacturers, retailers, and consumers.

Fraud Prevention: Blockchain technology can help prevent fraud in supply chain management by ensuring that all transactions are secure, transparent, and tamper-proof. This can help to prevent fraud at every stage of the supply chain, from production to distribution to retail.

Here are some examples of companies using blockchain technology in supply chain management:

Walmart: Walmart is using blockchain technology to track the journey of food products from farm to store. This helps to ensure the quality and safety of their products while also providing customers with more information about the products they buy.

Maersk: Maersk, the world's largest shipping company, is using blockchain technology to manage its global supply chain. The company has partnered with IBM to create a blockchain-based platform that tracks shipments in real-time, reducing paperwork and improving efficiency.

De Beers: De Beers, the world's largest diamond company, is using blockchain technology to track the journey of diamonds from mine to retailer. This helps to prevent conflict diamonds from entering the supply chain and also provides customers with more information about the origin and journey of their diamonds.

In conclusion, blockchain technology has the potential to revolutionize supply chain management by providing greater transparency, security, and efficiency. As more companies adopt this technology, we can expect to see significant improvements in supply chain operations, leading to better quality products, lower costs, and increased customer satisfaction.

Discussion of the potential for decentralized autonomous organizations (DAOs) to change the way businesses are structured and operated

Decentralized Autonomous Organizations (DAOs) have been hailed as a potential game-changer in the world of business. DAOs are organizations that are run through code, and they are designed to operate without the need for human intervention. These organizations are built on blockchain technology, which provides a decentralized and transparent platform for managing business operations.

DAOs have the potential to change the way businesses are structured and operated. Here are some of the ways in which DAOs can impact the business world:

Decentralization: DAOs are built on a decentralized platform, which means that there is no central authority that controls the organization. Instead, decisions are made through a consensus mechanism, where all members of the organization have a say in the decision-making process. This means that there is no hierarchy, and everyone has an equal say in how the organization is run.

Transparency: DAOs are built on blockchain technology, which provides a transparent platform for managing business operations. All transactions are recorded on a public ledger, which means that anyone can see how the organization is operating. This provides a level of transparency that is not possible with traditional business structures.

Efficiency: DAOs are designed to operate through code, which means that they can automate many business processes. This can lead to greater efficiency and cost savings, as there is no need for human intervention in many cases.

Trust: DAOs are built on a trustless platform, which means that trust is built into the system. All members of the organization have a stake in the success of the organization, which provides a level of trust that is not possible with traditional business structures.

Access: DAOs are designed to be accessible to anyone, regardless of their location or background. This means that anyone can participate in the organization, regardless of their social status or financial means.

While DAOs have the potential to change the way businesses are structured and operated, there are also some potential risks and challenges. Here are some of the risks and challenges associated with DAOs:

Legal and regulatory challenges: DAOs operate in a legal gray area, as they are not regulated in the same way as traditional businesses. This can lead to legal and regulatory challenges, which can impact the operation of the organization.

Security risks: DAOs are built on blockchain technology, which is not immune to security risks. There have been instances of DAOs being hacked, which has resulted in significant losses for members of the organization.

Governance challenges: DAOs are designed to operate through a consensus mechanism, which can lead to governance challenges. It can be difficult to reach consensus on important decisions, which can impact the operation of the organization.

Scalability challenges: DAOs are still in their early stages of development, and there are scalability challenges associated with these organizations. As the number of members in the organization grows, it can be difficult to maintain the same level of efficiency and transparency.

In conclusion, DAOs have the potential to change the way businesses are structured and operated. These organizations are built on blockchain technology, which provides a decentralized and transparent platform for managing business operations. While there are some risks and challenges associated with DAOs, they are still in their early stages of development, and there is significant potential for these organizations to revolutionize the business world. Students interested in the legal implications of DAOs will need to stay informed about the evolving regulatory landscape, while business owners and entrepreneurs should consider the potential benefits of DAOs for their own operations.

CHAPTER 45: LEGAL AND REGULATORY ISSUES

As blockchain and cryptocurrency continue to gain traction in the global economy, regulators and policymakers around the world are grappling with how to address the legal and regulatory issues that arise from these new technologies. While blockchain and cryptocurrency offer exciting opportunities for innovation and growth, they also pose significant risks to investors and consumers.

In this chapter, we will examine the legal and regulatory issues surrounding blockchain and cryptocurrency, including the challenges of regulating a decentralized and borderless technology, the risks associated with investing in cryptocurrencies, and the impact of blockchain on traditional legal and regulatory frameworks.

We will begin by exploring the regulatory landscape for blockchain and cryptocurrency around the world. We will examine the different approaches taken by countries such as the United States, China, and Japan, and discuss the challenges of regulating a technology that transcends national borders.

Next, we will examine the risks associated with investing in cryptocurrency. We will explore the different types of cryptocurrencies, the factors that determine their value, and the potential for fraud and market manipulation. We will also discuss the challenges of enforcing existing securities laws in the context of cryptocurrency investments.

Finally, we will discuss the impact of blockchain on traditional legal and regulatory frameworks. We will examine how blockchain can be used to create decentralized autonomous organizations (DAOs) and smart contracts that can automate traditional legal processes. We will also discuss the challenges of adapting existing legal frameworks to accommodate these new technologies.

Overall, this chapter will provide a comprehensive overview of the legal and regulatory issues surrounding blockchain and cryptocurrency. We will examine the challenges of regulating a decentralized and borderless technology, the risks associated with investing in cryptocurrencies, and the impact of blockchain on traditional legal and regulatory frameworks. By the end of this chapter, you will have a deep understanding of the legal and regulatory challenges associated with blockchain and cryptocurrency, as well as the opportunities and potential benefits these new technologies offer.

Overview of how existing laws and regulations are being challenged by the emergence of Web 3 and Bitcoin

The emergence of Web 3 and Bitcoin has created numerous legal and regulatory challenges. While the decentralized and open nature of these technologies has led to unprecedented levels of innovation and growth, it has also raised questions about how existing legal frameworks can accommodate them.

Bitcoin, for example, has been the subject of intense regulatory scrutiny since its inception. Governments around the world have struggled to determine how to classify and regulate Bitcoin, which has been variously described as a currency, a commodity, and a security.

Similarly, Web 3 technologies, such as decentralized applications (dApps) and smart contracts, are challenging traditional legal concepts like contractual capacity and liability. As these technologies continue to evolve, they are likely to have an increasingly significant impact on the legal and regulatory landscape.

Legal Challenges for Bitcoin

One of the primary legal challenges posed by Bitcoin is its status as a currency. While some countries have recognized Bitcoin as legal tender, many others have taken a more cautious approach. For example, in the United States, the IRS has classified Bitcoin as property, rather than currency, for tax purposes. This means that any gains or losses from Bitcoin transactions must be reported as capital gains or losses.

Bitcoin's status as a commodity is also a matter of debate. In 2015, the US Commodity Futures Trading Commission (CFTC) declared that Bitcoin was a commodity under the Commodity Exchange Act. However, this classification has not been universally accepted, and other regulators have taken a more cautious approach.

Perhaps the most significant legal challenge posed by Bitcoin is its potential use in illegal activities. Because Bitcoin transactions are anonymous and decentralized, they have been used to facilitate a range of criminal activities, including drug trafficking, money laundering, and tax evasion. Governments around the world have responded by implementing a range of regulations aimed at preventing Bitcoin from being used for illicit purposes.

Regulatory Challenges for Web 3

Web 3 technologies, such as dApps and smart contracts, are also raising a range of legal and regulatory challenges. For example, because dApps are often decentralized and run on a blockchain, it can be difficult to determine who is responsible for them in the event of a dispute or malfunction. Similarly, smart contracts raise questions about contractual capacity and the enforceability of contracts.

The legal challenges posed by Web 3 technologies are particularly acute in the financial sector. Decentralized finance (DeFi) platforms, which use Web 3 technologies to create financial products and services, are challenging traditional financial regulations. For example, because DeFi platforms are often decentralized and do not have a central authority, it can be difficult to determine who is responsible for ensuring compliance with regulations.

In addition to these legal and regulatory challenges, Web 3 technologies are also raising broader questions about the role of governments in regulating emerging technologies. Some proponents of Web 3 argue that decentralized technologies have the potential to fundamentally transform the way that society is governed, and that traditional legal and regulatory frameworks are ill-equipped to deal with these changes.

Conclusion

The emergence of Web 3 and Bitcoin has created numerous legal and regulatory challenges. While these technologies have the potential to drive unprecedented levels of innovation and growth, they are also challenging existing legal frameworks and raising questions about the role of governments in regulating emerging technologies. As these technologies continue to evolve, it is likely that they will continue to have a significant impact on the legal and regulatory landscape. As such, it is essential for lawyers, policymakers, and regulators to stay abreast of these developments and adapt their approaches accordingly.

Discussion of the challenges of regulating decentralized systems and their impact on traditional legal frameworks

The emergence of decentralized systems, such as blockchain and cryptocurrencies, has presented new challenges for traditional legal frameworks. Decentralized systems operate differently from traditional centralized systems, where there is a central authority that governs and controls the system. In decentralized systems, there is no central authority, and decision-making is distributed among the network's participants. This presents unique challenges for regulators and lawmakers as they try to apply existing legal frameworks to these emerging technologies. This section will discuss the challenges of regulating decentralized systems and their impact on traditional legal frameworks.

Challenges of Regulating Decentralized Systems

One of the main challenges of regulating decentralized systems is the lack of a central authority. Decentralized systems operate on a peer-to-peer network, where every participant has an equal say in decision-making. This means that traditional regulatory approaches, which rely on a central authority to enforce regulations and monitor compliance, are ineffective in decentralized systems.

Another challenge is the cross-border nature of decentralized systems. Blockchain and cryptocurrencies operate on a global scale, and it is challenging for regulators to enforce regulations in different jurisdictions. This has led to a patchwork of regulations, with different countries taking different approaches to regulate decentralized systems. Some countries have banned cryptocurrencies, while others have embraced them.

Decentralized systems also pose challenges for traditional legal frameworks. For example, cryptocurrencies can be used for illicit activities, such as money laundering and financing terrorism. This has led to concerns about the use of cryptocurrencies and the need to develop new legal frameworks to address these issues. Additionally, decentralized systems challenge traditional notions of property and ownership. Smart contracts, which are self-executing contracts that are stored on a blockchain, can transfer ownership of assets without the need for intermediaries, such as banks or lawyers.

Impact on Traditional Legal Frameworks

Decentralized systems and blockchain technology have the potential to disrupt traditional legal frameworks. Smart contracts, for example, can be used to automate legal processes, such as property transfers and the execution of wills. This reduces the need for intermediaries and makes legal processes more efficient and cost-effective.

However, the use of smart contracts also poses challenges for traditional legal frameworks. Smart contracts are code-based, and there may be errors or bugs in the code that could lead to unintended consequences. Additionally, smart contracts are immutable, meaning that once they are executed, they cannot be changed. This poses challenges for dispute resolution and the enforcement of legal contracts.

Decentralized systems also challenge traditional notions of governance and decision-making. In decentralized systems, decision-making is distributed among the network's participants, and there is no central authority to enforce regulations or resolve disputes. This makes it challenging for regulators to ensure compliance with regulations and to enforce legal contracts.

Conclusion

Decentralized systems and blockchain technology present unique challenges for regulators and lawmakers. The lack of a central authority and the cross-border nature of decentralized systems make it challenging to enforce regulations and monitor compliance. Additionally, decentralized systems challenge traditional legal frameworks, such as property and ownership rights, and the use of smart contracts poses challenges for dispute resolution and contract enforcement. However, decentralized systems also have the potential to disrupt traditional legal frameworks by making legal processes more efficient and cost-effective. Regulators and lawmakers will need to find ways to balance the benefits of decentralized systems with the need to protect consumers and ensure compliance with regulations.

Analysis of potential legal and regulatory approaches to address these challenges

As discussed in the previous section, regulating decentralized systems such as Web 3 and Bitcoin presents unique challenges to traditional legal frameworks. In this section, we will examine potential legal and regulatory approaches to address these challenges. We will first discuss the current state of regulation and its limitations, followed by an analysis of potential approaches to regulation.

Current State of Regulation

Currently, most regulatory approaches to Web 3 and Bitcoin are focused on traditional legal frameworks. This is because the technology is relatively new, and regulators are still trying to understand its potential risks and benefits. The primary regulatory bodies overseeing decentralized systems are financial regulators, such as the SEC and CFTC, as well as international organizations such as the FATF.

While financial regulators have implemented some regulatory measures, such as registration and licensing requirements for cryptocurrency exchanges, these measures are often inadequate in addressing the unique challenges of decentralized systems. For example, it can be challenging to regulate anonymous peer-to-peer transactions and smart contracts that are not tied to a specific legal entity.

Moreover, different countries have taken different regulatory approaches, with some countries, such as China, banning cryptocurrencies outright, while others, such as the United States, have taken a more hands-off approach. This lack of consistency in regulatory approaches across jurisdictions can lead to confusion and uncertainty for businesses and consumers alike.

Potential Approaches to Regulation

Given the challenges of regulating decentralized systems, traditional legal frameworks may not be sufficient to address these challenges. Therefore, regulators may need to consider alternative approaches to regulation, such as technology-based solutions.

One potential approach is to use smart contracts as a regulatory mechanism. Smart contracts are self-executing contracts with the terms of the agreement directly written into lines of code. By enforcing regulatory requirements through smart contracts, regulators can ensure compliance with regulatory requirements in a decentralized manner. However, this approach requires careful consideration of the potential risks and limitations of smart contracts.

Another potential approach is to use decentralized autonomous organizations (DAOs) as regulatory bodies. DAOs are decentralized organizations that operate using smart contracts and can execute business logic autonomously. By creating DAOs to oversee decentralized systems, regulators can ensure compliance with regulatory requirements without relying on traditional legal frameworks. However, this approach raises questions about the accountability and transparency of such organizations.

Finally, regulators could also consider working directly with decentralized systems' communities to develop self-regulatory measures. By working with the community, regulators can gain a better understanding of the unique challenges of decentralized systems and develop solutions that work for everyone. However, this approach requires significant cooperation and coordination between regulators and the community.

Conclusion

Regulating decentralized systems presents unique challenges to traditional legal frameworks. While current regulatory approaches are focused on traditional legal frameworks, they may be inadequate in addressing these challenges. Therefore, regulators may need to consider alternative approaches to regulation, such as technology-based solutions or working directly with decentralized systems' communities to develop self-regulatory measures. Ultimately, finding the right regulatory approach will require careful consideration of the potential risks and limitations of each approach, as well as significant cooperation and coordination between regulators and the community.

In conclusion, the emergence of Web 3 and Bitcoin has significantly challenged existing legal frameworks and regulatory approaches. The decentralized and autonomous nature of these systems poses several challenges, including jurisdictional issues, legal uncertainty, and enforcement challenges. Moreover, the potential risks and benefits associated with Web 3 and Bitcoin are yet to be fully understood, and the legal and regulatory approaches to address these challenges are still evolving.

Despite these challenges, there are several potential legal and regulatory approaches that could address these issues. These include developing clear legal definitions and frameworks for decentralized systems, enhancing cross-jurisdictional cooperation and harmonization, and utilizing new technologies and tools such as smart contracts and blockchain analysis. Additionally, there is a need for greater collaboration between regulators, industry players, and other stakeholders to ensure that the legal and regulatory frameworks are aligned with the changing landscape.

It is important for business owners, entrepreneurs, corporate executives, legal departments, compliance officers, financial analysts, investment bankers, accountants, regulators, and business lawyers to stay informed about these developments and the potential legal and regulatory implications for their businesses. As the landscape continues to evolve, it is critical to remain adaptable and proactive in addressing these challenges and opportunities.

In conclusion, the emergence of Web 3 and Bitcoin is changing the financial landscape and laws as they relate to business in significant ways. While these changes pose several challenges, they also offer exciting opportunities for innovation and growth. By embracing these changes and working collaboratively to develop effective legal and regulatory frameworks, businesses can position themselves to thrive in this rapidly evolving landscape.

Call to action for businesses and policymakers to stay informed and engaged with these changes.

As we have seen in this part, the emergence of Web 3 and Bitcoin has brought about significant changes to the financial landscape, including changes to traditional legal frameworks and regulations. While these changes bring about new opportunities for businesses and individuals alike, they also pose significant challenges for regulators and policymakers.

It is therefore crucial for businesses and policymakers to stay informed and engaged with these changes to ensure that they are able to adapt to the evolving landscape and make informed decisions that will benefit all stakeholders.

One way businesses can stay informed is by keeping up with the latest developments in the field through attending conferences, workshops, and seminars. They can also engage with industry experts and thought leaders, who can provide valuable insights into the changing landscape and offer guidance on how to navigate these changes.

Another way businesses can stay engaged is by actively participating in industry associations and working groups. These groups provide a platform for businesses to collaborate with other stakeholders and influence the direction of regulatory policy in a manner that is conducive to the interests of all stakeholders.

For policymakers, staying informed and engaged requires a deep understanding of the technologies and business models that underpin Web 3 and Bitcoin. This can be achieved through ongoing education and training, as well as engaging with industry experts and stakeholders to gather insights into the potential impact of these technologies.

Policymakers can also work with industry associations and other stakeholders to create a regulatory framework that is both supportive of innovation and protective of the interests of all stakeholders. This requires a balance between providing regulatory clarity and certainty, while also allowing for flexibility and adaptability in response to changing circumstances.

In conclusion, the emergence of Web 3 and Bitcoin has brought about significant changes to the financial landscape, which require businesses and policymakers to stay informed and engaged with these changes. By doing so, they can ensure that they are able to adapt to the evolving landscape and make informed decisions that will benefit all stakeholders.

Final thoughts and recommendations for further study

In conclusion, the financial landscape is rapidly changing due to the emergence of Web 3 and Bitcoin. The traditional banking system is being challenged, and new players are entering the market with innovative ideas and technologies. As businesses and policymakers navigate these changes, it is important to stay informed and engaged with the latest developments.

First, businesses should consider the benefits of incorporating Web 3 and Bitcoin into their operations. These technologies can offer increased security, efficiency, and transparency in financial transactions. For example, businesses can use smart contracts to automate payment processing and reduce the risk of fraud. By staying up-to-date with these developments, businesses can stay competitive and better serve their customers.

Second, policymakers should consider the regulatory implications of these changes. As new technologies emerge, there may be a need for new laws and regulations to protect consumers and ensure market stability. However, policymakers should also be careful not to stifle innovation or create unnecessary barriers to entry for new players. It is important to strike a balance between protecting the public interest and fostering innovation in the financial sector.

Third, further research is needed to fully understand the implications of Web 3 and Bitcoin on the financial landscape. This research should consider both the legal and technical aspects of these changes, and should involve collaboration between experts in finance, law, and technology. Some key areas for future research include:

The legal and regulatory challenges of integrating Web 3 and Bitcoin into the existing financial system.

The potential impact of Web 3 and Bitcoin on financial inclusion and access to banking services for underserved populations.

The potential benefits and risks of decentralized finance (DeFi) platforms, which use Web 3 and blockchain technologies to offer financial services without the need for traditional banks.

The role of cryptocurrencies in money laundering and other illicit activities, and the potential for new technologies to combat these activities.

By conducting this research and staying informed about the latest developments in the financial landscape, policymakers and businesses can better navigate these changes and ensure a smooth transition to a new era of finance.

In summary, the emergence of Web 3 and Bitcoin is changing the financial landscape in profound ways. These changes offer opportunities for businesses to increase efficiency and transparency, while also posing new challenges for policymakers and regulators. By staying informed and engaged with these changes, businesses and policymakers can work together to ensure a stable and innovative financial system for years to come.

APPENDIX:

Alphabetical Appendix of Key Words or Terms for Part 1: Introduction to Business Law

Basic Legal Concepts and Principles: Fundamental concepts and principles that form the foundation of the legal system, such as the rule of law, due process, equity, fairness, and justice.

Business Law: The body of law that governs business transactions and activities, including contracts, corporate governance, securities regulation, and intellectual property.

Breach of Contract: A violation of the terms of a contract, which occurs when one party fails to perform their obligations as specified in the agreement.

Contract Law: The branch of law that deals with the formation, interpretation, and enforcement of contracts between parties.

International Business Law: The legal rules, regulations, and practices that govern business transactions and activities that cross national borders.

Legal Systems: The different methods of organizing and administering justice in a society, such as common law, civil law, religious law, and customary law.

Legal Issues in Business Transactions: The legal problems and challenges that arise in the course of conducting business, including contract disputes, intellectual property infringement, and regulatory compliance.

Sources of Law: The origins of legal authority, including constitutions, statutes, regulations, judicial decisions, and customary practices.

Types of Legal Systems: The different approaches to organizing and administering the law, including common law, civil law, religious law, and customary law.

Alphabetical Appendix of Key Words or Terms for Part 2: Business Organizations and Corporate Governance

Board of Directors: A group of individuals elected by a corporation's shareholders to oversee the management and decision-making processes of the corporation.

Corporate Finance: The area of finance that deals with the financial decisions made by corporations, including capital structure, investment decisions, and financial risk management.

Corporate Governance: The system of rules, practices, and processes by which a company is directed and controlled. It involves balancing the interests of a company's many stakeholders, such as shareholders, management, customers, suppliers, financiers, government, and the community.

Executive Compensation: The financial compensation and non-financial benefits given to a corporation's top executives, such as CEOs, CFOs, and COOs, for their services and performance.

Securities Law: The area of law that regulates the issuance, trading, and sale of securities, such as stocks, bonds, and other financial instruments. Securities laws are intended to protect investors from fraud and other unfair practices.

Types of Business Organizations: The legal structures that businesses can take, including sole proprietorships, partnerships, corporations, and limited liability companies (LLCs). Each type of business organization has its own advantages and disadvantages in terms of liability, taxation, and management.

Alphabetical Appendix of Key Words or Terms for Part 3: Contract Law and Legal Issues in Business Transactions

Breach of Contract: A breach of contract occurs when one party fails to perform its obligations under a valid agreement without a legal excuse. It gives the non-breaching party the right to pursue remedies to recover losses caused by the breach.

International Business Law: International business law is the set of legal rules and regulations that govern business activities between parties from different countries. It includes international trade law, foreign investment law, and commercial law, among others.

Legal Issues in Business Transactions: Legal issues in business transactions refer to the legal considerations that arise in the course of conducting business transactions, such as mergers and acquisitions, financing arrangements, and contractual agreements. These issues may involve issues of contract law, intellectual property law, securities regulation, and other areas of law.

Remedies: Remedies refer to the legal means available to a party to enforce its rights or recover damages caused by a breach of contract or other legal violation. Remedies can be legal, such as monetary damages or specific performance, or equitable, such as injunctive relief or rescission.

Formation of Contracts: Formation of contracts refers to the process by which a valid contract is created between parties. It typically requires an offer, acceptance, consideration, and the intention to create legal relations. Additionally, the contract must not violate any laws or public policy.

Alphabetical Appendix of Key Words or Terms for Part 4: Securities Laws and Regulations

Exemptions: In the context of securities law, exemptions refer to provisions that allow certain securities offerings to be exempt from full SEC registration requirements. Exemptions can be used for a variety of reasons, including to reduce costs and regulatory burdens for small businesses.

Regulation A: Regulation A is a set of SEC rules that allow companies to raise up to $50 million in a 12-month period from the public through a simplified registration process. This regulation is often used by small and medium-sized businesses that want to raise capital from the public without undergoing the full registration process required by the SEC.

Regulation A+: Regulation A+ is an amendment to Regulation A that was implemented as part of the JOBS Act of 2012. This amendment allows companies to raise up to $75 million in a 12-month period through a streamlined registration process, making it easier for companies to raise capital from the public.

Securities Act of 1933 and 1934: The Securities Act of 1933 and the Securities Exchange Act of 1934 are two landmark pieces of securities legislation in the United States. The Securities Act of 1933 requires companies that issue securities to provide full disclosure of all material information to investors, while the Securities Exchange Act of 1934 established the SEC and gave it broad regulatory authority over the securities industry.

Securities Exchange Act of 1934: The Securities Exchange Act of 1934 is a federal law that regulates the trading of securities in the United States. The act established the Securities and Exchange Commission (SEC) and gave it broad authority to regulate and oversee the securities industry. The act also requires companies that trade securities on public exchanges to make periodic disclosures of financial and other information to investors.

Alphabetical Appendix of Key Words or Terms for Part 5: Securities Regulation Agencies and Enforcement

Financial Industry Regulatory Authority (FINRA): A non-governmental organization that regulates broker-dealers in the United States. FINRA oversees securities firms and professionals, enforces regulations and standards, and provides education and training to industry members.

Securities and Exchange Commission (SEC): A US government agency responsible for enforcing federal securities laws and regulating securities markets. The SEC's primary mission is to protect investors and maintain fair and efficient markets.

State Securities Regulators: State-level agencies responsible for regulating securities offerings and securities professionals operating within their respective states. Each state has its own securities regulator, often known as the State Securities Commission, that enforces state securities laws and regulations.

Alphabetical Appendix of Key Words or Terms for Part 6: Disclosure Requirements and Liability

Disclosure Requirements: These are the rules and regulations that require companies to disclose certain information about their business and financial condition to the public. For example, public companies are required to file periodic reports with the SEC that include financial statements, management discussion and analysis, and other disclosures.

Liability for Securities Fraud: This refers to the legal responsibility that individuals and companies may face if they make false or misleading statements or omissions in connection with the sale of securities. Securities fraud can lead to civil and criminal liability, including fines, imprisonment, and the requirement to pay damages to investors who have been harmed.

Securities Violations: This refers to any action or omission that violates securities laws and regulations. Examples of securities violations include insider trading, market manipulation, and making false or misleading statements in connection with the sale of securities.

Insider Trading: This refers to buying or selling securities on the basis of material non-public information. Insider trading is illegal and can lead to civil and criminal liability, including fines and imprisonment.

Alphabetical Appendix of Key Words or Terms for Part 7: Public Offering Process and Requirements

IPO Alternatives: IPO alternatives refer to other means by which a company can go public without going through the traditional initial public offering (IPO) process. These alternatives include direct listings, special purpose acquisition companies (SPACs), and reverse mergers.

IPO Considerations: IPO considerations refer to the factors a company should take into account when deciding whether or not to go public. These considerations include market conditions, the company's financial performance, the costs and benefits of going public, and the potential risks and challenges.

Pre-IPO Planning: Pre-IPO planning refers to the preparation a company must undergo before launching an IPO. This includes preparing financial statements, conducting due diligence, establishing a governance structure, and selecting underwriters.

Public Scrutiny: Public scrutiny refers to the level of attention and scrutiny a company can expect to receive once it becomes a publicly traded company. This includes increased media attention, regulatory oversight, and shareholder activism.

Alphabetical Appendix of Key Words or Terms for Part 10: Financial Statement Analysis and Reporting

Balance sheets: A balance sheet is a financial statement that provides a snapshot of a company's financial position at a specific point in time. It summarizes a company's assets, liabilities, and equity, and provides information on what the company owns and owes.

Cash flow statements: A cash flow statement is a financial statement that shows how much cash a company has generated and used during a specific period. It includes cash inflows and outflows from operating activities, investing activities, and financing activities.

Financial statement analysis: Financial statement analysis is the process of reviewing and analyzing a company's financial statements to gain insights into its financial health and performance. This can include analyzing the income statement, balance sheet, and cash flow statement, as well as calculating and interpreting financial ratios.

Income statements: An income statement (also known as a profit and loss statement) is a financial statement that shows a company's revenues and expenses over a specific period of time, typically a quarter or a year. It provides information on a company's profitability by subtracting expenses from revenues to determine net income or net loss.

Ratio analysis: Ratio analysis involves calculating and interpreting financial ratios, which are relationships between different financial statement items. Examples of commonly used financial ratios include the current ratio, quick ratio, debt-to-equity ratio, and return on investment ratio.
Ratio analysis can help identify trends and patterns in a company's financial performance, and can be used to compare a company's performance to industry benchmarks or competitors.

Alphabetical Appendix of Key Words or Terms for Part 11: Financial Reporting Standards and Regulations

Generally Accepted Accounting Principles (GAAP): These are the set of accounting standards, guidelines, and principles that are used in the preparation of financial statements in the United States. GAAP provides a common framework for financial reporting that helps ensure consistency, comparability, and transparency across companies and industries.

International Financial Reporting Standards (IFRS): These are the global accounting standards that are used in many countries around the world, including the European Union, Australia, Canada, and Japan. IFRS provides a common language for financial reporting that helps ensure consistency, comparability, and transparency across countries and industries.

Sarbanes-Oxley Act of 2002: This is a US federal law that was passed in response to a number of high-profile accounting scandals in the early 2000s, such as the Enron scandal. The Sarbanes-Oxley Act (SOX) introduced a number of measures to improve corporate governance, financial reporting, and auditing, including the creation of the Public Company Accounting Oversight Board (PCAOB) and the requirement for companies to implement internal controls over financial reporting.

Alphabetical Appendix of Key Words or Terms for Part 12: International Accounting Standards

Global Financial Reporting Standards: these are a set of accounting standards and principles that are used to prepare financial statements for companies that operate internationally. They are designed to provide consistency and transparency in financial reporting across different countries and regions.

International Financial Reporting Standards (IFRS): these are a set of global accounting standards that provide guidelines for the preparation and presentation of financial statements. IFRS is used by many countries around the world, although some countries, such as the United States, have not fully adopted them.

US GAAP: this stands for Generally Accepted Accounting Principles, which are a set of accounting standards and procedures that are used by companies in the United States to prepare financial statements. These principles are established by the Financial Accounting Standards Board (FASB).

Challenges and Opportunities in Global Financial Reporting: these refer to the issues and opportunities that arise in the process of preparing financial statements for companies that operate across different countries and regions. Some of the challenges include dealing with different accounting standards, currency conversions, and cultural differences. Some of the opportunities include gaining access to new markets and investors, and the ability to achieve greater economies of scale.

Alphabetical Appendix of Key Words or Terms for Part 13: How Web 3 and Bitcoin are Changing the Financial Landscape and Laws as it Relates to Business

Bitcoin: the first and most well-known cryptocurrency, created in 2009 by an unknown person or group using the name Satoshi Nakamoto.

Blockchain: a decentralized digital ledger that records transactions across many computers so that the record cannot be altered retroactively without the alteration of all subsequent blocks and the consensus of the network.

Central bank digital currency (CBDC): a digital version of a country's fiat currency issued and regulated by its central bank.

Cryptocurrency: a digital or virtual currency that uses cryptography for security and operates independently of a central bank.

Cryptocurrency mining: the process of verifying and adding transactions to a blockchain by solving complex mathematical algorithms, which requires significant computational power and energy consumption.

Decentralization: the process of transferring control of an activity or organization from a centralized authority to a distributed network.

Digital identity: the online representation of an individual's identity that can be verified and authenticated using various digital technologies.

Distributed ledger technology (DLT): a digital system for recording transactions that is maintained by a network of computers rather than a centralized authority.

Initial coin offering (ICO): a type of crowdfunding using cryptocurrencies that offer investors a digital token or coin representing a share in a project or company.

Privacy coins: cryptocurrencies that offer enhanced privacy and anonymity to their users.

Regulatory sandboxes: programs run by regulatory authorities that allow fintech companies to test their products and services in a controlled environment with reduced regulatory requirements.

Smart contracts: self-executing contracts with the terms of the agreement directly written into code, which are stored on a blockchain.

Stablecoins: cryptocurrencies that are pegged to the value of a stable asset, such as a fiat currency or commodity, to reduce price volatility.

Tokenization: the process of converting an asset, such as real estate or a company share, into a digital token that can be traded on a blockchain.

Web 3.0: the next generation of the internet, characterized by decentralization, distributed ledgers, and advanced machine learning algorithms.

ANSWER KEY :

Chapter 10: Breach of Contract and Remedies:
Exercise 1: Calculation of Damages

What is the first step in calculating damages in a breach of contract case?
a) Determine the value of the performance received
b) Determine the value of the promised performance
c) Determine the contract price
d) Calculate the difference
Answer: c) Determine the contract price

What does the injured party need to prove in order to calculate damages in a breach of contract case?
a) The fact of the loss
b) The amount of the loss suffered
c) Both a and b
d) None of the above
Answer: c) Both a and b

How is the difference between the promised performance and the actual performance received calculated?
a) By subtracting the contract price from the value of the promised performance
b) By subtracting the value of the performance received from the contract price
c) By adding the value of the promised performance to the value of the performance received
d) By multiplying the value of the performance received by the contract price
Answer: b) By subtracting the value of the performance received from the contract price

Exercise 2: Remedies

What is specific performance?
a) A monetary remedy that compensates the injured party for their losses
b) An order requiring the breaching party to perform their contractual obligations
c) A court order preventing the breaching party from performing their contractual obligations
d) A remedy that cancels the contract and restores the parties to their pre-contractual position

Answer: b) An order requiring the breaching party to perform their contractual obligations

What is an injunction?
a) A monetary remedy that compensates the injured party for their losses
b) An order requiring the breaching party to perform their contractual obligations
c) A court order preventing the breaching party from performing their contractual obligations
d) A remedy that cancels the contract and restores the parties to their pre-contractual position
Answer: c) A court order preventing the breaching party from performing their contractual obligations

What is rescission?
a) A monetary remedy that compensates the injured party for their losses
b) An order requiring the breaching party to perform their contractual obligations
c) A court order preventing the breaching party from performing their contractual obligations
d) A remedy that cancels the contract and restores the parties to their pre-contractual position
Answer: d) A remedy that cancels the contract and restores the parties to their pre-contractual position

Exercise 3: Breach of Contract

What is a material breach of contract?
a) A breach that does not affect the essence of the contract
b) A breach that affects the essence of the contract
c) A breach that is trivial or minor
d) A breach that occurs before the contract is formed
Answer: b) A breach that affects the essence of the contract

What is a repudiatory breach of contract?
a) A breach that does not affect the essence of the contract
b) A breach that affects the essence of the contract
c) A breach that is trivial or minor
d) A breach that occurs before the contract is formed
Answer: b) A breach that affects the essence of the contract

What is an anticipatory breach of contract?

a) A breach that occurs before the contract is formed
b) A breach that occurs after the contract is formed
c) A breach that is trivial or minor
d) A breach that is threatened or declared before the time for performance
Answer: d) A breach that is threatened or declared before the time for performance.

What is the purpose of an equitable remedy?
A) To punish the breaching party
B) To compensate the injured party
C) To prevent future breaches
D) Both B and C
Answer: D) Both B and C

Which of the following is an example of an equitable remedy?
A) Damages
B) Specific performance
C) Rescission
D) All of the above
Answer: D) All of the above

What is the purpose of an accounting remedy?
A) To punish the breaching party
B) To compensate the injured party
C) To prevent future breaches
D) Both A and B
Answer: B) To compensate the injured party

When is specific performance typically granted?
A) When the contract involves the sale of goods
B) When the subject matter of the contract is unique
C) When the injured party can easily obtain damages
D) Both A and B
Answer: D) Both A and B

Which of the following is an example of an injunction remedy?
A) Ordering the breaching party to perform its contractual obligations
B) Ordering the breaching party to cease a certain activity
C) Ordering the breaching party to pay damages
D) None of the above
Answer: B) Ordering the breaching party to cease a certain activity.

Chapter 43

What is the main advantage of the decentralized model of currency introduced by Bitcoin and other cryptocurrencies?
a. Eliminates the need for intermediaries
b. Provides greater privacy and security
c. Allows for greater financial inclusion
d. All of the above
Answer: d. All of the above.

The decentralized model of currency introduced by Bitcoin and other cryptocurrencies has several advantages. Which of the following is not one of these advantages?
a. Eliminates the need for intermediaries
b. Provides greater privacy and security
c. Allows for greater financial inclusion
d. Is controlled by a central authority
Answer: d. Is controlled by a central authority.